# THIN-FILM SILICON SOLAR CELLS

ENGINEERING SCIENCES Micro- and Nanotechnology

# THIN-FILM SILICON SOLAR CELLS

Arvind Shah, Editor

The main authors of *Thin-Film Silicon Solar Cells* are Christophe Ballif, Wolfhard Beyer, Friedhelm Finger, Horst Schade, Arvind Shah, and Nicolas Wyrsch, with additional contributions by Jean-Eric Bourée, Corinne Droz, Luc Feitknecht, Daniel Oppizzi, Martin Python, Julian Randall, Ricardo Rüther, Michael Stückelberger, and Reto Tscharner.

EPFL Press
A Swiss academic publisher distributed by CRC Press

**Taylor and Francis Group, LLC**
6000 Broken Sound Parkway, NW, Suite 300,
Boca Raton, FL 33487

**Distribution and Customer Service**
orders@crcpress.com

**www.crcpress.com**

Library of Congress Cataloging-in-Publication Data
A catalog record for this book is available from the Library of Congress.

This book is published under the editorial direction of
Professor Jürgen Brugger (EPFL).

The authors and publisher express their thanks to the Swiss Federal Institute of Technology in Lausanne (EPFL) for the generous support towards the publication of this book.
The editor and authors also kindly acknowledge the financial and logistical support received during the preparation of the manuscript from the PV-Lab of the Institute of Microtechnology (IMT) of the EPFL.

We would like to thank the management of Flexcell (VHF-Technologies SA) for providing us with a photograph of the silicon solar cells featured on the cover of this book.

# EPFL Press

is an imprint owned by Presses polytechniques et universitaires romandes, a Swiss academic publishing company whose main purpose is to publish the teaching and research works of the Ecole polytechnique fédérale de Lausanne (EPFL) and other universities and institutions of higher learning.

Presses polytechniques et universitaires romandes
EPFL – Rolex Learning Center
Post office box 119
CH-1015 Lausanne, Switzerland
E-Mail: ppur@epfl.ch
Phone: 021 / 693 21 30
Fax: 021 / 693 40 27

**www.epflpress.org**


ISBN 978-2-940222-36-0 (EPFL Press)
ISBN 978-1-4200-6674-6 (CRC Press)

Printed in Italy

# PREFACE

Photovoltaic solar energy constitutes one of the main hopes for solving the future energy problems of Mankind, together with other important forms of renewable energy, such as hydroelectricity, wind energy, solar thermal generation of electricity, biomass-based generation and geothermal energy.

When we started our work in Neuchâtel on solar cells in the early 1980s, we fully recognized this potential but were faced with a few difficult choices: In which solar cell material should we invest? What should be the focus of our work? Which should be the milestones and goals towards which we should direct our plan of research?

In 1983 we selected amorphous silicon as semiconductor material; we decided to focus our attention on the reduction of manufacturing costs; and we defined the following goals:

(1) To master all the fabrication steps of an amorphous silicon module, and especially the deposition of the silicon layer itself;

(2) To reduce the fabrication costs by at least a factor of three, and to reduce the production energy for solar modules decisively;

(3) To contribute to increase the overall conversion efficiency of commercial amorphous silicon modules three-fold, to over 9 %.

These three goals have now been attained, almost 30 years after we started our work. It took almost three times longer than foreseen to arrive at this point. Even though our Laboratory in Neuchâtel has indeed contributed some essential building blocks to this success, it was only thanks to the fact that many other laboratories world-wide (especially in Japan, Europe and in the USA) made complementary contributions towards these milestones that these goals were finally reached.

During this thirty-year period, our laboratory had to weather many difficulties: Time and again, our hopes were shattered by results that were not up to our expectations. Time and again we had to modify our equipment, our process parameters, and our cell design. Time and again we strived in vain to find industries interested in collaborating with us and in continuing our work, towards the development of commercial products. Thanks to the unflinching support of Gerard Schriber, initiator of the research program at the Swiss Federal Department of Energy and of Jean-Daniel Perret, head of University affairs, at the Republic and Canton of Neuchâtel, we were able to continue our work, year after year, until we finally were successful. The lesson we learnt thereby is that it is not at all a straightforward task to develop better solar cells; it is a task, which can not be easily planned ahead and which needs long-term financing. It is a tribute to all the women and men, who made up our research team, that in spite of all the odds they continued with enthusiasm and drive. From the very beginning of our work, their commitment and motivation remained intact. Their efforts have been decisive in attaining our goals.

In 1983, our choice of amorphous silicon as solar cell material was motivated by the fact that amorphous silicon appeared at that moment as the only solar-cell material, with a true potential for massive cost reduction, and with a solid basis of reliable and proven mass production tools. Today, thanks to the pioneering work of our research group in Neuchâtel, microcrystalline silicon has emerged as "novel" photovoltaic material and is used together with amorphous silicon to implement the so-called *micromorph* modules and to achieve thereby stabilized commercial total-area efficiencies over 10 %. The field of thin-film silicon solar cells has, thus, been considerably extended.

In a sense we have reached a turning point in the development of silicon solar-cell technology, and we feel that the timing is appropriate for a book to summarize current understanding and to provide a comprehensive overview on which the present and future generation of researchers will be able to set their own goals. To students and to all other novices wishing to enter the field, we would like to formulate the *following recommendations for the use of this book*: They may start with Chapter 1, which provides a general introduction to the field of thin-film silicon modules. They should then look through Chapters 2 and 3, which cover both amorphous and microcrystalline silicon, in considerable detail. They need not go through the whole of these two Chapters before studying the rest of the book. In a first phase, they may just read Sections 2.1 to 2.3 as well as Sections 3.1 and 3.2. They should above all study in depth the central Chapter of this book, namely Chapter 4: here they will find a complete description of how a solar cell functions – a description which is not limited to the classical *pn*-type solar cell as used for "conventional" modules, but mainly devoted to the understanding of *pin*-type solar cells, on which thin-film silicon modules are based. Chapters 5, 6, 7 and 8 can be taken on later, as and when desired or needed. Let us just state that Chapter 6 is another key Chapter of the book, especially for all those concerned with the fabrication and testing of modules.

As editor of this book and as author of a large part of the text, I wish to express my thanks to all those who encouraged me to continue to assemble and to filter data, to write-up chapters and revise them, and to incorporate new representations, to recruit co-authors, and to introduce additional texts. It is quite impossible in this preface to mention all of those who have helped me here, during the thirty months it took to prepare this book. In addition to those mentioned in the *About the Authors* section just after this preface, and all those who are explicitly acknowledged for providing smaller contributions of text or figures throughout the book, I would like to specifically thank Dr. Richard Bartlome and Dr. Corsin Battaglia of IMT's PV Lab, Dr. Stefano Bengali of OERLIKON SOLAR, Milan Vanacek of the Academy of Sciences in Prague, and Professor Hans Beck of the University of Neuchâtel, who helped with proofreading, corrections and calculations. I also thank Dr. Frederick Fenter, of the EPFL Press, who accompanied my efforts during this long process. Some very special words of thanks go to my wife Brigitte and my youngest daughter Annjali, who encouraged me to bring this book to completion. I remember very well our summer holidays on the

Island of Corsica, where I wrote the most difficult parts of the book – on the beach, in glaring sunlight. Truly an appropriate setting to write a text on solar energy! It has remained my dream that some future readers will also be taking this book with them to the beach, on their own summer vacations, and be able to follow all that the book contains without having to refer to other texts.

We may hope that by the year 2040, i.e. 30 years from now, renewable energy will be able to provide one third of the world's electricity and that photovoltaic (PV) solar energy may contribute a third of that, i.e. about 10% of the world's total electricity requirements. This may at first not appear to be a very high share, but it is almost as much as provided by nuclear energy today. To attain such a level, PV module production and installation will have to continue to grow by 30 % per year – just as it has done during the last three decades. This means PV module production will have to grow from just under 10 $GW_p$ in 2010 to over 10,000 $GW_p$ in 2040. This is optimistic, but not impossible.

Of all the PV technologies, however, only thin-film silicon will be able, during the next three decades, to provide and sustain such a growth. All other technologies are either too energy–intensive and inherently costly (like crystalline wafer-based silicon), or based on rare and partly toxic raw materials (like CIGS and CdTe), or not yet ripe for extensive field use (like dye-sensitized and polymer cells).

The production of thin-film silicon modules will, on the other hand, have to be thoroughly revised: Novel production tools, costing only a fraction of the present machinery, will have to be developed; the use of toxic and dangerous gases in production avoided; module reliability and diagnosis improved; module design and structure better adapted to the requirements of the end users.

To sum up, let us simply state: Thin-film silicon PV modules have a central role to play in providing an important share of the world's electricity, particularly in remote or tropical zones, in the medium-term future. PV is probably the only reasonable way of providing electricity for these zones (e.g. Mongolia, Tibet, Arabia, Andes, Amazonia, Sahel). Thin-film silicon PV modules have here three significant advantages over other PV technologies: They can operate with little loss in efficiency at temperatures up to 80 °C; they can be produced in a more robust, safe and reliable way; they can be delivered as light-weight, flexible and unbreakable rolls (like those shown on the front cover of this book). We therefore have no doubt that thin-film silicon PV modules will find vast applications in about 10 years from now – when PV becomes the economically best solution for providing electricity to large parts of the world. The present book is our contribution to assist all those involved in this process.

*Arvind V. Shah,*
*Founder,*
*Photovoltaic Laboratory at the Institute of Microtechnology*
*(IMT PV Lab)*
*Neuchâtel (Switzerland) June 2010*

# ABOUT THE AUTHORS

## CHRISTOPHE BALLIF

Christophe Ballif is director of the Phototovoltaics and Thin Film Electronics Laboratory at the Institute of Microtechnology (IMT) Neuchâtel, which is part of the Swiss Federal Institute of Technology (EPFL) in Lausanne since 2009. The lab is active in the fields of thin-film silicon, high efficiency heterojunction crystalline cells and module technology. He graduated as a physicist from the EPFL in 1994, where he also obtained his Ph.D. degree in 1998 working on novel photovoltaic materials. He conducted postdoctoral research at NREL (Golden, US) on compound semiconductor solar cells; he has also worked at the Fraunhofer Institute for Solar Energy Systems (Germany), where he studied crystalline silicon photovoltaics until 2003, and at EMPA in Thun (CH), before becoming full professor at the IMT in 2004, taking over the chair of Professor Arvind Shah. Professor Ballif is author or co-author of over 180 publications.

## WOLFHARD BEYER

Wolfhard Beyer obtained his Ph.D. degree in physics at the University Marburg, and starting in 1979 was a scientific staff member at Forschungszentrum Jülich (Research Center Jülich, Germany), where he worked in the fields of surface physics and vacuum techniques, thin film and ion technology, and photovoltaics in various Institutes. Wolfhard Beyer has conducted research on amorphous silicon and related materials for the past 40 years with a focus on material issues of thin-film silicon solar cells. Areas of work include the deposition of thin-film materials, the investigation of (micro-) structural and electronic properties, hydrogen incorporation and hydrogen dynamics, and of doping effects.

## JEAN-ERIC BOURÉE

Dr Jean-Eric BOUREE is a CNRS researcher in the Laboratoire de Physique des Interfaces et des Couches Minces of Ecole Polytechnique (France). After graduating in solid state physics from the University of Orsay, he defended a third-cycle thesis on solid state physics at the University of Orsay in 1970, followed by a thesis (Doctorat d'Etat ès Sciences Physiques) at the University of Paris in 1975, on magnetic insulators. Dr Bourée has been active in the field of photovoltaics since 1976, within the framework of numerous French, European and international research programs, and he has been involved in the realization of different types of solar cells.

**CORINNE DROZ**

A native of Neuchâtel, Switzerland, Corinne Droz obtained a M.Sc. degree in microtechnical engineering at the EPFL in 1997. After one year spent at the NEC Corporation research center in Japan, she continued her research activities in 1998 in the Photovoltaics Laboratory of the IMT, where she obtained a Ph.D. degree in 2003 in thin-film PV-cell characterization under the supervision of Professor Arvind Shah. Since then, she has worked as energy delegate for the city of La Chaux-de-Fonds; starting in 2007 she works at Pasan SA, a company producing module testers for the photovoltaics industry.

**FRIEDHELM FINGER**

Friedhelm Finger is a member of the Institute of Energy Research – Photovoltaic in the Forschungszentrum Jülich, where he heads the "Materials and Solar Cells" section. After studying physics in Marburg (Germany) and Dundee (Great Britain), he received his Ph.D. in Marburg before joining the Institute of Microtechnology at the University Neuchâtel, where he started his work on microcrystalline silicon and VHF-PECVD processes. In September 1991 he moved to the Forschungszentrum Jülich. His main topics of interest are deposition processes for thin-film silicon alloys, in particular microcrystalline silicon, defect states in these materials, and applications in thin-film solar cells.

**MARTIN PYTHON**

Martin Python graduated with an M.Sc. degree in physics at Ecole Polytechnique Fédérale de Lausanne (EPFL). He received his Ph.D. from University of Neuchâtel for his thesis entitled *Microcrystalline silicon solar cells: growth and defects*. He currently works as a research engineer in thin-film department at Saint-Gobain Recherche in Paris.

**JULIAN RANDALL**

Julian Randall has been an FP7 National Contact Point since 2006 at Euresearch. His present responsibilities include the Energy Thematic that also covers photovoltaics. Before that he was a postdoctoral fellow at ETHZ in the Institute for Electronics where he conducted research on photovoltaics in wearable applications. In 2003 he obtained a doctorate from EPFL in photovoltaic systems, focusing on indoor applications. The resulting thesis formed the basis of a book he authored entitled *Designing Indoor Solar Products: Photovoltaic Technologies For AES*, published by J. Wiley & Sons. He has also worked for the University of Neuchâtel PV-Lab on the industrialization of a novel PV technology.

## HORST SCHADE

Dr. Horst Schade received his Ph.D. in solid-state physics from the Swiss Federal Institute of Technology (ETH), Zurich, Switzerland. Throughout his career he has been engaged in R&D, mainly dealing with semiconductor and surface physics. In the early eighties, while still at RCA Laboratories, Princeton, New Jersey, USA, he specialized in photovoltaics based on amorphous silicon. He continued to be active in that field for 15 years with Phototronics Solartechnik (now SCHOTT Solar Thin Film GmbH), Putzbrunn (Munich), Germany, both in research, and in various issues of upscaling to production levels. Upon his retirement in 2005, he is a scientific consultant in silicon-based thin-film photovotaics.

## NICOLAS WYRSCH

Nicolas Wyrsch studied physics at the ETH Zurich, obtaining his diploma in 1984, after which he joined the group of Professor Arvind Shah at the Institute of Microtechnology (IMT) of in Neuchâtel to contribute to the development of a-Si:H solar cells, focusing on the study electronic properties of the material. After a stay at Princeton University he obtained his Ph.D. from the University of Neuchatel in 1991. Since then he is team leader in the laboratory of photovoltaics (PV-LAB) of IMT and responsible for several national and international projects on the development of thin-film silicon solar cells and devices.

# CONTENTS

CHAPTER 1

# INTRODUCTION

*Christophe Ballif*

The direct conversion of sunlight to electricity, the photovoltaic effect, has always fascinated scientists. Photovoltaics (PV) is a multidisciplinary field that requires an understanding of physics, chemistry, material science and production technologies. The objective of people involved in this field is usually threefold: (1) to improve the conversion efficiency of the devices, (2) to develop processes that will allow for lower production costs, (3) to ensure that module performances will be maintained for several decades in outdoor conditions, thereby providing much more energy than used in production. Photovoltaics is also controversial: while it is easy to prove that, technically, the full energy needs of humanity can be fulfilled by such a renewable solar energy, in the past and still today, detractors have argued that the manufacture of large-area low-cost semiconductor panels is virtually an impossible task. They now appear to have been wrong, as solar PV electricity at a few cents per kWh has proved to be a realistic target in sunny areas. This opens bright perspectives for producing a substantial part of our electricity needs with PV. Challenges linked to the intermittence of solar energy must still be overcome for the vision of a "solar energy world" to become true. This should happen with the development of clever energy management systems, and by providing better ways to "store" electricity, be it chemically (batteries, $H_2$, or other "solar fuels") or as mechanical energy (hydroelectric storage, compressed air,...). In such a context, thin-film silicon technology will most likely play a key role, as it is today the only technology combining the advantage of virtually infinite resources (Si and Zn in the most simple form) with low material usage.

## 1.1 A STRONG MARKET GROWTH FROM 1999 TO 2008

The last decade has been particularly exciting for the field of photovoltaics. Different governmental programs, especially in Japan, Germany and Spain, have allowed a strong growth of the PV market. In particular the cleverly designed German Renewable Energy Sources Act (Erneuerbare-Energien-Gesetz, EEG), started in 2000, ensured a strong but controlled growth of the internal German market. Figure 1.1(a) shows how solar cell production worldwide increased from 200 MW in 1999 to over 7 GW

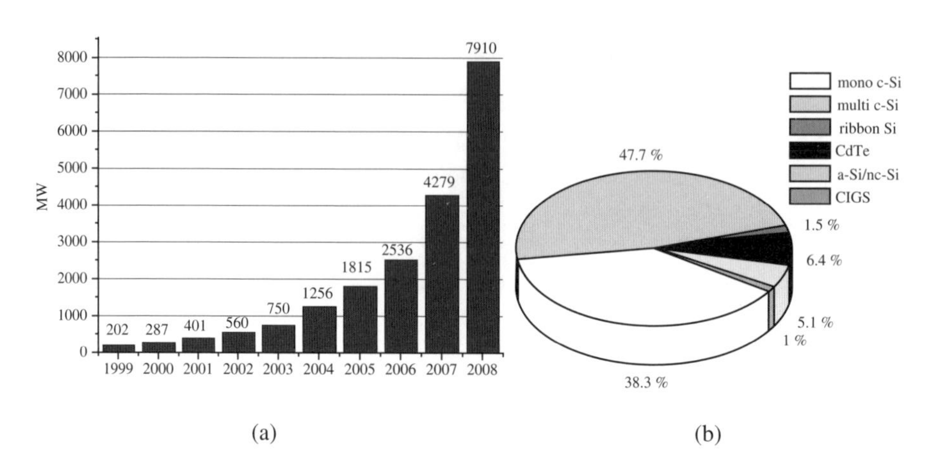


**Fig. 1.1** (a) Annual PV production (both crystalline cells and thin-film modules) over the last decade. Due to delay between cell manufacturing and module installation, the installed capacities in the equivalent year can be a factor 30 to 40 % lower, especially in the last three years. The unit W, Wp or $W_{peak}$ corresponds to 1 W of power delivered under a standard illumination of 1000 W/m$^2$. 1 $W_{peak}$ of module delivers between 1 and 2 kWh annually depending on the location (central or southern Europe). The reported 7.9 GW for 2009, if installed in central Europe, would deliver on average as many kWh as a 1 GW thermal or nuclear power plant with an uptime of 90 %. It would correspond to two such plants, if installed in the sunniest place in Europe. (b) Percentages of production for the various types of technologies. Source: [Photon -2009]

in 2008. The years 2005 to 2008 will be remembered as the golden years for photovoltaics. Not only did the production volume increase every year by 40 to 70 %, but hundreds of new companies entered the field of photovoltaics and announced the launching of new manufacturing capacities. This "explosion" was stimulated by three factors:

- the perspective of short-term profit on markets with high feed-in tariffs;
- the proof that manufacturing costs could be decreased significantly year after year, following typical learning curves of high-tech manufacturing thus paving the way for a huge long-term market for photovoltaics, even without any subsidies or supporting schemes;
- the perception that green technologies will play an increasingly important role in a society plagued by excessive consumption of fossil fuels and faced with major climatic changes induced by human activity.

## 1.2 A TECHNOLOGY COMING TO MATURITY: CRYSTALLINE SILICON

Until now, crystalline silicon (c-Si) PV technology, based on the use of monocrystalline or multicrystalline wafers, has dominated the photovoltaic market and will probably continue to do so for several more years. In 2008, 88 % of the cell and module production was based on c-Si (Fig. 1.1(b)). This technology could initially benefit indirectly from past research and investments made by the semiconductor industry before developing into a sector in itself, with dedicated processes and equipment. An advantage of c-Si technology is the possibility of splitting the complex manufacturing chain from "sand to module" into several separate steps: these include polysilicon production, ingot casting, wafering, cell processing and module manufacturing.

Several companies were successful in the early 2000's at focusing initially on a specific step in the chain, such as solar cell processing (e.g. Q-cells, Suntech, Yingli), wafering (e.g. LDK), modules (e.g. Solon), whereas some other companies looked for a more complete process integration (e.g. REC, Solarworld). Starting in 2007, an increasing number of companies have attempted to achieve a higher degree of integration, in order to capture the full potential of cost reduction, as basically achievable for c-Si.

In the years 2000 to 2008, most of the c-Si growth was achieved based on the "standard" solar cell manufacturing process used for both *p*-type mono- and multicrystalline (mc-Si) silicon wafers or even for Si-ribbons (for which the saw losses of silicon material are reduced). The cell processing step is crucial as it will determine the cell efficiency and, finally, the module efficiency; it consequently influences other related costs (e.g. use of Si in g/W, the module costs in $/W), as less material and surface are used for the same rated power. The standard c-Si cell process is simple. The typical process flow is illustrated in Figure 1.2 and the typical device structure is illustrated in Figure 1.3(a). Using this process, cell efficiency is in the range of 15-16 % for mc-Si and 16-17 % for mono-Si; this translates into module efficiencies of 13 to 15 %. Considered globally, the "standard" crystalline silicon technology can be regarded as robust and reliable. For solar cell processing, production cost reductions of 5 to 8 % per year have been reported by several manufacturers. This was obtained by reducing the wafer thickness, by reaching higher throughputs, a better yield, and by moderate gains in efficiency as well as by moving to larger production quantities. Furthermore, the availability of several turnkey equipment suppliers has allowed newcomers to enter the field with minimal technological risks. Further potential for cost reduction is possible and several studies show that c-Si technology has the potential to reach production costs of 1.2-1.5 $/W_p$ at the module level, within the next 5 years [Canizo-2009].

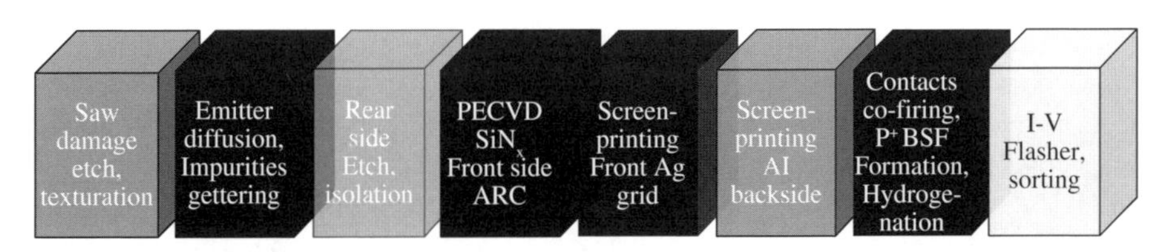


**Fig. 1.2** Standard process flow for mono- and multi-crystalline silicon solar cells.

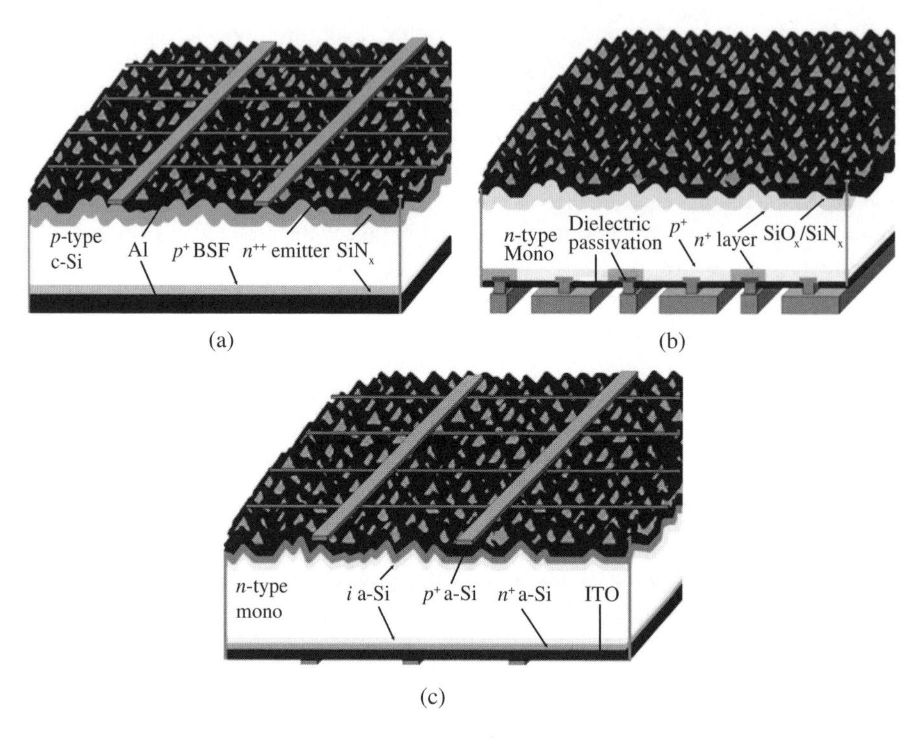


**Fig. 1.3** (a) Standard c-Si solar cell; (b) Sunpower's back-contacted solar cell; (c) HIT solar cell. The three schemes are represented here with "random pyramids" texture.

## 1.3 HIGH-EFFICIENCY CRYSTALLINE SILICON SOLAR CELLS

A few companies have developed alternative solutions to the standard cell manufacturing process, thus permitting significantly higher efficiencies to be reached. For instance, the company Sunpower manufacturers fully back-contacted solar cells with interdigitated electrodes, which reach up to 22 % conversion efficiency in production, at the cost of a significantly more complex manufacturing procedure. Sunpower delivers (status 2009) the world's highest efficiency modules with efficiencies between 17 and 19.6 %. Using a completely different "low temperature approach", Sanyo fabricates its so-called HIT solar cells, which combine the best of the c-Si and thin film Si layers, as discussed in this book. In this case, an $n$-type wafer is "sandwiched" with thin intrinsic layers of amorphous silicon (typically 2-5 nm) on both sides, which provide a quasi-perfect electronic passivation of the wafer. The emitter and backside contacts are ensured by depositing doped amorphous or microcrystalline $p$- and $n$-layers, respectively. The lateral conduction is ensured by transparent conducting electrodes. In HIT concepts, the surface passivation is decoupled from the metallized area and no alignment is necessary for the electrical grid contacting scheme. This concept is remarkable because it allows for the fabrication of devices with record voltages (up to

739 mV for a thin 85 $\mu$m wafer), which translates into a favorable temperature coefficient compared to standard crystalline Si technology. Both Sunpower and Sanyo have reported large cells (>100 cm$^2$) with efficiencies above 23 % at the laboratory level.

In parallel, several companies are now introducing additional steps or modifying the "standard" process to increase further the cell efficiencies: privileged ways include, for instance, introducing a selective emitter (reinforced doping below the contacting finger), improving the backside passivation (e.g. using a $SiO_x/SiN_y$ stack with local contact opening) or improving the metallization (e.g. by fine line printing with subsequent plating of contacts). Finally we note that the physics of c-Si solar cells based on *pn* homojunctions is covered by several textbooks [Green-1982, Goetzberger-1997, Ricaud-1997], is well understood, and the performance of high efficiency devices can be more easily assessed (optically and electrically) than the thin-film silicon cells treated in this book.

## 1.4 THE SILICON FEED-STOCK ISSUE: A TRIGGER FOR THIN-FILM DEPLOYMENT

From 2006 to 2008, the growth of the crystalline silicon solar cell industry was limited by the availability of the so-called "Si feedstock" or purified polysilicon, which was typically required at a level of 9-12 g/W$_p$ [Müller-2006]. Indeed from 2005 onwards the PV industry has been using more purified polysilicon than the semiconductor industry[1]. The companies which did not secure enough silicon faced large risks as they had to buy silicon at very high spot market prices or engage in long-term binding contracts with silicon suppliers. This situation has triggered massive investments in new polysilicon plants as well as attempts to use lower grade silicon such as UMG silicon (upgraded metallurgical silicon). Because of the technical challenges associated with polysilicon production and with the modifications thereof, most of the new plants will need several years to be brought up to speed.

This lack of purified silicon was one of the important factors that opened up a window of opportunity for thin-film PV technologies in the years 2005 to 2008. It enabled firms to exploit the advantages inherent in thin-film technologies. The quantity of active materials required is limited, as layers with thicknesses of only a few hundreds of nanometers or a few micrometers are used, and the total number of processing steps is strongly reduced compared to the full production chain for crystalline silicon. In contrast with-wafer based technology where cells have to be series interconnected by soldering metal ribbons, monolithic integration series interconnection (Fig. 1.4(c)) by laser or mechanical scribing can be applied directly during the module manufacture. Together with the low material usage, this gives the prospect of very low production costs; it should, in principle, allow for the extension of the $/W learning curve towards production costs as low as 0.5 $/W, which is necessary to make solar electricity competitive with base electricity prices in highly insulated areas. Besides that, thin-film technologies offer the prospect of significantly lower energy pay back times when

[1] Note, however, that the level of purity for purified Si for c-Si cells can be slightly lower than the level for microelectronics, simplifying for instance the $SiHCl_3$ distillation process.

compared to crystalline silicon [Fthenakis-2008a and b], coming already close to one year at the system level in sunny areas (with system lifetimes > 20-25 years). In addition, the more aesthetic aspects of thin-film modules and the possibility of manufacturing flexible, unbreakable modules, open up a full range of new PV solutions. As nature never gives anything for free, it turns out that the large-scale manufacture of low-cost thin-film solar modules is far from trivial and several companies have faced major challenges in the past years whilst up-scaling lab processes to real "mass production".

The "classical" thin-film PV technologies can be grouped in three large categories, based on different active absorber materials: copper-indium-gallium-selenide and variations thereof (hereafter named "CIGS"), cadmium telluride (CdTe) and, finally, hydrogenated amorphous silicon (a-Si:H)/microcrystalline silicon (μc-Si:H) and their alloys, regrouped in the category "thin-film silicon" in this book. Approaches based on the crystallization of amorphous thin silicon films are not considered here: on one

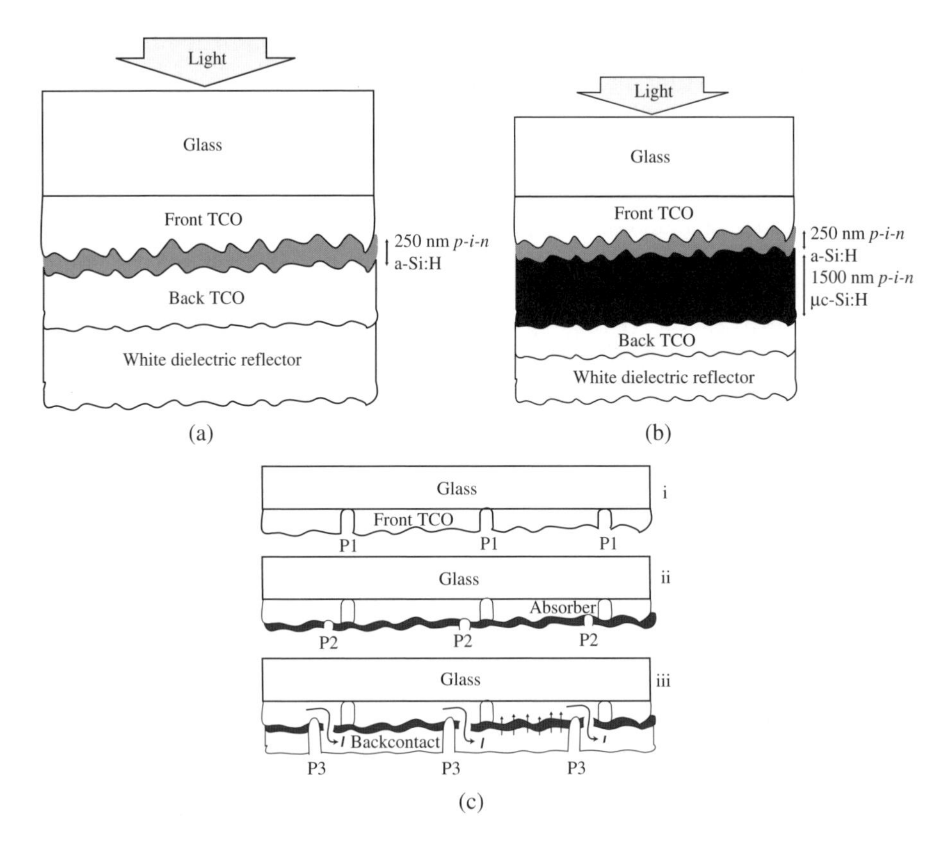


**Fig. 1.4** Sketch of typical thin-film silicon solar cells: (a) amorphous and (b) "micromorph" tandem. These cells are fabricated by depositing thin layers (300 nm a-S:Hi, 1.5 μm μc-Si:H) at low temperatures (200 °C) on a transparent conductive oxide. (c) Cross-section view of the laser scribing sequences (P1, P2, P3) for the monolithic series interconnection. Through the P2 openings, the back-contacts connect the front contact of the next segment. The same current *I* flows from one cell segment to the next. The voltage of the module is the sum of the voltage of the individual cells or segments.

side, the physics of the devices is closely related the properties of ordered crystalline silicon materials and, on the other, they are still in the research or R&D phase. Certified records [Green-2009], typical commercial efficiencies as well as general strengths and weaknesses of the three categories are highlighted in Table 1.1.

Solutions based on the use of chalcopyrites (CIGS and derived compounds) are attractive, because of the record efficiencies achieved here: they are, at the laboratory scale, close to those of crystalline silicon. One of the major challenges is to control properly the deposition of the quaternary alloy used as absorber material in a reproducible way under mass production conditions. Also, for very large production volumes, indium availability might become a concern. Cadmium telluride technology also benefits from the medium to high efficiencies achieved at the laboratory scale and has the advantage of the very fast and relatively "easy" deposition techniques used for the CdTe absorber. Concerns about the availability of Te as well as acceptance issues linked to an intensive use of a cadmium-containing compound could become critical factors for this technology. Finally, thin-film silicon technologies (amorphous silicon, SiGe alloys, microcrystalline silicon), even though they have not so far not demonstrated similarly high efficiencies at the laboratory scale, do have the advantage of plentiful raw material supply and of strong possible synergies with other industrial sectors, such as the flat panel display sector.

**Table 1.1** Short summary status for commercialized thin-film PV technologies (2009). Record efficiencies are taken from [Green-2009]. In this reference, a confirmed stable efficiency of 12.5 % for a-Si/μc-Si/μc-Si cell of smaller dimension, 0.27 $cm^2$, is also reported. Company names are indicative of the status in 2008 and are subject to change. Production quantities based on [Photon - 2009].

| | CdTe | Chalcopyrites CIGS $Cu(In,Ga)Se_2$ $Cu(In,Ga)\ S_2$ | Si, SiGe (amorphous microcrystalline)* |
|---|---|---|---|
| Record lab cell efficiency (>1 $cm^2$) | 16.7 % (1 $cm^2$) | 19.4 % (1 $cm^2$) | 11.7 % (mini-module) (14 $cm^2$) |
| Commercial module efficiency | 8-11 % | 8-12 % | 6-9 % |
| Production quantity 2008 | 510 MW | 50 MW | 400 MW |
| Main producers 2008 (>20 MW) | First Solar | Würth Solar, Honda Solar | United Solar Ovonic, Kaneka, MHI, Sharp, Trony, Ersol, SCHOTT Solar |
| Strength | Fast and "easy process" for absorber deposition | Efficiency close to c-Si | Synergy with flat panel display sectors, ressources unlimited |
| Weakness | Concern on Te availability, reliability of back contact, acceptance linked to Cd | Concern on In availability, sensitive process for absorber deposition, Cd used for high efficiency. devices | Medium efficiency |

*Stabilized efficiencies are given here (after the initial phase of light-induced degradation).

Even though thin-film PV technologies have long held the promise of very low production costs, only a few companies have succeeded, until 2005-2006, in up-scaling production technologies, and this was mostly based on in-house developments. This includes the impressive case of First Solar which succeeded after almost 20 years of development to successfully scale up its production of CdTe modules. By performing smart copies of its first efficient production lines and by continuous efficiency and productivity improvements, it was also the first company who reported ($1^{st}$ quarter 2009) production costs below 1 \$/W. With CIGS, only a few companies have succeeded so far to scale up production capacities, but to a much lower extent than First Solar, and also without demonstrating yet the low \$/Wp capacities. Several tens of companies worldwide are now attempting to ramp up production capacities based on CdTe and CIGS technologies and a few equipment suppliers have announced that they will develop turn key solutions. In the thin-film silicon field, Unisolar, Kaneka, Mitsubishi Heavy Industries (MHI) and Sharp were the first companies to expand capacities in the years 2004-2006 up to several tens of megawatts.

In 2008, close to 1 GW of thin-film PV modules were produced, corresponding to PV market shares of 6.4 % for CdTe, 5 % for thin thin-film silicon and 0.5 % for GIGS. Considering the announced production capacities in the different sectors, most experts expect thin films to play an increasingly important role in PV module production, whereas it is also expected that if the market develops to really large dimensions (>100 GW), each of these three technologies, as well as wafer-based crystalline silicon (c-Si) modules and concentrator systems, will find their own applications, depending on the geographical location (hot-cold, sunny-cloudy) and the type of installation (facades, roofs, solar parks, ...).

## 1.5 THIN-FILM SILICON: A UNIQUE THIN-FILM TECHNOLOGY WITH A "LONG" HISTORY

Thin-film silicon occupies a special place within the field of thin-film photovoltaics for several major reasons: (1) The fact that it is based on abundant materials (Si and Sn or Zn in the simplest version of a single-junction amorphous silicon solar cell with $SnO_2$ or ZnO contact layers) is essential. For a market size reaching several hundred $GW_p$ of PV modules annually, thin-film silicon technology is currently the only viable option among the proven thin-film PV technologies [Wadia-2009]. (2) At such a market level, approaches based on a substantial use of In or Te would become constrained in growth. Finally, (3) no hazardous materials (such as Cd or Pb) are required in thin-film silicon technology, making it, if not safer, at least free from acceptance problems and/or regulatory risks (e.g. ban on Cd).

Another reason for the uniqueness of thin-film silicon is that the basics are well established: this includes the control of doping levels, as well as reproducible fabrication processes for depositing device-quality layers. A strong synergy with another industrial sector, the flat panel display industry, is present. Even though the efficiencies are a little lower than those obtained with the two other competing thin-film technologies, the low temperature coefficient of down to –0.2 %/°C and the self-annealing effect of the amorphous silicon material should ensure that, for high module operating temperatures, the annual energy yield is superior by 5 to 10 % when compared to other types of thin-film or crystalline-silicon modules. Eventually, and as will be described in Chapter 7, broad field experience exists for thin-film silicon modules, so that thin-film silicon can be considered to be a "reliable" technology.

The history of thin-film silicon solar technology started over 30 years ago. Following the first demonstration, in 1976, by Carlson and Wronski [Carlson-1976] of efficient a-Si based devices, and shortly thereafter the demonstration that hydrogenated amorphous silicon (a-Si:H) undergoes a so-called "light-induced degradation (LID)" effect also called the "Staebler-Wronski effect" [Staebler-77], several attempts were made in the 80's to commercialize thin-film silicon modules (Solarex building up on RCA, AFG and Chronar, Arco Solar). The first large-area commercial a-Si:H modules were released in 1986 by Arco Solar (Genesis G-400). At that time, however, the small market size, unsatisfactory control of the reliability aspects, as well as a strong LID effect, did not allow for a successful large-scale deployment of these modules. Based on the simplicity of their manufacturing process and their relatively good performance at low light intensities, a-Si:H cells were in the '80s widely used for consumer applications and, in particular, for calculators. A few companies, derived from the initial ones, with annual production capacities in the 1-2 $MW_p$ range, continued to sell and instal products for outdoor applications; they demonstrated that intrinsically the long-term reliability of the absorber material is ensured. In the late 80's and in the 90's, improvements were made at controlling the a-Si:H material stability by adapting the plasma-deposition process, for instance by introducing very high frequency (VHF) plasma deposition, which allows for higher deposition rates [Curtins-1987]. In Chapter 2, the detailed physical aspects of amorphous silicon will be described; the device physics will be treated in Chapter 4. Plasma fabrication processes are described in Chapter 6.

## 1.6 AMORPHOUS SILICON, MICROCRYSTALLINE SILICON AND "MICROMORPH" DEVICES

The demonstration in the mid 90's that efficient microcrystalline solar cells could be fabricated in a relatively simple way providing that external impurities (e.g. O) were carefully managed [Meier-1996], revived interest in thin-film silicon solar cells, as it paved the way for higher efficiency tandem devices, which make a better use of the solar spectrum. Whereas amorphous silicon (a-Si:H) has a typical bandgap of 1.75 eV and absorbs mostly visible light, microcrystalline silicon (μc-Si:H) has a bandgap of 1.1 eV and absorbs also the near-infrared light (Fig. 1.5(a)). Device-grade μc-Si:H can be fabricated with a similar PECVD process as a-Si:H, based on $H_2$ and $SiH_4$ input gases. μc-S:H is a mixed-phase material composed of Si nanocrystals through which carriers can percolate, embedded in a passivating a-Si:H matrix. The properties of μc-S:H (described in Chapter 3) depend, as for a-Si:H, strongly on the deposition conditions. Hydrogenated μc-Si material is also almost unaffected by the Staebler-Wronski effect. The beauty of the microcrystalline/amorphous or "micromorph" tandem cell concept lies in combining a-Si:H with μc-Si:H within a single device, allowing stabilized efficiencies in the range of 12 % to be reached now. The potential maximum efficiency is substantially higher, if one considers the ideal bandgaps of a-Si:H and μc-Si:H within a tandem structure. Figure 1.4

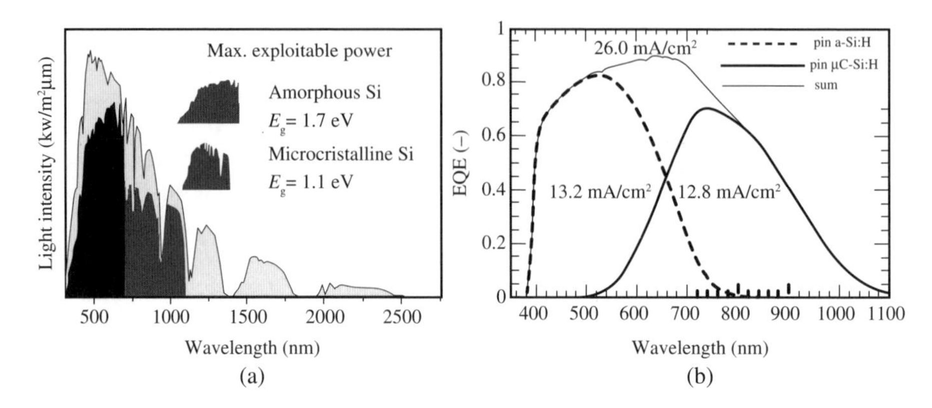


**Fig. 1.5** (a) Solar spectrum compared with the maximum power that can be generated using amorphous and microcrystalline silicon. (b) External quantum efficiency of a micromorph solar cell (Source: IMT, Neuchâtel).

shows typical device constructions for single-junction a-Si:H and tandem a-Si:H/μc-Si:H devices. Figure 1.5(b) shows, for a state of the art "micromorph" device, the ratio of collected electrons versus incident photons, i.e. the external quantum efficiency (EQE) in short-circuit conditions. A large fraction of the light from 400 to 1000 nm is collected. Such tandem and multi-junction devices will be covered in detail in Chapter 5.

In the time frame 1990 to 2005, intense research activities took place in various academic and industrial laboratories (see, e.g., [Shah-2004]). Record stabilized efficiencies were confirmed: 9.5 % for a-Si:H (1 cm², IMT), 10.3 % for μc-Si:H (1 cm², Kaneka) and 11.7 % for "micromorph" tandems (14 cm², Kaneka) [Green-2009], Even higher stabilized efficiencies were reported for smaller triple-junction cells (e.g. 12.3 % for a-Si/a-SiGe/a-SiGe and 12.5 % for a-Si/μc-Si/μc-Si for ~0.25 cm² devices from United Solar Ovonic [Green-2009]). Japanese companies were the first to start mass production of amorphous silicon modules on glass (Sharp, Kaneka, MHI); Kaneka also introduced the first pilot production for "micromorph" modules in 2001. Most companies carried out in-house development of the production equipment for module manufacturing. For instance, Mitsubishi Heavy Industries (MHI) developed a novel electrode design that allows for record deposition rates in industrial reactors. The US company United Solar Ovonic was the first firm to commercialize triple-junction a-Si/a-SiGe/a-SiGe cells on flexible substrates. In parallel, several other companies started pilot lines for flexible modules based on amorphous silicon (e.g. VHF-Technologies, Powerfilm).

The process flow for fabricating a thin-film silicon solar module is simple; it is illustrated, for modules prepared on glass substrates, in Figure 1.7. After glass washing, a textured transparent conductive oxide (e.g. ZnO, as illustrated in Figures 1.6 and 1.9) is deposited, with typical sheet resistances in the order of 5-15 ohm/square. As an alternative, a $SnO_2$-coated glass can also be used. The surface texture, with features in the range of 0.1 to 1 $\mu$m, allows for light scattering into the absorber layer, thereby enhancing the solar cell's current, as compared to the case of a flat surface, thus, making, the use of a low absorber layer thickness possible. A first laser scribe divides the TCO into several segments. In the next step, the active silicon layers are deposited by plasma-enhanced chemical vapor deposition (PECVD), at typical excitation frequencies between 13.56 and 60 MHz. Boron-doped, intrinsic and phosphorus-doped layers are deposited. Depending on the process conditions, amorphous or microcrystalline layers can be grown (typically with an increased H flow and/or an increased plasma power, microcrystalline layers result). When tandem devices are prepared, an amorphous *pin* sub-cell is deposited first followed by a microcrystalline *pin* sub-cell. The two sub-cells are connected by a tunnel junction. A second laser scribe is then performed, followed by deposition of the electrical back contact. A typical back contact should be highly reflective and is usually of the type "thin ZnO/Ag" (or thin "ZnO/Al" for amorphous silicon); it can also be constituted by a thick conductive ZnO layer covered, later, by a white dielectric reflector. A third scribe terminates the monolithical integration (Fig. 1.7). Finally, the coatings are removed on the module edges, electrical contacts are applied, and a final lamination step takes place. Chapter 6 will present a detailed description of the various processes used for the manufacturing of thin-film silicon PV modules.

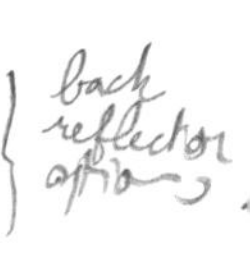

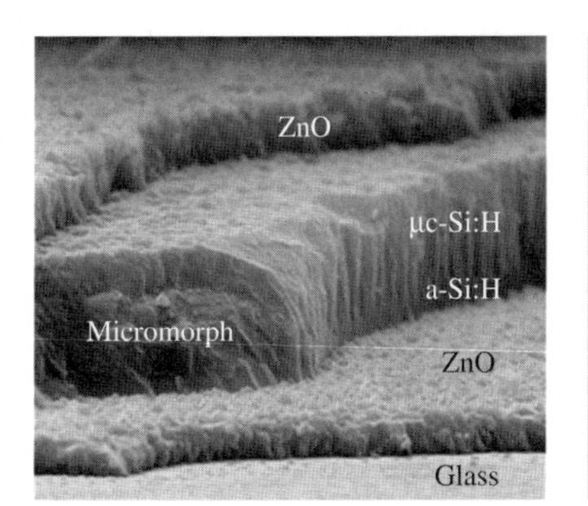


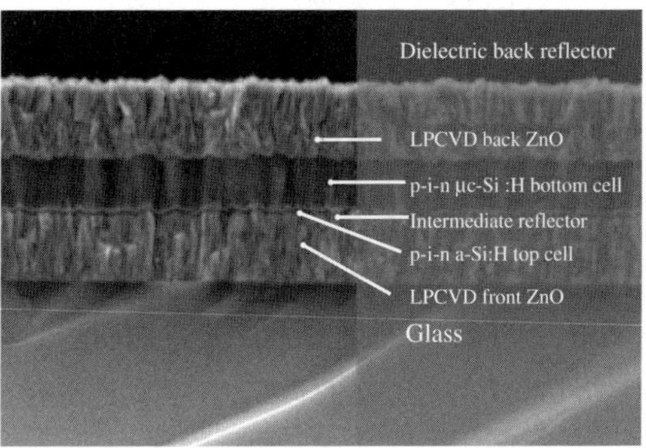


**Fig. 1.6** Two different SEM views of "micromorph" solar cells (Source IMT, [Dominé-2009]). The cells are deposited on a glass coated with ZnO by low pressure chemical vapor deposition (LPCVD).

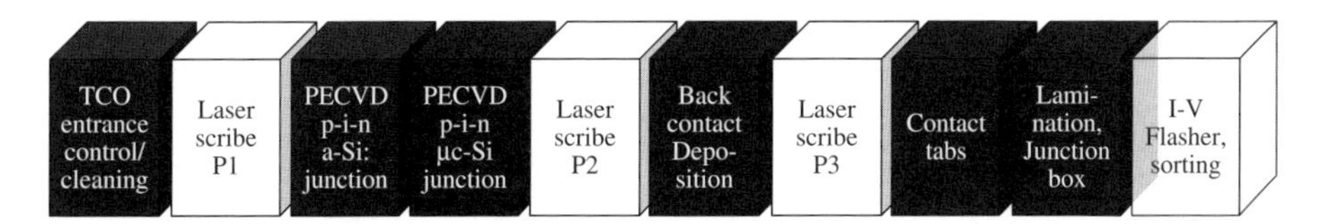


**Fig. 1.7** Diagram flow for the manufacture of complete thin-film silicon modules based on a-Si:H and μc-Si:H.

## 1.7 SYNERGY WITH THE DISPLAY SECTOR AND EMERGENCE OF A LARGE PV SECTOR

The flat panel display sector also developed in an impressive manner in the years 1995 to 2005. In the manufacture of liquid crystal displays (LCD) over 10 million square meters had already been coated in 2004 with thin amorphous silicon layers, both doped and undoped, and with silicon nitride layers used for the fabrication of thin-film transistors. This topic will be the object of Chapter 8. The core equipment used for such deposition processes is based on Radio-Frequency (RF) PECVD. The flat panel display industry invested massively in production-scale fabrication tools and demonstrated a continuous decrease of their production costs (20 % annually per/square meter). It was hence natural that equipment makers identified thin-film silicon solar technology as a high potential growth market for their products, as similar PECVD equipments can be used for the absorber layer deposition. This led companies such as Oerlikon, Applied Films (later purchased by Applied Material) and Ulvac to enter the solar PV market, transferring to mass production (as illustrated in Fig. 1.8) several of the processes developed in research laboratories.

The possibilities for module production companies to purchase either turnkey manufacturing lines or, at least, professional equipment provided by high tech equipment suppliers, coupled with the difficulties of crystalline silicon technology to deal with the feedstock problem, led to a large number of announcements (over 50) of companies entering the thin-film silicon module manufacturing business from 2006 to 2008. At the beginning of 2010, the leading companies are able to manufacture single-junction or tandem amorphous silicon modules with total-area stabilized efficiencies of 6.5 to 7 %, whereas the best "micromorph" products reach 8 to 9 % total-area stabilized efficiency (on glass). There is still an efficiency gap of about one-third compared to the best laboratory results, but this should be reduced over the coming years. The best flexible PV modules (with triple-junction cells based on a-Si:H and on a-SiGe;H alloys) reach up to 7.4 % stable efficiency (aperture area).

Table 1.2 gives a partial overview of the companies working in the field of thin-film silicon, as of the start of 2010. Indeed, the various announcements of established and new companies could culminate in as much as 4 to 6 $GW_p$/year by 2012. Noticeably, the development of the thin-film silicon sector has been impressive: this sector has been able to rapidly expand, with mass production facilities installed or announced in more than 15 countries.

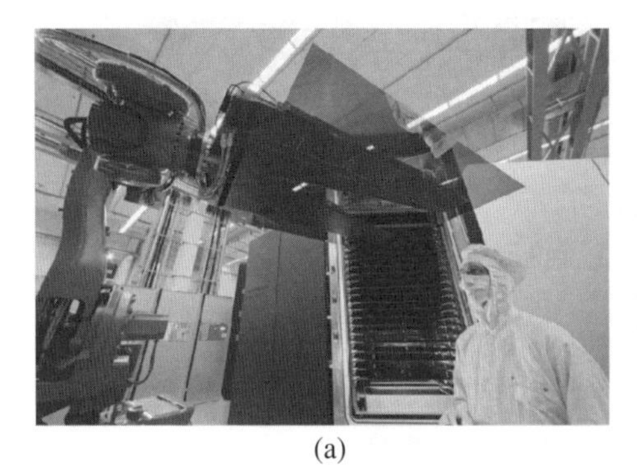

(a)

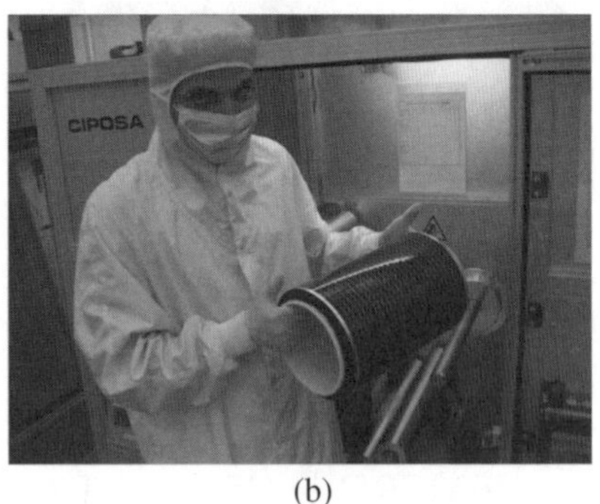


(b)

**Fig. 1.8** (a) KAI-production tools of Oerlikon Solar for a-Si:H and μc-Si:H deposition (Trübbach, CH). (b) Roll of a-Si:H cells on low-cost polymer at VHF-Technologies "Flexcell" (Yverdon, CH).

**Table 1.2** Overview of some of the major industrial actors in the thin-film silicon PV sector, status at the beginning of 2010. Names of companies and their future development are subject to major change.

| Thin-film silicon module producers | Entering the market 2008-2010 | Equipment suppliers (PECVD) | Pilot-line and/or consumer products |
|---|---|---|---|
| Sharp ++ | Pramac++ | Oerlikon++ | Sanyo~~ ++ |
| Kaneka ++ | Heliosphera++ | Applied materials++ | VHF-Technologies ~~ |
| Mitsubishi HI ++ | Gadir | Leybold optics++ | Power film~~ |
| United Solar | Astroenergy++ | Ulvac++ | Nuon ~~ |
| Ovonic** ~~ | GET | TEL++ | Xunlight ** ~~ |
| Fuji** ~~ ,Bosch ++ | Masdar PV | Jusung | ICP |
| Trony, Sunwell | XinAo | Ishikawa HI++ | Amelio |
| Inventux ++ | Nexpower | Thin Film Solar | SOLEMS |
| Auria Solar ++ | Amplesun | Energo Solar | Bengbu PB |
| Sunfilm ++ ENN ++ | Dupont Apollo | ...(EPV) | various small ... |
| Schott Solar | Masdar PV | | |
| Boading Tianwai | Suntech various | | |
| Moser Baer, Malibu | a-Si:H ... | | |
| T-Solar Global | | | |
| Signet | | | |
| VHF-Technologies ~~ | | | |
| EPV, | | | |
| BSC | | | |
| various a-Si:H ... | | | |

++ indicates a-Si/μc-Si technology, ~~ indicates flexible products, ** indicates use of a-Si,Ge:H alloys. Several amorphous Si plants, mostly in Asian countries or on first-generation equipment are not shown here.

## 1.8 PERSPECTIVES AND CHALLENGES FOR THIN-FILM SILICON TECHNOLOGY

In 2009-2010, the situation has become more challenging for many companies active in photovoltaics. It is likely that not all expansion plans and capacity increases will be realized, including those in the thin-film silicon sector. A combination of factors creates a difficult situation. On one side, lured by the prospect of high feed-in tariff especially in Spain, many PV companies have increased their production capacity. The virtual disappearance of the Spanish market, where an inappropriately high feed-in tariff led first to an explosion of installed capacity at high prices, followed by the introduction of a very low annual "cap" ( = maximum new capacity eligible for feed-in tariff) for newly installed capacity, as well as the slower than expected market growth in countries supportive of photovoltaic electricity, created a situation of oversupply. This forced a sharp drop in module prices and caused cash flow and debt problems for many companies. The financial crisis reinforced this effect: investments in photovoltaics were reduced and only the lowest cost and/or most established products made it to the end customer. First Solar succeeded earlier than others with CdTe thin-film modules in bringing down its cost and in bringing its efficiency to close to 10 %, therefore also puts strong pressure on the PV market.

However, considering the huge market of hundreds of GWs that could be deployed, in conjunction with solutions for energy storage, transformation and transportation, fantastic promise remains for PV modules based on amorphous silicon, microcrystalline silicon and their alloys. As mentioned earlier, this prospect is based on the abundance of the materials involved, the possibility of strongly reducing manufacturing costs, at very low levels in terms of $/m$^2$, and further increasing efficiencies. Most companies entering production have so far purchased first-generation equipment issued and adapted from the flat panel display industry, but the prospect of a new generation of fabrication solutions should guarantee further improvement of the technology. Altogether, long-term success will be linked to addressing the major challenges for the thin film silicon sector, including the following:

- Better substrates and textured front electrodes, with a "proper nanostructure", that allows for both strong light confinement and for proper growth of the silicon layers, will have to be produced industrially. This aspect is certainly as important as those linked to the fabrication of the silicon absorbers. Figure 1.9 illustrates the various shapes and structures used in today's state-of-the-art devices [Dominé-2008], [Dominé-2009] at IMT Neuchâtel.
- A new generation of equipment allowing state-of-the-art deposition of μc-S absorber material at lower costs needs to come on the market. This can be achieved with equipment allowing higher throughput (e.g. through higher deposition rates) and/or by designing equipment with lower capital expenditure ("capex") and running costs.
- Environmentally friendly and cost-effective solutions for reactor cleaning have to be introduced, moving away from the typical $SF_6$ or $NF_3$ gases used today in some production lines, as this leads to sub-optimum $CO_2$ equivalent emission in g/kWh [Fthenakis-2008a]. Well-controlled $NF_3$ or $SF_6$ use, without release at the gas production site and with perfect gas abatement at thin-film silicon plants, or $F_2$ with on-site (at module plant) production and use, would allow better lifecycle performances.
- Improvement in device efficiencies will have to be continuously demonstrated, in order to secure the future of the technology. At the level of laboratory devices, stable efficiencies over 13-14 % should be achieved in the near future with tandem junction devices; within about a further decade, over 16 % should be attained by introducing additional junctions. The quest for stable high (2 eV)

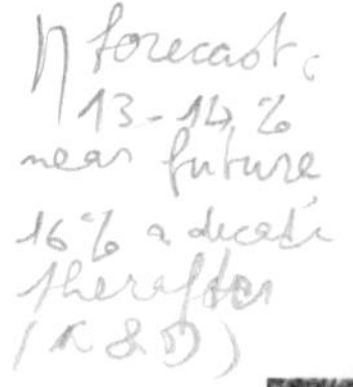

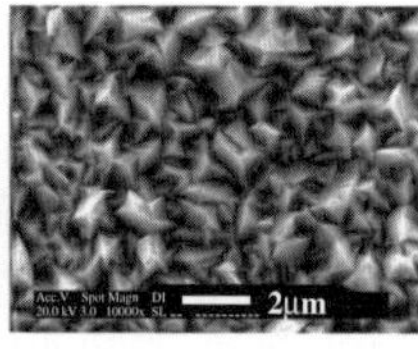

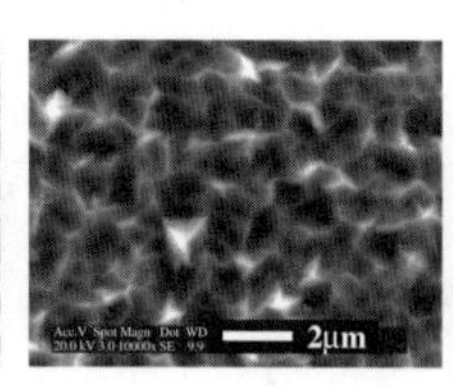

**Fig. 1.9** Various surface textures of TCO surfaces, all from LP-CVD ZnO deposition (Source IMT, [Bailat-2006, Dominé-2009]). An extended and complete review of ZnO for solar cell applications can be found in [Ellmer-2008].

and mid-bandgap (1.5 eV) material, if successful, could improve the technology even further.

- Confidence has to be fostered in future customers that thin-film silicon technology is reliable and robust and that investment in this technology is a must. This implicates low-cost but ultra-reliable module technology, as well as a proper communication of the achieved durability performances.

Provided that the right efforts are made by the various industries, academic and research institutions involved, thin-film silicon technology will play a key role in renewable electricity and energy production in the future. The topic of energy resources is of too high an importance for the thin-film silicon road not to be fully explored and brought to full maturity!

## 1.9 REFERENCES


[Bailat-2006] Bailat J., Dominé D., Schlüchter R., Steinhauser J., Faÿ S., Freitas F., Bücher C., Feitknecht L., Niquille X., Tscharner R., Shah A., Ballif C. "High-efficiency p-i-n microcrystalline and micromorph thin film silicon solar cells deposited on LPCVD ZnO coated glass substrates" (2006), *Proceedings of the 4th World Conference on Photovoltaic Energy Conversion*, pp. 1533-1436

[Canizo-2009] C. del Cañizo, G. del Coso, and W.C. Sinke, "Crystalline silicon solar module technology: towards the 1 € per watt-peak goal", (2009), *Progress in Photovoltaics: Research and Applications Vol* 17, pp. 199–209

[Carlson-1976] Carlson D.E., Wronski C.R., "Amorphous silicon solar cells", *Applied Physics Letters*, (1976), Vol. **28,** pp 671-673.

[Curtins-1987] Curtins H., Wyrsch N.; Shah A. "High-rate deposition of amorphous hydrogenated silicon: effect of plasma excitation frequency", (1987), *Electronic Letters*, Vol. **23**, pp. 228-230

[Dominé-2009] Dominé, D., "The role of front electrodes and intermediate reflectors in the optoelectronic properties of high-efficiency micromorph solar cells", *Ph. D. thesis, Faculté des Sciences, University of Neuchâtel*, in print (2009)

[Dominé-2008] Dominé D., Buehlmann P., Bailat J., Billet A., Feltrin A., Ballif C., "High-efficiency micromorph silicon solar cells with in-situ intermediate reflector deposited on various rough LPCVD ZnO", (2008) *Proceedings of the 23rd European Photovoltaic Solar Energy Conference*, pp. 2091-2095

[Ellmer-2008] Ellmer, K., Klein, A., Rech, B. Editors, "Transparent conductive zinc oxide: basics and applications in thin film solar cells", *Springer Series in Materials Science*, **104**, (2008). ISBN: 9783540736110

[Fthenakis-2008a] Fthenakis V., Gualtero S.; van der Meulen, R, Kim, H.C., "Comparative life-cycle analysis of photovoltaics based on nano-materials: A proposed framework", (2008), *Materials Research Society Symposium Proceedings,* Vol. **1041**, pp. 25-32

[Fthenakis-2008b] Fthenakis, V.M., Kim, H.C., Alsema, E., "Emissions from Photovoltaic Life Cycles", (2008), *Environmental Science and Technology,* Vol. **42**, pp 2168–2174

[Goetzberger-1997] Goetzberger A., "Sonnenergie: Photovoltaik", Teubner Verlag, Stuttgart (1997)

[Green-1982] Green M.A., "Solar Cells" Volumes 1, 2, 3, Prentice Hall, Englewood Cliffs, NJ (1982)

[Green-2009] Green M.A., Emery K., Hishikawa Y., Warta W., "Solar Cell Efficiency Tables (Version 34)",(2009), *Progress in Photovoltaics: Research and Applications*, Vol. **17**, pp. 320–326

[Meier-1996] Meier J., Torres P., Platz R., Dubail S., Kroll U., Selvan J.A., Pellaton-Vaucher N., Hof C., Fischer D., Keppner H., Shah A., Ufert K.-D., Giannoulès P., Koehler J., "On the way towards high-efficiency thin film silicon solar cells by the 'micromorph' concept", (1996), *Materials Research Society Symposium Proceedings,* Vol. **420,** pp. 3–14

[Müller-2006] Müller, A., Ghosh, B., Sonnenschein, R., Woditsch, P. "Silicon for photovoltaic applications", (2006), *Materials Science and Engineering: B*, Vol. **134**, pp. 257–262

[Ricaud-1997] Ricaud A., "Photopiles solaires", (1997), *Cahiers de chimie*, PPUR, pp. 1-352 (1997). ISBN: 2-88074-326-5

[Photon-2009] *Photon International*, March 2009 pp 170-183.

[Shah-2004] Shah A., Schade H., Vanecek M., Meier J., Vallat-Sauvain E., Wyrsch N., Kroll, U., Droz C., Bailat J., "Thin-film Silicon Solar Cell Technology", (2004), *Progress in Photovoltaics: Research and Applications*, Vol. **12**, pp. 113–142

[Staebler-1977] Staebler D.L., Wronski C.R., "Reversible conductivity changes in discharge-produced amorphous Si", (1977), *Applied Physics Letters*, Vol. **31**, pp. 292–294 (1977).

[Wadia-2009] Wadia C., Alivisatos A.P., Kammen D.M., "Materials Availability Expands the Opportunity for Large-Scale Photovoltaics Deployment", (2009), *Environmental Science and Technology*, Vol. **43**, pp. 2072–2077

CHAPTER 2

# BASIC PROPERTIES OF HYDROGENATED AMORPHOUS SILICON (a-Si:H)

*Arvind Shah, with the collaboration of Wolfhard Beyer*

## 2.1 INTRODUCTION

Pure silicon thin films can be deposited by vacuum deposition techniques, like evaporation, sputtering, etc. They are either amorphous, i.e. they consist of a disordered atomic structure, or they can be microcrystalline, i.e. they contain small crystallites with diameters around 1 μm or less. In general, such thin layers of silicon contain a very high density of dangling bond defects (around $10^{19}$ defects per $cm^3$) and cannot be used to build semiconductor devices.

On the other hand, if thin silicon layers are deposited by Plasma-Enhanced Chemical Vapor Deposition (PECVD) from silane ($SiH_4$), or preferably from a mixture of silane and hydrogen, the hydrogen atoms will passivate a large part of the defects and the resulting thin films can, if suitably optimized, be useful for forming semiconductor devices. This specific method of fabricating thin silicon layers was pioneered by Walter Spear, Peter Le Comber and their research group at the University of Dundee in the first half of the 1970's. In a landmark paper published in 1975 [Spear-1975] (see also [Spear-1976]), they demonstrated that amorphous silicon layers deposited from silane by PECVD could be doped by adding either phosphine ($PH_3$) or diborane ($B_2H_6$) to the plasma discharge. They showed that the conductivity of these thin amorphous silicon layers could be increased by several orders of magnitude. They based their investigations on earlier, only partially published work by R.C. Chittick et al. [Chittick-1969]. The publications of Spear and Le Comber opened up the use of amorphous silicon layers for fabricating diodes and, thus, solar cells, as well as thin-film transistors, which can be used for the active addressing matrix in liquid crystal displays.

## 2.1.1 Structure of amorphous silicon

***Structural disorder: bond angle and bond length distributions***

Amorphous solids possess short-range order: the nearest atomic neighbors are almost placed in the same position as in the corresponding crystal. However, over a longer distance the periodic structure or "lattice", which is characteristic of crystalline material, is not maintained and the deviations become very large. This is shown schematically (for the simplified two-dimensional case) in Figure 2.1a.

Due to the short range order, many physical properties of a material are similar in the amorphous and crystalline forms. Other properties, like bandgap, optical absorption coefficients, electrical conductivities, undergo, in general, significant changes (see Sects. 2.3 to 2.5).

In Figure 2.1b, the concept of structural disorder is illustrated, again for the simplified two-dimensional case, but for a rather extreme situation. (This situation does not correspond to the one found in amorphous silicon, but gives us a sort of "caricature", which the reader may easily understand.)

The atomic structure of an amorphous material is characterized by a certain amount of disorder. Indeed an amorphous material can possess more or less disorder. The disorder has an effect on the bond angle and bond length distributions. Let us first look at bond angles: in the case of crystalline silicon the angle between two adjacent Si-Si bonds is fixed and is exactly 109° 28′, as represented in Figure 2.2. For amorphous silicon, we have a distribution of bond angles with a standard deviation

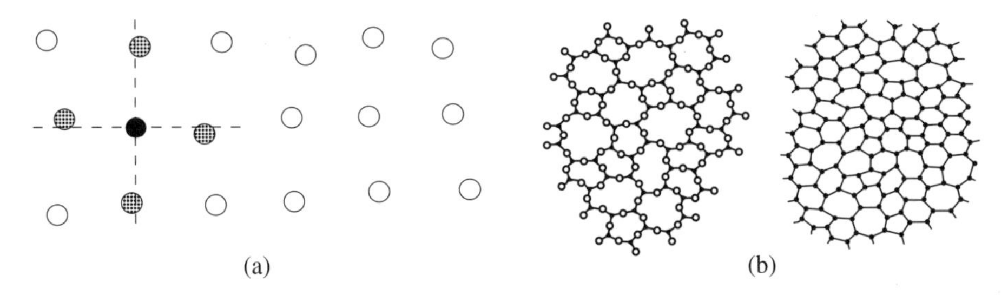


**Fig. 2.1** Schematic two-dimensional representations of the material structure in an amorphous solid: (a) diagram indicating that the atoms (small circles) have nearest neighbors, which are almost in the same position as in a crystalline structure (symbolized by the broken lines); and (b) diagram with a large disorder in the short-range atomic configuration. (according to [Zallen-1983])

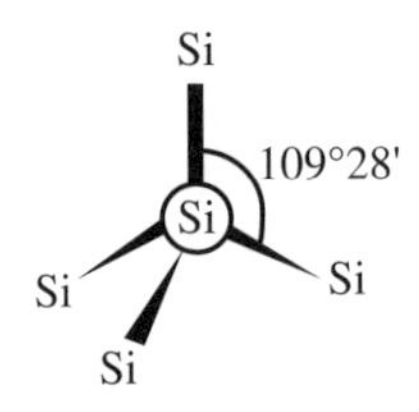


**Fig. 2.2** Atomic model for crystalline silicon, indicating the bond angle formed between two adjacent bonds. In the case of amorphous silicon, this angle has a distribution of values.

of 6° to 9° around this value. Now let us look at bond lengths: in the case of crystalline silicon the bond length is fixed and is approximately 0.235 nm. For amorphous silicon, one finds a random distribution of bond lengths around this value (see e.g. [Singh-2004]).

If the distributions are broad, i.e. if the bond angles and bond lengths vary a lot within the material, we consider the material to be highly disordered. It will then also have very pronounced bandtails (see Sect. 2.2.1). Very often such highly disordered material will have, at the same time, a high density of "mid-gap defects" or "dangling bonds" (see Sect. 2.2.2). Such a material will not be suitable for fabricating semiconductor devices.

What we would like to have for applications is an amorphous material with a minimum amount of disorder. However, it turns out that when we try to reduce the disorder below a certain threshold (corresponding to a standard deviation in the bond angle distribution of around 6°), we are no longer able to deposit amorphous layers, and the material crystallizes.

### *Dangling bonds*

An amorphous material cannot extend over larger distances without the formation of defects in the structure: this means that "point defects" will occur and that, in fact, certain bonds will be totally missing. Such a missing bond is also called a "dangling bond". A silicon atom with a dangling bond is only bonded to three neighboring atoms instead of four, as shown in Figure 2.3a. This means that such an atom has bonds which are not saturated and the corresponding electronic states lie near the midgap and act as recombination centers. This will be explained in Sections 2.2.2 and 2.4.4.

The dangling bonds are "amphoteric", which means that they can have either positive or negative charge (or be neutral). In fact, due to capture kinetics, they will generally assume, in the majority, the same charge state as the free carriers (holes or electrons) in the amorphous semiconductor. For instance, in a *p*-type semiconductor material the dangling bonds will mostly be positively charged.

In amorphous silicon layers, as deposited by Plasma-Enhanced Chemical Vapor Deposition (PECVD) from silane ($SiH_4$), a large proportion (over 99 %) of the

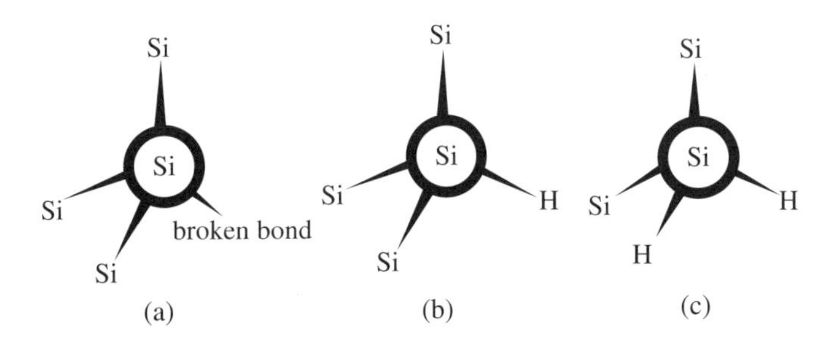


**Fig. 2.3** Model for silicon atom with (a) unpassivated dangling bond (acting as recombination center or mid-gap state); (b) dangling bond "passivated" by a hydrogen atom (and no longer acting as a dangling bond); (c) two hydrogen atoms connected to it. ($SiH_2$-configuration)

original dangling bonds are "passivated" by hydrogen. These "passivated" dangling bonds have a hydrogen atom sitting on them, as represented in Figure 2.3b. These "passivated" dangling bonds do not act as recombination centers and do not constitute gap states; they will therefore not be counted as dangling bonds in the following discussion in Section 2.2.2.

One may note the following points:

The density of "unpassivated" dangling bonds increases under the influence of light, until it asymptotically reaches a substantially higher, "stabilized" value: this is the so-called "light-induced degradation" or "Staebler-Wronski effect", which will be described in Section 2.2.3.

Some of the silicon atoms in hydrogenated amorphous silicon (a-Si:H) are bonded to two hydrogen atoms (Figure 2.3c). In this configuration, the energy for Si-H bond rupture is less than the energy required for breaking the Si-H bonds in Figure 2.3b, where the silicon atom is bonded to a single hydrogen atom. Therefore this configuration (also called $SiH_2$-configuration) may contribute in a more pronounced manner to light-induced degradation.

The amount of hydrogen atoms incorporated in usual plasma-deposited amorphous silicon layers is found to be much larger than what would be quantitatively required to "passivate" all the "original" dangling bonds. Indeed the hydrogen content in usual plasma-deposited amorphous silicon layers has been found by measurements to be about 5 to 10 atomic %, whereas less than 1 % of hydrogen atoms would certainly be sufficient to "passivate" all the dangling bonds. There is therefore certainly a large amount of hydrogen atoms incorporated at positions interrupting Si-Si bonds. These "additional" hydrogen atoms are probably connected with the light-induced degradation effect described hereunder. Experiments have revealed cases where as one increases the hydrogen density in a-Si:H layers, one also increases the degradation effect.

Additionally, it is in practice impossible to "passivate" all dangling bonds. There is always a residual amount of "unpassivated" dangling bonds.

***Microstructural properties***

Thin films deposited by chemical vapor deposition (and also by other methods) are never fully uniform, but always posses a particular "microstructure" (see Figures 2.4 and 2.5).

The first few atomic layers are in general deposited on a substrate constituted by another material; therefore there will be an increased density of defects in this region, due to mismatch between the lattices (atomic structures). One can well imagine that this leads to increased mechanical stress and to an increased density of broken bonds (and, also, of weak bonds) in both the substrate and in the layer which is being deposited.

The surface of the layer is another region where the defect density is greater than in the bulk. If we look at a free surface (covered only by air), we expect to have here a major discontinuity in the structure; the whole atomic matrix comes to an end with a whole "sheet" or "plane" of broken bonds; most of these will (at least in the case of hydrogenated amorphous silicon, a-Si:H) be passivated by hydrogen (or oxygen) atoms, but even so the remaining surface defect density will be rather large.

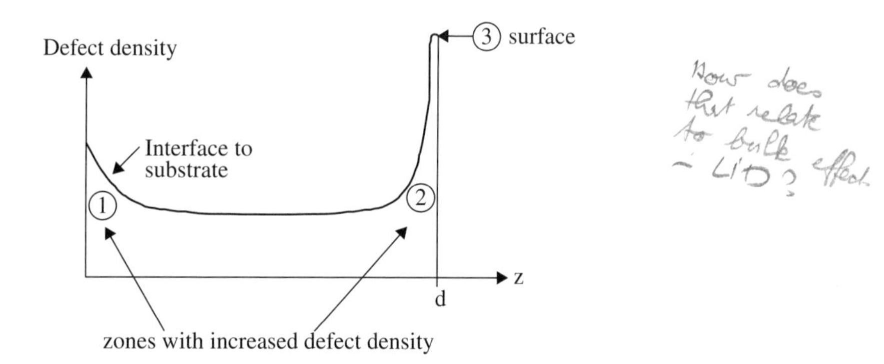


**Fig. 2.4** Schematic diagram showing increased defect density in the initial and final regions of a thin film of thickness *d*.

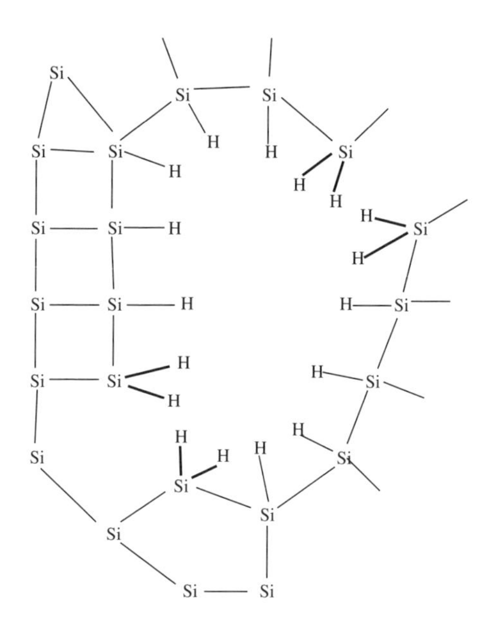


**Fig. 2.5** Schematic diagram showing micro-voids as commonly formed within a thin-film silicon layer.

Often another layer will be deposited on top of the semiconductor layer (in the case of solar cells, this will often be a metallic contact layer or a transparent conductive oxide layer); in this case we have a similar situation as described above, for the initial part of the layer.

Micro-voids can be formed within the layer due to irregularities in the deposition process; such microvoids invariably lead to the formation of internal surfaces; they can be assumed to have, therefore, a larger density of broken bonds (and of weak bonds) than the rest of the layer.

The microstructural properties are of great importance when assessing the electronic quality of a semiconductor layer. In both amorphous and microcrystalline silicon layers, they determine the quality of the resulting devices. In both cases it is quite decisive to work with layers that are as dense as possible (i.e. have relatively few microvoids) and where the lattice mismatch problems on both sides of the layers have been reduced, as far as possible. It is not always evident how to obtain such a situation.

In the case of amorphous silicon, the density of the layers has a direct relationship with the light-induced degradation effect (so-called Staebler-Wronski effect, see below): porous layers, which have a high density of micro-voids and therefore many weak bonds at the "internal surfaces" of these micro-voids, tend to have very pronounced degradation behavior.

In microcrystalline silicon, micro-voids and cracks/fissures are also observed; they can lead to reversible changes in the conductivity (see Sects. 3.4 and 3.5) and – if they are very pronounced – can also result in very poor solar cell performance (see Sect. 4.5.11).

### 2.1.2 "Free" and "trapped" carriers (electrons and holes); mobility gap

We shall now look at the behavior of charge carriers, i.e. of electrons and holes within an amorphous semiconductor. As we shall see in Chapter 4, the performance of all solar cells is determined by three main effects, which, in turn, depend on the semiconductor used for fabricating the solar cell:

(1) The conversion of photons into electron-hole pairs within the semiconductor. This effect depends on the bandgap of the semiconductor (Sect. 4.1.1).

(2) The separation of electrons and holes (Sect. 4.1.2).

(3) The transport of electrons and holes to the electrical contacts of the solar cell. This effect depends on the mobilities of electrons and holes within the semiconductor, and on the recombination process between the electrons and holes. It is treated in Section 4.3 for "classical" *pn*-type solar cells as used for example in the case of crystalline silicon wafers; here transport is mainly by diffusion of charge carriers. It is treated in Sections 4.5 to 4.11 for *pin*- and *nip*-type solar cells, as generally used with thin-film silicon (with both amorphous and microcrystalline silicon); here, transport is mainly by drift, i.e. it occurs under the influence of an electric field.

Now, in amorphous semiconductors an overwhelming part of the charge carriers (electrons and holes) are not free to move under the influence of an electric field, but are "trapped" carriers. This is a direct effect of the amorphous nature of the material and it is referred to as "localization":

Considerations based on quantum physics (i.e. relatively complex theoretical and numerical calculations) show the following: In a periodic atomic structure (such as is the case in crystalline materials) an electron can basically possess an extended wave-function, i.e. is basically present in the whole crystal; this means that at a temperature of 0 K it would be free to move with a velocity/mobility that is theoretically infinite, and at higher temperatures with a velocity/mobility limited only by lattice vibrations. However, in practice the velocity/mobility of electrons in crystalline silicon is mainly limited by crystal defects and impurities.

On the other hand, in an amorphous material, a large part of the electrons are "localized" and confined ("trapped") within a short perimeter. In a semiconductor this applies not only to the states with the lowest energy in the conduction band, i.e. those occupied by electrons, but also to the states with the highest energy in the valence band, i.e. to those states that are normally considered to be occupied by electrons with a positive charge or by "holes". Even the small fractions of electrons and holes that are free have a mobility that is reduced by the amorphous nature of the material. This leads to a situation as shown in Figure 2.6, where there is no real energy gap anymore, but where the energy levels between the valence and conduction band are filled with additional states: with bandtail states due to "localization" or "trapping" and midgap states due to the dangling bonds mentioned above.

In contrast to a crystalline semiconductor where a real bandgap exists, i.e. where we have a range of energies with virtually no electronic states at all, we have in amorphous semiconductors a "mobility gap", i.e. a range of energies, where there are electronic states but these states are "localized" and have zero-mobility associated with them. An additional effect is that the mobility gap of hydrogenated amorphous silicon has a larger value (typically between 1.7 and 1.8 eV) than the bandgap of crystalline silicon (approx. 1.1 eV). Furthermore, the bandgap of amorphous silicon depends on the parameters of the deposition process. For example, a low deposition temperature generally results in a higher mobility gap. Another noteworthy effect is the variation

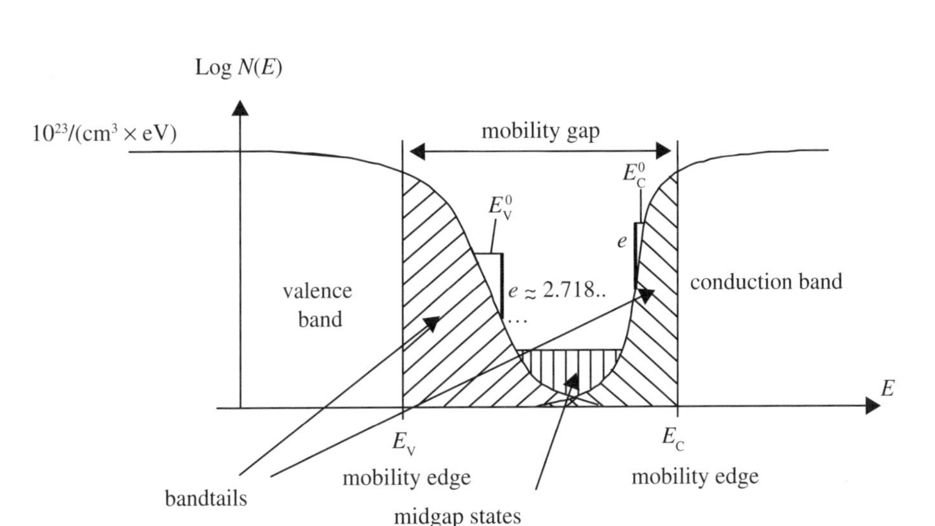


**Fig. 2.6** Diagram showing schematically the electronic "gap states", which are a result of the amorphous nature of the semiconductor. The density of states $N(E)$ is represented logarithmically and as the ordinate (on the vertical axis); the energy $E$ of the corresponding electronic state is represented as the abscissa (on the horizontal axis). The "gap states" lie in the mobility gap, i.e. in the energy range between the valence band edge $E_V$ and the conduction band edge $E_C$. Note that the band edges are simply considered as "frontiers" separating localized states (containing "trapped" carriers) from extended states (containing "free" carriers), see also Section 2.4.

of the bandgap when hydrogen is added to the silane in the deposition plasma. If the hydrogen dilution is increased over a certain threshold, the layers are no longer completely amorphous, but become (at least partly) microcrystalline and their bandgap tends to be approximately equivalent to that of crystalline silicon: see Chapter 3. Depending upon the exact deposition conditions, one can deposit layers which are still (almost) completely amorphous, but have more short-range and medium-range order than the "standard" type of amorphous silicon layers: these layers are called "protocrystalline" layers (see [Kol-1988], [Collins-2003]); they generally also have a somewhat higher bandgap [Pearce-2009] than "standard" amorphous silicon layers, deposited with no or little hydrogen dilution. If not only the hydrogen dilution but also other deposition conditions are modified, one may obtain "polymorphous" layers, which are characterized by the inclusion of very small crystallites [Fontcuberta-2001]; these layers also have a slightly higher bandgap [Fontcuberta-2004], [Soro-2008].

If the hydrogen dilution is increased even further, one obtains (as just mentioned) microcrystalline layers (see Chapter 3).

## 2.2 GAP STATES

### 2.2.1 Bandtail states

In amorphous semiconductors, disorder creates bond angle and bond length variations, as discussed above. Due to these variations, bandtails are formed at the edge of both valence and conduction bands. The density of states in these bandtails decreases exponentially as one moves away from the band edge towards the middle of the "bandgap" or, to be more precise, towards the middle of the "mobility gap". Thereby, the conduction band is limited by the mobility edge, i.e. by the energy level, which separates free electrons (which have a certain finite mobility and can therefore move about in the semiconductor) and trapped electrons (which have zero mobility and therefore remain at a fixed location). In a similar way, there exists also a mobility edge that limits the valence band at higher energies.

Group IV semiconductors, such as silicon and germanium, have a valence bandtail that is much more pronounced than their conduction bandtail. This means that the exponential function, which describes the density of states in the bandtail, falls off more sharply at the edge of the conduction band than at the edge of the valence band. (In the case of amorphous selenium and other chalcogen/chalcogenide semiconductors, one suspects the opposite to be true, i.e. the conduction bandtail is more pronounced than the valence bandtail).

As a quantitative value to describe the exponential function involved, one uses the energy constant $E^0$: this is the energy required for the exponential function to fall off by a factor of ***e*** (= 2.718…), where ***e*** is the base of the natural logarithm.

This is shown schematically in Figure 2.6 above.

Both bandtails have an exponential density of states *N(E)*, as a function of the energy level *E*. In the valence bandtail, *N*(*E*) is proportional to $\exp\{(E_V - E)/E_V^0\}$, and in the conduction bandtail *N*(*E*) is proportional to $\exp\{-(E_C - E)/E_C^0\}$, where $E_V$ and $E_C$ are the energy levels at the mobility edges of the valence and conduction bands, respectively; $E_V^0$ and $E_C^0$ are therefore energy constants characterizing the exponential functions.

In "device-quality" amorphous silicon, the valence bandtail has a value $E_V^0$ of approximately 45 to 50 meV and the conduction bandtail a value $E_C^0$ of approximately 25 meV. The values of $E_V^0$ and $E_C^0$ depend on the disorder present in the amorphous network: if the amorphous network or "matrix" is highly disordered, then many bandtail states will be present and both $E_V^0$ and $E_C^0$ will be large.

Bandtail states can capture both electrons and holes. They can also basically re-emit, by thermal emission, both electrons and holes.

If we have thermal equilibrium, i.e. if the amorphous semiconductor is in the dark and if, furthermore, no current is injected, then the occupation of the bandtail states is governed by equilibrium statistics (similar to the Fermi-Dirac statistics that govern the occupation of the valence and conduction bands in all semiconductors). Under the influence of light (or under the influence of current injection), equilibrium statistics do not apply anymore and the occupation of the bandtail states is governed by the kinetics of the capture and thermal emission processes.

As we will describe in more detail later in Section 2.4.4, the states in the conduction bandtail will capture mainly electrons, and the captured electrons in the bandtail can also be easily re-emitted thermally into the conduction band. It can be reasonably assumed that these two transfer processes are much more probable (at least for the upper part of the bandtail, which is adjacent to the conduction band) than the capture of a hole or the thermal emission of a hole from the conduction bandtail into the valence band. This consideration is (partly) based on the fact that the probability of thermal emission is proportional to $\exp(-\Delta E/kT)$, where $\Delta E$ is the energy to be invested during the emission process, $k$ the Boltzmann constant and $T$ the absolute temperature. (Note that $kT = 26$ meV for ambient temperature, 300K.) In the case of thermal emission from the bandtail to the adjacent band, $\Delta E$ is a relatively small value (in the order of 20 to 100 meV) and the probability of thermal emission is, thus, relatively large.

Similarly, the states in the valence bandtail will capture mainly holes, and the captured holes in the bandtail can also be easily re-emitted thermally into the valence band. This is certainly true for the lower part of the valence bandtail, i.e. the part adjacent to the valence band itself.

One therefore describe the states in the bandtail (especially those states which are very near to the bands) as "*traps*". They do not act as recombination centers, but just as "trapping centers", where a free carrier is captured and later on re-emitted into the same band.

In an operating device such as a solar cell, the bandtails are thereby in constant exchange with the corresponding band: i.e. the valence bandtail is continuously exchanging charge carriers with the valence band and the conduction bandtail is continuously exchanging carriers with the conduction band. This is shown schematically in Figure 2.7.

In steady-state conditions under illumination (these are the operating conditions that are usually of importance for a solar cell), the two processes (capture and re-emission) are in equilibrium: there are exactly the same amounts of free holes, which are captured, and of trapped holes, which are released by thermal emission. As capture and thermal emission exactly balance each other, one need not consider

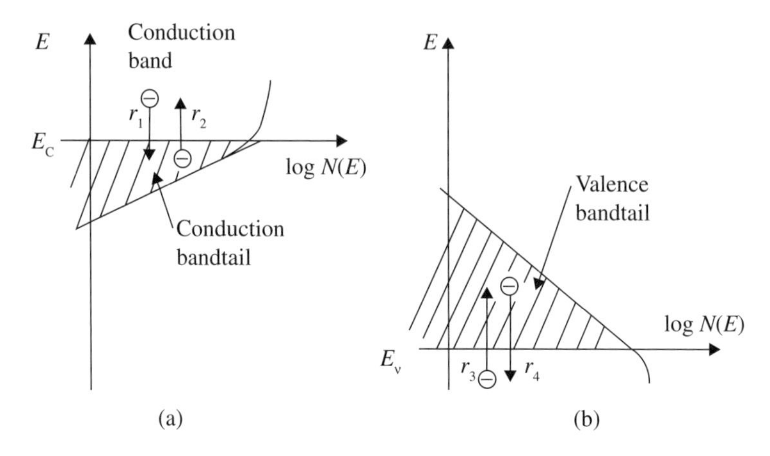


**Fig. 2.7** Exchange between trapped carriers from the conduction and valence bandtails and free carriers from the corresponding bands; under steady-state conditions these two processes are in equilibrium and do not affect the current in the bands: the capture rate $r_1$ of free electrons from the conduction band is equal to the thermal emission rate $r_2$ of trapped electrons from the conduction bandtail; similarly the capture rate $r_3$ of free holes from the valence band is equal to the thermal emission rate $r_4$ of trapped holes from the valence bandtail.

these processes at all when one looks at the currents in a solar cell. On the other hand, one certainly has to look at the trapped charge stored in the bandtail when one considers the total charge stored in an amorphous layer. In fact, for every free (or "mobile") hole in the valence band (i.e. for every hole that contributes to the current), one has $1/\theta_p$ trapped holes in the valence bandtail. ( $\theta_p$ is called the "trapping factor" and gives the relationship between the density $p_f$ of free carries and the density $p_t$ of trapped carriers: (see Sect. 4.5.4.)

$$\theta_p = p_f/(p_f + p_t) \approx p_f/p_t ,$$

In practice one finds, for device quality material with $E_V^0 \approx 50$ meV, that $\theta_p \approx 10^{-2}$ to $10^{-3}$. This means that for every free hole that contributes to the flow of current, there are a few hundred trapped holes that only contribute to the positive charge. It is as if there were a huge cloud of immobilized positive charge in the valence bandtail, which is simply necessitated by the flow of free carriers. This huge amount of positive charge will deform the internal electric field in the intrinsic layer (*i*-layer) of *pin* solar cells, as will be described in Section 4.6. Now, if the electric field is deformed, the collection of photogenerated carriers in the solar cell will be made more difficult and the solar cell performance reduced.

For the intrinsic layer (*i*-layer) of solar cells, the energy constant $E_V^0$ should therefore be as small as possible, in order to keep this "huge cloud" of trapped holes reasonably small. Because of the exponential distribution of states in the bandtail, a decrease of $E_V^0$ by even a few meV will lead to a substantial diminution of the factor $1/\theta_p$ and, thus, also to a substantial reduction in the density of trapped holes. This is one of the reasons why, for amorphous silicon solar cells, intrinsic layers with the lowest value $E_V^0$ that can be obtained should be used.

The situation is similar for the conduction band. However, as the conduction bandtail is much less pronounced and as the corresponding energy constant $E_C^0$ is only about half the value 25 meV (instead of 45 to 50 meV), the effect of trapping is also less pronounced and one assumes that for every free electron there are approximately 10 trapped electrons ($\theta_n \approx 10^{-1}$).Therefore the effect of trapped electrons on the field deformation in the *i*-layer of *pin* solar cells is not significant.

The bandtail states or "traps" also play a large role when a device, be it a solar cell or a thin-film transistor, is operated in the "transient mode", i.e. when it is switched on or switched off. Whenever the device is switched on, the "trap states" of the bandtails will have to be filled with a certain quantity of charge carriers: i.e. the valence bandtail will have to be filled with $1/\theta_p \approx 10^2$ to $10^3$ trapped holes for every free hole in the valence band. Similarly, the conduction band will have to be filled with $1/\theta_n \approx 10$ trapped electrons for every free electron in the conduction band. Whenever the device is switched off, all these trapped carriers will have to be evacuated again. This makes the switching process very slow, much slower than in classical crystalline semiconductor devices. This does not play a large role for solar cells, because the illumination conditions do not change so rapidly. One should, however, remember this effect when one conducts measurements on solar cells, thereby changing the state of the solar cell often in a relatively rapid manner.

Thin-film transistors (TFT) used as switches (e.g. in liquid-crystal displays) suffer from the fact that the trap states have to be partly filled up and then emptied again, whenever the transistor is switched on or off, respectively. For the usual *n*-channel amorphous silicon TFT, the relevant parameter is the electron field effect mobility $\mu_n^{TFT}$; this quantity is therefore very low, i.e. it is equal to the band mobility $\mu_n^0$ times the trapping factor $\theta_n$ for electrons, i.e. it is only about one-tenth of the band mobility. Because the trapping factor $\theta_p$ for holes is very much smaller (only about $10^{-2}$ to $10^{-3}$, as stated above), *p*-channel amorphous silicon TFTs are not at all viable. Trapping by bandtails is less pronounced in microcrystalline silicon, and for this reason both *n*-channel and *p*-channel TFTs are feasible with microcrystalline silicon.

### 2.2.2 Midgap states: dangling bonds

In amorphous semiconductors there are (as mentioned above) always dangling bonds. Even though the hydrogen atoms present in hydrogenated amorphous silicon layers "passivate" a large part of the dangling bonds (usually well over 99.9 % of the "original" dangling bonds), there always remains a certain amount of "unpassivated" dangling bonds, which are not linked to a hydrogen atom and remain as broken bonds. Their density in device-quality hydrogenated amorphous silicon (a-Si:H) is somewhere between $10^{14}$ and $10^{17}$ cm$^{-3}$. (The value $10^{17}$ dangling bonds per cm$^3$ refers to a-Si:H layers after the light-induced degradation effect; the value $10^{14}$ dangling bonds per cm$^3$ refers to the bulk of the very best a-Si:H layers in the as-deposited or annealed state, i.e. before light-induced degradation, see below.) The following discussion will be concerned exclusively with these "unpassivated" dangling bonds. (The dangling bonds that are passivated by hydrogen atoms are not associated with localized states in the bandgap, i.e. they no longer count as dangling bonds as far as their electronic properties are concerned.)

These "unpassivated" dangling bonds are associated with electronic states that are situated more or less in the center of the mobility gap. Originally, a covalent bond between two silicon atoms has also two bonding electrons in the orbital with the highest energy. These two electrons are shared between the two atoms that are bonded together. When such a bond is broken because of the amorphous nature of the atomic network, the basic situation is that each of the separated atoms now has a dangling bond, with just one bonding electron. In this situation the dangling bond is neutral and it is denoted by the symbol $D^0$. Evidently, the neutral dangling bond $D^0$ can capture a free hole from the valence band and this hole will *recombine* with the single electron that was sitting originally on the dangling bond, creating thereby a positively charged dangling bond $D^+$. (In a much rarer process, the single electron sitting originally on the neutral dangling bond $D^0$ can be released by thermal emission to the conduction band, leaving behind again a positively charged dangling bond $D^+$.) After these processes, by which the dangling bond has become positive (symbol $D^+$), it is no longer occupied by an electron, but it can be considered to be occupied by a positive charge carrier, i.e. by a hole. In this state it can capture a free electron from the conduction band (or, in a much rarer process, release a hole by thermal emission to the valence band) and enter once more into the neutral state $D^0$. The transitions from $D^0$ to $D^+$ and back from $D^+$ to $D^0$ occur at a certain energy level $E^{+/0}$.

On the other hand (according to the Pauli principle), every electronic state can have in total two electrons (with two different spins); i.e. the neutral dangling bond $D^0$ can also capture an electron (or, in a much rarer process, release a hole by thermal emission to the valence band); it will then become a negatively charged dangling bond $D^-$. Again, once it is in this state, the negatively charged dangling bond $D^-$ can now capture a hole, which *recombines* with one of the two electrons (or, in a much rarer process, it can release an electron, by thermal emission to the conduction band) and become again a neutral dangling bond $D^0$. The transitions from $D^0$ to $D^-$ and back from $D^-$ to $D^0$ occur at a certain energy level $E^{0/-}$.

What is generally assumed (and has partly been substantiated by experimental evidence) are the following hypotheses:

The energy level $E^{0/-}$ at which transitions between a neutral dangling bond $D^0$ and a negatively charged dangling bond $D^-$ will occur is slightly higher than the energy level $E^{+/0}$ at which transitions will occur between a positively charged dangling bond $D^+$ and a neutral dangling bond $D^0$; the difference between these two energy levels is called the "correlation energy" or "Hubbard energy" $U$; one has: $U = E^{0/-} - E^{+/0}$

In amorphous silicon, the value of the correlation energy $U$ is, thus, assumed to be positive; experiments tend to indicate that it is somewhere around 0.3 eV; the fact that $U$ is positive can be intuitively understood, because the two electrons sitting on the same electronic state have the same negative charge and there is therefore a charge repulsion (electrostatic repulsion) between them; this charge repulsion has to be overcome when the second electron is added to the dangling bond;

Under usual device-operating conditions (especially in the case of solar cells under illumination), the densities of electrons and holes in the conduction and valence bands are much higher than the densities prevailing in conditions of thermal equilibrium (i.e. in dark conditions and without current injection). In thermal

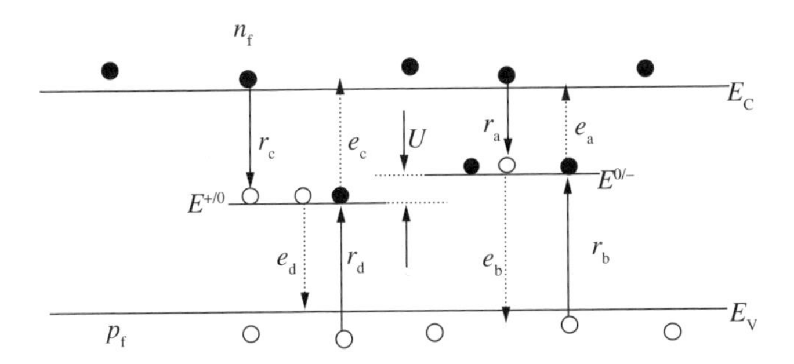


**Fig. 2.8** Schematic representation showing the energy levels associated with dangling bonds in an amorphous silicon layer: the energy level $E^{+/0}$ for the transition from a positively charged to a neutral dangling bond; the energy level $E^{0/-}$ associated with the transition from a neutral dangling bond to a negatively charged dangling bond; the correlation energy or "Hubbard energy" $U$, being the difference of energy between these two levels. Also shown are the capture processes $r_a$, $r_b$, $r_c$ and $r_d$, as well as the thermal emission processes $e_a$, $e_b$, $e_c$ and $e_d$.

equilibrium, according to the thermodynamic principle of "detailed balance", there exists a balance between thermal emission and capture: the transition rates have to be equal in both directions between the dangling bond states and any one of the bands (and vice-versa). (This means that for thermal equilibrium we will have, in Figure 2.8: $r_a = e_a$; $r_b = e_b$; $r_c = e_c$; $r_d = e_d$.) If the densities of free electrons and holes are now strongly increased, the capture rates for these free carriers will also be strongly increased. The rates for thermal emission from dangling bonds will, on the other hand, hardly be modified. Therefore, in illuminated solar cells, the probabilities for thermal emission of holes and electrons from dangling bond states are very much smaller than the probabilities of capturing holes and electrons from the corresponding bands. Thermal emission will therefore be neglected when describing the operation of solar cell devices under illumination.

The capture of a charge carrier by a dangling bond of the opposite polarity (i.e. the capture of an electron by a positively charged dangling bond $D^+$ or the capture of a hole by a negatively charged dangling bond $D^-$) is assisted by electrostatic attraction and is therefore very much more probable than the capture of an electron or of an hole by a neutral dangling bond.

The energy levels $E^{+/0}$ and $E^{0/-}$, the correlation energy $U$, the capture processes $r_a$, $r_b$, $r_c$ and $r_d$ (where a free hole or a free electron is captured by a dangling bond), as well as the thermal emission processes $e_a$, $e_b$, $e_c$ and $e_d$ (where an electron or a hole is emitted from a dangling bond into the corresponding band) are schematically shown in Figure 2.8.

It is, in fact, easy to convince oneself that thermal emission from dangling bond states into the bands is a process with a low probability, because it means that the charge carrier (electron or hole), which is thermally emitted, has to acquire a large amount of energy. As already mentioned in Section 2.2.1, the probability $e_i$ of thermal emission is proportional to $\exp(-\Delta E/kT)$, where $kT \approx 26$ meV for ambient temperature ($T = 300$K); this probability is therefore extremely small for large values of $\Delta E$. Therefore if we neglect the thermal emission processes and only look at the capture processes shown in Figure 2.8, we can immediately see that these processes involve for each of the two paths ($r_c/r_d$ and $r_a/r_b$) a recombination between an electron and a hole. It is therefore easy to intuitively accept that dangling bonds generally act as *recombination centers* within the intrinsic layers of amorphous silicon solar cells. For this reason it is very important to keep the density of dangling bonds low within device-quality intrinsic layers. The recombination process will be studied in more detail in Section 2.4.4.

The density of dangling bonds that are in the neutral state, i.e. which have only one electron, can be measured by *the Electron Spin Resonance (ESR)* technique; see [Brodsky-1969], [Stutzmann-1987], [Stutzmann-1988], [Stutzmann-1989a,b]; as well as Section 3.2.2, where the ESR measurement technique is described in more detail, in the context of microcrystalline silicon. It is generally found that, in the best amorphous intrinsic layers, one has a dangling bond density of approximately $10^{14}$ cm$^{-3}$. However, in the zones near the substrate/silicon interface and in the zones near the surface of the layer, one has very much higher dangling bond densities. If the layers measured are not fully intrinsic but are slightly doped (e.g. through residual quantities of dopant impurities, like O, B, P, etc.), then the dangling bonds will already be charged in the state of thermal equilibrium, i.e. without injection of additional carriers, by photo-generation or by current injection. This is due to the requirement of charge neutrality and to the corresponding shift of the Fermi level within the semiconductor. Indeed if we have an excess of impurities which are donors, like O and P (i.e. which tend to give an *n*-doped character to the material), then the donor atoms themselves will be positively ionized ($O^+$, $P^+$), and a large part of the dangling bonds will be negatively charged ($D^-$). On the other hand, if the impurities have acceptor-like character, like B, then a large part of the dangling bonds will be positively charged ($D^+$). If we do have a high density of charged dangling bonds, then the ESR measurement method will no longer give a correct value for the dangling bond density, because these charged dangling bonds have either 0 spin or 2 spins and no longer participate in spin resonance. These charged dangling bonds show up, on the other hand, in sub-bandgap absorption measurements, such as those done with PDS (Photothermal Deflection Spectroscopy) or CPM (Constant Photocurrent Method) (see Sect. 2.3.6). However, one needs a calibration factor to translate the absorption coefficient (usually taken for amorphous silicon at 1.2 eV) into a dangling bond density. This calibration factor will be discussed in Section 2.3.6.

### 2.2.3 Light-induced degradation (Staebler-Wronski effect)

#### *Basic phenomena*

When the amorphous silicon layer is exposed to light (or when an electric current is injected into the amorphous silicon layer), a degradation effect takes place. The weaker bonds within the amorphous silicon network are broken and thereby new dangling bonds are created. This is the light-induced degradation effect or "Staebler-Wronski effect (SWE)", first observed by D. L. Staebler and C. R. Wronski, [Staebler-1977]. By annealing the layer at 150 °C (for a few hours) or at 250 °C (for a few minutes), the effect can be reversed and the newly created dangling bonds are "passivated" again.

We then reach the "annealed" state, which is very similar to the state of the layer before light-induced degradation.

In a rather simplified model, there would, even at lower temperatures, be an equilibrium between the creation of new dangling bonds and the "suppression" or "healing" (passivation) of dangling bonds. In this simplified model it follows that: the higher the temperature, the higher is the rate of healing; the higher the light intensity, the higher is the rate of creating new dangling bonds. Evidently, there would be, under these circumstances, an equilibrium point where an equal number of new dangling bonds are created and existing dangling bonds are suppressed. This equilibrium point may be reached at room temperature after several hundred hours. This is why one generally says that after a time of about 1000 hours of exposure to light, amorphous silicon layers "stabilize", i.e. they are supposed to asymptotically have reached a new and higher value of dangling bond density, which is now considered to be more or less stable. This value is, for the best material made to date, about a factor 10 to 100 higher than the initial value and therefore lies in the range of $10^{16}$ to$10^{17}$ dangling bonds per $cm^3$. It is, however, often observed that degradation still continues, albeit at a lower rate, even after 1000 hours of light exposure. In fact, there are certain amorphous silicon layers where light-induced degradation has been observed to continue, even in a pronounced manner, for a very long time, and other layers where apparent stabilization takes place relatively soon. If the temperature is higher, the apparent stabilization will take place much sooner.

For reasons that are not fully clear to date, stabilization would appear to be more pronounced in amorphous silicon solar cells (see Sect. 4.10) than in individual amorphous silicon layers. (One of the reasons that one suspects as being responsible for this difference is the fact that solar cells are sealed against the influence of humidity and atmospheric oxygen, whereas individual layers are not sealed [Vaneček-2010].)

The real criterion for evaluating degradation would be to observe the increase in dangling bond density, by performing ESR measurements or by determining the sub-bandgap absorption (see Sect. 2.3.5). Such an evaluation is, however, time-consuming. It is therefore customary to observe degradation in layers by plotting the value of the photoconductivity $\sigma_{photo}$ as a function of time (see Fig. 2.9).

photoconductivity stabilises at ~ 8x10-5 Ω-1 cm-1

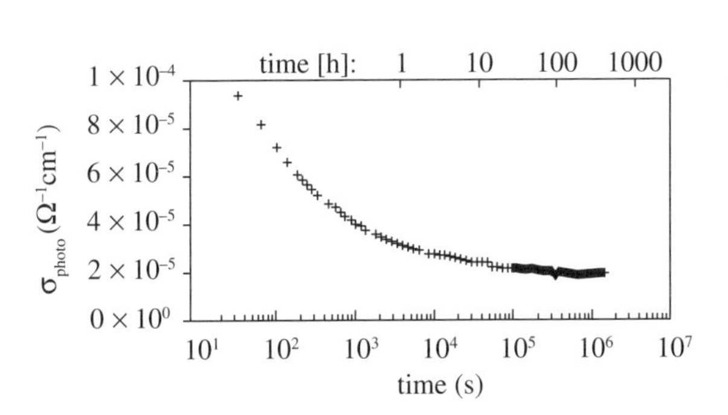


**Fig. 2.9** Typical curve for the decrease of photoconductivity under the influence of light-soaking, taken, with permission, from [Hof-1999]. One remarks an apparent saturation of the degradation effect at about $10^6$ seconds, i.e. at a few hundred hours.

As will be mentioned in Section 2.4.5, the value of the photoconductivity can be taken as a coarse indication for $1/N_{db}$. It is important to specify the temperature and the light intensity at which one conducts the light-soaking experiments. (In solar cells, it has become customary to perform light-induced degradation experiments at a temperature of 50 °C and with a light beam that corresponds approximately to the AM 1.5 spectrum (see Sect. 4.1.1d) and has, thus, a light intensity of about 100 mW/cm$^2$.)

A considerable amount of research work has been carried out in attempting to elucidate the nature of the Staebler-Wronski effect and also to find methods for eliminating or at least reducing this degradation phenomenon (see e.g. [Shimizu-2004], for a relatively recent review). Up to now the exact microscopic nature of the Staebler-Wronski effect is, however, still under debate.

**Note** that one also talks about "*metastability*", when referring to light-induced degradation: This is motivated by the fact that an a-Si:H layer can remain in the initial state or in any of the degraded states for an arbitrarily long time, if it is not subjected to an "external perturbation" – either to light exposure (which will cause the layer to degrade further) – or to heat (which will cause the layer to return towards the "initial" state, or at least towards the "annealed" state). *The "annealed" state can be considered, for all practical purposes, to be identical with the "initial" state.*

***Methods of reducing light-induced degradation***

There exist so far only two methods, which clearly lead to a certain reduction of light-induced degradation: (1) an increase in the *deposition temperature* (see [Platz-1998]); and (2) using *hydrogen dilution*, i.e. one deposits the amorphous silicon layers from a mixture of silane plus hydrogen [Tsu-1997], [Guha-2003]. In the latter case, the plasma used for the deposition of amorphous silicon has a higher concentration of atomic hydrogen than if a pure silane plasma is used.

Note that whereas all industrial manufacturers of solar cells use, at present, exclusively plasma deposition for depositing amorphous silicon layers, a considerable amount of research work has been done on an alternative method of fabricating amorphous silicon layers: by Hot-Wire Deposition (see Sect. 6.2). In Hot-Wire Deposition both factors that lead to a reduction in light-induced degradation are re-united: an increase in deposition temperature and also an increase in the concentration of atomic hydrogen (see [Mahan-1991] and [Vaneček-1992]).

The hydrogen dilution method is currently used for the industrial production of virtually all amorphous silicon solar cells. The dilution ratios used, i.e. the ratios between the hydrogen and the silane gas flows as fed-in into the deposition reactor, vary strongly, from very low values of hydrogen dilution (1:5 to 1:1) where the optical properties of the layers are virtually unchanged and the deposition rate is hardly reduced, up to high values of hydrogen dilution (4:1 or even more) where the optical bandgap of the layers is slightly increased and the deposition rate typically reduced. The appropriate hydrogen dilution ratio depends on the deposition pressure, deposition temperature, the deposition reactor used and on other parameters of the deposition process.

If the hydrogen to silane dilution ratio is strongly increased, one may obtain the "protocrystalline" and "polymorphous" silicon layers mentioned in Sect. 2.1.2. These layers are indeed reported to be more stable than "standard" amorphous silicon layers, but within single-junction amorphous silicon solar cells this improvement in stability is, in general, somewhat compensated by an increase in the bandgap and by a corresponding decrease in the absorption of the solar spectrum and, thus, a decrease in the electron-hole pairs generated by sunlight. This fact necessitates (at least in the single-junction case) thicker solar cells, which are then again less stable.

There has of late been a substantial amount of research invested into solar cells made with "protocrystalline" silicon (see e.g. [Pearce-2000]) and "polymorphous" silicon (see e.g. [Poissant-2003] [Soro-2008]). On the other hand, very little is known on the industrial production of solar cells and modules with such layers. Protocrystalline silicon (i.e. amorphous silicon containing traces of very small nanocrystals, which can be seen in high-resolution electron microscopy) does appear to be of some interest for solar cell production, whereas polymorphous silicon has not yet attracted much interest from solar module manufacturers, possibly because of problems with powder formation within the deposition reactor.

For even higher values of hydrogen dilution, one obtains microcrystalline silicon (also called "nanocrystalline silicon"; see Chapter 3).

### *Light-induced degradation and deposition rate*

As already stated, increasing the hydrogen dilution generally leads to a decrease in the deposition rate of the layers [Kroll-1998]. What about the relation between the deposition rate and the dangling bond density? The "stabilized" density of dangling bonds depends, in a very sensitive way, on the deposition conditions and on the microstructure of the layer that is deposited, as follows.

In general, if the deposition rate is markedly increased, the density of dangling bonds is also correspondingly increased, both in the initial and in the "stabilized" state, and the resulting layers are not at all usable in solar cells. At least this can be evaluated immediately after deposition.

However, the authors have, when increasing the deposition rate (by increasing the plasma power), often encountered a situation where they obtained amorphous silicon layers with a relatively low density of dangling bonds in the *initial* state, but with a high density of dangling bonds in the *degraded* or "stabilized" state. These layers are, of course, also not usable in solar cells. These layers were usually layers with a *high density of micro-voids* and, thus, also a high proportion of hydrogen atoms bonded to the "internal surfaces" formed by these micro-voids (see Fig. 2.5).

### *Fourier Transform Infrared Spectroscopy (FTIR): used as a tool for detecting a-Si:H layers with potentially high degradation*

The easiest manner to *evaluate the density of micro-voids* is by Fourier Transform Infrared Spectroscopy (FTIR), a non-destructive and relatively rapid technique that allows one to probe the bonds between atoms by inducing vibrations of bonds. Infrared spectroscopy works as follows: depending upon the wavelength of the infrared light that enters the sample, one can induce different vibrational modes of (polar) bonds between two neighbouring atoms. If a bond vibration is triggered, the material sample that one is investigating will absorb part of the infrared light. If the density of the specific atomic bonds that are vibrating is large, the absorption of infrared light will be high. Instead of using a monochromator and scanning the infrared spectrum, one prefers today to analyze an interferogram which contains information on the whole spectral range of infrared light and thereafter use the Fourier transform technique to numerically decompose the absorption signal into its different spectral components. This method of performing infrared spectroscopy is much faster and more precise than the original method with a monochromator. One obtains, thus, a FTIR spectrum, as shown in Figure 2.10 for typical amorphous silicon samples. Si-H bonds situated within the amorphous matrix have a vibrational mode (Low Stretching Mode, LSM) at a wavenumber of 2000 cm$^{-1}$; Si-H bonds on external and internal surfaces have a vibrational mode (High Stretching Mode, HSM) at a wavenumber of 2070-2090 cm$^{-1}$. ("Wavenumber" is a unit used in infrared spectroscopy and is defined as the reciprocal of the wavelength of the radiation in vacuum; it is generally given in cm$^{-1}$.) The difference in stretching frequency is attributed to a different force constant of the Si-H oscillator when located at a void surface or embedded within a dielectric medium [Wagner-1983], [Cardona-1983]. For amorphous silicon layers, one defines [Bhattacharya-1988], [Mahan-1987], [Mahan-1989] a microstructure factor.

$$r = I_{2090} / (I_{2090} + I_{2000}),$$

where $I_{2000}$ and $I_{2090}$ are the intensities of the FTIR signal (i.e. of the infrared absorption) at 2000 cm$^{-1}$ and around 2070-2090 cm$^{-1}$, respectively. One obtains these intensities $I_{2000}$ and $I_{2090}$ by deconvolution of the FTIR spectrum (see e.g. [Müllerová-2008], [Smets-2003]). The microstructure factor $r$ has been shown to increase when the concentration of micro-voids increases. On the other hand, if the density of the micro-voids

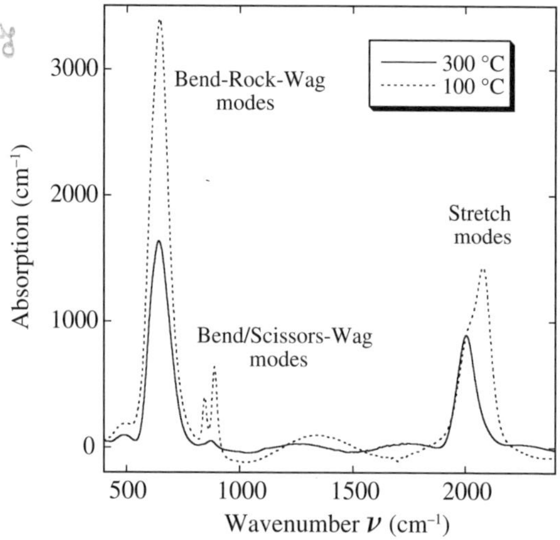

**Fig. 2.10** Typical infrared absorption spectra obtained by FTIR (Fourier Transform Infrared Spectroscopy) for amorphous silicon layers: straight line, for a device-quality intrinsic amorphous silicon layer deposited at 300 °C; dashed line, for an inferior-quality intrinsic amorphous silicon layer deposited at 100 °C – this layer has a pronounced peak around 2070 to 2090 cm$^{-1}$ indicating a very large density of micro-voids leading (probably) to a very pronounced light-induced degradation effect; reproduced with permission from [Kroll-1995], and revised by Dr. Kroll.

is high [Bhattacharya-1988], one generally also finds that the light-induced degradation effect will be very pronounced. For this reason, the microstructure factor $r$ can be used as a fast and convenient first indication of the relative stability of the material. An amorphous silicon layer, which can be accepted as the intrinsic layer of a solar cell, should have a microstructure factor around 0.2 (or preferably lower) [Kroll-1998]. If within an intrinsic amorphous silicon layer one finds a value of $r$ that is larger than 0.25 [Bhattacharya-1988], one may, with some confidence, predict that the dangling bond density after degradation will be too high for application within solar cells.

Note: FTIR is a diagnostic tool that is commonly used for various investigations in materials research within the field of thin-film silicon, for example it is often used in amorphous (and microcrystalline) silicon to evaluate the total amount of (bonded) hydrogen within the sample by integrating the signal under the rocking-wagging mode around 640 $cm^{-1}$. The atomic percentage of bonded hydrogen can be calculated from this mode by using the procedure described for example in [Kroll-1996]. One thereby has to integrate the quantity $\alpha(\nu)/\nu$ (where $\alpha$ is the infrared absorption coefficient at the wavenumber $\nu$) and then multiply the result by the proportionality factor $A_w$ = $1.6 \times 10^{19}$ $cm^{-2}$ and divide it by $N_{Si}$, which is the the atomic density of pure silicon. If the hydrogen content is very high (much higher than 10 at %), this is generally considered to be detrimental to the quality of an intrinsic amorphous layer.

Furthermore, FTIR is also used, specifically, to detect $SiH_2$ bonds (2 hydrogen atoms bonded on the same silicon atom (see Figure 2.3c, above), and also to detect $SiH_3$ bonds (3 hydrogen atoms bonded on the same silicon atom). $SiH_2$ bonds lead to vibrational modes (bend/scissors and wag modes) in the range of 800 to 900 $cm^{-1}$; their density can therefore also easily be evaluated qualitatively by FTIR spectroscopy. If there is a high proportion of $SiH_2$ bonds we must consider the material quality to be poor. In fact, a high hydrogen concentration usually results in an increase of $SiH_2$ vibrations and, as discussed in Section 2.6, in an increase of the microstructure factor $r$. $SiH_3$ bonds lead to vibrational modes (stretch modes) around 2400 $cm^{-1}$. $SiH_3$ bonds are detected if the material is porous and thus of very poor quality.

Finally, FTIR can be used to detect $SiO_x$ bonds and, thus, to evaluate the percentage of oxygen atoms within an amorphous silicon sample (this can be done by looking at the relatively strongly absorbing $SiO_x$ mode in the 1100 to 1000 $cm^{-1}$ region; it is only viable if the oxygen content exceeds about 0.5 atomic %).

These applications of the FTIR technique are typically employed in advanced materials research, whereas the microstructure factor $r$ (as given above) is a convenient and practical technique for the optimization of deposition processes in the production of amorphous silicon solar cells (see Sect. 6.1.3).

## 2.3 OPTICAL ABSORPTION: OPTICAL GAP AND SUB-BANDGAP ABSORPTION

**Note:** When working with plots involving optical parameters, one often has to go from a representation as a function of the wavelength $\lambda$ of light to a representation as a function of photon energy $E_{ph}$, and vice-versa. Here the following conversion formula will be useful:

$$E_{ph}[\text{eV}] = \frac{hc}{q}\frac{1}{\lambda} \approx \frac{1.240}{\lambda[\mu\text{m}]}[\text{eV}],$$

where $h$ is Planck's constant, c the velocity of light, $q$ the charge of an electron (unit charge).

### 2.3.1 Absorption coefficient plot

Absorption measurements are an important way of characterizing amorphous silicon layers. The absorption coefficient for amorphous silicon generally follows the behavior shown in Figure 2.11: there are three distinct regions A, B and C, which can be clearly identified.

*Region A*: this part of the absorption curve is given by band-to-band transitions, i.e. the photon is absorbed and generates a hole in the valence band and an electron in the conduction band; by extrapolating from the absorption curve in region A to lower energies, one finds what is called the "optical gap" $E_{og}$ of the amorphous layer (see Sect. 2.3.4). It is this part of the absorption curve that is useful for photo-generation of free carriers (pairs of holes and electrons) within solar cells and other photoelectronic devices. Because of the amorphous nature of the material, the absorption coefficient in this region is about a factor of 10 higher than the absorption coefficient for crystalline silicon. On the other hand, the optical gap $E_{og}$ of amorphous silicon has a value

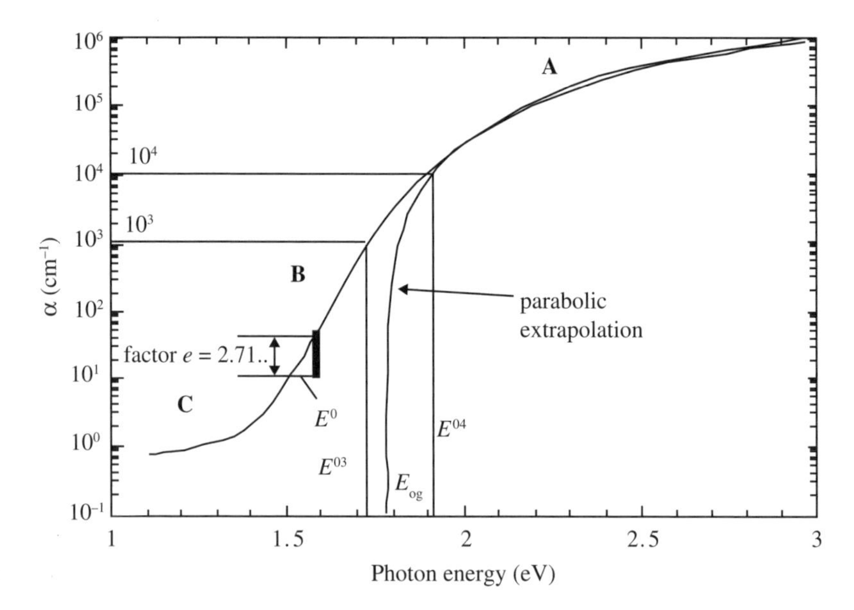


**Fig. 2.11** Typical curve for the absorption coefficient as a function of photon energy. Note that the absorption coefficient is plotted in a logarithmic scale and varies here over 7 orders of magnitude.

between 1.65 and 1.85 eV (and, in most cases, between 1.7 and 1.75 eV), i.e. it is considerably higher than the bandgap of crystalline silicon. Note that instead of determining the optical gap by an extrapolation procedure, as described below, one sometimes uses the value $E^{03}$ (photon energy for $\alpha = 10^3\,\mathrm{cm}^{-1}$) or the value $E^{04}$ (photon energy for $\alpha = 10^4\,\mathrm{cm}^{-1}$) as an approximate value for the optical gap.

*Region B*: this part of the absorption curve is given by transitions between the bandtails and the opposite band, i.e. between the valence bandtail and the conduction band, as well as between the conduction bandtail and the valence band (see Sect. 2.2.1).

Because in amorphous silicon the valence bandtail is much more pronounced than the conduction bandtail, one actually mainly "sees" here the valence bandtail. One can measure the energy constant $E_V^0$ in this way: $E_V^0$ is approximately equal to the exponential decrease of the optical absorption coefficient; the latter is called the "Urbach energy" and denoted by the symbol $E^0$:

$$E_V^0 \approx E^0$$

The name "Urbach energy" is actually borrowed from crystalline semiconductors; however, the reason for having an exponential fall-off of the absorption coefficient at photon energies below the energy of the bandgap is quite different in crystalline semiconductors (where this dependence is caused by excitons) than in amorphous semiconductors (where this fall-off is a consequence of an exponential distribution of bandtail states).

*Region C*: this part of the absorption curve is given by transitions from the midgap states (i.e. essentially from the states generated by the dangling bonds) to the valence and conduction bands. The absorption in region C is therefore a measure of the density of dangling bonds within the layer. During degradation, this part of the absorption curve will be enhanced by 1 to 2 orders of magnitude. (Such an enhancement of defect absorption by 1 to 2 orders of magnitude is seen in the majority of high-quality layers; on the other hand, one does find layers whose degradation behaviour is quite different.)

Note that in the above discussion we have not taken into account the transitions between gap states themselves, but only the transitions between gap states, on one hand, and states in the band, on the other hand. This is justified because the density of states in the bands is very much larger than the density of states in the gap – therefore transitions involving states in the band are much more likely than transitions involving only gap states. {There is yet another reason why transitions involving only gap states are not very probable: electrons (or holes) sitting on gaps states are "localized" and therefore cannot transit to another gap state, unless it is in virtually the same geometric position.}

Generally speaking, the absorption coefficient $\alpha(E_{\mathrm{photon}})$ is given by summing up the probabilities of all the transitions within the semiconductor where an electron acquires an energy $E$; the probability for each transition is basically proportional to the product of the density of occupied states at the original energy level $E_{\mathrm{initial}}$ of the electron and the density of free states at the final energy level $E_{\mathrm{final}}$ of the electron: $E_{\mathrm{final}} = E_{\mathrm{initial}} + E_{\mathrm{photon}}$. Therefore, the plot of the absorption coefficient is basically a convolution between the lower energy part of the density of states (the part below the Fermi level referring to states that are assumed to be almost fully occupied) and the higher energy part of the density of states (the part above the Fermi level pertaining to states, that are assumed to be almost all free). This is explained below in more detail.

## 2.3.2 Link between density of states and absorption coefficient

If we take a very intuitive viewpoint, we can say that the absorption coefficient is given by the probability of absorbing a photon and transferring the energy of the photon $E_{photon} = h\nu$ to an electron within the amorphous layer. (Here, $h$ is Planck's constant and $\nu$ the angular frequency of the incoming light.) Now for every energy level $E$ within the amorphous layer there is a certain density of states $N(E)$, as shown schematically in Figure 2.12.

Furthermore, these states may be occupied or empty and it is their occupation function $f(E)$ that will govern the occupation. For a transition to occur, the initial electronic state, from where the electron is taken, has originally to be occupied, and the final electronic state, where the electron will find itself after the transition has taken place, has first to be empty.

If we take an arbitrary value $E_{initial}$ for the original energy of the electron, then the electron will find itself, after the transition, at the energy value $E_{final} = E_{initial} + E_{photon}$. At the energy value $E_{initial}$ there will be $N(E_{initial})$ states in total and $N_i = N(E_{initial}) \times f(E_{initial})$ occupied states, where $f(E)$ is the occupation function of the states. At the energy value $E_{final}$ there will be $N(E_{final})$ states in total and $N_f = N(E_{final}) \times \{1 - f(E_{final})\}$ empty states. Thus, it is reasonable to assume that the probability of a transition from the states situated at $E_{initial}$ to states situated at $E_{final}$ is proportional to the product $N_i \times N_f$. Thereby, we will assume that all electronic states have an equal probability of participating in the transition. (In terms of quantum physics, this means that we are assuming the corresponding "optical matrix elements" to be all equal.) We will furthermore assume that there is a weighting factor $W(E_{photon})$, which is associated with the transition probability; i.e. depending upon the energy absorbed, the transition will be more or less

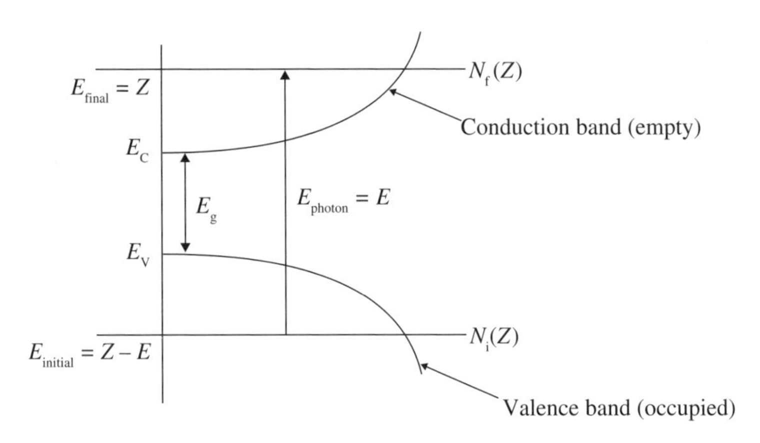


**Fig. 2.12** Diagram of energy (vertical axis) versus the density of states (horizontal axis), as used for the calculation of the absorption coefficient; the states in the gap are not shown. We are looking here at the absorption of a photon of energy $E_{photon} = E$, which induces a transition of an electron from a state at the energy level $E_{initial} = Z–E$ to a state at the energy level $E_{final} = Z$; the density of empty states at the final energy level $Z$ is $N_f(Z)$ and the density of occupied states at the initial energy level $Z–E$ is $N_i(Z–E)$.

probable. We then obtain the following relationship, by integrating over all possible transitions, i.e. over all possible values for $E_{\text{initial}}$:

$$\alpha(E) = W(E)\int_{-\infty}^{+\infty} \underbrace{N(Z-E)f(Z-E)}_{N_i} \times \underbrace{N(Z)\big(1-f(Z)\big)}_{N_f} dZ \tag{2.1}$$

(In order to simplify the expression we have substituted $Z$ for $E_{\text{final}}$ and $E$ for $E_{\text{photon}}$; $E_{\text{initial}}$ then becomes $Z$–$E$.)

In the conduction and valence bands, the occupation function is given by the Fermi-Dirac function. The Fermi-Dirac function is based on statistical considerations, and is valid for electrons and holes that are independent and move freely within a band. For the localized states in the bandtails and at midgap, the occupation functions will therefore differ slightly from the Fermi-Dirac function. However, to simplify we can just make the following statement: generally, the states below the Fermi level $E_F$ will (almost) all be occupied and the states above the Fermi level will (almost) all be empty. This is a rather coarse approximation and neglects the few states below the Fermi level $E_F$ which are empty, and the few states above the Fermi level $E_F$ which are occupied. But this approximation will hardly affect the value of the absorption coefficient computed by Equation (2.1). By making this approximation we "ensure" that for a transition to take place, one has to have $E_{\text{final}} \geq E_F$ and $E_{\text{initial}} \leq E_F$. Indeed, the electron will then always be transiting from a state below the Fermi level $E_F$ to a state above the Fermi level $E_F$.

Now, take a look at Figure 2.12 and at Equation (2.1): we are integrating in Equation (2.1) over $Z = E_{\text{final}}$. Therefore, the lower boundary of the integral can be specified as $E_F$; the upper boundary of the integral can be specified as $E_F + E$.

We, thus, obtain:

$$\alpha(E) \approx W(E)\int_{E_F}^{E_F+E} N(Z-E)N(Z)dZ \tag{2.2}$$

Thereby, we are not explicitly considering the necessity for the conservation of momentum, when the electron is shifted from a lower energy to a higher energy; if the momentum has to undergo a change and requires interaction with a phonon to do so (see Sect. 4.1.1, Fig. 4.7), this will simply result in a lower value of the weighting function.

Now we have to look more closely at the weighting function $W(E)$; this weighting function is linked to the optical matrix elements. The optical matrix elements are given by the probabilities for transitions from one state to another. In general, they not only depend on the energy absorbed during the transition, but they depend individually on the initial and on the final state. One has, however, no clear indication of what to assume for the optical matrix elements in the case of amorphous semiconductors. Therefore the simplest possible assumption is used [Cody-1984], [Jackson-1985]: one assumes all the optical matrix elements to be equal. Now, there are two types of optical matrix elements: dipole matrix elements and momentum matrix elements. If the dipole matrix elements are assumed to be equal, then $W(E)$ becomes a constant; if the momentum matrix elements are assumed to be equal, then $W(E) \times E$ becomes a constant.

### 2.3.3 Exponential density of states in bandtails and Urbach energy in plot of absorption coefficient

Let us first look at transitions between one of the two bandtails and the other band, i.e. at transitions between the valence bandtail and the conduction band, as well as transitions between the conduction bandtail and the valence band. These two types of transitions will contribute to region B in the plot of the absorption coefficient as a function of photon energy (Figure 2.11). Region B of the absorption plot is, as mentioned above, characterized by an exponential function with an energy constant $E^0$, also called the "Urbach energy", which is, *in amorphous silicon, more or less equivalent to the energy constant of the valence bandtail* $E_V^0$ (Figure 2.6).

This can be seen with the following considerations: both bandtails have an exponential density of states *N(E)* as a function of the energy level *E*. We therefore have the following expressions (see Figure 2.6)

$$N(E) = N_V^0 \exp\{(E_V - E)/E_V^0\}, \text{ for the valence bandtail}$$

$$\text{and } N(E) = N_C^0 \exp\{-(E_C - E)/E_C^0\}, \text{ for the conduction bandtail,}$$

where $N_V^0$ and $N_C^0$ are the density of states at the mobility edges of the valence band and the conduction band; $E_V$ and $E_C$ are the energy levels at the mobility edges of the valence and conduction bands, respectively; $E_V^0$ and $E_C^0$ are the energy constants characterizing the exponential functions.

On the other hand, in the conduction and valence bands, the density of states is not given by exponential functions, but possibly by parabolic functions, as formulated in Equation 2.4 below. As a very coarse approximation, we may, however, in the present computation assume that the density of states in conduction and valence bands is given by a constant.

Now, the absorption coefficient $\alpha(E)$ is given by Equation 2.2, which has the form of a convolution function between the density of states *N(Z)* and the density of states *N(Z-E)* for energy levels shifted by the photon energy $E = E_{photon}$. *The absorption coefficient in the region of the Urbach absorption edge results from transitions between the valence bandtail and the conduction band plus transitions between the valence band and the conduction bandtail.* It is therefore clearly given by the superposition of two exponential functions. Indeed, detailed calculations show that it is given by an expression of the following form:

$$\alpha(E) = c_1 \exp\left\{(E - E_g)/E_V^0)\right\} + c_2 \exp\left\{(E - E_g)/E_C^0)\right\}$$

Thus, $\alpha(E)$ is given by the sum of two exponential functions with two different energy constants $E_V^0$, $E_C^0$. A similar sum of two exponential functions is found in Electronic Circuit Theory, when a pulse waveform (a step waveform, to be precise) is deformed by two cascaded RC-lowpass filters. Instead of "energy constants" $E_V^0$, $E_C^0$, we are dealing there with "time constants" $\tau_1$, $\tau_2$ . It has been found there empirically [Millman-1965] that such a sum of two exponential functions can be, with an error that is less than 5 %, replaced by a single exponential function, with the resulting time constant being $\tau_{resulting} = \sqrt{\tau_1^2 + \tau_2^2}$. In analogy with this we can write for the case of optical absorption:

$$E^0 = \sqrt{\left(E_C^0\right)^2 + \left(E_V^0\right)^2} \tag{2.3}$$

### 2.3.4 Determination of the optical gap

There are basically two ways of determining the optical gap:

In *the first, empirical method*, we simply set the absorption coefficient $\alpha$ to a certain value; e.g. we set $\alpha = 10^3$ cm$^{-1}$ or $\alpha = 10^4$ cm$^{-1}$ and determine the corresponding photon energy, i.e. we determine $E^{03}$ or $E^{04}$ (see Fig. 2.11 above). This method has, however, two significant disadvantages:

(a) The values of $E^{03}$ and (to a lesser extent) of $E^{04}$ are influenced not only by the conduction and valence bands themselves, but also by the bandtails – the more pronounced the bandtails, the lower these values become (the absorption curve becomes higher and the points for $E^{03}$ and $E^{04}$ are shifted to the left).

(b) We need to know the thickness $d$ of the layer, in order to determine the absorption coefficient. The thickness $d$ will directly influence the result.

Therefore this method is not currently used today, whenever precise measurements of the optical gap are required. It is, however, a rapid and practical method for assessing the approximate value of the bandgap and for distinguishing, in a coarse manner, between amorphous and microcrystalline layers.

In the *second, improved method,* we *extrapolate* the curve A of Figure 2.11 and find out where this extrapolation reaches down to the horizontal axis; i.e. where, in the extrapolated curve, $\alpha$ would become zero. Region A in the absorption coefficient plot is the result of band to band transitions. To make an extrapolation of curve A, we have therefore to take care to assume a realistic function for the density of states in the bands and also an appropriate weighting function $C(E)$ in the convolution function, which links the density of states to the absorption coefficient, according to Equation 2.2.

The most common extrapolation procedure is that which was empirically introduced by Tauc [Tauc-1966] and is given by Equation 2.5, below.

This procedure can be justified theoretically if one assumes that the density of states in the valence and conduction bands of amorphous silicon are, just as is the case in crystalline silicon, parabolic functions of the energy difference $\Delta E$ between the energy level $E$ considered and the band edges (mobility edges) $E_V$ and $E_C$, respectively. This means that we will have:

$$N(E) = (\Delta E_{C,V})^{1/2}, \tag{2.4}$$

with $\Delta E_V = -E + E_V$ and $N(E) = 0$, for $E > E_v$
or $\Delta E_C = E - E_C$, and $N(E) = 0$, for $E < E_C$, respectively

Now one can assume that the momentum matrix elements are constant, so that $W(E) \times E = C,$ where $C$ is a constant ([Tauc-1966], [Cody-1982], [Cody-1984],

[Street-1991]). By additionally substituting the $N(E)$ values from Equation 2.4 into Equation 2.2, one finds the integral:

$$\alpha(E) \approx \frac{C}{E} \int_{E_F}^{E_F+E} \Delta E_V{}^{1/2} \Delta E_C{}^{1/2} \mathrm{d}Z$$

now with:

$$\Delta E_V = -Z + E + E_V$$

and:

$$\Delta E_C = Z - E_C$$

and, for the energy of the gap:

$$E_g = E_C - E_V$$

The above integral is of the type $\int_0^g \sqrt{a + bx + cx^2}\,\mathrm{d}x$ and yields an arcsine-function; the boundary conditions are such that a very simple result follows:

$$\alpha(E) \approx \frac{C}{E} \times \frac{\pi}{8} (E - E_g)^2$$

One obtains here:

$$\alpha(E)E \propto \left(E - E_g^T\right)^2 \tag{2.5}$$

where $E_g^T$ now denotes what is generally called "the Tauc optical gap". In order to find the value of the Tauc gap $E_g^T$, one can plot $\sqrt{\alpha(E)E}$ versus $E$, as shown in Figure 2.13 for a typical amorphous silicon layer.

The Tauc optical gap is the most common method used for characterizing the gap of amorphous silicon layers. Several other researchers have, however, criticised this method and suggested alternative methods for determining the optical gap on amorphous silicon layers. As an example, [Cody-1982] suggest using the assumption that the dipole matrix elements are constant, i.e. that $W(E)$ = constant; they retain the assumption of a parabolic density of states in the conduction and valence bands; this leads them to the relationship:

$$\alpha(E)/E \propto \left(E - E_g^T\right)^2 \tag{2.6}$$

This means using a plot of $\{\alpha(E)/E\}^{1/2}$ instead of a plot of $\{\alpha(E)E\}^{1/2}$. The resulting value of the optical gap is about 50 to 100 meV lower than the value obtained with Tauc's procedure.

In a comprehensive, recent publication [Mok-2007], the experimental and theoretical factors that play a role in the determination of the optical gaps according to Tauc and Cody have been presented in detail. The authors especially studied the

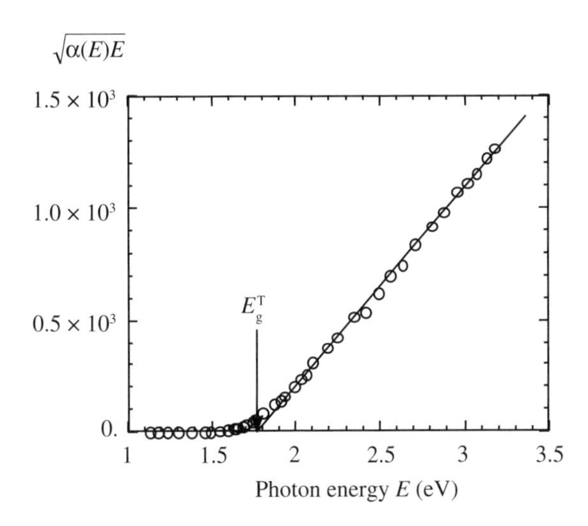


**Fig. 2.13** Plot of $\{\alpha(E)E\}^{1/2}$ versus photon energy $E$ (with in $cm^{-1}$ and $E$ in eV) as used for the determination of the Tauc optical gap $E_g^T$.

influence of film thickness on the values obtained; one can infer from this work that a more precise explanation becomes very complex.

However, it is probably not really so important which procedure one uses; what is important is to compare only values for the optical gap, that are obtained by the same method. If one compares optical gaps from films of different thicknesses, additional errors will be introduced, especially for thin films (thickness below 2 μm).

The value which has a real physical meaning is not the optical gap, but the mobility gap. However, it is rather difficult to determine the mobility gap; if one had "truly" intrinsic layers with no contamination and therefore with the Fermi level in the center of the "mobility gap", this may, in principle, be possible by evaluating the activation energy of the dark conductivity (see Sect. 2.4.3). However, in practice, one rarely has such layers. Other methods, like low-temperature (10 K) photocurrent spectra measurements [Vaneček-1985] and internal photoemission [Wronski-1989], have been used and give values which are about 100 to 160 meV higher than the optical gap according to Tauc's procedure. These methods are rather time-consuming. One therefore almost never utilizes the "mobility gap" to characterize the properties of amorphous silicon layers, but almost always the "optical gap". Thereby, one simply assumes that the "mobility gap" is about 0.1 eV larger than the "optical gap".

### 2.3.5 Relationship between sub-bandgap absorption and defect density

Transitions between midgap defects (also called "deep defects" and usually composed mainly of dangling bonds) and the two bands give rise to region C in the absorption plot. There are two methods of evaluating the density of midgap defects from region C of the plot of the absorption coefficient $\alpha(E)$ versus $E$, given in Figure 2.11. The first method, which goes back to Amer and Jackson [Amer-1984], consists of integrating

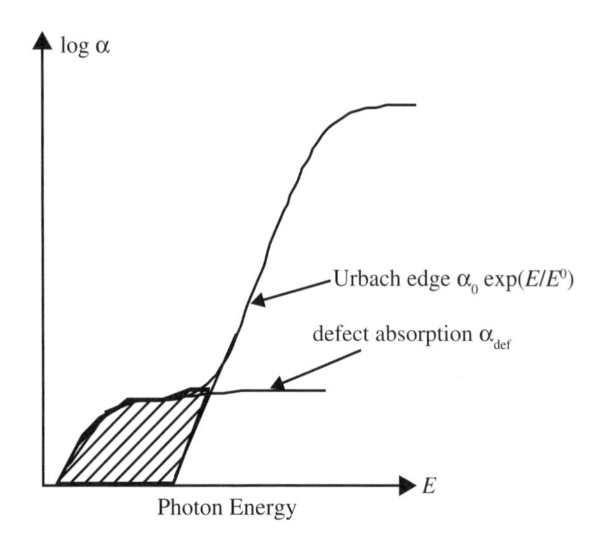


**Fig. 2.14** Plot of log $\alpha$ versus photon energy $E$ used for evaluating Urbach energy $E^0$, defect absorption $\alpha_{def}$ and mid gap defect density $N_{def}$ (in principle equal to the dangling bond density $N_{db}$).

the curve $\alpha(E)$ to the left side of the Urbach edge (exponential fall-off) in the plot of log$\alpha$ versus $E$, as shown by the hatched area in Figure 2.14.

The second method (which has become the preferred method) is to simply evaluate the value of the absorption coefficient $\alpha$ at an adequate value of photon energy, which for amorphous silicon layers is usually set to 1.2 eV.

To translate the sub-bandgap absorption into the deep defect density or dangling bond density, one needs a calibration factor. Such a calibration factor can be obtained by performing absorption measurements (PDS or CPM measurements), on one hand, and ESR measurements, on the other hand, on identical samples (i.e. on co-deposited intrinsic layers; the density of defects in the material can be easily varied by degradation, thermal annealing, ion bombardment, etc.). According to [Wyrsch-1991a], the calibration factor is different for PDS and CPM, which are the two methods commonly used for evaluating sub-bandgap absorption in amorphous silicon layers (see also Sect. 2.3.6): an absorption of 1 $cm^{-1}$ at 1.2 eV (as measured by CPM) or of 2 $cm^{-1}$ at 1.2 eV (as measured by PDS) correspond to a deep defect density between 2.4 and $5 \times 10^{16}$ $cm^{-3}$. One notes that it is not possible to determine the deep defect density or dangling bond density with great precision. The values found in the literature depend very much on the measurement technique used and (if applicable) on the calibration factor employed.

For most practical purposes, one therefore expresses the quality of amorphous silicon layers, not in terms of defect density or of dangling bond density, but in terms of defect absorption $\alpha$ (1.2 eV) at a photon energy of 1.2 eV.

## 2.3.6 Measurement of sub-bandgap absorption

For thin-film silicon devices one usually has layer thicknesses around 1 to 2 μm or 1 to $2 \times 10^{-4}$ cm; this means that an absorption coefficient of 1 $cm^{-1}$ leads to an absorbance value of 0.1 to 0.2 ‰, i.e. only about $10^{-4}$ of the incoming light is absorbed by the thin silicon layer. It is not possible to measure such low absorbance values with the usual, conventional type of photospectrometers, which are based on transmission and reflection measurements; such a low value of absorbed light will be totally masked by the measurement errors. For this reason, various methods have been developed in order to directly measure the absorption *within* the layer, independently of the reflections occurring at either side of the layer. At the moment three different techniques are in use as follows.

*Photothermal deflection spectroscopy (PDS):* Here one measures the very slight heating up of the thin-film silicon layer as induced by the absorption of a monochromatic light beam. The layer is placed in a liquid (e.g. $CCl_4$) which has a refractive index that varies strongly with temperature. Thus, the temperature increase is translated into the deviation of a probe beam (here, a He-Ne laser beam). This deviation can easily be evaluated by a photoelectric position detector. Thereby, lock-in techniques are used to increase the sensitivity (Fig. 2.15). The PDS technique is a relative technique for the measurement of optical absorption: the form of the curve that one obtains is correct, but absolute values can only be assigned if one calibrates the curve at a point where the absorption is sufficiently high that it can be measured by the conventional optical measurement technique, i.e. by using a transmission/reflection meter. Furthermore, if

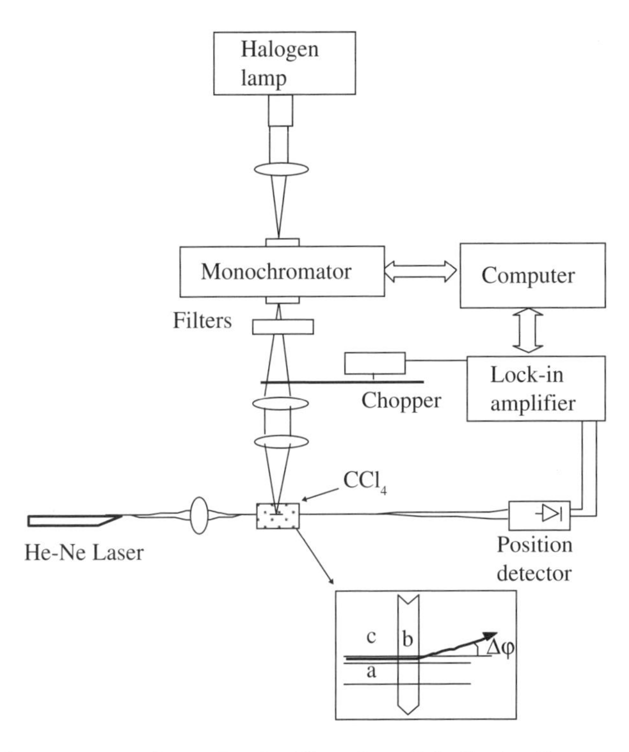


**Fig. 2.15** Experimental set up for performing Photothermal Deflection Spectroscopy (PDS) measurements, in order to evaluate sub-bandgap absorption, i.e. absorption for photon energies $h\nu$ smaller than the energy of the bandgap (or rather of the mobility gap) $E_g$. Expanded box shows the substrate (a), monochromatic light (b), and the $CCl_4$ bath (c).

the layer that one wants to measure is very thin (1 μm or less) and, at the same time, has a smooth surface, one invariably has interference fringes in the PDS spectrum; these fringes make a precise interpretation quite difficult.

The PDS technique evaluates the total absorption of the light in the whole layer, including the interface between the substrate and the silicon layer and including also the surface of the silicon layer; these regions are, however, regions with increased defect density as was discussed in Section 2.1.1 (refer to Fig. 2.4). Therefore, the PDS measurement method gives us a sub-bandgap absorption value (and consequently, an indication of the defect density) that is too high, especially for thin layers (2 μm and lower). There are three ways to overcome this difficulty: (1) measure the thickness of a series of layers with the PDS technique and extrapolate to find the "bulk" defect density [Favre-1987], [Curtins-1988]; (2) measure very thick layers where the interface/surface effects do not play a role; (3) compare PDS defect densities always for the same layer thickness, say for 1 μm – this method (3) is probably the most appropriate one as, within solar cells, the intrinsic layers also suffer from an increased density of interface defects.

*Constant photocurrent method (CPM):* Here, one measures the photoconductivity that has been induced by the absorption of the light at different wavelengths: under certain circumstances, the value of the photoconductivity of a semiconductor layer can be considered to be roughly proportional to the light absorbed and to the density of charge carriers generated within the layer. By means of an electronic regulation circuit, the photocurrent is kept constant and one correspondingly modulates the amplitude of the monochromatic light for which one wishes to measure the absorption. If the absorption coefficient is low at that specific wavelength, then a high light intensity is required to obtain the fixed value of the photocurrent; if, on the other hand, the absorption coefficient is high, then a low value of light intensity will be sufficient. As one keeps the photocurrent (i.e. the density of charge carriers) constant, it is not necessary to have an exact proportionality between the photoconductivity and the density of charge carriers. As both electrons and holes contribute to the photoconductivity, albeit in a slightly different manner, the CPM technique depends, to a slight extent, on the position of the Fermi level $E_F$. The basic CPM technique is explained in more detail in [Vaneček-1981].

In a similar way as for PDS, the basic CPM technique just gives a relative curve and has to be calibrated by an optical measurement using a "conventional" transmission/reflection meter. The calibration is done at an appropriate value of photon energy, where the absorption is sufficiently high that such conventional techniques allow for a precise measurement.

Interference fringes are also a problem for the basic version of the CPM technique. An improved version of CPM (that does not require calibration with additional transmission/reflection measurements at higher photon energies and also avoids the interference fringes) is given in [Vaneček-1995].

With CPM one is "looking at" photoconductivity, i.e. one is basically "looking at" the less defective parts of the layer, i.e. at those parts which contribute the most to photoconductivity. Interface and surface defects therefore hardly show up in the CPM method. CPM has also been successfully used to evaluate defect densities of intrinsic layers within *pin*- and *nip*-solar cells [Holovskú-2008].The CPM method is about an order of magnitude (factor 10) more sensitive than the PDS method. On the other hand, it is an indirect method for measuring absorption and therefore may be subject to artefacts and misinterpretation. In general it is, however, considered to be very reliable [Vaneček-2010].

*Fourier transform photocurrent spectroscopy (FTPS):* This method is based, just like CPM, on measurement of the photocurrent; one uses here a white bias light (rather than a monochromatic light with a variable wavelength); spectral resolution is now obtained with the help of a Fourier transformation that can be conveniently performed on a commercial FTIR instrument (see Sect. 2.2.3). FTPS is based on similar physical principles as CPM, but it is much faster and more sensitive [Vaneček-2002]. It requires, however, linearity between the photoconductivity and the charge carrier density, which is not necessarily guaranteed in amorphous silicon. One therefore generally prefers to use it mainly for microcrystalline silicon, although [Holovský-2008] shows that excellent results are, in practice, also obtained for amorphous silicon, both for layers and for cells (see also Sects. 3.2.2 and 4.8.3).

All methods described above (PDS, CPM and FTPS) are basically just "relative" methods for measuring absorption and need to be calibrated, e.g. with an additional absorption measurement by transmission/reflection (at the high end of the $\alpha(E)$ versus $E$ curve, so as to obtain absolute values of the absorption coefficient. An exception is the "improved CPM" method mentioned above, and introduced as "absolute constant photocurrent method" in [Vaneček-1995].

## 2.4 TRANSPORT, CONDUCTIVITY AND RECOMBINATION

### 2.4.1 Transport model

When one looks at scientific literature on the transport of charge carriers within amorphous silicon, one may at first become rather confused by the variety and the complexity of models studied. In fact, carrier transport at lower temperatures (well below 0 °C) in hydrogenated amorphous silicon layers is usually governed by various forms of "hopping", whereas in non-hydrogenated amorphous silicon layers, hopping even plays a role at room temperature. "Hopping" is transport by tunnelling from one localized state to the other. Various forms of "hopping" have been described: "Variable-range hopping", which is prevalent at very low temperatures, and "Nearest-neighbour hopping", which can be identified at somewhat higher temperatures. All these phenomena have indeed been extensively studied in the literature, because they give physical insight into transport mechanisms and material structure.

However, at room temperature and with device-quality material, electronic transport in hydrogenated amorphous silicon layers (and also in hydrogenated microcrystalline silicon) becomes, as far as most devices are concerned, very similar to transport in classical, crystalline semiconductors.

(A notable anomaly in these amorphous layers is the Hall Effect, where the "wrong" sign is obtained. This anomaly has so far not been explained in a satisfactory way. For the purpose of the present book, we will just state that the Hall Effect should not be used to evaluate amorphous layers. For microcrystalline silicon layers, however, the Hall Effect has no such anomaly.)

It can therefore be well described by the so-called "standard transport model". In this model, both free electrons, i.e. electrons within the conduction band, as well as free holes, i.e. holes within the valence band, travel by drift and/or by diffusion from one place to the other. On the other hand, localized electrons and holes (i.e. electrons and holes trapped in their respective bandtails) do not contribute to the current. Thus, one has for the respective electrical current densities

$$J_{\mathrm{n}} = +q\mu_{\mathrm{n}}^{0}\, n_{\mathrm{f}}\, E + D_{\mathrm{n}}\, \partial n_{\mathrm{f}} / \partial x$$

and

$$J_{\mathrm{p}} = +q\mu_{\mathrm{p}}^{0}\, p_{\mathrm{f}}\, E - D_{\mathrm{p}}\, \partial p_{\mathrm{f}} / \partial x,$$

where $q$ is the unit charge (charge of an electron), $\mu_{\mathrm{n}}^{0}$ is the band mobility of electrons, $\mu_{\mathrm{p}}^{0}$ the band mobility of holes, $n_{\mathrm{f}}$ the density of free electrons, $p_{\mathrm{f}}$ the density of free holes, $E$ the electric field prevailing at the point considered, $D_{\mathrm{n}}$ the diffusion constant of electrons, and $D_{\mathrm{p}}$ the diffusion constant of holes.

## 2.4.2 Measurement of conductivity in a co-planar configuration

Let us now look at a uniform layer of amorphous (or microcrystalline) silicon, with two "ohmic contact pads" separated by a distance $l$ of 1 to 2 cm, as shown in Figure 2.16. Note that such "ohmic" contacts can be obtained by evaporating for example a few hundred nanometres of aluminium on the amorphous layer and then annealing the whole structure for about an hour at 180 °C. We can now apply an

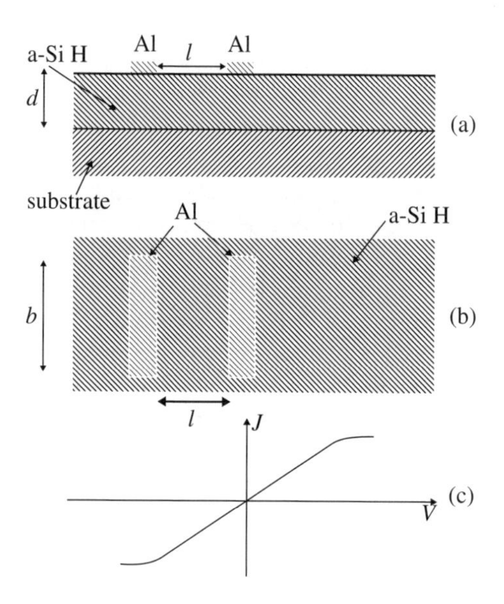


**Fig. 2.16** Experimental structure for co-planar conductivity measurements on amorphous and microcrystalline silicon layers: (a) cross-sectional view; (b) view from above; (c) typical $J(V)$ characteristic obtained.

electric voltage $V$ between the two contact pads and measure the resulting electrical current $I = A \times J$, where $J$ is the current density and $A$ the cross-section of the amorphous silicon layer through which the current is flowing. (We are assuming here, for simplicity's sake, that the current density is uniform over the whole cross-section; in a crude approximation, we may also consider $A$ to simply be the product of the thickness $d$ of the amorphous silicon layer and of the side length $b$ of the rectangular contact pad, see Figure 2.16.)

Within a uniform amorphous silicon layer, the density of carriers will everywhere be identical and there will be, thus, no diffusion. On the other hand, the voltage $V$ applied to the contacts will create, provided the contacts are ohmic (i.e. provided the $J(V)$-curve given in Figure 2.16 is linear), an electric field $E = V/l$. This will lead to a current density:

$$J = \sigma E = J_n + J_p$$

Or:

$$J = (q\mu_n^0 n_f + q\mu_p^0 p_f)E \tag{2.7}$$

i.e.

$$\sigma = (q\mu_n^0 n_f + q\mu_p^0 p_f) = \sigma_n + \sigma_p \tag{2.8}$$

Here, $\sigma$ is the conductivity of the amorphous silicon layer, with $\sigma_n$ being that part of the conductivity due to electrons and $\sigma_p$ that part of conductivity due to holes.

### 2.4.3 Dark conductivity $\sigma_{dark}$

The measurement of the dark conductivity of thin-film silicon layers is a basic and important evaluation technique. The value $\sigma_{dark}$ of the dark conductivity is generally found to vary exponentially with the absolute temperature $T$ (in Kelvin), i.e. we generally have a temperature dependence of the form

$$\sigma_{dark} = \sigma_0 \exp[-E_{act}/kT],$$

where $\sigma_0$ is called the "conductivity prefactor" and $E_{act}$ is denoted as "activation energy".

The values of $\sigma_0$ and $E_{act}$ give us a first indication about possible impurities in the layer. Necessary (but not sufficient) conditions for an a-Si:H layer to be considered "uncontaminated" are: $\sigma_{dark} < 10^{-11}(\Omega\text{cm})^{-1}$ and $E_{act} \approx E_g/2$ i.e. $E_{act} > 0.75$ eV. In μc-Si:H layers the conditions read: $\sigma_{dark} < 10^{-5}(\Omega\text{cm})^{-1}$ and $E_{act} > 0.45$ eV (see Chapter 3). Those a-Si:H and μc-Si:H layers, that do not fulfil these conditions can be considered to be contaminated, in most cases by oxygen. Indeed, oxygen acts within a-Si:H and μc-Si:H as *n*-type dopant, shifting the Fermi level towards the conduction band edge.

**Note** that the determination of $\sigma_{dark}$ in *amorphous* silicon layers is a delicate task, because of the very low conductivity values. It generally has to be done in a shielded measurement setup and with very sensitive conductivity measuring equipment.

#### *Theoretical derivation of the expression for* $\sigma_{dark}$

In the dark, i.e. without any illumination, the densities $n_f$ and $p_f$ correspond to the values $n_{f0}$ and $p_{f0,}$ i.e. to the densities prevalent at thermal equilibrium; they are given by the Fermi-Dirac distribution. Now, the Fermi level is usually quite far away from the band edges $E_C$ and $E_V$, so that the Fermi-Dirac distribution can be well approximated by the Boltzmann distribution, i.e. by a simple exponential function. We thus have

$$n_{f0} = 2\int_{E_C}^{\infty} N(E)\exp\{-(E-E_F)/kT\}dE \tag{2.9}$$

and

$$p_{f0} = 2\int_{0}^{E_V} N(E)\exp\{-(E_F-E)/kT\}dE \tag{2.10}$$

where $N(E)$ is the density of states, $k$ the Boltzmann constant, and $T$ the absolute temperature. The factor 2 is introduced because each state can be occupied with two electrons, of opposite spins.

The integration should actually be executed between the limits of the corresponding bands, i.e. between the lower and the upper band edge of the conduction band for Equation 2.9, and between the lower and the upper band edge of the valence band for Equation 2.10. However integrating up to $\infty$ in Equation 2.9 or from 0 on in Equation 2.10 will introduce only a very minor error. Furthermore, the occupation function (here the Boltzmann distribution) drops off very rapidly when moving away from the band edges, so that $N(E)$ can be replaced by the values $N_C$ and $N_V$, respectively, of the density of states at the band edges. Thus, we obtain for the density $n_{f0}$ of free electrons in thermal equilibrium (i.e. in the dark):

$$n_{f0} \approx 2N_C\, kT\exp\{-(E_C-E_F)/kT\} \tag{2.11}$$

And similarly for the density $p_{f0}$ of free holes in thermal equilibrium (i.e. in the dark):

$$p_{f0} \approx 2N_V\, kT\exp\{-(E_F-E_V)/kT\} \tag{2.12}$$

This leads to:

$$\sigma_n = 2qkT\, N_C\mu_n^0\exp\{-(E_C-E_F)/kT\} \tag{2.13}$$

And to:

$$\sigma_p = 2qkT\, N_V\,\mu_p^0\exp\{-(E_F-E_V)/kT\} \tag{2.14}$$

In amorphous silicon, the band mobility $\mu_n^0$ of electrons is considered to be around 1 to 20 $cm^2$ $(Vs)^{-1}$ and the band mobility $\mu_p^0$ of holes around 0.1 to 5 $cm^2$ $(Vs)^{-1}$(see e.g. [Zeman-2006], p.184, and Tables 4.3 and 4.4). Furthermore, undoped amorphous silicon very often has a slightly $n$-type character, due to the presence of oxygen and other undesired impurities, so that $(E_C - E_F) < (E_F - E_V)$ . Thus, in many cases, $\sigma_n >> \sigma_p$ and $\sigma_{dark} \approx \sigma_n$.

We will therefore, from now on, only consider $\sigma_n$, i.e. at that part of the conductivity that is given by electrons. By replacing the electron mobility $\mu_n^0$ with the electron diffusion constant $D_n$, according to Einstein's relation $\mu_n^0 = D_n / kT$, we obtain for $\sigma_{dark}$ the formula given below in Equation 2.15.

***Expression for $\sigma_{dark}$ and its practical consequences***

The following expression for the dark conductivity $\sigma_{dark}$ is generally used:

$$\sigma_{dark} \approx \sigma_0 \exp\{-(E_C - E_F)/kT\} \tag{2.15}$$

where $\sigma_0 = 2qN_C D_n$ is called the "prefactor" of the conductivity and is assumed, in this simple model, to be a material constant that does not depend on temperature (see [Overhof-1989, p.26 and p.77]).

The quantity $\sigma$, which we have calculated here, in the calculations leading up to Equation 2.15, is the dark conductivity and is in practice usually denoted by the symbol $\sigma_{dark}$, in order to distinguish it from the photoconductivity $\sigma_{photo}$.

There are two consequences to be drawn from Equation 2.15 as follows:

The first consideration relates to the *absolute value of the dark conductivity* $\sigma_{dark}$ *at room temperature*. Because of the relatively large negative value in the argument of the exponential function, i.e. $-(E_C - E_F)/kT \approx -20\ldots-30$, the dark conductivity $\sigma_{dark}$ of undoped amorphous silicon layers is very small at temperatures of about 25 to 30 °C and it may be as low as $10^{-12}$ $(\Omega\text{cm})^{-1}$, if the layer has a low level of impurities. (Because $e^2 \approx 10$ and $2kT \approx 50$ meV at room temperature, we can conclude that the room temperature dark conductivity increases by a factor of 10 for every 50 meV that the Fermi level $E_F$ is pushed nearer to the conduction band, due to impurities or due to voluntary doping.)

The second consideration relates to the *temperature dependence of the dark conductivity* $\sigma_{dark}$, when plotted on a logarithmic scale with $k \times \ln(\sigma)$ as a function of $(1/T)$. Such a plot is shown in Figure 2.17. The plot is a linear function with a negative slope. The slope is called the activation energy $E_{act}$ of the dark conductivity $\sigma_{dark}$. One obtains from Equation 2.15:

$$\ln\sigma \approx \ln\sigma_0 - (E_C - E_F)/kT \tag{2.16}$$

Assuming the Fermi level $E_F$ remains fixed when the temperature varies, the plot of $k \times \log\sigma$ versus $(1/T)$ would have a slope of $-(E_C - E_F)$ and the activation energy $E_{act}$ would be equal to $(E_C - E_F)$.

The activation energy $E_{act}$ of the dark conductivity does, indeed, give a rough indication of the position of the Fermi energy. If the activation energy $E_{act}$ is $\geq 0.75$ eV for amorphous silicon (and $\geq 0.45$ eV for microcrystalline silicon), one can assume that the Fermi level $E_F$ is in the middle of the gap. If, for layers that are supposed to be intrinsic, the activation energy $E_{act}$ is lower, then this is generally an indication of contamination by impurities, especially by oxygen.

In reality, the Fermi level $E_F$ does not remain fixed when the temperature varies: it shifts, especially in the case of *doped layers*. This effect is called the *statistical shift* (see [Overhof-1989], Sect. 8.1.1); it occurs because charge neutrality has to be maintained in the layer even when the temperature varies. If the density of states $N(E)$ is not constant, but is a function of energy $E$, as is the case in the region of the bandtails, the Fermi level $E_F$ will have to shift in order to maintain charge neutrality. The net result for *n-type doped layers* is that the measured activation energy $E_{act}$ has to be corrected (reduced) by about 100 to 200 meV in order to obtain the quantity $(E_C - E_F)$,

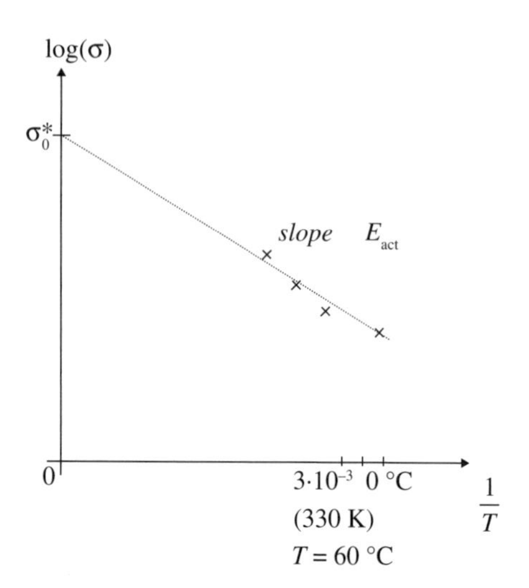


**Fig. 2.17** Plot of the dark conductivity $\sigma_{dark}$ (on a logarithmic scale and multiplied by the Boltzmann constant $k$) as a function of the inverse of the absolute temperature, i.e. as a function of (1/T); this plot usually gives a straight line. In a first approximation, the extrapolation of this line to $T = 0$ gives us the conductivity prefactor $\sigma_0$, and the slope of this line, also called the "activation energy $E_{act}$ of the dark conductivity", gives us the value $E_C–E_F$ , or the value $E_F –E_V$, i.e. the position of the Fermi level $E_F$, with respect to the nearest band edge. This procedure can be applied to intrinsic and near-intrinsic layers; for doped layers a correction factor has to be introduced because of the "statistical shift" of the Fermi level (see below).

i.e. in order to obtain the position of the Fermi energy. This correction is valid for "standard" (usual) hydrogenated amorphous silicon layers; the correction due to the statistical shift is probably much lower for hydrogenated microcrystalline layers.

For *intrinsic and near intrinsic layers*, there is virtually no correction to be made and the activation energy $E_{act}$ directly gives us the approximate position of the Fermi level $E_F$.

For *p-type doped layers*, by the measurement of $E_{act}$, one assesses the quantity $(E_F – E_V)$; here again one has to correct for the statistical shift; the correction is certainly higher than in the case of *n*-doped layers, because the valence bandtail is more pronounced than the conduction bandtail. It can be estimated that the measured value of $E_{act}$ has to be reduced by some 150 to 250 meV to obtain the quantity $(E_F – E_V)$.

To recapitulate: by plotting the measured values $k \times \ln \sigma_{dark}$ versus $(1/T)$, we obtain (more or less) a straight line with the slope $E_{act}$. If the value of $E_{act}$ is relatively high (around 800 meV for a-Si:H, around 500 meV for μc-Si:H), then we have material which is "truly" intrinsic. If the slope is somewhat lower, then we have material which is contaminated by impurities (like oxygen) causing involuntary doping effects. If the slope is much lower (i.e. if it is only 200 to 300 meV for a-Si:H, or even almost 0 meV for μc-Si:H) then we have layers which are strongly doped. However, the value $E_{act}$ of the slope does not give us directly the position of the Fermi level $E_F$ for strongly doped a-Si:H layers, but tends to underestimate the distance of the Fermi level $E_F$ from the nearest band edge. Still, the value $E_{act}$ can be used as a *relative* measure to evaluate the efficiency of doping procedures in shifting the Fermi level $E_F$.

## 2.4.4 Recombination

### *Recombination via dangling bonds*

In hydrogenated amorphous silicon layers (and probably also in hydrogenated microcrystalline silicon layers), recombination mainly occurs by free electrons and free holes being captured by dangling bond states and recombining there. Thus, the main recombination centers are the dangling bonds. The recombination processes via dangling bonds have already been briefly described in Section 2.2.2; they are schematically shown again in Figure 2.18., under the simplifying assumption that the thermal emission of electrons and holes from dangling bonds into the conduction and valence bands, respectively, is negligibly small, because of the large differences in energy involved. The corresponding arrows have therefore been omitted. Furthermore, the capture processes $r_b$ and $r_c$, that have large capture cross-sections have been denoted with a thick arrow, whereas the capture processes $r_a$ and $r_d$, that are associated with small capture cross-sections have been drawn with thin arrows.

For steady-state conditions, the occupation of the dangling bonds has to remain constant, so that

$$r_a = r_b \text{ and } r_c = r_d \tag{2.17}$$

In a similar way as in classical semiconductor recombination theory, introduced by Shockley, Hall and Read, the rates for the four capture processes can be expressed

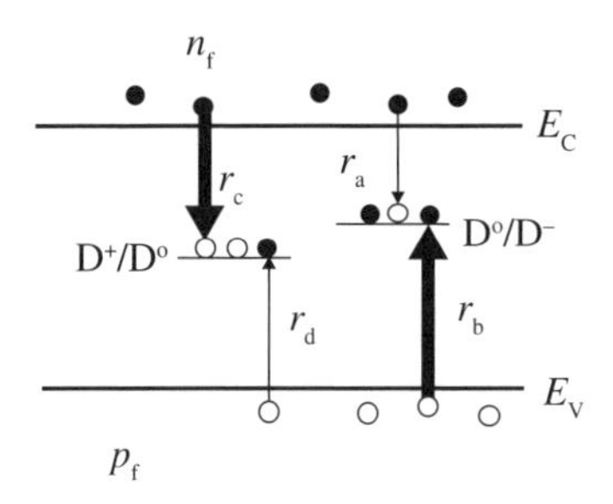


**Fig. 2.18** Schematic representation showing the capture of free holes and free electrons by a dangling bond; $r_a$ is the rate of capture of a free electron by a neutral dangling bond $D^0$; $r_b$ is the rate of capture of a free hole by a negatively charged dangling bond $D^-$; $r_c$ is the rate of capture of a free electron by a positively charged dangling bond $D^+$; $r_d$ is the rate of capture of a free hole by a neutral dangling bond $D^0$; the capture processes described by $r_b$ and $r_c$ are assisted by the attraction between two opposite charges and are therefore basically much more rapid than the capture processes described by $r_a$ and $r_d$, which involve capture by a neutral dangling bond. Thermal emission of holes and electrons from the dangling bond are not shown in this diagram, as they are considered to be much less probable than the corresponding capture processes under normal operating conditions for the intrinsic layer within solar cell devices.

as the products of the density of free carriers $n_f$, $p_f$ which are being captured, of the density $N_{db}^{+-0}$ of dangling bonds in the specific state, of the respective capture cross-section $\sigma_{n,p}^{+-0}$, and of the thermal velocity of the free carriers $n_f$ and $p_f$, i.e. of $v_{th.n}$ or $v_{th.p}$, respectively:

$$r_a = v_{th.n}\,\sigma_n^0\,n_f\,N_{db}^0, \tag{2.18a}$$

$$r_b = v_{th.p}\,\sigma_p^-\,p_f\,N_{db}^-, \tag{2.18b}$$

$$r_c = v_{th.n}\,\sigma_n^+\,n_f\,N_{db}^+ \tag{2.18c}$$

$$r_d = v_{th.p}\,\sigma_p^0\,p_f\,N_{db}^0 \tag{2.18d}$$

In accordance with part of the research literature [Wyrsch-1991b], [Vaillant-1986], [Spear-1985], [Klimovsky-2002], we will assume that $\sigma_n^+$ and $\sigma_p^-$ are almost two orders of magnitude larger than $\sigma_n^0$ and $\sigma_p^0$. Furthermore, the occupation functions $f^+$, $f^0$and $f^-$ are introduced to denote the occupation of the dangling bonds, such that:

$$N_{db}^+ = f^+ N_{db}, \tag{2.19a}$$

$$N_{db}^0 = f^0 N_{db}, \tag{2.19b}$$

$$N_{db}^- = f^- N_{db}, \tag{2.19c}$$

Evidently one can write: $f^+ + f^0 + f^- = 1$; from this last relationship and from the steady-state condition, Equation 2.17, we can calculate the occupation functions and express them in terms of the density of free carriers $n_f$, $p_f$. This enables us to calculate the densities of positively charged, neutral and negatively charged dangling bonds, $N_{db}^+$, $N_{db}^0$ and $N_{db}^-$, respectively, and insert them into Equations (2.18a)-(2.18d). Thus, the capture rates $r_a$, $r_b$, $r_c$ and $r_d$ can be determined. Now, as in the Shockley-Hall-Read theory used in classical crystalline semiconductors, we can calculate the recombination function $R$ in a similar way. The recombination function is given by the probability of an electron being captured and, subsequently, to recombine with a hole: $R = r_a + r_c$. It is also given by the probability of a hole being captured and, subsequently, to recombine with an electron: $R = r_b + r_d$. We can, thus, write the "double equation":

$$R = r_a + r_c = r_b + r_d \tag{2.20}$$

The difference from classical crystalline semiconductors is that that here there are, because of the amphoteric nature of the dangling bonds with their three charge states, not only one recombination path, but *two* parallel recombination paths, as shown in Figure 2.18. The calculations are rather complicated and the resulting general expressions are "bulky" and provide little insight. They are given in [Hubin-1992].

Thereby, one replaces the capture cross-sections $\sigma_n^+$, $\sigma_p^-$, $\sigma_n^0$ and $\sigma_p^0$ with the corresponding "capture times" $\tau_{n,p}^{+\,0-}$, such that:

$$\frac{1}{\tau_n^+} = v_{th.n}\,\sigma_n^+\,N_{db} \tag{2.21a}$$

$$\frac{1}{\tau_p^-} = v_{th.p}\,\sigma_p^-\,N_{db} \tag{2.21b}$$

$$\frac{1}{\tau_n^0} = v_{th.n}\,\sigma_n^0\,N_{db} \tag{2.21c}$$

$$\frac{1}{\tau_p^0} = v_{th.p}\,\sigma_p^0\,N_{db} \tag{2.21d}$$

The "capture times" $\tau_{n,p}^{+\,0-}$ are similar to lifetimes in classical crystalline semiconductors; they already contain information about the material quality via the density of dangling bonds $N_{db}$. Note that

$$\tau_{n,p}^{+-} << \tau_{n,p}^{0} \tag{2.22}$$

Equation 2.22 just states once again that capture which is assisted by Coulomb attraction (attraction of two opposite charges) is much more probable than capture by a neutral dangling bond; the corresponding charged-assisted capture times are therefore much smaller.

Evidently the actual *lifetimes* of the free carriers $n_f$ and $p_f$ will depend on the prevailing occupation of the dangling bonds, i.e. on $f^+$, $f^0$ and $f^-$, and these in their turn depend, for thermal equilibrium conditions, on the position of the Fermi level $E_F$; for other conditions, within actual operating devices, on the kinetics of all the capture processes.

For our further considerations we will now assume that

$$v_{th.n}\,\tau_n^{+0} \approx v_{th.p}\cdot\tau_p^{-0} \tag{2.23}$$

and that

$$\sigma_{n,p}^{\pm} \approx \varsigma \times \sigma_{n,p}^{0};\varsigma = 10...100 \tag{2.24}$$

Furthermore, we will just look at three special cases:

1. *The majority of dangling bonds are neutral*: $N_{db}^0 >> N_{db}^-, N_{db}^+$

This condition occurs, for the assumptions just given in Equations 2.23 and 2.24, (and for $\varsigma = 10$) if the ratio $n_f/p_f$ is between 0.1 and 10. It is a condition generally prevailing in the intrinsic (*i*) layer of the usual amorphous and microcrystalline silicon solar cells, which are generally of *pin*-type (see Sect. 4.9). We obtain then the following recombination function:

$$R = n_f / \tau_n^0 + p_f / \tau_p^0 \tag{2.25}$$

Both recombination paths in Figure 2.18 are now active and it is, for every recombination path, the rate-limiting (i.e. smaller) capture time (i.e. $r_a$ for the right-side recombination path and $r_d$ for the left-side recombination path), which determines the recombination function $R$. Interestingly enough, Equation 2.25 is a very simple linear expression, whereas in classical crystalline semiconductors the corresponding recombination function for intrinsic layers is non-linear!

2. *The majority of dangling bonds are positive*: $N_{db}^{+} \gg N_{db}^{0}, N_{db}^{-}$
This condition occurs, for the assumptions just given in Equations 2.23 and 2.24, if the ratio $n_f/p_f$ is less than 0.1. It is a condition, that generally prevails in $p$–type doped layers, and also within the intrinsic ($i$) layer of the usual amorphous and microcrystalline silicon solar cells, towards the $p/i$ –interface. We obtain then the following recombination function:

$$R = n_f / \tau_n^{+} \tag{2.26}$$

Equation 2.26 is formally similar to the equation for the recombination function $R$ in the $p$-type doped layers of classical crystalline semiconductors.

3. *The majority of dangling bonds are negative*: $N_{db}^{-} \gg N_{db}^{0}, N_{db}^{+}$
This condition occurs, for the assumptions just given in Equations 2.23 and 2.24, if the ratio $n_f/p_f$ is larger than 100. It is a condition, that generally prevails in $n$–type doped layers, and also within the intrinsic ($i$) layer of the usual amorphous and microcrystalline silicon solar cells, towards the n/$i$ -interface. We obtain then the following recombination function:

$$R = p_f / \tau_p^{-} \tag{2.27}$$

Equation 2.27 is formally similar to the equation for the recombination function $R$ in the $n$-type doped layers of classical crystalline semiconductors.

We will use Equations 2.25 to 2.27 in Section 4.5.9 when we look at recombination within $pin$-type amorphous and microcrystalline silicon solar cells.

***Bandtail States: separation between "traps" and "recombination centers"***

Bandtail states constitute a distribution of gap states spread out over a large range of energies. J.G. Simmons and R.W. Taylor have studied [Simmons-1972] such a situation of distributed gap states. They come to the conclusion that one can separate such states into two categories: "traps" (which exchange carriers only with the nearer of the two bands) and "recombination centers" (which capture carriers from both bands, leading thus to the recombination of these carriers). The demarcation lines between these two different kinds of states are given by what J.G. Simmons and R.W. Taylor call the "quasi-Fermi level for trapped electrons $E_{tn}$" and the "quasi-Fermi level for trapped holes $E_{tp}$".

The analysis of J.G. Simmons and R.W Taylor [Simmons-1972] is based on the kinetics of capture and thermal emission processes; it is relatively cumbersome. All we need to retain here for practical application to solar cells and other devices are the following principles (see Figure 2.19):

The quasi-Fermi levels for trapped carriers $E_{tn}$ and $E_{tp}$ are approximately equivalent to the "usual" quasi-Fermi levels $E_{Fn}$ and $E_{Fp}$, respectively. The quasi-Fermi levels are a purely computational concept used to illustrate the effect of illumination

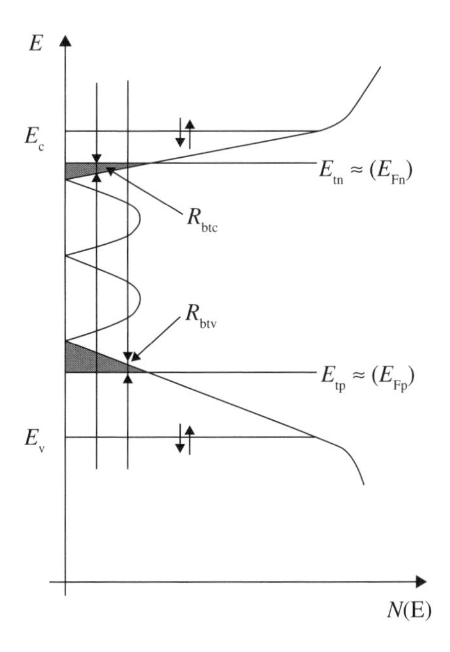


**Fig. 2.19** Schematic representation showing the "quasi-Fermi levels for trapped electrons $E_{tn}$" and the "quasi-Fermi level for trapped holes $E_{tp}$". According to J.G. Simmons and R.W. Taylor [Simmons-1972] these levels constitute the "frontiers" between "traps" (on the outside of these two levels, i.e. towards the respective bands) and "recombination centers" (between the two levels, i.e. generally in the center of the bandgap). "Traps" are gap states, which above all interact with states in the nearby bands, by capture and thermal emission of the charge carrier concerned; "recombination centers" are states which above all capture charge carriers of opposite polarity and, thus, lead to recombination. As the densities of free carriers $n_f$ and $p_f$ increase, the "frontiers" move away from the middle of the gap towards the band edges $E_c$ and $E_v$, and a larger part of the bandtails become recombination centers, i.e. the zones $R_{btc}$ and $R_{btv}$ become larger.

on the density of free electrons and holes. They are thereby defined as follows: they are the levels at which the Fermi level would have to be placed in order to obtain in the respective bands the same density of free electrons and free holes, respectively, as in the case of thermal equilibrium (i.e. without light).

As the density $n_f$ of free electrons and the density $p_f$ of free holes increase, the levels $E_{tn}$ and $E_{tp}$ move away from the center of the gap towards the bands, and an increasing part of the lower conduction bandtail states ($R_{btc}$ in Figure 2.19) and also an increasing part of the upper valence bandtail states ($R_{btv}$ in Figure 2.19) become recombination centers; this means that the density of the recombination centers increases with the increase of $n_f$ and $p_f$. This effect is one of the factors which intervenes in explaining the non-linear variation of photoconductivity with the illumination level.

On the other hand, if we keep the quasi-Fermi levels at a fixed position, by using a strong bias light, we typically obtain linear behavior for an additional, weak AC light source, when evaluating, in this way, the photocurrent spectrum. This effect is used when performing CPM and FTPS measurements.

In order to understand the basic functioning of optoelectronic devices, such as solar cells, it is not really necessary to take into consideration the additional recombination centers constituted by these bandtail states situated near the center of the gap.

### 2.4.5 Photoconductivity

We shall now look at the conductivity of an amorphous (or microcrystalline) silicon layer in the co-planar configuration of Figure 2.16 under illumination, when the densities of free carriers $n_f$ and $p_f$ are no longer given by the thermal equilibrium values $n_{f0}$ and $p_{f0}$, but are significantly larger. We assume, as in Section 2.4.2 that the contacts are "ohmic", i.e. that they do not constitute barriers and let the current of both carriers circulate freely.

Note that in the literature, this situation is sometimes referred to as "secondary photoconductivity". If one makes the distinction between "primary" and "secondary" photoconductivity, then the term "primary photoconductivity" is used to designate the photocurrent $J_{ph}$ that can be obtained in devices with blocking contacts, i.e. in a photodiode or a solar cell.

In the case of "primary photoconductivity" the current one measures is given (in the optimal case, when current collection is perfect) by the photons absorbed in the material and the electron-hole pairs generated as a result of this absorption process.

In the case of "secondary photoconductivity", on the other hand, the current flowing in the photoconductive layer is not limited by the generation rate of electron-hole pairs (as is the case in "primary photoconductivity"), because the "ohmic" contacts can supply or absorb additional electrons and holes, as needed for transporting a current density that is given by $J = \sigma \times E$ (at least as long as one remains in the linear part of the $J(V)$ characteristic shown in Fig. 2.16c).

In the present book, we do not use the term "primary photoconductivity" at all, so that in our case when we speak of "photoconductivity", we are implicitly always talking of "secondary photoconductivity".

When we have ohmic contacts, then there will be an uniform distribution of charge carriers (i.e. of both electrons and holes) in the whole amorphous (or microcrystalline) silicon layer, as shown schematically in Figure 2.20. Under the influence of an electric field $E$, holes will be transported in the direction of the electric field and electrons in the opposite direction. The current density and the conductivity will still be given by Equations (2.7) and (2.8), only now, due to photogeneration, the densities of free electrons and free holes will be much larger than in the case of thermal equilibrium (as was applicable in the dark). Indeed we have now $n_f >> n_{f0}$ and $p_f >> p_{f0}$.

The mechanism is now the following: Due to photogeneration, $n_f$ and $p_f$ increase to relatively high levels which are limited only by recombination, as we will discuss below; these high levels of $n_f$ and $p_f$ determine according to Equation (2.7) the current density $J$ for a given value of the electric field $E$. The current will flow from one contact to the other. In the case shown in Figure 2.20, holes will flow from left to right and electrons in the opposite direction; "fresh" carriers will be continuously supplied by the contacts ("fresh" holes by the contact on the left and "fresh" electrons by the contact on the right). After travelling to the opposite electrode, electrons and holes

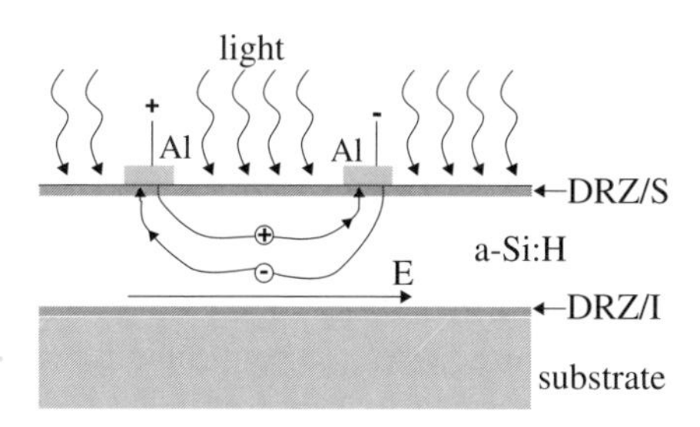


**Fig. 2.20** Schematic representation of a cross-section through an amorphous (or microcrystalline) silicon layer, showing charge transport by electrons and holes in the case of "secondary photoconductivity". The current flows between the two contacts, largely avoiding the defect-rich zones (DRZ) at substrate (S) and interface (I). The aluminum (Al) contacts are "ohmic", i.e. they can supply or absorb free electrons and holes as needed for current continuity and ensure a corresponding electric current in the connecting wires. The density $n_f$ of free electrons and the density $p_f$ of free holes are given here by the equilibrium between photogeneration and recombination processes; these densities define the conductivity $\sigma$. If the electric field $E$ is large, then the current density $J = \sigma \times E$ will also be large and may indeed become larger than the generation rate of electrons and holes.

will be absorbed there. The continuity of the electrical current is ensured by a corresponding flow of electrons in the wires connected to the electrodes.

There is now, contrary to the case of "primary photoconductivity", no direct connection between the generation rate of electron-hole pairs due to the incoming light, on one hand, and the flow rate of electrons and holes from one electrode to the other, on the other.

The current density is given by the conductivity and, in its turn, the conductivity is given by the free carrier densities $n_f$ and $p_f$, which are determined by assuming that in the steady-state (considered here) the generation rate $G$ and the recombination rate $R$ are equal. Thereby, one may neglect thermal generation of carriers (by thermal emission), provided that the illumination level is sufficiently high. This is what we will assume in the following. Here again (as in the calculation of dark conductivity) we assume that electrons contribute much more to the total current than holes.

There are three reasons that can be given for this assumption in the case of photoconductivity:

The first is the difference in the values of the mobilities, $\mu_n^0 > \mu_p^0$, already mentioned above. Secondly, there is the effect of trapped charge in the charge balance sheet: there has to be charge neutrality and therefore, even if the layer is ideally intrinsic and has no charged (ionized) impurities whatsoever, we will have the relation $n_f + n_t = p_f + p_t$. Thereby, charge neutrality is mainly maintained by the large amount of trapped charge $n_t \approx p_t$. To convince oneself of the latter, remember that $n_f = \theta_n\, n_t$ and $p_f = \theta_p\, p_t$; but remember also that $\theta_p << \theta_n << 1$. Now, for the total charge one has:

$$Q^- = n_f + n_t \text{ and } Q^+ = p_f + p_t \text{ with } Q^- = Q^+$$

$$\text{i.e. } \Theta_n n_t + n_t = \Theta_p p_t + p_t$$

This means that:

$$n_t \approx p_t$$

and that

$$n_f >> p_f$$

Thirdly, if there are impurities in the layer, which have an *n*-type doping effect, and are therefore positively ionized, like oxygen, this will increase even more the density $n_f$ of free electrons over the density $p_f$ of free holes; this last effect is, however, only relatively modest in the case of photoconductivity. The reason for this is that the densities of free electrons and holes are now much larger than in the case of dark conductivity, so that the density of ionized impurities only changes the charge balance sheet in a relatively moderate way.

We may now formally write for the photoconductivity $\sigma_{photo}$:

$$\sigma_{photo} = q\mu_n^0 n_f, \tag{2.28}$$

with $n_f$, such that

$$R = n_f/\tau_f = G \tag{2.29}$$

In a very simplified manner one may set $\tau_f \approx \tau_n^0$; then, according to Equation (2.21c), $\tau_f$ would be a material constant that is indirectly proportional to the dangling bond density: $\tau_f = k_{photo}(N_{db})^{-1}$, where $k_{photo}$ stands for $(v_{th.n}\sigma_n^0)^{-1}$. If this were really the case, the photoconductivity could be used directly as a quality parameter for the amorphous silicon layer.

However, $\tau_f$ is now not really a constant, but will depend in a complex manner on both $n_f$ and $p_f$, i.e. on the illumination level (expressed by the generation rate $G$), as well as on the trapping effects (expressed by $\theta_n$ and $\theta_p$). The essential point is that there will be charge neutrality in an uniform layer; therefore $\tau_f$ will also depend on the Fermi level within the material, i.e. on the ionized impurities in the layer.

By combining Equation (2.28) and Equation (2.29), we obtain for the photoconductivity:

$$\sigma_{photo} = q\mu_n^0 \tau_f G \tag{2.30}$$

There are two practical consequences of Equation (2.30). Firstly, $\sigma_{photo}$ is, as just mentioned, sometimes used as quality parameter for amorphous and microcrystalline silicon layers; however, $\sigma_{photo}$ is not a reliable quality parameter: the reason is that $\tau_f$ is not really a constant, but depends on the whole charge balance sheet for free and trapped carriers and for ionized impurities; indeed $\sigma_{photo}$ will become larger if for example the oxygen contamination is increased.

Secondly, $\sigma_{photo}$ should therefore only be used as a quality parameter if one is sure of having layers with the Fermi level $E_F$ in the middle of the gap, i.e. layers that have a correspondingly low value of dark conductivity $\sigma_{dark}$; these are also layers with a correspondingly high value of the dark conductivity activation energy $E_{act}$.

It has been proposed to use the ratio ($\sigma_{photo}$:$\sigma_{dark}$), which is called "photosensitivity", as a quality parameter for amorphous and microcrystalline silicon layers, but this ratio is not a true quality parameter either, although it can be taken as a first criterion for discriminating "potentially good" layers from "bad" layers: if this ratio is higher than $10^5$, in the case of amorphous silicon and higher than $10^3$ in the case of microcrystalline silicon, the layers being measured are potentially (but not surely) "device-quality" layers.

For white light of approximately 100 mW/cm$^2$ energy intensity, the values of $\sigma_{photo}$ are around $1 \times 10^{-5}\ \Omega^{-1}\ cm^{-1}$ for amorphous silicon and around $1 \times 10^{-4}\ \Omega^{1}\ cm^{-1}$ for microcrystalline silicon.

It has been shown (for amorphous silicon by [Beck-1996] and for microcrystalline silicon by [Goerlitzer-1996] that the "small-signal" photoconductivity (called "SSPC" or "steady-state photoconductivity) can be used in "midgap layers" (layers with the Fermi level $E_F$ in the middle of the gap) as a quality parameter to obtain the mobility × lifetime product ($\mu^0\tau^0$ - product), which is useful for characterizing recombination and transport within the *i*-layers of *pin*- and *nip*-type solar cells.

The "small-signal" photoconductivity is a sort of differential or incremental photoconductivity; it is obtained by measuring, with a lock-in amplifier, the additional current provoked by a chopped white light source of small amplitude, when the layer is already subject to a white bias light, with an intensity corresponding to AM 1.5 illumination (100 mW/cm$^2$, see Sect. 4.1.1).

If the layers are not "midgap layers", but have for example an *n*-type character due to the presence of oxygen contamination, then the measured value of the "small-signal" photoconductivity will be too high and not reflect correctly the "true" $\mu^0\tau^0$ - product; it will have to be combined with the measurement of the ambipolar diffusion length $L_{amb}$ in order to obtain the "true" $\mu^0\tau^0$ - product. For more information, the reader is referred to the papers by [Beck-1996] for amorphous silicon, and by [Goerlitzer-1996] for microcrystalline silicon. In practice the determination of the "true" $\mu^0\tau^0$ - product in this manner is a rather cumbersome process, so that it is preferable to perform absorption measurements in order to evaluate layer quality.

Alternatively, if one is sure of having layers with the Fermi level $E_F$ in the middle of the gap, the photoconductivity $\sigma_{photo}$ can indeed be conveniently used as a quality criterion.

## 2.5 DOPING OF AMORPHOUS SILICON LAYERS

Amorphous silicon layers without hydrogen cannot be easily doped and therefore cannot be used to fabricate devices.

On the other hand, amorphous silicon layers deposited by plasma deposition from silane contain hydrogen atoms, which "passivate" a large part of the dangling bonds (see Sect. 2.1.1); such layers can be doped by mixing phosphine ($PH_3$) with the silane ($SiH_4$) in order to obtain *n*-type doped layers, or by mixing diborane ($B_2H_6$) with the silane in order to obtain *p*-type doped layers . This was first shown by R.C. Chittick [Chittick-1969] and, later, in a systematic study by W. E. Spear and P. G. Le Comber, see e.g. [Spear-1975] [Spear-1976]. What one experimentally finds is, however, that the doping efficiency of phosphorous and boron atoms in hydrogenated amorphous silicon layers is much lower than in crystalline silicon layers.

As the dopant level increases, the doping efficiency decreases even more. It is not possible by doping to push the Fermi level $E_F$ really near to the band edges $E_C$ and $E_V$. This can be explained by two observations:

Firstly, there is the presence of bandtail states: as the Fermi level $E_F$ moves upwards towards $E_C$ or, especially, as it moves downwards towards $E_V$, it runs into the bandtails; this signifies that the electrons or holes liberated by the ionized dopants will largely be used to fill up these bandtail states and will only, to an increasingly smaller proportion, become free electrons and holes.

The second observation is the creation of additional dangling bonds by the incorporated dopant atoms: as the doping level increases, one finds that the density of dangling bonds also increases.

The *atomic model* used for explaining the second observation is shown in Figure 2.21. One assumes that there is a transformation process which changes an electronically inactive, neutral P atom, which is connected to a Si atom with four activated bonds, into an ionized P atom connected now to a Si atom with a dangling bond and vice versa. The transformation of the P atom from the neutral form (left side of Fig. 2.21) to the non-ionized form (right side of Fig. 2.21) is in equilibrium with the reverse process, and the law of mass action applies here. This leads, under certain simplifying assumptions [Street-1991], to the relationship:

$$\frac{[P_{active}]}{[P_{total}]} \approx C \frac{1}{\sqrt{[P_{total}]}}, \tag{2.31}$$

where $[P_{active}]$ denotes the concentration of active, ionized P atoms, and $[P_{total}]$ the total concentration of P atoms, whereas C is a constant. Equation 2.31 signifies that the doping efficiency decreases proportionally to $1/\sqrt{[P_{total}]}$. On the other hand, one also finds that the density of additional dangling bonds provoked by doping increases proportionally to $\sqrt{[P_{total}]}$. Similar reasoning can be applied to doping by boron (B) atoms.

W. E. Spear and P. G. Le Comber have measured, in their landmark paper [Spear-1975], the (room temperature) dark conductivity $\sigma_{dark}$ and the activation energy $E_{act}$ of the dark conductivity, for a series of amorphous silicon samples deposited in an RF-plasma from a mixture of silane and a dopant gas (diborane $B_2H_6$ or phosphine $PH_3$). The result is still valid to date; it is given in Figure 2.22 (thick, full line). It is possible to increase the dark conductivity $\sigma_{dark}$ of the layer substantially by adding a dopant gas to

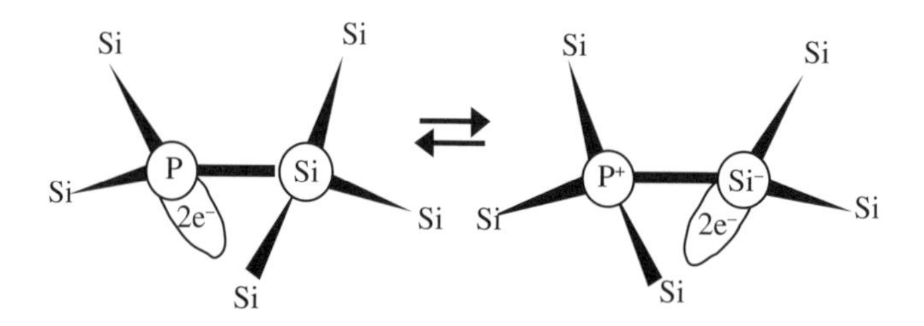


**Fig. 2.21** Conversion process from neutral to ionized doping atom (here P atom) and vice-versa. Through the conversion process a dangling bond is created on the neighboring Si atom.

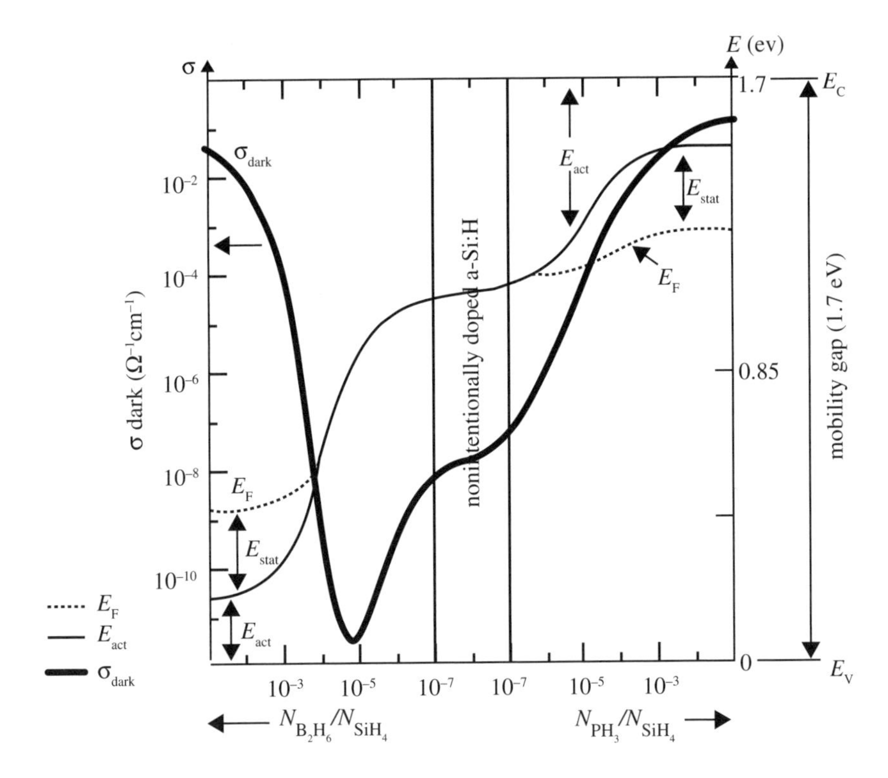


**Fig. 2.22** Measured value of dark conductivity $\sigma_{dark}$ (full, thick line); measured value of the dark conductivity activation energy $E_{act}$ (full, thin line; plotted in the Figure as distance from the corresponding mobility edge $E_C$, $E_V$) and estimated position of the Fermi energy $E_F$ (dotted line), for amorphous silicon layers, produced by PECVD on glass, as a function of the gas phase doping ratio $N_{PH3}/N_{SiH4}$ (for *n*-type layers) and $N_{B2H6}/N_{SiH4}$ (for *p*-type layers). Values of $\sigma_{dark}$ and $E_{act}$ are from [Spear-1975]. To obtain the curve for the Fermi level $E_F$, the statistical shift $E_S$ has additionally been taken into account, with certain reasonable assumptions (see Text). The equivalent bandgap of a-Si:H (or "mobility gap" as it is called here), is taken to be 1.7 eV, while drawing the graph; this corresponds to generally published values. Figure adapted from [Shah-2004].

the silane plasma; however, after reaching a certain doping level this effect saturates: it is not possible to reach conductivities much higher than $10^{-2}$ $(\Omega cm)^{-1}$ (for *p*-type layers) or much higher than $10^{-1}$ $(\Omega cm)^{-1}$ (for *n*-type layers). The measured values of the activation energy $E_{act}$ are also plotted in Figure 2.22 (thin, full line): for *n*-type samples we have plotted (just as in the original publication of [Spear-1975]), $E_{act}$ as the distance extending below the mobility edge $E_C$ of the conduction band; for *p*-type samples we have plotted $E_{act}$ as the distance extending above the mobility edge $E_V$ of the valence band curve. In our Figure 2.22, we have, additionally, corrected the curve for $E_{act}$, by taking into account the "statistical shift $E_{stat}$", a concept that was briefly introduced in Section 2.4.3. We thereby obtained the dotted line in Figure 2.22, a line that shows the estimated position of the Fermi level for these samples. The correction used here is based on the method given by P. Thomas and H. Overhof (see [Overhof-1989], sect. 8.1.1), assuming a constant density for the deep states in the gap of $10^{16}$/cm$^2$ eV and an exponential bandtail as shown in their

Figure 8.3 (full line), Note that here an identical correction has been assumed both for *n*-type and for *p*-type layers (which is, of course, only a very coarse approximation as the valence bandtail is much more pronounced than the conduction bandtail). The curve for $E_F$ in Figure 2.22 is therefore only approximate, as the density of gap states may not really correspond to the above hypotheses, and the statistical shift may, thus, be slightly different. However, the main message from Figure 2.22 is that one is not able, in amorphous silicon layers, to shift the Fermi level $E_F$ more than about half the way from the middle of the gap to the band edges (or mobility edges) $E_C$ and $E_V$.

## 2.6 HYDROGEN IN a-Si:H (by Wolfhard Beyer)

### 2.6.1 Introduction

From the preceding sections it is evident that hydrogen (H) plays a beneficial role in amorphous silicon, because it "passivates" a large part of the defects (dangling bonds). Indeed, when amorphous silicon (a-Si) layers are deposited without hydrogen (e.g. by vacuum evaporation or sputtering) they have a high defect density (around $10^{19}$ defects/cm$^3$) and are largely unusable as device material.

*Hydrogenated* amorphous silicon (a-Si:H) on the other hand, can have defect densities of $10^{16}$ defects/cm$^3$ and less. This involves in particular the a-Si:H material prepared by plasma decomposition of silane ($SiH_4$) [Chittick-1969] and disilane ($Si_2H_6$). Other methods for a-Si:H deposition include the decomposition of silane by a hot wire and sputtering of Si in an argon-hydrogen atmosphere. In all these cases, the deposition process results in hydrogen concentrations of several atomic percent or more. The H content of a-Si materials of low hydrogen concentration can be enhanced by in-diffusion of hydrogen from a flow of atomic H to the silicon surface and by hydrogen ion implantation.

It turns out that all the properties of hydrogenated amorphous silicon (a-Si:H) layers (not only defect density and electronic transport properties, but also the optical properties and the density of the layers) very much depend on the hydrogen content, on its profile within the layer and on the precise way in which hydrogen is incorporated into the layer (e.g. whether it is bonded to the silicon atoms or is trapped, as $H_2$ molecule, in voids and interstitial sites). All these characteristic quantities linked to hydrogen vary not only with the parameters of the deposition process itself, but also with various post-deposition treatments, like thermal annealing, post-hydrogenation, etc.

Looking specifically at solar cells, the crucial questions on the application of a-Si:H layers is the degree of dangling bond passivation by hydrogen and the stability of the bonded hydrogen. Another issue of hydrogen presence in a-Si:H is its role in light-induced degradation (Staebler-Wronski effect). And finally, the effect of hydrogen content on the bandgap of the layer and on its optical absorption coefficient plot (as a function of wavelength) is another key factor when designing solar cells.

As hydrogen plays a central role in amorphous silicon layers and in amorphous silicon solar cells, we must understand more about hydrogen behavior if we want to understand more about a-Si:H layers and solar cells. For all these reasons, in Section 2.6 a rather detailed picture on the role of hydrogen in a-Si:H will be presented.

While various methods have been applied to characterize and quantify the incorporated hydrogen in a-Si:H, including nuclear magnetic resonance, infrared absorption and the modelling of hydrogen incorporation by theoretical techniques, we focus here primarily on hydrogen effusion and hydrogen diffusion studies which characterize directly hydrogen stability.

### 2.6.2 Hydrogen incorporation

Within the a-Si:H material and at its surfaces, various incorporation sites of hydrogen are conceivable. One needs to distinguish between hydrogen bonded to individual Si atoms, hydrogen molecules ($H_2$) and atomic hydrogen which may be free (in voids) or dissolved in the silicon network (interstitial hydrogen). Only Si-bonded hydrogen will show up in infrared absorption spectra.

Si-bonded hydrogen can be incorporated in the amorphous silicon network in different manners, e.g.: (a) as an isolated Si-H bond (attached to a single Si atom and at a certain distance from the other H atoms); (b) as a pair of hydrogen atoms interrupting a weak Si-Si bond and forming an isolated Si-H H-Si site, at a certain distance from other silicon-bonded H atoms; (c) in the form of *hydrogen clusters* (where the SiH, $SiH_2$ bonds and the Si-H H-Si sites are grouped together). Case (c) occurs preferentially at the surfaces of voids (empty spaces) of various sizes.

Hydrogen incorporation according to (a) and (b) seems generally beneficial for the a-Si:H layers, as it leads to a relatively "stable passivation" of the dangling bonds and of weak Si-Si bonds. Hydrogen incorporation according to (c), i.e. in the form of hydrogen clusters, is, on the other hand, considered "harmful" for the a-Si:H layers, because it may lead to a situation, where hydrogen bonding becomes unstable and additional silicon dangling bonds and weak Si-Si bonds are generated. The presence of clustered hydrogen (often termed "dihydride, $SiH_2$" in literature and evaluated by the intensity of the 2100 $cm^{-1}$ infrared absorption line, see Section 2.2.3 d) has, in fact, been associated with an enhanced light-induced degradation (see [Nishimoto-2002] and Sects. 2.6.9 to 2.6.11, hereunder). According to the experience of IMT Neuchâtel and of other laboratories, a-Si:H layers deposited at low temperatures are often porous and have a high density of silicon-hydrogen clusters; they also exhibit a particularly strong light-induced degradation effect [Shah-2010].

From studies of hydrogen incorporation in crystalline silicon, various sites for the incorporation oh H are known, both from experiments and from model calculations. But it is not clear so far how these sites are related to the hydrogen sites in a-Si:H. Some of these sites may not exist at all in a disordered material. These sites (omitting complexes of hydrogen with impurities) include: vacancies of different sizes (mono-vacancies, di-vacancies, tri-vacancies etc.), the $H_2$* configuration, H platelets and H in bond-centered positions (Si-H-Si) [Van de Walle-1991]. The latter position can also be considered as an interstitial site for atomic hydrogen. In crystalline silicon, other possible interstitial sites for atomic H are also known, like the tetrahedral interstitial site.

Considering hydrogen which is not bonded to silicon, we can first look at molecular hydrogen ($H_2$): It may, in a-Si:H be incorporated in interstitial sites or in voids. Its presence seems to depend strongly on the density of the material, i.e. on whether the molecular H can diffuse out at the deposition temperature or not. The presence of unbonded atomic H in voids is unlikely because of its high reactivity with silicon, as evidenced, e.g., by etching effects [Wanka-1997].

The predominant configuration for hydrogen incorporation will depend on chemistry (binding energies), chemical reactions, the amorphous/crystalline state, microstructure, solubility effects and other factors. The binding energy (measured for diatomic molecules) lies for Si-H in the range of 3 to 3.6 eV, for Si-Si in the range of 2 – 2.5 eV and is for H-H about 4.5 eV. Accordingly, in chemical equilibrium, silicon will bind to atomic hydrogen rather than to another silicon atom. This explains the high etching rates of Si achieved with a flow of atomic H [Wanka-1997].

On the other hand, in plasma and hot wire deposition of a-Si:H, the films are commonly considered to grow by the polymerization of $SiH_3$ and $SiH_2$ type radical species under the release of $H_2$, governed by the polymerization reaction

$$\mathrm{SiH_x} + \mathrm{SiH_x} \leftrightarrow \text{Si-Si} + x\mathrm{H_2}.$$

According to this reaction, H-free Si should grow if molecular hydrogen can leave the system and if the temperature is high enough. Indeed, the hydrogen concentration in a-Si:H (see Figure 2.23) decreases strongly with rising substrate temperature so that material deposited near 600 °C contains only about 0.1 atomic % H.

According to the film growth model by polymerization, H in (plasma and hot-wire grown) a-Si:H arises primarily because of incomplete cross-linking and, as a consequence, H is predominantly incorporated at positions interrupting Si-Si bonds [Beyer-1985a]. This incorporation concept of hydrogen can also explain why at high hydrogen concentration (above 10 atomic %) hydrogen-related voids are present. Such voids are expected to form if per silicon atom more than one bond to other silicon atoms

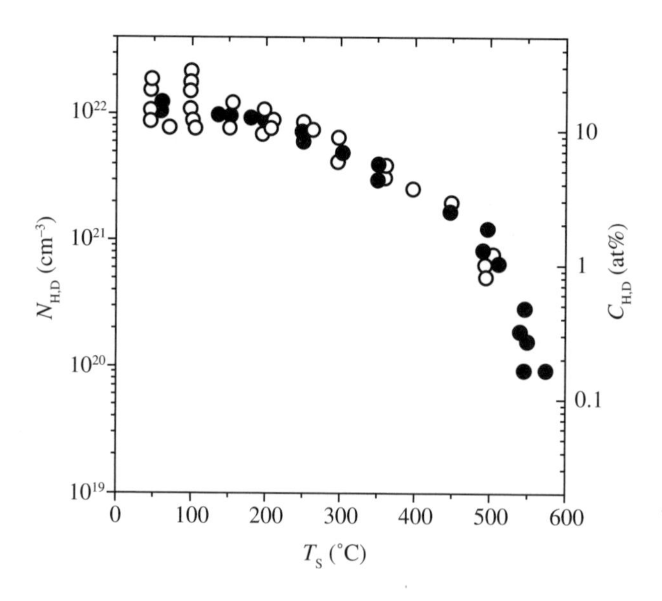


**Fig. 2.23** Density $N_{H,D}$ and concentration $C_{H,D}$ of hydrogen and deuterium determined by effusion experiments for various series of undoped plasma-grown a-Si:H (open symbols) and a-Si:D films (closed symbols) as a function of substrate temperature $T_S$ (data from [Beyer-1999]).

is interrupted by hydrogen, since the Si network thus loses connectivity. Accordingly, at hydrogen concentrations $\geq$ 25 at.% hydrogen-related voids will generally form, while at lower H concentration hydrogen-related voids still form when hydrogen is incorporated statistically or in clusters (like as $SiH_2$, $SiH_3$, polysilane or multiple Si-HH-Si sites). Indeed, the best quality plasma grown a-Si:H films have hydrogen concentrations $\leq$ 10 atomic %. Also easily explained by this growth model is the generally observed increase in hydrogen concentration with increasing deposition rate [Street-1991], as newly arriving Si radicals will inhibit hydrogen release from the film growth surface. Concerning the other H incorporation sites mentioned above, one may conclude from data of H solubility in interstitial sites in c-Si [Van Wieringen-1956] that the concentration of interstitial H is negligible up to the crystallization temperature of a-Si:H (about 600 °C). Also quite negligible (in device-grade a-Si:H) is the concentration of H molecules ($H_2$), which is of the order of one percent of the total H content according to calorimetric measurements [Von Löhneysen-1984], [Graebner-1984].

### 2.6.3 Hydrogen dilution during deposition

Hydrogen dilution is widely applied to produce a-Si:H of improved quality or to deposit microcrystalline Si. Instead of using pure silane as a process gas, gas mixtures of silane and hydrogen are employed [Guha-1981],[Guha-2003]. It seems that the major effect of hydrogen dilution is some change of chemistry of the film growth surface. In particular, the concentration of atomic hydrogen near the growing surface is enhanced. This can result in various surface effects like the abstraction of hydrogen or of $SiH_4$ groups from the growth surface and an enhanced surface mobility of $SiH_x$ ($x = 0\ldots3$) species, both leading to a denser structure of the a-Si:H films. Furthermore, gas phase reactions of process gas molecules leading to larger molecules/radicals are inhibited by gas dilution. Again this has presumably positive influences on the film microstructure. Often, hydrogen dilution leads to a decrease in hydrogen concentration [Beyer-1999]. At very high hydrogen dilution one obtains microcrystalline silicon. It has been proposed that the deposition of a-Si:H at or near the onset of crystallinity (proto-crystalline material) is a criterion to achieve best quality (stable) a-Si:H [Collins-2003].

### 2.6.4 Hydrogen effusion and hydrogen surface desorption

Depending on the microstructure of the Si:H material and other parameters, hydrogen may not remain in a stable location but will migrate through the amorphous silicon network, especially at higher temperatures. Thus, hydrogen stability is affected by diffusion and desorption processes. A highly informative experiment to get information about these effects is the measurement of hydrogen effusion. In such an experiment, a sample is usually heated under vacuum at constant heating rate and the gases leaving the sample are detected by a quadrupole mass analyzer. Hydrogen is found to leave the film in the form of $H_2$, not for example as atomic H or $SiH_4$. In Figure 2.24(a) the hydrogen effusion rate $dN/dt$ is plotted as a function of temperature for a-Si:H films grown by plasma-deposition at different substrate temperatures $T_S$. Since hydrogen is predominantly bound to silicon, the hydrogen effusion rate may be limited by three processes (see Fig. 2.24b), the primary Si-H bond rupture, the diffusion

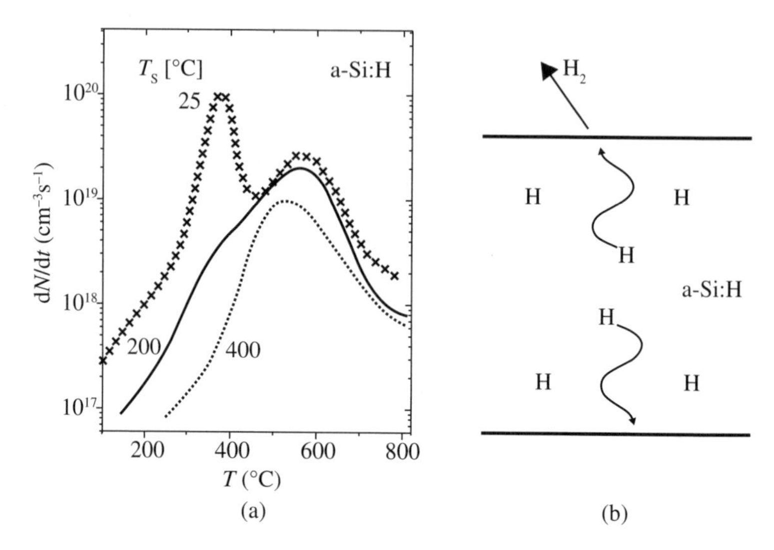


**Fig. 2.24** (a) Hydrogen effusion rate d$N$/d$t$ as a function of temperature for a-Si:H films of different substrate temperature $T_S$ (and hydrogen content). The film thickness is approx. 1 $\mu$m, according to [Beyer-1985a]. (b) Schematic illustration of hydrogen effusion from an a-Si:H film.

of atomic or molecular hydrogen to the film surface and, in the case of the diffusion by hydrogen atoms, a surface recombination step of atomic hydrogen to form $H_2$.

A detailed analysis using layered films of hydrogenated and deuterated material and a series a of films of different thickness shows that the diffusing hydrogen species (atomic or molecular H) depends strongly on the microstructure of the material [Beyer-1983], [Beyer-1985a], [Beyer-1991a]. No major difference is found in the effusion curves if hydrogen is replaced by the isotopes deuterium or tritium [Beyer-1988], [Kherani-2008].

In a ***compact (dense) material***, H diffuses in the atomic form and H diffusion limits the effusion rate. The resulting H effusion peak depends on film thickness. In Figure 2.24(a), the H effusion peak at 500-600 °C is caused by this effusion mechanism and, assuming an Arrhenius dependence of the H diffusion coefficient ($D = D_0$ exp $(-E_D/kT)$), one obtains $D_0 \approx 10^{-2}$ cm$^2$s$^{-1}$ and $E_D \approx 1.5$ eV for undoped Si:H with a hydrogen concentration of about 10 atomic %.

In a ***void-rich material***, on the other hand, hydrogen diffuses predominantly as a molecule. In this case, the hydrogen effusion rate is usually not limited by diffusion but by the primary rupture of Si-H bonds at the void surfaces, i.e. by a surface desorption process. For undoped a-Si:H, this hydrogen effusion process shows up as an effusion peak near 400 °C (see Figure 2.24(a)), which is largely independent of film thickness. The dependence of the hydrogen effusion rate d$N$/d$t$ on temperature can be described by the surface desorption formula

$$d(N/N_0)/dt = \nu_{ph} (1 - N/N_0)^n \exp(-\Delta G/kT)$$

with $\nu_{ph}$ the phonon frequency, $N_0$ and $N$ the original and effused hydrogen concentrations, $n$ the order of reaction, $k$ the Boltzmann constant and $\Delta G$ the free energy of desorption. For undoped a-Si:H with a H concentration exceeding 10 %, $\Delta G \approx 1.95$ eV and $n = 1$

is obtained [Beyer-1985a]. This free energy of desorption has been associated with the simultaneous rupture of two Si-H bonds and the formation of $H_2$, i.e. $\Delta G = 2\ E(\text{Si-H}) - E(\text{H-H})$. With literature data of the H-H binding energy of $E$(H-H) of 4.5 eV, one obtains $E(\text{Si-H}) \approx 3.2$ eV in good agreement with literature data for SiH molecules ($E$(Si-H) = 3.06 eV ) [Huber-1972]. In agreement with the only small isotope effects in the Si-H binding energy [Huber-1972], the low temperature effusion peaks for hydrogen, deuterium and tritium are observed at nearly the same temperature [Beyer-1988], [Kherani-2008].

For typical plasma-grown a-Si:H material, the hydrogen effusion curves (see Fig. 2.24(a)) show, furthermore, that the surface desorption related peak of hydrogen-rich material (curves 1 and 2) is followed by a diffusion related peak at higher temperature. This result demonstrates that the films become compact after desorption of hydrogen from void surfaces, presumably because of cross-linking of neighboring silicon dangling bonds. Such structural change of plasma-grown a-Si:H by annealing-induced hydrogen release was confirmed by thickness and other measurements [Beyer-1983], [Beyer-2004b] and supports the concept of predominant H incorporation in plasma-grown a-Si:H due to incomplete cross-linking. Note that under special deposition conditions, in particular for columnar growth, such densification by annealing does not take place [Beyer-1999].

Both H diffusion and H surface desorption are found to depend on doping, as shown in Figure 2.25. A detailed analysis shows that both the free energy of desorption of hydrogen and the hydrogen diffusion energy depend on the Fermi level and decrease considerably when the Fermi level is shifted (by $p$-type doping using $B_2H_6$) towards the valence band. For $n$-type ($PH_3$) doping, the effect is smaller. This doping dependence of H diffusion and desorption energies has been associated with simultaneous electronic transitions during rupture of Si-H bonds [Street-1987], [Beyer-1989a].

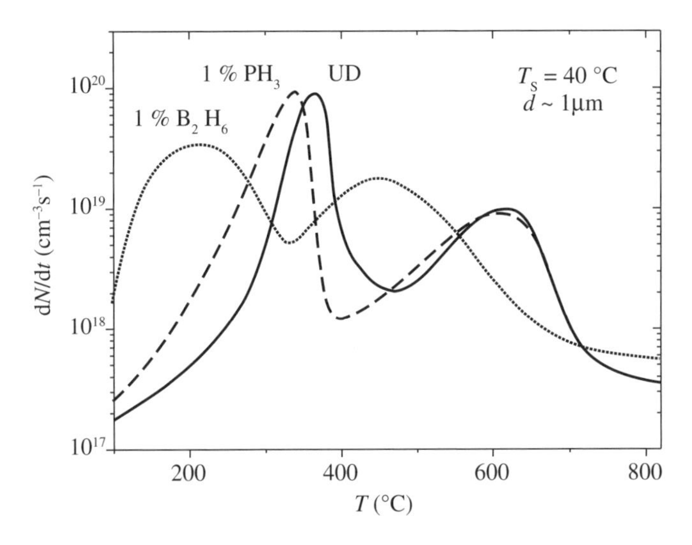


**Fig. 2.25** Hydrogen effusion rate d$N$/d$t$ for hydrogen-rich undoped (UD) and doped [with 1 % diborane ($B_2H_6$) and 1 % phosphine ($PH_3$)] a-Si:H deposited at a substrate temperature $T_S = 40$ °C (adapted from [Beyer-1991a]).

## 2.6.5 Hydrogen diffusion

Hydrogen diffusion, i.e. the migration of hydrogen atoms at a given temperature through and out of the material, can be measured not only by hydrogen effusion but also by evaluation of depth profiles of hydrogen prior to and after annealing. A highly sensitive method is the SIMS (secondary mass spectroscopy) measurement of hydrogen and deuterium inter-diffusion in layered structures of hydrogenated and deuterated material (Figure 2.26a) as first applied by Carlson and Magee [Carlson-1978]. Another important experiment to study hydrogen diffusion is given by SIMS measurements of depth profiles of deuterium introduced by a deuterium plasma (Figure 2.26b).

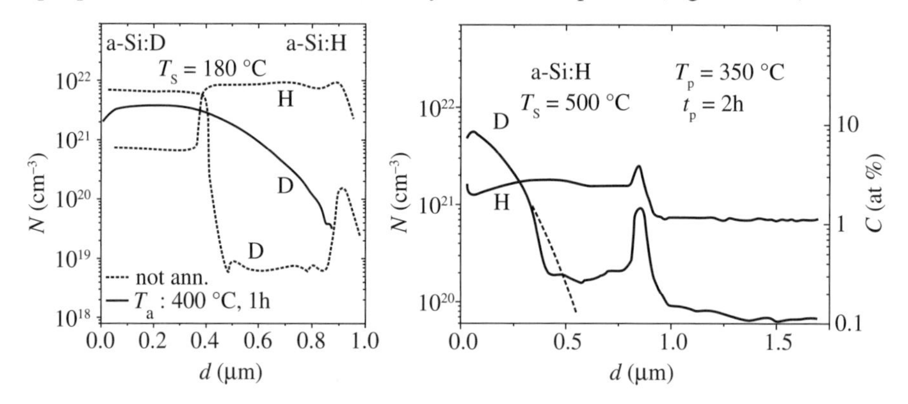


**Fig. 2.26** SIMS depth profiles of hydrogen and deuterium: (a) in layered structure of a-Si:H and a-Si:D (deposited at $T_s$ = 180 °C); for further explanations see caption of Figure 2.29. Dotted line: as deposited, full line: deuterium, after annealing at 400 °C for one hour. (b) For in-diffusion of deuterium (at a temperature $T_p$ = 350 °C for two hours) from a deuterium plasma. (adapted from [Beyer-2003]).

Such types of measurements showed that the diffusion of atomic H proceeds in a trapping process with silicon bonds acting as traps. The diffusion parameters ($D_0$, $E_D$) are not fixed but are found to depend on various other parameters, like H concentration in the investigated layer as well as in a H-source layer, doping, illumination and on other parameters [Street-1991], [Beyer-1999], [Beyer-2003]. As an explanation it is assumed that H diffusion involves not only Si-H rupture but also Si-Si bond reconstruction effects [Beyer-1999]. A linear relation between log $D_0$ and $E_D$ is observed involving a wide range of different Si:H materials (a-Si:H and μ-Si:H) and H-diffusion experiments (see Fig. 2.27). Such linear dependences between prefactor and activation energy in Arrhenius dependences have been reported for various activated physical phenomena, like conductivity in amorphous semiconductors and dielectric relaxation in polymers [Yelon-1990], [Street-1991], [Beyer-1999].

## 2.6.6 Hydrogen solubility effects

In materials science, solubility characterizes the ability of a material to incorporate foreign atoms. Solubility may limit the incorporation of such foreign atoms or, if foreign atoms are incorporated in concentrations above a solubility limit, structural

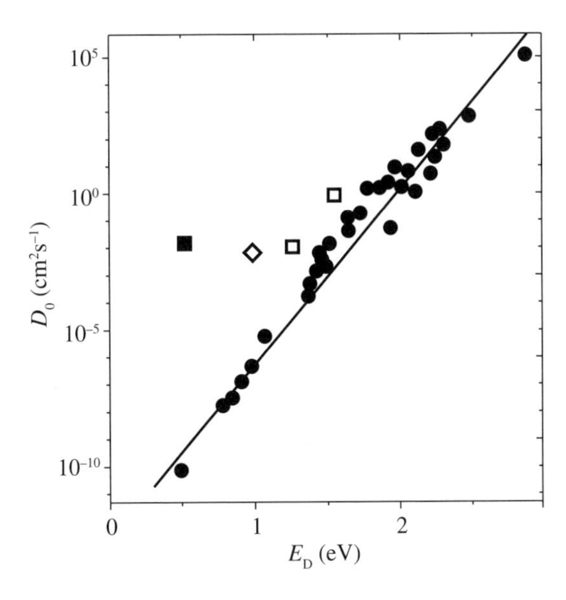


**Fig. 2.27** Experimental diffusion prefactor $D_0$ as a function of hydrogen diffusion energy $E_D$ for amorphous, microcrystalline and single-crystal silicon. The full dots (●) show the diffusion results as measured as D-H interdiffusion on many different kinds of silicon layers: undoped a-Si:H (both plasma-CVD and Hot Wire technique), μc-Si:H and H implanted c-Si and phosphorus-doped a-Si:H.The data points also include illumination enhanced H diffusion and H in-diffusion from a plasma. All these H diffusion data exhibit a *linear dependence* of the prefactor on the diffusion energy $E_D$ (the so called Meyer-Neldel rule). Exceptions to this rule are only found for single-crystal silicon (■), amorphous silicon of columnar growth (◊) and highly B-doped material (□).The figure is based on data in [Beyer-2003].

changes in a material are expected. Two solubility effects of hydrogen in a-Si:H have been discussed extensively. The first solubility effect characterizes the transition from compact to void-rich material when the H concentration exceeds about 10 atomic %. This effect is demonstrated in Figure 2.28 for measurements of H effusion. It is seen that for three series of samples (plasma grown films based on $SiH_4$, $Si_2H_6$ as well as sputtered films) hydrogen at concentrations below about 15 atomic % is incorporated in dense material (high temperature effusion peak only), while at higher H concentration a void-rich material (low temperature effusion peak) is always present. This solubility effect is attributed to a loss of connectivity in the a-Si network when more than one Si-Si bond per silicon atom is interrupted by hydrogen incorporation [Beyer-1983], [Beyer-1985a].

A second solubility effect shows up when a H- (D-) rich layer is deposited on H- (D-) poor material. As seen in Figure. 2.29, deuterium diffuses into the layer with less hydrogen only (approximately) on the level of this lower H concentration. This effect has been explained by the presence of a limited concentration of trapping sites for hydrogen in a-Si:H, since in strong Si-Si bonds with a bonding distance of 2.35 Å (as in crystalline silicon) Si-H with a bond length of 1.6 Å does not fit in. Si-H only fits into weak Si-Si bonds which are present in a-Si:H from the deposition process and

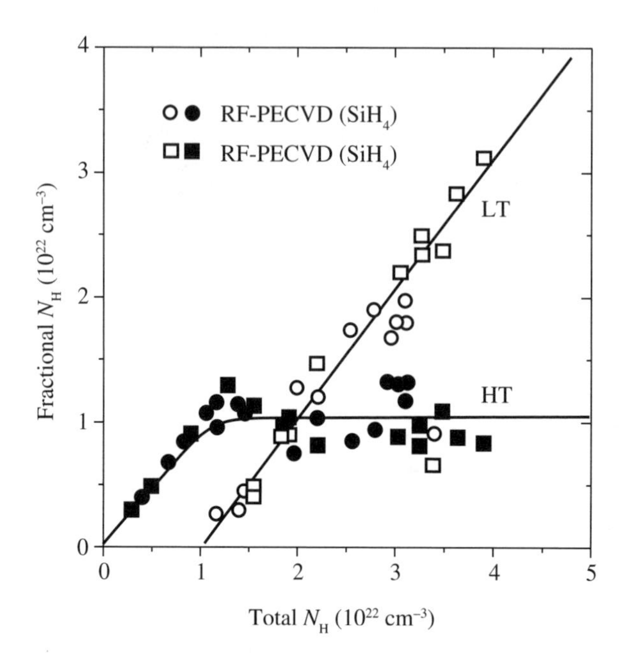


**Fig. 2.28** Low-temperature (LT, void-related, open data points) and high-temperature (HT, bulk related, closed data points) fraction of hydrogen effusion as a function of total hydrogen density (sum of LT and HT effusion) for differently prepared a-Si:H, adapted from [Beyer-1985a]. See also Chapter 6 for explanations of the PECVD deposition method.

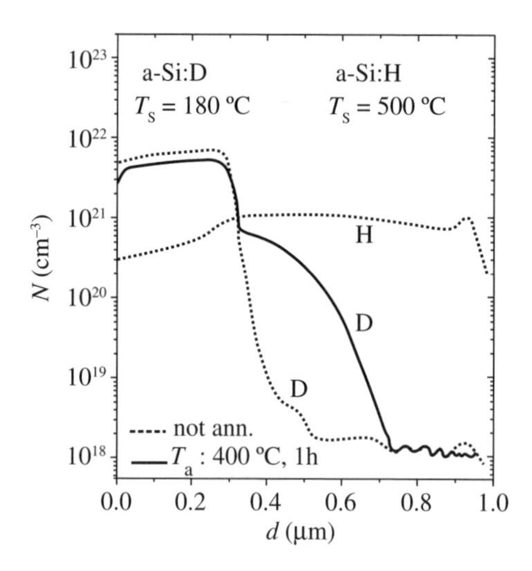


**Fig. 2.29** SIMS depth profiles of hydrogen and deuterium prior to and after annealing for layered structure of deuterium-rich a-Si:D (deposited at $T_s$ = 180 °C) on top of H-poor a-Si:H (deposited at $T_s$ = 500 °C). For the annealed state, only the profile of deuterium is given, as the annealing step hardly alters the profile of hydrogen. Deuterium diffuses into the hydrogenated material only at the level of hydrogen concentration in the hydrogen-poor material, according to [Beyer-2003] .

are usually interrupted by incorporated hydrogen. Thus, the H solubility of a given a-Si:H material is basically given by its hydrogen concentration $C_H$. Only for small hydrogen concentrations, $C_H \leq 1$ atomic %, does the H solubility attain a constant value of approximately 1 atomic % for plasma grown a-Si:H. This is an indication of the presence of structural defects in H-poor a-Si:H, such as unhydrogenated weak Si-Si bonds or isolated dangling bonds. Apparently, these defects in a concentration of about 1 atomic % act as hydrogen traps [Beyer-1997], [Beyer-1999].

### 2.6.7 Hydrogen effects on optoelectronic properties

Hydrogen incorporation influences the electronic properties of a-Si:H in various ways: (a) increase of the bandgap, (b) saturation of dangling bonds, (c) the generation of voids (see Sect. 2.6.6) and (d) causing metastability effects and material equilibration by diffusing H. These effects will be discussed in Sections 2.6.8 to 2.6.11.

### 2.6.8 Effect of hydrogen incorporation on the bandgap of a-Si:H

The influence of H incorporation on the optical bandgap of a-Si:H is demonstrated in Figure 2.30. The optical gap is seen to vary from about 1.5 eV to about 2 eV when the hydrogen concentration varies between 0 and about 50 atomic %. This effect has been used in a-Si:H based solar cells to change the spectral sensitivity.

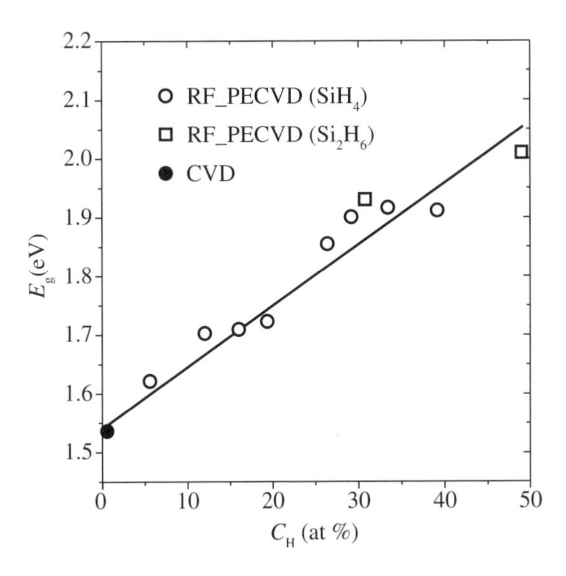


**Fig. 2.30** Optical bandgap $E_g$ of a-Si:H as a function of hydrogen concentration $C_H$ for films deposited by (conventional) thermal chemical vapor deposition (CVD) at a substrate temperature near 600 °C as well as by plasma-enhanced chemical vapour deposition (PECVD), from silane and disilane process gases at various substrate temperatures (adapted from [Beyer-1985a]). See Chapter 6 for explanations of the various deposition methods.

### 2.6.9 Stability of dangling bond passivation

The dangling bond passivation stability in a-Si:H depends strongly on the material microstructure, i.e. whether H leaves a Si-H bonding site by desorption or diffusion processes.

For a void-rich material, H leaves the Si-H bonding sites primarily by H surface desorption. Thus, for undoped material, the surface desorption formula (see Sect. 2.6.4)

$$dN/dt = \nu_{ph}(N_0 - N)\exp(-1.9\text{eV}/kT)$$

(with $\nu_{ph}$ the phonon frequency, $N_0$ and $N$ the original and effused hydrogen concentrations) can be used to estimate the hydrogen desorption rate. For a typical process temperature of 200 °C a desorption rate of about $10^{14}$ $H_2$ molecules/cm$^3$s$^{-1}$ results. Thus, for a device (solar cell) process time of 1000 s or more, this hydrogen release process can lead to high defect concentrations. Even higher H desorption will occur when the material is doped (see Figure 2.25). Thus, good quality material with defect (dangling bond) densities < $10^{16}$/cm$^3$ requires an absence (or low concentration) of voids.

In compact material, on the other hand, the diffusion of hydrogen atoms is not expected to create additional dangling bond defects, as long as the hydrogen remains within the sample. In the H diffusion process, Si-H bonds are broken and restored. Defects may be generated, however, in parts of the a-Si:H material near the surface or near the interfaces to other layers of the device and at void surfaces due to H out-diffusion. Defects may also be generated, as discussed in 2.6.11, by illumination.

### 2.6.10 Hydrogen and material microstructure

A fairly straightforward way to detect microstructure in a-Si:H is the measurement of Si-H vibrational modes in a-Si:H by infrared absorption. The Si-H stretching modes near 2090 cm$^{-1}$ and 2000 cm$^{-1}$ are considered to arise from surface bound (clustered) and bulk incorporated Si-bonded hydrogen, respectively [Cardona-1983], [Wagner-1983]. Therefore, the microstructural characterization of a-Si:H using the infrared microstructure parameter $r = I_{2090} / (I_{2000} + I_{2090})$ [Mahan-1987] is widely applied, particularly as this is a rapid non-destructive method. Thereby, $I_{2000}$ and $I_{2090}$ are the integrated absorptions values $I = \int \omega^{-1}\, \alpha(\omega)\, d\omega$ near $\omega$ =2000 and 2090 cm$^{-1}$, respectively, with $\alpha(\omega)$ the absorption coefficient at the frequency $\omega$, see also Section 2.2.3. It should be noted, however, that processes like hydrogen surface desorption or oxidation of void surfaces may lead to erroneous results [Beyer-2008]. More reliable measurement techniques of material microstructure involve small angle scattering techniques [Williamson-1998] and effusion of implanted rare gas atoms [Beyer-2004a].

### 2.6.11 Role of hydrogen in light-induced degradation

An important consequence of H diffusion at elevated temperature is that migrating H is expected to eliminate any dangling bond states in a-Si:H, such as those produced by (e.g.) illumination (Staebler-Wronski effect) or ion implantation.

For a given sample, the critical temperature $T_E$, above which diffusing hydrogen will saturate dangling bonds may be estimated by the temperature at which one-hour annealing results in a hydrogen diffusion length of about 1 Å (i.e. an atomic distance). This means that the H diffusion coefficient $D$ in a given material is approximately $10^{-20}$ cm$^2$/s at the temperature $T_E$.

Figures 2.31 a) and b) show $T_E$ as a function of hydrogen concentration and of the Fermi level position relative to the midgap, i.e. as a function of doping. These data were obtained by extrapolation of experimental Arrhenius dependences of H diffusion to lower temperatures [Beyer-2009].

For the doping dependence of $T_E$, there is a striking agreement between $T_E$ values found from H diffusion and $T_E$ data from evaluation of kinks in the $1/T$ dependence of conductivity, as shown in Table 2.1 [Street-1991]. These kinks were explained by the transition from a thermal equilibrium state of the material at higher $T$ to a frozen-in state of the material at lower $T$ [Street-1991]. The agreement between H diffusion data and the results of electronic properties of a-Si:H is a strong argument for the hypothesis that hydrogen diffusion causes the equilibrium state of the a-Si:H material at $T>T_E$ and, thus, basically causes metastability.

Note that, according to the results of Figure 2.31(a), no light-induced degradation in the *i*-layer of a-Si solar cells is expected in a dense material at high hydrogen concentration and temperatures ≥ 80 °C.

Apart from the role of hydrogen in saturating dangling bonds and, thus, removing light-induced dangling bonds at enhanced temperatures, hydrogen may also directly be involved in light-induced defect generation which takes place at

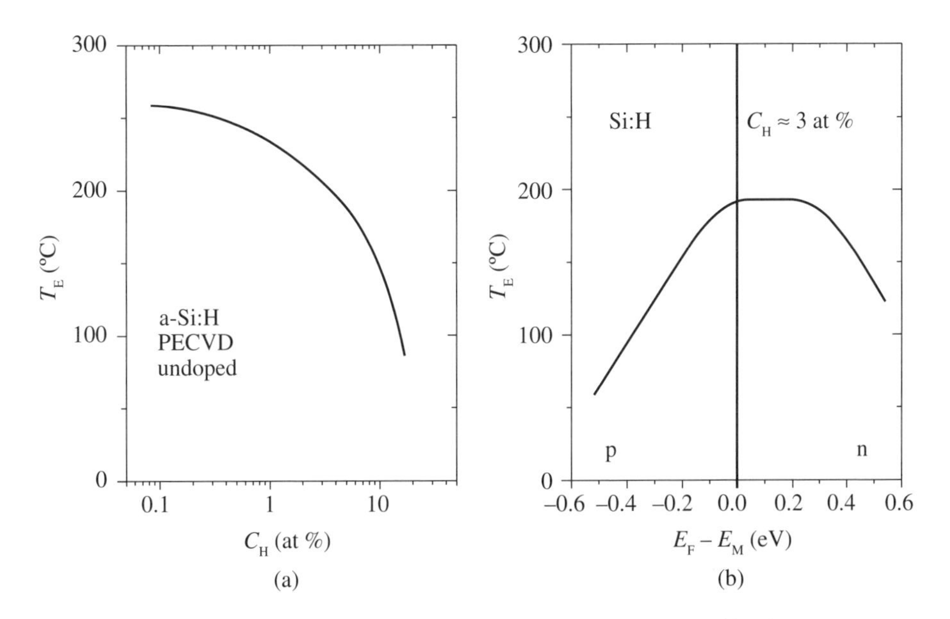


**Fig. 2.31** Temperature $T_E$ at which the H diffusion coefficient $D$ equals $10^{-20}$ cm$^2$/s as a function of (a) H concentration and (b) position of the Fermi level $E_F$ relative to midgap $E_M$; $(E_F-E_M)<0$ refers to *p*-type material, $(E_F-E_M) > 0$ refers to *n*-type material [Beyer-2009].

**Table 2.1** Comparison of measured and calculated temperatures $T_E$ for different types of doping; according to [Street-1991]

| Material | $T_E$ [°C] (measured by conductivity) | $T_E$ [°C] (calculated for $D = 10^{-20}$ cm²/s) |
|---|---|---|
| *p*-type | 80 | 70 |
| *n*-type | 130 | 110 |
| Undoped | 200 | 160 |

$T < T_E$. Indeed, an increase of light-induced solar cell degradation with increasing dihydride concentration was reported [Nishimoto-2002]. Also reported was a decrease of light-induced degradation when hydrogen is replaced by deuterium [Sugiyama-1997]. In the models by Dersch et al. [Dersch-1981] and Stutzmann et al. [Stutzmann-1985] bond switching of hydrogen is assumed to take place stabilizing the light-induced rupture of weak Si-Si bonds. Various other models [Branz-1999], [Tuttle-2004] assume the generation of highly mobile H by light. However, from the H-diffusion data there is little evidence so far for such highly mobile H near room temperatures. Yet there is considerable experimental evidence for an enhancement of H diffusion by illumination [Weil-1988], [Santos-1991], i.e. through illumination hydrogen is induced to move from one silicon bonding site to another. This light-induced motion of hydrogen could play an important role in the observed light-induced defect generation. Hydrogen atoms induced to move could both saturate existing dangling bonds and break (weak) Si-Si bonds, resulting in the generation of additional dangling bonds. Clearly this highly complex field requires future work for clarification.

## 2.7 AMORPHOUS SILICON-GERMANIUM AND SILCON-CARBON ALLOYS (by Wolfhard Beyer)

### 2.7.1 Introduction

Amorphous alloys of silicon with two other elements of group IV A (carbon group) of the periodic table (see Fig. 2.32), namely with carbon and germanium, have been extensively studied for use within thin-film solar cells. Some attention has also been given to alloys of amorphous silicon with oxygen, nitrogen and tin. The reason for these investigations was the possibility of substantially varying the bandgap of the resulting thin film layer:

- towards higher values, by alloying amorphous silicon with carbon (a-Si,C:H), oxygen (a-Si,O:H) or nitrogen (a-Si,N:H);
- towards lower values by alloying silicon with germanium (a-Si,Ge:H) or tin (a-Si, Sn:H).

In the present section, we will only deal with alloys of silicon and carbon (a-Si,C:H) and alloys of silicon and germanium (a-Si,Ge:H). These alloys are, in general, fabricated by plasma-enhanced chemical deposition, just like unalloyed

| | | | | | 2 He |
|---|---|---|---|---|---|
| 5 B | 6 C | 7 N | 8 O | 9 F | 10 Ne |
| 13 Al | 14 Si | 15 P | 16 S | 17 Cl | 18 Ar |
| 31 Ga | 32 Ge | 33 As | 34 Se | 35 Br | 36 Kr |
| 49 In | 50 Sn | 51 Sb | 52 Te | 53 I | 54 Xe |
| 81 Tl | 82 Pb | 83 Bi | 84 Po | 85 At | 86 Rn |

**Fig. 2.32** Periodic table of elements (partial).

amorphous silicon layers are. Thereby, in addition to silane ($SiH_4$) (or disilane ($Si_2H_6$)) and hydrogen ($H_2$), the following process gases are commonly used for producing alloys:

- methane ($CH_4$) for producing a-Si,C:H alloys;
- germane ($GeH_4$) for producing a-Si,Ge:H alloys.

Alloys that are used today in commercial thin-film silicon solar cells and modules for the photogenerating intrinsic layers (*i*-layers), in *pin* and *nip* solar cells are:

- a-Si,Ge:H in the bottom and middle sub-cells of the triple-junction cells of Uni-Solar (see Sect. 5.4.2);
- a-Si,C:H in the top sub-cell of the tandem cells of EPV, Bangkok Solar, and several other companies producing modules according to the recipes of the CHRONAR company (see again Sect. 5.4.2 and, specifically, the reference therein to [Delahoy-1989])

For *p*-type window layers in *pin* and *nip* solar cells, a-Si,C:H layers are used.

The following Sections deal briefly with the fabrication, structure, defects and the electronic and optical properties of a-Si,Ge:H and a-Si,C:H alloys, as produced by plasma-CVD.

### 2.7.2 Fabrication

As mentioned above, a-Si,Ge and a-Si,C alloys are usually fabricated by plasma-enhanced CVD. One major advantage of the plasma deposition process is the fact that it is very easy to grow a-Si alloys: As an example: just by adding, germane or methane to silane, one obtains a-Si,Ge and a-Si,C alloys.

For a-Si,C:H, various gas mixtures of silicon- and carbon-bearing hydrides like silane, disilane as well as methane, ethane, and acetylene are used, as well as gases containing Si-C bonding configurations, like silylmethane or methylsilane [Beyer-1987a], [Beyer-1989b]. For a-Si,Ge:H the gases silane, disilane, germane, digermane as well as silylgermane and others are employed [EMIS-1989], [EMIS-1998], [Finger-1998a]. Fluorinated gases like $GeF_4$ are also applied, in particular for a-Si,Ge alloys [Finger-1998a].

A modification of conventional plasma deposition called "cathodic deposition" [Wickboldt-1997] has been successfully used for the deposition of a-Si,Ge:H and a-Si,C:H. In "cathodic deposition", the substrates are placed on the powered electrode of a capacitive PECVD apparatus. Within an RF-plasma, this electrode charges up negatively and is therefore termed "cathode". Due to the positive charge of most ions in the plasma, "cathodic" samples grown with an enhanced ion bombardment (see also Chapter 6, Sects. 6.1.1 and 6.1.2).

One method extensively (and very successfully) studied is the deposition of a-Si,Ge:H and a,Si,C:H by the hydrogen dilution method (see Sect. 2.6.3) [Guha-1981], [Guha-2003], [Matsuda-1987]. Also used for the deposition of a-Si,Ge:H alloys was the helium (He-) dilution method [Tsuo-1991], [Middya-1994], [Jones-2000]. Another deposition method, which has been investigated with some success for the deposition of a-Si,Ge:H and a-Si,C:H, is the hot-wire method (see Sect. 6.2).

Figure 2.33 shows the alloy composition $y$ as a function of gas phase composition $x$ for a-Si,Ge:H and a-Si,C:H alloys prepared by plasma deposition. Hot wire data for a-Si,Ge:H are also included, for comparison. As one can see, no unique relation

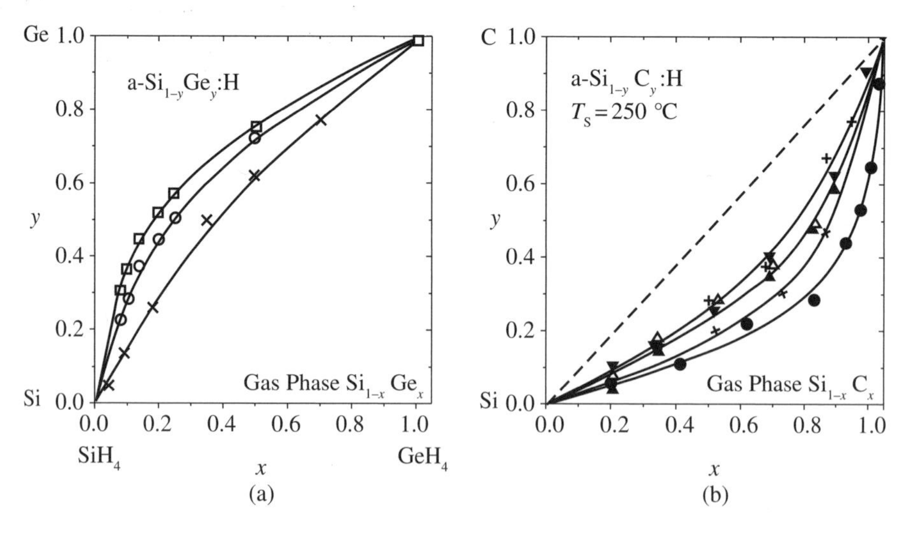


**Fig. 2.33** Solid-state composition parameter $y$ as a function of gas phase composition $x$ for (a) a-$Si_{1-y}Ge_y$:H alloys ($T_S \approx 270$ °C, plasma-deposited, with He-dilution (□), plasma-deposited with H-dilution (o), data from [Tsuo-1991]; deposited with the hot wire method ($T_S \approx 360$ °C) (x), data from [Nelson-1998]; and for (b) a-$Si_{1-y}C_y$:H alloys (plasma-deposited, using undiluted gas mixtures : $SiH_4$-$CH_4$ (•), $SiH_4$-$C_2H_6$ (▲), $SiD_4$-$C_2D_6$ (Δ), $SiH_4$-$C_2H_4$ (▼), $Si_2H_6$-$CH_4$ (x), $Si_2H_6$-$C_2H_4$ (+): data from [Beyer-1987a]).

between $y$ and $x$ exists. Indeed, the incorporation of atoms from the gas phase into a solid involves various steps, like the fractionation of gas species into radicals, the adsorption of radicals, as well as desorption and etching effects at the growing film surface. Thus, deposition parameters such as substrate temperature, pressure, gas dilution, plasma power, etc. are expected to play a role in the film composition, as well as the respective binding energies.

If $p$ is the probability that a given gas species enters the solid, the probability ratio which defines a preferential incorporation can be calculated [Stutzmann-1984] according to $p(\text{Si})/p(\text{Ge,C}) = (1-(1/y))/(1-(1/x))$. The data in Figure 2.33 yield $p(\text{Si})/p(\text{Ge}) \approx 0.25$ for a-Si,Ge deposition based on (He-diluted) $SiH_4$-$GeH_4$ gas mixtures, and $p(\text{Si})/p(\text{C}) \approx 6$ for a-Si,C deposition based on a $SiH_4$-$CH_4$ gas mixture.

### 2.7.3 Structure of a-Si:Ge:H and a-Si:C:H alloys

An ideal alloy of silicon with germanium or carbon would consist entirely of Si-Ge or Si-C bonds, respectively, i.e. it would be atomically ordered (see Figure 2.34 (a)). Of course, such a situation can only apply to stochiometric alloys. In reality, in particular if the alloy is fabricated at rather low deposition temperatures, Ge and Si atoms (or Si and C atoms) are incorporated in a random or clustered manner (see Figure 2.34 (b)). One reason is that the binding energies of Si-Ge, Ge-Ge and Si-Si bonds are quite similar, lying between 2.8 and 3.1 eV [Huber-1972]. In case of a-Si,C alloys, the binding energy of Si-C (4.6 eV) exceeds that of Si-Si bonds, but C-C double bonds are much stronger. Thus, the formation of Si-C or Si-Ge bonds is barely favored by binding energy effects.

An improvement in the film structure is expected and found using Si-Ge and Si-C -bearing gases for not too high plasma powers [Beyer-1989b], as in that case Si-C and Si-Ge bonds in the molecules may be directly transferred into the films. Improved microscopic bonding was demonstrated e.g. by measurement of the (infrared) Si-C absorption near 800 $cm^{-1}$ which is expected to attain its highest values for the highest concentration of Si-C bonds [Beyer-1987a].

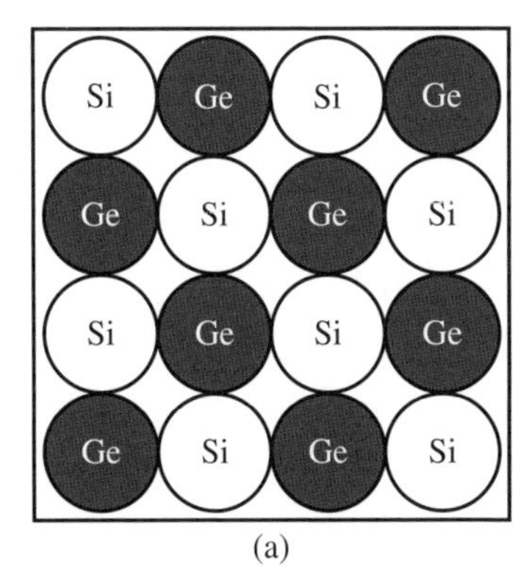


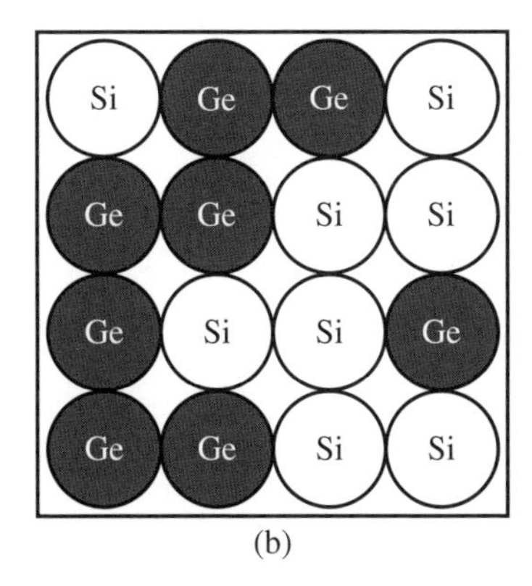


**Fig. 2.34** (a) Atomically ordered Si,Ge alloy, (b) statistical/clustered Si and Ge incorporation.

Besides the difficulties in attaining atomic ordering, a central problem of a-Si,Ge and a-Si,C alloy growth is the formation of a void-rich structure [Williamson-1998]. This means that empty spaces in the films of different sizes exist, from isolated vacancies to interconnected void clusters.

For both a-Si,Ge:H and a-Si,C:H, significant success in structural improvement has been achieved by the method of hydrogen dilution [Guha-1981], [Matsuda-1987], [Alvarez-1992].

For a-Si,Ge:H, the use of a higher substrate temperature [Yang-2009] than applied for a-Si:H (typically 200 °C) is likely to lead to a more dense microstructure, as known for a-Ge:H deposition [Beyer-2004b]. The deposition on the cathode side (powered electrode) of the plasma reactor was also reported to result in structurally improved a-Si.Ge:H alloys [Wickboldt-1997].

### 2.7.4 Hydrogen incorporation, effusion, surface desorption and diffusion

The binding energies of hydrogen to Ge, Si and C atoms are listed in Table 2.2.

According to the concept of H incorporation into Si-H alloys by incomplete polymerization (see Sect. 2.6.2), one would expect an increase of hydrogen incorporation when hydrocarbons are applied as a process gas and a decrease in H concentration when germane is used, as the binding energy of hydrogen to C is higher and to Ge is lower than to Si. As demonstrated in Figure 2.35, such an increase in H concentration upon C incorporation and a corresponding decrease upon Ge incorporation are indeed found [Beyer-1985a]. Thus, for a-Si,C:H alloys there are often problems with a "too high" hydrogen concentration (mostly due to $CH_3$ incorporation) and for Si-Ge alloys with a "too low" concentration of hydrogen.

In a similar way as in a-Si:H, bonded hydrogen in a-Si alloys can be detected by infrared absorption measurements [by FTIR-measurements, see Sect. 2.2.3]. Vibrational frequencies of hydrogen in bonds with alloy atoms are listed in Table 2.3. From both infrared absorption and H effusion experiments, it was

**Table 2.2** Binding energies of hydrogen to silicon, germanium and carbon:

| | Binding energies [eV] | | |
|---|---|---|---|
| Bond | Si-H | Ge-H | C-H |
| Literature [Huber-1972] | 3.06 | 2.95 | 3.47 |
| From H effusion [Beyer-1991] | 3.20 | 2.97 | 3.85 |

**Table 2.3** Vibrational frequencies (expressed as wavenumbers) of hydrogen bond stretching modes (see Sect. 2.2.3), as used for the evaluation of FTIR data.

| Bond | Si – H | Ge – H | C – H |
|---|---|---|---|
| Wavenumber [$cm^{-1}$] | 2000-2100 | 1900-2000 | 2900-3000 |

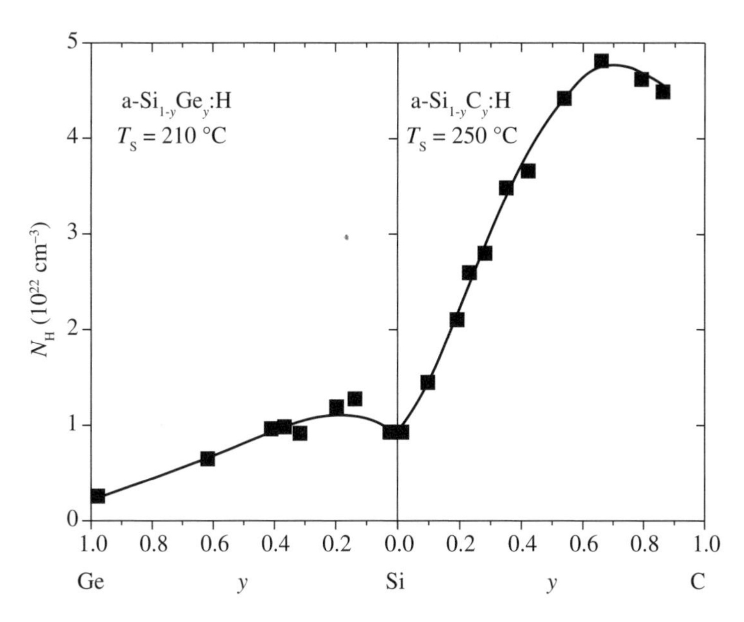


**Fig. 2.35** Hydrogen density $N_H$ in a-Si,Ge:H and a-Si,C:H alloys as a function of solid-state composition $y$. Substrate temperature is $T_S = 210$ and 250 °C, respectively. Adapted from [Beyer-1985b].

concluded that in the case of carbon-rich a-Si,C:H alloys there is an incorporation of high amounts of hydrogen in the configuration of methyl (-$CH_3$) groups [Beyer-1987a].

H effusion from a-Si alloys shows qualitatively a similar temperature dependence as for a-Si:H [see Sect. 2.6.4, Fig. 2.24(a)], i.e. there are usually one or two effusion peaks. As for a-Si:H, the low temperature effusion peak is attributed to H-surface desorption and the high temperature peak to the diffusion of hydrogen in a (more or less) compact material [Beyer-1987a]. The bond rupture energy $\Delta G$ for hydrogen surface desorption varies considerably as the composition changes, because bonded hydrogen atoms with considerably different binding energies are involved. Temperatures of low-temperature hydrogen effusion peaks and corresponding hydrogen bond rupture energies $\Delta G$ are listed in Table 2.4 for undoped material [Beyer-1991a].

**Table 2.4** Temperature of low-temperature hydrogen effusion peak as well as free energy of hydrogen desorption $\Delta G$ for (undoped) a-Si:H, a-Ge:H and a-C:H.

| Material | Temperature [K] of low-temperature hydrogen effusion peak, for a heating rate of 20°/min | $\Delta G$ [eV] |
|---|---|---|
| a-Si:H | 645 | 1.95 |
| a-Ge:H | 480 | 1.44 |
| a-C:H | 1005 | 3.2 |

For certain materials with dense structure the H diffusion coefficients were measured. As shown in Figure 2.36 (a), the H diffusion coefficient (at $T = 400$ °C) in the Si-Ge alloy system increases by an order of magnitude when the composition is changed from 0 to 20 % at.% Ge, and remains constant at higher Ge concentrations [Beyer-1991b]. For sputtered a-Si,C:H (with a nearly constant H concentration) a hydrogen diffusion coefficient decreasing with increasing carbon content was reported, while for a-Si,O:H alloys an increase in the H diffusion coefficient was observed (see Figure 2.36 (b)). This latter increase was attributed to an increase in hydrogen concentration [Beyer-2000].

### 2.7.5 Microstructural effects (voids)

In a similar way as in a-Si:H, a major defect in a-Si alloys is the formation of voids of various sizes. These voids are primarily related to:

- film growth;
- alloy formation;
- hydrogen incorporation above the hydrogen solubility limit.

Film growth results in voids primarily at low surface mobility, with the consequence of island formation and self-shadowing effects. Alloy formation may lead to the presence of voids, e.g., due to the different binding energies of the alloy atoms to hydrogen. This leads to a preferential attachment of hydrogen to those alloy atoms which have the stronger hydrogen bonding [Paul-1981] and, as

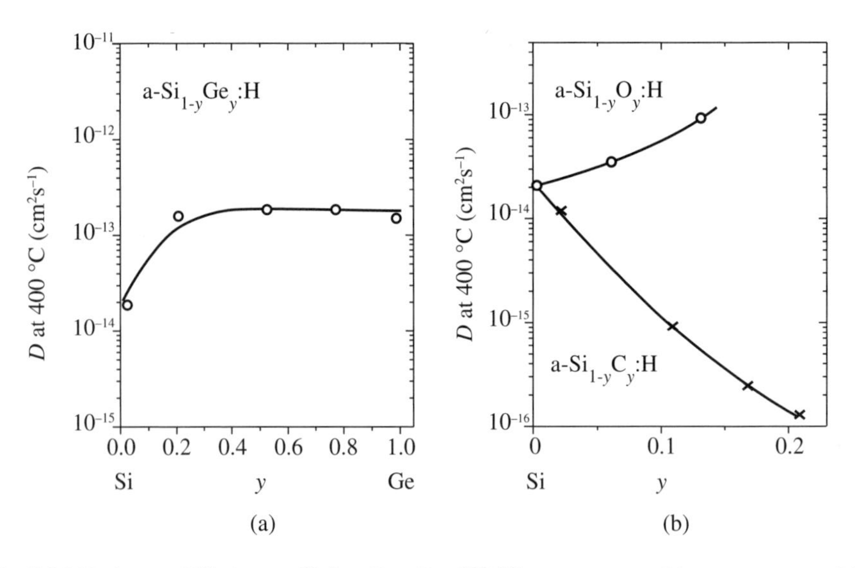


**Fig. 2.36** Hydrogen diffusion coefficient $D$ at T = 400 °C versus composition parameter $y$: (a) for a-Si,Ge:H alloys [Beyer-1991b] and (b) a-Si,C:H and a-Si,O:H alloys; from (adapted from [Beyer-2000]).

a consequence, to dangling bonds and voids near atoms of less strong hydrogen bonding.

For a-Si,Ge:H prepared with undiluted gases and on the standard (anode) side of plasma reactors, void formation related to film growth and to alloy formation seems to prevail. In the case of a-Si,C:H, void formation is mostly caused by the high hydrogen incorporation, largely related to the incorporation of $CH_3$ groups [Williamson-1998].

## 2.7.6 Dangling bonds, density of defect states

Upon alloying, quite generally an increase in defect density (i.e. in dangling bond concentration) is observed, as shown in Figure 2.37 [Beyer-1987b], [Cohen-1998].

Various reasons for this increase in defect concentration have been proposed; the most straightforward one appears to be the formation of void-rich materials where H is unstable at void surfaces and desorbs, resulting in dangling bonds with states near the midgap. The preferential bonding of hydrogen to the alloy component with higher hydrogen binding energy has also been proposed as a possible source of defects in a-Si,Ge:H alloys [Paul-1981]. Matsuda and Ganguly [Matsuda-1995] argued that formation and incorporation of short-lived $GeH_x$ ($x = 0\text{-}2$) radicals may be a reason for the increased defect density.

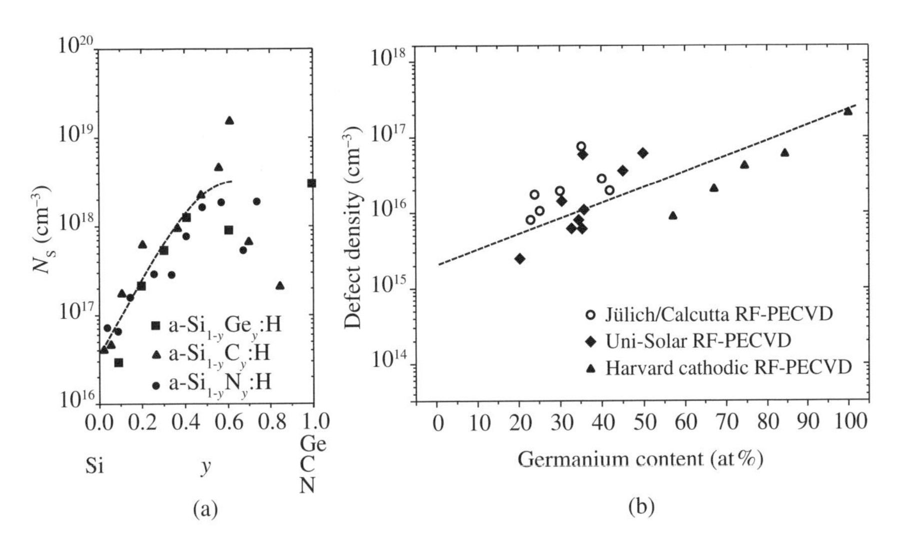


**Fig. 2.37** (a) Spin density $N_S$ (from ESR-measurements, used here as a measure of defect density, of various a-Si alloys as a function of composition $y$ (adapted from [Beyer-1987b]), and (b) defect density versus germanium content of a-Si,Ge:H alloys prepared at various laboratories (according to [Cohen-1998]). "Cathodic RF-PECVD" is a deposition method with relatively strong ion bombardment on the growing layer; refer to the remarks at the beginning of Section 2.7.2 as well as to [Wickboldt-1997]. See also Chapter 3, Section 3.2.2 for a detailed description of the ESR measurement technique and Chapter 6, Sections 6.1.1 and 6.1.2, for further explanations on the PECVD deposition method and on the ion bombardment issue.

### 2.7.7 Hydrogen stability versus alloy composition

In a similar way as in a-Si:H, hydrogen stability in a-Si alloys will depend on the material structure, i.e. on the ability of hydrogen to diffuse out of the material as a molecule or atom. In the case of void-rich materials, hydrogen can diffuse out rapidly as a molecule. In this case, hydrogen stability is determined by the reaction converting hydrogen atoms bonded to Si, C, etc. to molecular hydrogen. H stability can then be estimated with the surface desorption formula in Section 2.6.4, using the values of the free energy of desorption of Table 2.4 (or interpolations). Much higher H stability is expected and observed for compact material, where (slow) diffusion of atomic hydrogen dominates H stability. While diffusion of atomic hydrogen (in the bulk) is unlikely to create new defects, surface desorption of $H_2$ from internal and external surfaces may result in defect creation.

### 2.7.8 Doping effects

*n*- and *p*-type doping has been reported for a-Ge [Spear-1976], for the full range of a-Si,Ge alloys [Stutzmann-1986] , as well as for a-Si,C alloys up to a bandgap of about 2.3 eV [Tawada-1982]. A Fermi level dependence of H desorption (and diffusion) was observed for both alloy systems [Beyer-1991c].

### 2.7.9 Light-induced degradation

Light-induced degradation has been reported for a-Ge:H [Su-2004], for a-Si,Ge alloys [Stutzmann-1985] as well as for silicon-rich a-Si,C:H alloys [Chen-1984], [Dawson-1996]. Assuming that the annealing out of light-induced defects proceeds by H diffusion, the temperature of this annealing process can be approximated by the temperature $T_E$ at which the H diffusion coefficient attains a value of $D \approx 10^{-20}$ cm$^2$/s (see Sect. 2.6.11). In Figure 2.38, such data is shown, based on an extrapolation from H diffusion results [Beyer-1991b].

### 2.7.10 Optical absorption

The influence on the bandgap of alloy formation of silicon with Ge and C is demonstrated in Figure 2.39.

It is often assumed that the bandgap variation is caused solely by the incorporation of a different semiconductor material. One needs to note, however, that in these alloy systems the hydrogen concentration also varies (at fixed deposition conditions), in such a way that the decreased bandgap for a-Ge:H is partially due to a decreased hydrogen content, while the enhanced bandgap in a-Si,C is caused, to a great extent, by an increased hydrogen content. This influence of the hydrogen concentration on the bandgap $E_g$ is demonstrated in Figure 2.40. Here, the (Tauc) bandgap $E_g$ for a-Si:H as well as for 1:1 gas mixtures of $SiH_4$ with $GeH_4$, $CH_4$ and $NH_3$ is plotted as a function of hydrogen density $N_H$ – the latter is varied by changing the substrate temperature [Beyer-1987b]. For a-Si,Ge:H alloys, similar results were reported by Terakawa et al. [Terakawa-1993].

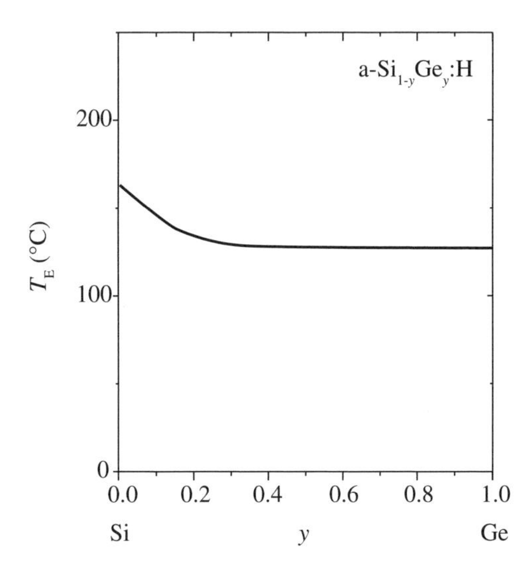


**Fig. 2.38** Temperature $T_E$ as a function of composition $y$ of (undoped) a-Si,Ge:H alloys; (adapted from [Beyer-2009]).

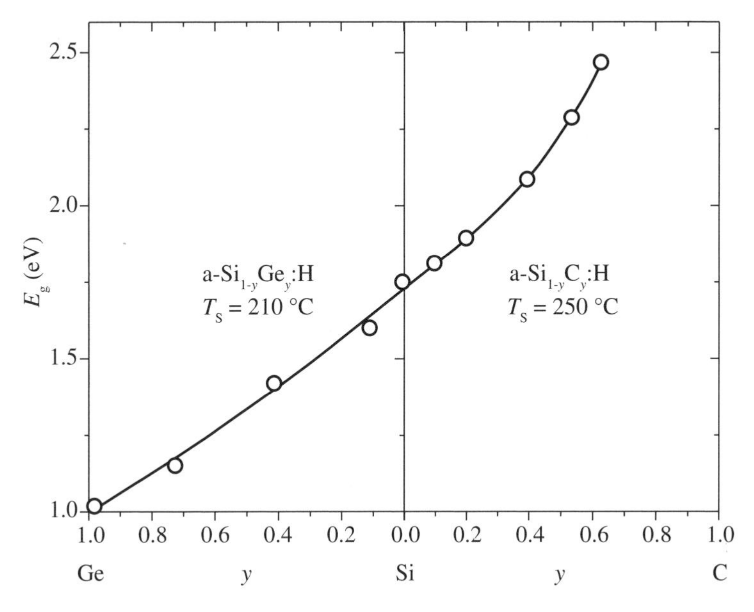


**Fig. 2.39** Optical gap $E_g$ of (plasma grown) a-Si,Ge:H and a-Si,C:H alloys as a function of composition $y$. Substrate temperature was $T_S \approx 210$ °C for a-Si,Ge:H and 250 °C for a-Si,C:H; (adapted from [Beyer-1985b]).

### 2.7.11 Electronic transport properties

A significant difference in the electronic conduction process in the dark between a-Si:H and a-Si alloys is not expected (as long as one can use the "standard transport

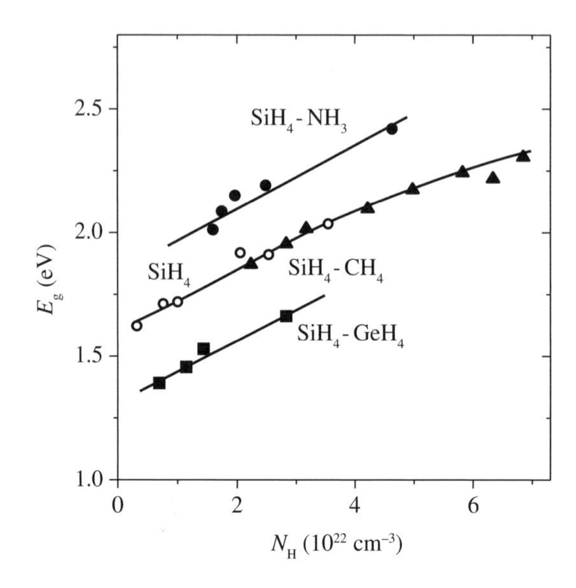


**Fig. 2.40** Optical gap $E_g$ (Tauc gap) of films based on (undiluted) $SiH_4$ as well as 1:1 gas mixtures of $SiH_4$ with $GeH_4$, $CH_4$ and $NH_3$ as a function of hydrogen density $N_H$, varied by changing the substrate temperature; (adapted from [Beyer-1987b]).

model" and can neglect transport via localized states, see Sect. 2.4.1), since the conductance at the mobility edge is defined by the minimum metallic conductivity [Street-1991]. For near-intrinsic material, the variation of dark conductivity σ according to the equation $\sigma \approx \sigma_0 \exp[-E_g / 2kT]$ is expected and has been observed.

## 2.7.12 Slope of the valence bandtail; Urbach energy

For both a-Si:H and a-Si alloys, the slope of the valence bandtail as determined by sub-bandgap absorption measurements is considered to be a measure of disorder [Street-1991], [Cohen-1998] (see Sect. 2.2.1): If the material is highly disordered, one obtains a valence bandtail, with a "gentle" slope and a high value of the characteristic energy $E_V^0$. Such "highly disordered" material will correspondingly have a high density of bandtail and defect states. The presence of such states may be detrimental to the functioning of solar cells (see Sect. 4.5.5).

The characteristic energy $E_V^0$ of the valence bandtail is determined optically by sub-bandgap absorption measurements (see Sect. 2.2.2). $E_V^0$ is approximately equal to the slope $E^0$ of the absorption edge (if the absorption coefficient is plotted on a logarithmic scale, as in Figure 2.11), which is called the "Urbach energy". The dependence of $E^0$ on alloy composition has been studied by various groups, both for a-Si,C:H and for a-Si,Ge:H alloys. While "device-quality" a-Si:H films show values of $E_V^0$ around 45 meV, much higher values of $E_V^0$ were initially observed for a-Si,C:H [Boulitrop-1985] and a-Si,Ge:H [Stutzmann-1989c]. More recently, much "better" films with values of $E_V^0$ near 45 meV were obtained for the whole a-Si,Ge:H alloy system [Finger-1998b], [Cohen-1998]. Improved a-Si,C:H films have also been reported [Baker-1990].

### 2.7.13 Strategies for obtaining good quality alloys

In particular with regard to the microstructure of the currently used a-Si alloys, a considerable potential for improvement seems to be present. Strategies for improving these alloys are derived both from empirical work and from physical/technical considerations.

Analysis of experimental data by Cohen [Cohen-1998] in terms of the bond-breaking model by Stutzmann [Stutzmann-1989b] suggests that the key to obtain material of low defect density are deposition processes leading to improved network order as witnessed by a steep valence bandtail (and as measured by a low Urbach energy $E^0$).

Various recipes have been proposed to improve film quality including (a) the use of hydrogen dilution and in the case of a-Si,Ge alloys the application of gas mixtures of disilane and germane [Guha-1981], [Matsuda-1995], [Lundszien-1997], and (b) an increase in substrate temperature [Yang-2009]. The success obtained by depositing good quality a-Ge:H with the "cathodic" deposition method (see Sect. 2.7.2) indicates that controlled ion bombardment is an important factor for obtaining a better alloy quality.

In all Si-based alloys, the hydrogen density must be kept below a limit value and excessive hydrogen incorporation leads to voids (see Sect. 2.6.6).

On the other hand, the Fermi-level dependence of hydrogen diffusion and desorption results in, for boron doping at substrate temperatures $\geq$ 200 °C, hydrogen content that is often too low and, thus, in an optical gap which is too low. This is true for *p*-type window layers of a-Si:H solar cells fabricated from (unalloyed) a-Si:H. Alloying with carbon increases the optical gap (see Fig. 2.39) and stabilizes the hydrogen incorporation (see Fig. 2.36(b)). However, carbon incorporation at concentrations >10 % also leads to hydrogen clustering and (possibly) C-C double bond formation (see Sect. 2.7.3): these two effects are detrimental for material quality, mostly because the boron doping effect and thus the conductivity is decreased. The latter is undesirable for window layers in solar cells.

While a large amount of work has already been done in this field, future substantial work appears worthwhile for achieving high quality solar cell materials.

## 2.8 CONCLUSIONS

Whereas wafer-based crystalline silicon is a material that is very well known and largely mastered in all its engineering properties, hydrogenated amorphous silicon (a-Si:H) is still a research topic and possesses many "mysterious" effects and properties, which are not fully understood. For example:

- the effect of hydrogen on its bandgap;
- the effect of structural disorder on layer properties;

- the electronic transport properties (specifically, the values of electron and hole mobilities);
- the relationship between fabrication parameters and material properties;
- substrate and interface effects;
- the exact nature of the microstructure (especially of micro-voids) always present in a-Si:H;
- the physical background of optical absorption in a-Si:H;
- the doping behavior of a-Si:H layers;
- the effect of various impurities on a-Si:H layers;
- the exact physical mechanisms underlying the light-induced degradation effect (Staebler-Wronski effect).

There is no doubt that hydrogenated amorphous silicon (a-Si:H) is a much more complex material than wafer-based crystalline silicon. The fact that we are dealing in a-Si:H with thin films and not with a bulk material makes the situation even more complex. Nevertheless, we do possess a whole arsenal of experimental tools with which we can reliably characterize amorphous silicon for device applications:

- the evaluation of defect density (dangling bond density) through sub-bandgap absorption;
- the evaluation of structural disorder by measuring the optical absorption edge;
- the evaluation of microstructure by infrared spectroscopy (FTIR) and other methods;
- the determination of the approximate position of the Fermi level, by measuring the dark conductivity of the layers and their activation energy.

With the help of these experimental tools it is possible to produce amorphous silicon layers of "standard" device quality, which can be reliably used in solar cells and other engineering devices. The fact that a-Si:H is subject to a strong light-induced degradation effect (Staebler-Wronski effect) is particularly disturbing from an applications point of view. It also warrants particular care when measuring amorphous silicon layers.

Many attempts have been made to eliminate the light-induced degradation effect, but all have so far failed. Still, through hydrogen dilution (during plasma deposition of a-Si:H layers) one is able to reduce the magnitude of the degradation. The same result is obtained by moderately increasing the deposition temperature.

Hydrogen plays an absolutely central role in a-Si:H: On one hand, it passivates the midgap defects (or dangling bonds), which act as recombination centers and have a detrimental effect in a-Si:H solar cells. In this way, hydrogen atoms saturating silicon dangling bonds are essential in making the amorphous silicon layers "usable" for solar cells and other, similar devices. However, incorporated hydrogen can be differently distributed within the film, as there can be the isolated incorporation of Si-H or Si-HH-Si groups in a compact material, the clustered incorporation of hydrogen in the form of Si-H, Si-$H_2$ at surfaces of voids of different sizes and the incorporation as molecular hydrogen. In particular, the incorporation of silicon-hydrogen clusters is likely to be harmful for the material, as hydrogen bonding becomes unstable and additional silicon dangling bonds and weak Si-Si bonds may be generated. The presence of clustered hydrogen has been associated with an enhanced light-induced degradation. As an example, the a-Si:H layers, which are deposited at low temperatures, are often porous and have a high density of silicon-hydrogen clusters; they also exhibit a particularly strong light-induced degradation effect.

Hydrogen is usually incorporated into the a-Si:H layer during deposition; therefore the hydrogen content of the layer will be determined by the deposition parameters. But the hydrogen content and the nature of its incorporation within the amorphous network can also be modified after deposition, e.g. by thermal annealing of the layer, or by post-hydrogenation, i.e. exposure of the layer, after deposition, to a hydrogen plasma. It is therefore useful to have a solid understanding of hydrogen behavior in a-Si:H layers. The experimental observations that enable us to gain this understanding have therefore been extensively described in this chapter.

One has now over 30 years experience of working with a-Si:H – this means that a-Si:H has become a reasonably predictable material. It means that there exists a "standard type" of device-quality a-Si:H layers, which are generally used in all devices to date. It may be possible to obtain even "better" layers, but this remains very much a research topic.

When it comes to alloys of amorphous silicon with carbon or germanium, the situation is even more complex and difficult to understand. Here much more research is called for, and there is considerable scope for a further improvement in layer "quality".

## 2.9 REFERENCES

[Alvarez-1992] Alvarez, F., Sebastiani, M., Pozzilli, F., Fiorini, P., Evangelisti, F., "Influence of hydrogen dilution on the optoelectronic properties of glow discharge amorphous silicon carbon alloys", (1992), *Journal of Applied Physics* Vol. **71**, pp. 267-272

[Amer-1984] Amer, N. M., Jackson, W. B., "Optical Properties of Defect States in a-Si:H" in "Hydrogenated Amorphous Silicon, Part B Optical Properties", ed. Pankove, J. I., Vol. **21B** of *Semiconductors and Semimetals*, Academic Press, Orlando, 1984

[Baker-1990] Baker, S. H., Spear, W. E., Gibson, R. A. G., "Electronic and optical properties of a-$Si_{1-x}$ $C_x$ films prepared from a $H_2$ diluted mixture of $SiH_4$ and $CH_4$", (1990), *Philosophical Magazine B*, Vol. **62**, pp. 213-223

[Beck-1996] Beck, N., Wyrsch, N., Hof, Ch., Shah, A., "Mobility lifetime product—A tool for correlating a-Si:H film properties and solar cell performances", (1996), *Journal of Applied Physics*, Vol. **79**, pp. 9361-9368

[Beyer-1983] Beyer, W., Wagner, H., "The role of hydrogen in a-Si:H - results of evolution and annealing studies", (1983), *Journal of Non-Crystalline Solids*, Vol. **59-60**, pp. 161-168

[Beyer-1985a] Beyer, W., "Hydrogen incorporation in amorphous silicon and processes of its release", in "*Tetrahedrally-Bonded Amorphous Semiconductors*" (1985), eds. Adler, D., Fritzsche, H., Plenum Press, New York, 1985, pp.129-146

[Beyer-1985b] Beyer, W., Wagner, H, Finger, F., "Hydrogen evolution from a-Si:C:H and a-Si:Ge:H alloys", (1985), *Journal of Non-Crystalline Solids*, Vol. **77 - 78**, pp. 857-860

[Beyer -1987a] Beyer, W., Mell, H., "Composition and thermal stability of glow-discharge a-Si:C:H and a-Si:N:H alloys" (1987) in: *Disordered Semiconductors*, eds. Kastner, M.A.,Thomas, G.A., Ovshinsky, S.R., Plenum, New York, pp.641-658

[Beyer-1987b] Beyer, W., "Structural and electronic properties of silicon-based amorphous alloys", (1987), *Journal of Non-Crystalline Solids*, Vol. **97 & 98**, pp. 1027-1034

[Beyer-1988] Beyer, W., Herion, J., Mell, H., Wagner, H., "Influence of boron-doping on hydrogen diffusion and effusion in a-Si:H and a-Si alloys", (1988), *Materials Research Society Symposium Proceedings,* Vol. **118,** pp 291-296

[Beyer-1989a] Beyer, W., Herion, J., Wagner, H., "Fermi energy dependence of surface desorption and diffusion of hydrogen in a-Si:H", (1989), *Journal of Non-Crystalline Solids*, Vol. **114**, pp. 217-219

[Beyer-1989b] Beyer, W., Hager, R., Schmidbaur, H., Winterling, G., "Improvement of the photoelectric properties of amorphous $SiC_x$:H by using disilylmethane as a feeding gas", (1989), *Applied Physics Letters,* Vol. **54**, pp.1666-1668

[Beyer-1991a] Beyer, W.,"Hydrogen effusion: a probe for surface desorption and diffusion", (1991), *Physica B*, Vol. **170**, pp.105-114

[Beyer-1991b] Beyer, W., Weller, H. C., Zastrow, U., "Hydrogen diffusion in the a-Si,Ge alloy system", (1991), *Journal of Non-Crystalline Solids,* Vol. **137-138,** pp 37-40

[Beyer-1991c] Beyer, W., Herion, J, Wagner, H., Zastrow, U., "Hydrogen stability in hydrogenated amorphous silicon-based alloys", (1991), *Material Research Society Symposium Proceedings*, Vol. **219**, pp. 81-84

[Beyer-1997] Beyer, W., "New Insights into Processes of Hydrogen Incorporation and Hydrogen Diffusion in Hydrogenated Amorphous Silicon", (1997), *Physica Status Solidi A*, Vol. **159**, pp. 53-63

[Beyer-1999] Beyer W., "Hydrogen Phenomena in Hydrogenated Amorphous Silicon", in "Introduction to Hydrogen in Semiconductors II", (1999), ed. Nickel, N. H., Vol. **61** of *Semiconductors and Semimetals*, Academic Press, San Diego, 1999

[Beyer-2000] Beyer, W., "Infrared absorption and hydrogen effusion of hydrogenated amorphous silicon-oxide films", (2000), *Journal of Non-Crystalline Solids,* Vol. **266-269**, pp. 845-849

[Beyer-2003] Beyer, W., "Diffusion and evolution of hydrogen in hydrogenated amorphous and microcrystalline silicon", (2003), *Solar Energy Materials and Solar Cells*, Vol. **78**, pp. 235-267

[Beyer-2004a] Beyer, W., "Characterization of microstructure in amorphous and microcrystalline silicon and related alloys by effusion of implanted helium", (2004), *Physica Status Solidi C*, Vol. **1**, pp. 1144-1153

[Beyer-2004b] Beyer, W, "Microstructure characterization of plasma-grown a-Si:H and related materials by effusion of implanted helium", (2004), *Journal of Non-Crystalline Solids,* Vol. **338-340,** pp 232-235

[Beyer-2008] Beyer, W., Carius, R., Lennartz, D., Niessen, L., Pennartz, F., "Microstructure effects in hot-wire deposited undoped microcrystalline silicon films", (2008), *Materials Research Society Symposium Proceedings,* Vol. **1066,** pp 179-184

[Beyer-2009] Beyer, W., unpublished data (2009)

[Bhattacharya-1988] Bhattacharya, E., Mahan, A. H., "Microstructure and the light-induced metastability in hydrogenated amorphous silicon", (1988) *Applied Physics Letters*, Vol. **52**, pp. 1587-1589

[Boulitrop-1985] Boulitrop, F., Bullot, J., Gauthier, M., Schmidt, M. P., Catherine Y., "Disorder and density of defects in hydrogenated amorphous silicon-carbon alloys", (1985), *Solid State Communications*, Vol. **54**, pp. 107-110

[Branz-1999] Branz, H. M., "Hydrogen collision model: Quantitative description of metastability in amorphous silicon", (1999), *Physical Review B,* Vol. **B-59** pp. 5498-5512

[Brodsky-1969] Brodsky, M. H., Title, R. S., "Electron spin resonance in amorphous silicon, germanium, and silicon carbide", (1969), *Physical Review Letters*, Vol. **23**, pp. 581-585

[Brodsky-1977] Brodsky, M. H., Cardona, M., Cuomo, J. J., "Infrared and Raman spectra of the silicon-hydrogen bonds in amorphous silicon prepared by glow discharge and sputtering", (1977), *Physical Review B*, Vol. **16**, pp. 3556-3571

[Cardona-1983] Cardona, M., "Vibrational Spectra of Hydrogen in Silicon and Germanium", (1983), *Physica Status Solidi B,* Vol. **118**, pp. 463-481

[Carlson-1978] Carlson, D. E., Magee, C. W., "A SIMS analysis of deuterium diffusion in hydrogenated amorphous silicon", (1978), *Applied Physics Letters,* Vol. **33**, pp. 88-83

[Chen-1984] Chen, G., Zhang, F., Wang, Y., Zhang, Y., Xu, X., "Adsorbate and light-induced effects on the conductivity of GD a-$Si_xC_{1-x}$:H", (1984), *Solar Energy Materials and Solar Cells*, Vol. **11**, pp.281-282

[Chittick-1969] Chittick, R. C., Alexander, J. H., Sterling, H. F., "Preparation and properties of amorphous silicon", (1969), *Journal of Electrochemical Society*, Vol. **116**, pp 77-81

[Cody-1982] Cody, G. D., Brooks, B. G., Abeles, B., "Optical absorption above the optical gap of amorphous silicon hydride", (1982), *Solar Energy Materials and Solar Cells*, Vol. **8**, pp. 231-240

[Cody-1984] Cody, G. D., "The optical absorption edge of a-Si:H" in "Hydrogenated Amorphous Silicon, Part B Optical Properties", (1984), ed. Pankove, J.I., Vol. **21B** of *Semiconductor and Semimetals*, Academic Press, Orlando

[Cohen-1998] Cohen, J. D., "Electronic structure of a-Si:Ge:H", in "Amorphous Silicon and its Alloys" (1998), ed. Searle, T., Vol. 19 of EMIS Datareviews Series, INSPEC, London

[Collins-2003] Collins, R. W., Ferlauto, A. S., Ferreira, G. M., Chen, C., Koh, J., Koval, R. J., Lee, Y., Pearce, J.M., Wronski, C. R., "Evolution of microstructure and phase in amorphous, protocrystalline and microcrystalline silicon studied by real time spectroscopic ellipsometry", (2003), *Solar Energy Materials and Solar Cells,* Vol. **78**, pp.143–180

[Curtins-1988] Curtins, H., Favre, M., Ziegler, Y., Wyrsch, N., Shah, A.V., "Comparison of Light induced Degradation in Low and High-Rate Deposited VHF-GD a-Si:H: Effect of Film Inhomogeneities", (1988), *Materials Research society symposium prouedings*, Vol. **118**, pp. 159-166

[Dawson-1996] Dawson, R. M., A., Fortmann, C. M., "The Staebler-Wronski effect and the thermal equilibration of defect and free carrier concentrations", (1996), *Journal of Applied Physics*, Vol. **79**, pp. 3075-3081

[Delahoy-1989] Delahoy, A.E., "Recent developments in amorphous silicon photovoltaic research and manufacturing at Chronar corporation", (1989), *Solar Cells*, Vol. **27**, pp.39-57.

[Dersch-1981] Dersch, H., Stuke, J., Beichler, J, "Light-induced dangling bonds in hydrogenated amorphous silicon", (1981), *Applied Physics Letters,* Vol. **38,** pp. 456-458

[EMIS-1989] EMIS Datareview Series No 1, "*Properties of Amorphous Silicon*", INSPEC, London, (1989)

[EMIS-1998] EMIS Datareview Series No 19, "*Properties of Amorphous Silicon and its Alloys*", INSPEC, London, (1998)

[Favre-1987] Favre, M., Curtins, H., Shah, A.V., "Study of Surface/Interface and Bulk Defect Density in a-Si:H by Means of Photothermal Deflection Spectroscopy and Photoconductivity", (1987), *Journal of Non-Crystalline Solids*, Vol. **97-98**, pp. 731-734

[Fejfar-1996] Fejfar, A., Poruba, A, Vaneček, M., Ko ka, J., "Precise measurement of the deep defects and surface states in a-Si:H films by absolute CPM", (1996), *Journal of Non-Crystalline Solids*, Vol. **200**, pp. 304-308

[Finger-1998a] Finger, F., Beyer, W., "Growth of a-Si:Ge:H alloys by PECVD- gas sources, conditions in the plasma and at the interface" in "*Amorphous Silicon and its Alloys*" (1998), ed. Searle, T., Vol. 19 of EMIS Datareview Series, INSPEC, London

[Finger-1998b] Finger, F., Beyer, W., "Growth of a-Si:Ge:H alloys by PECVD- optimisation of growth parameters, growth rates, microstructure and material quality" in "*Amorphous Silicon and its Alloys*" (1998), ed. Searle, T., Vol. 19 of EMIS Datareview Series, INSPEC, London

Fontcuberta-2001] Fontcuberta i Morral, A., Roca i Caborrocas, P., "Shedding light on the growth of amorphous, polymorphous, protocrystalline and microcrystalline silicon thin films", (2001), *Thin Solid Films,* Vol. **383,** pp. 161-164

[Fontcuberta-2004] Fontcuberta i Morral, A., Roca i Caborrocas, P., Clerc, C., "Structure and hydrogen content of polymorphous silicon thin films studied by spectroscopic ellipsometry and nuclear measurements", (2004), *Physical Review B*, Vol. **69** Paper No 125307

Goerlitzer-1996] Goerlitzer, M., Beck, N., Torres, P., Meier, J., Wyrsch, N., Shah, A., "Ambipolar diffusion length and photoconductivity measurements on "midgap" hydrogenated microcrystalline silicon", (1996), *Journal of Applied Physics*, Vol. **80**, pp. 5111-5115

[Graebner-1984] Graebner, J. E., Golding, B., Allen, L. C., Biegelsen, D. K., Stutzmann, M., "Solid Hydrogen in Hydrogenated Amorphous Silicon". (1984), *Physical Review Letters*, Vol. **52**, pp. 553-556

[Guha-1981] Guha, S., Narasimhan, K., Pietruszko, S. M., "On light-induced effect in amorphous hydrogenated silicon", (1981), *Journal of Applied Physics,* Vol. **52**, pp. 859-860

[Guha-2003] Guha, S., Yang, J., Banerjee, A., Yan, B., Lord, K., "High-quality amorphous silicon materials and cells grown with hydrogen dilution", (2003), *Solar Energy Materials and Solar Cells*, Vol. **78**, pp. 329-347

[Hof-1999] Hof, C., "Thin Film Solar Cells of Amorphous Silicon: Influence of *i*-Layer Material on Cell Efficiency", *Ph. D. thesis, Faculté des Sciences, University of Neuchâtel*, UFO Atelier für Gestaltung & Verlag, Allensbach, Germany Vol. **379**, **ISBN** 3-930803-78-X, (1999)

[Holovskú-2008] Holovskú, J., Poruba, A., Purkrt, Z., Vaneček, M., "Comparison of photocurrent spectra measured by FTPS and CPM for amorphous silicon layers and solar cells", (2008), *Journal of Non-Crystalline Solids*, Vol. **354** pp. 2167-2170

[Huber-1972] Huber, K. P., "Constants of Diatomic Molecules" (1972) in: "*AIP Handbook of Physics*", ed Grey, D.E., McGraw Hill, New York

[Hubin-1992] Hubin, J., Shah, A., Sauvain, E., "Effects of Dangling Bonds on the Recombination Function in Amorphous Semiconductors", (1992), *Philosophical Magazine Letters*, Vol. **66**, pp. 115-125

[Jackson-1985] Jackson, W. B., Kelso, S. M., Tsai, C. C., Allen J. W., Oh, S. J., "Energy-dependence of the optical matrix element in hydrogenated amorphous and crystalline silicon, (1985), *Physical Review B*, Vol. **31**, pp. 5187-5198

[Jones-2000] Jones, S. J., Liu, T., Deng, X., Izu, M., "a-Si:H-based triple-junction cells prepared at i-layer deposition rates of 10 Å/s using a 70 MHz PECVD technique", (2000), *Proceedings 28$^{th}$ IEEE Photovoltaic Specialists Conf.*, pp. 845-848

[Kherani-2008] Kherani, N. P., Liu, B., Virk, K., Kosteski, T., Gaspari, F., Shmayda, W.T., Zukotynski, S., Chen, K. P., "Hydrogen effusion from tritiated amorphous silicon", (2008), *Journal of Applied Physics*, Vol. **103**, Paper No. 024906 1-7

[Klimovsky-2002] Klimovsky, E., Rath, J. K., Schropp, R. E. I., Rubinelli, F. A. "Errors introduced in a-Si:H-based solar cell modelling when dangling bonds are approximated by decoupled states", (2002), *Thin Solid Films*, Vol. **422** (1-2), pp. 211-219

[Koh-1998] Koh, J., Lee, Y., Fujiwara, H., Wronski, C.R., Collins, R.W. "Optimization of hydrogenated amorphous silicon p-i-n solar cells with two-step i layers guided by real-time spectroscopic ellipsometry", (1998), *Applied Physics Letters*, Vol. **73**, pp. 1526-1528

[Kroll-1995] Kroll, U., "VHF-Plasmaabscheidung von amorphem Silizium: Einfluss der Anregungsfrequenz, der Reaktorgestaltung sowie Schichtengenschaften", *Ph. D. thesis, Faculté des Sciences, University of Neuchâtel*, **ISBN** 3-89191-905-0, (1995)

[Kroll-1996] Kroll, U., Meier, J., Shah, A. Mikhailov, S., Weber, J., "Hydrogen in Amorphous and Microcrystalline Silicon Films prepared by Hydrogen Dilution", (1996), *Journal of Applied Physics,* Vol. **80**, pp. 4971-4975

[Kroll-1998] Kroll, U., Meier, J., Torres, P., Pohl, J., Shah, A., "From Amorphous to Microcrystalline Silicon Films Prepared by Hydrogen Dilution Using the VHF (70MHz) GD Technique", (1998), *Journal of Non-Crystalline Solids*, Vol. **227-230**, pp. 68-72.

[Lundszien-1997] Lundszien, D., Fölsch, J., Finger, F., Wagner, H., "Is there an optimization limit for hydrogenated amorphous silicon-germanium alloys for solar cell applications?", (1997), *Proceedings 14th European Photovoltaic Solar Energy Conference*, pp. 601-604

[Mahan-1987] Mahan, A.H., Raboisson, P., Willamson, D.L., Tsu, R., "Evidence for microstructure in glow-discharge hydrogenated amorphous Si-C alloys", (1987), *Solar Cells,* Vol. **21**, pp. 117-128

[Mahan-1989] Mahan, A. H., Williamson D. L., Nelson, B. P., Crandall, R. S., "Characterization of Microvoids in Device-Quality Hydrogenated Amorphous Silicon by small-angle X-Ray Scattering and Infrared Measurements", (1989), *Physical Review B*, Vol. **40**, pp. 12024-12027

[Mahan-1991] Mahan, A.., Vaneček, M., "Reduction in the Staebler-Wronski effect observed in low H content a-Si:H films deposited by the hot wire technique", *AIP Conference Proceedings* **234**, AIP , New York 1991, pp. 195-202

[Matsuda-1987] Matsuda, A., Tanaka, K., "Guiding principles for preparing highly photosensitive Si-based amorphous alloys", (1987), *Journal of Non-Crystalline Solids,* Vol. **97-98**, pp.1367-1374

[Matsuda-1995] Matsuda, A., Ganguly, G., "Improvement of hydrogenated amorphous silicon germanium alloys using low power disilane-germane discharges without hydrogen dilution", (1995), *Applied Physics Letters*, Vol. **67**, pp. 1274-1276

[Middya-1994] Middya, A. R., Hazra, S., Ray, S.,"Growth of device quality amorphous SiGe:H alloys with high deposition rate under helium dilution", (1994), *Journal of Applied Physics*, Vol. **76**, pp. 7578-7582

[Millman-1965] Millman, J., Taub, H., "*Pulse, Digital and Switching Waveforms*" McGraw Hill, New York (1965 and newer editions) Section 2.6 "The Low-Pass RC-cricuit (Exponential and Ramp Inputs"

[Mok-2007] Mok, T. M., O'Leary, S. K., "The dependence of the Tauc and Cody optical gaps associated with hydrogenated amorphous silicon on the film thickness: αl Experimental limitations and the impact of curvature in the Tauc and Cody plots", (2007), *Journal of Applied Physics*, Vol. **102**, Paper No.113525

[Müllerová-2008] Müllerová, J., Šutta, P., van Elzakker, G., Zeman, M., Mikula, M., "Microstructure of hydrogenated silicon thin films prepared from silane diluted with hydrogen", (2008), *Applied Surface Science* Vol. **254**, pp. 3690–3695

[Nelson-1998] Nelson, B. P., Yueqin, X., Webb, J.D., Mason, A., Reedy, R.C., Gedvilas, L.M., Lanford, W.A.,"Techniques for measuring the composition of hydrogenated amorphous silicon-germanium alloys", (1998), Journal of Non-Crystalline Solids , Vol. **266-269,** pp 680-684

[Nishimoto-2002] Nishimoto, T., Takai, M., Miyahara, H., Kondo, M., Matsuda, A., "Amorphous silicon solar cells deposited at high growth rate", (2002), Journal of Non-Crystalline Solids , Vol. **299-302,** pp 1116-1122

[Overhof-1989] Overhof, H., Thomas, P., "Electronic transport in hydrogenated amorphous semiconductors", *Springer Tracts in Modern Physics*, Vol. **114**, Springer Verlag, Berlin 1989

[Paul-1981] Paul, W., Paul, D. K., von Roedern, B., Blake, J., Oguz, S., "Preferential attachment of H in amorphous hydrogenated binary semiconductors and consequent inferior reduction of pseudogap state density", (1981), *Physical Review Letters*, Vol **46**, pp. 1016-1020

[Pearce-2000] Pearce, J. M., Koval, R. J., Ferlauto, A. S., Collins R.W., Wronski C. R., Yang, J., Guha, S., "Dependence of open-circuit voltage in hydrogenated protocrystalline silicon solar cells on carrier recombination in p/i interface and bulk regions", (2000), *Applied Physics Letters*, Vol. **77**, pp. 3093-3095

[Pearce-2009] Pearce, J., personal communication

[Platz-1998] Platz, R., Wagner, S., Hof, C., Shah, A., Wieder, S., Rech, B., "Influence of excitation frequency, temperature, and hydrogen dilution on the stability of plasma enhanced chemical vapor deposited a-Si : H", (1998), *Journal of Applied Physics*, Vol. **84**, pp. 3949-3953

[Poissant-2003] Poissant, Y., Chatterjee, P., Roca i Caborrocas, P, "Analysis and optimization of the performance of polymorphous silicon solar cells: Experimental characterization and computer modelling", (2003), *Journal of Applied Physics*, Vol. **94**, pp. 7305-7316

[Santos-1991] Santos, P. V., Johnson, N. M., Street, R. A., "Light-enhanced hydrogen motion in a-Si:H", (1991), *Physical Review Letters*, Vol. **67**, pp. 2686- 2689

[Shah-2004] Shah, A.V., Schade, H., Vaneček, M., Meier, J., Vallat-Sauvain, E., Wyrsch, N., Kroll, U., Droz, C., Bailat, J., (2004), "Thin-film Silicon Solar Cell Technology", *Progress in Photovoltaics: Research and Applications*, Vol. **12**, pp. 113-142

[Shah-2010] Shah, A.V, personal communication

[Shimizu-2004] Shimizu, T., "Staebler-Wronski effect in hydrogenated amorphous silicon and related alloy films", (2004), *Japanese Journal of Applied Physics,* Vol. **43**, pp.3257-3268

[Simmons-1972] Simmons, J. G., Taylor, G. W., "Basic equations for statistics, recombination processes, and photoconductivity in amorphous insulators and semiconductors", (1972), *Journal of Non-Crystalline Solids*, Vol. **8-10**, pp 940-946

[Singh-2004] Singh, R., Prakash, S., Shukla, N. N., Prasad, R., "Sample dependence of the structural, vibrational and electronic properties of a-Si:H: a density-functional-based tight-binding study, (2004) ), *Physical Review B*, Vol. **70**, paper No. 115213

[Smets-2003] Smets, A. H. M., Kessels, W. M. M., van de Sanden, M. C. M., "Vacancies and voids in hydrogenated amorphous silicon", (2003), *Applied Physics Letters*, Vol. **82**, pp. 1547-1549

[Soro-2008] Soro, Y. M., Abramov, A., Guenier-Farre, M. E., Johnson, E. V., Longeaud, C., Roca i Caborrocas, P., Kleider J. P., "Device- grade hydrogenated polymorphous silicon deposited at high rates", (2008), *Journal of Non-Crystalline Solids*, Vol. **354**, pp. 2092-2095

[Spear-1975] Spear W.E., Le Comber, P.G., "Substitutional doping of amorphous silicon", (1975), *Solid State Communications*, Vol. **17**, pp.1193-1196

[Spear-1976] Spear, W.E., Le Comber, P. G., "Electronic properties of substitutionally doped amorphous Si and Ge" (1976) *Philosophical Magazine B*, Vol. **33**, pp. 935-949

[Spear-1985] Spear, W.E., Howard, A, C., and Kinmond, S., "The temperature-dependence of the D and $D^+$ capture cross-section in a-Si", (1985), *Journal of Non-Crystalline Solids*, Vol. **77-78**, pp. 607-610

[Staebler-1977] Staebler, D. L., Wronski, C. R., "Reversible conductivity changes in discharge-produced a-Si", (1977), *Applied Physics Letters,* Vol. **31**, pp 292-294

[Street-1987] Street, R.A., Tsai, C. C., Kakalios J., Jackson, W.B., "Hydrogen diffusion in amorphous silicon", (1987), *Philosophical Magazine B,* Vol. **56**, pp. 305-320

[Street-1991] Street, R. A., "*Hydrogenated amorphous silicon*", Cambridge University Press, Cambridge, 1991

[Stutzmann-1984] Stutzmann, M., Nemanich, R. J., Stuke, J., "Electron-spin-resonance study of boron-doped $Si_xGe_{1-x}$:H alloys", (1984), *Physical Review B,* Vol. **30,** pp 3595-3602

[Stutzmann-1985] Stutzmann, M., Jackson, W. B., Tsai, C. C., "Light-induced metastable defects in hydrogenated amorphous silicon: A systematic study", (1985), *Physical Review B,* Vol. **32,** pp 23-47

[Stutzmann-1986] Stutzmann, M., "The doping efficiency in amorphous silicon and germanium", (1986), *Philosophical Magazine B*, Vol. **53**, pp. L15-L21

[Stutzmann-1987] Stutzmann, M., Biegelsen, D. K., Street, R. A., "Detailed investigation of doping in hydrogenated amorphous silicon and germanium", (1987), P*hysical Review B*, Vol. **35**, pp. 5666-5701

[Stutzmann-1988] Stutzmann, M. Biegelsen, D. K., "The microscopic structure of defects in a-Si:H and related materials" in "*Amorphous Silicon and Related Materials*" ed. H. Fritzsche, (1988), World Scientific Publishing Company, Singapore

[Stutzmann-1989a] Stutzmann, M., Biegelsen, D. K., "Microscopic nature of coordination defects in amorphous silicon", (1989), *Physical Review B*, Vol. **40**, pp. 9834-9840

[Stutzmann-1989b] Stutzmann, M, "The defect density in amorphous silicon", (1989), *Philosophical Magazine B,* Vol. **60**, pp. 531-546

[Stutzmann-1989c] Stutzmann, M., Street, R. A., Tsai, C. C., Boyce, J. B., Ready, S. E., "Structural, optical and spin properties of hydrogenated amorphous silicon-germanium alloys", (1989), *Journal of Applied Physics*, Vol. **66**, pp. 569-592

[Su-2004] Su, T., Taylor, P. C., Whitaker, J., "Metastable defects in a-Si:H and a-Ge:H: The role of hydrogen", (2004), *Materials Research Society Symposium Proceedings,* Vol. **808**, pp. 141-152

[Sugiyama-1997] Sugiyama, S., Yang, J., Guha, S., "Improved stability against light exposure in amorphous deuterated silicon alloy solar cell", (1996), *Applied Physics Letters,* Vol. **70,** pp 378-380

[Tauc-1966] Tauc, J., Grigorovici, A., Vancu, A., "Optical properties and electronic structure of amorphous germanium", (1966) , *Physica Status Solidi,* Vol. **15**, pp. 627-637

[Tawada-1982] Tawada, Y., Kondo, M., Okamoto, H., and Hamakawa, Y., "Hydrogenated amorphous silicon carbide as a window material for high efficiency a-Si solar cells", (1982), *Solar Energy Materials,* Vol **6** , pp.299-315

[Terakawa-1993] Terakawa, A., Shima, M., Sayama, K., Tarui, H., Tsuda, S., Nishiwaki, H., Nakano, S.,"Film Property Control of hydrogenated amorphous silicon germanium for solar cells", 1993, *Japan Journal of Applied. Physics,* Vol. **32**, pp. 4894-4899

[Tsu-1997] Tsu, D. V., Chao, B. S., Ovshinsky, S. R., Guha, S. Yang, J., "Effect of hydrogen dilution on the structure of amorphous silicon alloys", (1997), *Applied Physics Letters,* Vol. **71**, pp. 1317-1319

[Tsuo-1991] Tsuo, Y. S., Xu, Y., Balberg, I., Crandall, R. S., "Effects of helium dilution on glow-discharge depositions of a-$Si_{1-x}Ge_x$:H Alloys",(1991), *Proceedings 22nd IEEE Photovoltaic Specialists Conference*, pp. 1334-1337

[Tuttle-2004] Tuttle, B. R., "Theory of Hydrogen-Related Metastability in Disordered Silicon". (2004), *Physical Review Letters*, Vol. **93**, Paper No. 215504 1-4

[Vaillant-1986] Vaillant, F., Jousse, D., "Recombination at dangling bonds and steady-state photoconductivity in a-Si:H", (1986), *Physical Review B,* Vol. **34**, pp. 4088-4098

[Van de Walle-1991] Van de Walle, C. G., "Theoretical aspects of hydrogen in crystalline semiconductors", (1991), *Physica B,* Vol.**170,** pp 21-32

[Vaneček-1981] Vaneček, M., Kočka J., Stuchlik J., Triska A., "Direct measurement of the gap states and band tail absorption by constant photocurrent method in amorphous silicon", (1981), *Solid State Communications*, Vol. **39**, pp. 1199-1202

[Vaneček-1985] Vaneček. M., Stuchlik, J., Kocka, J., Triska, A., "Determination of the mobility gap in amorphous silicon from a low temperature photoconductivity measurement", (1985), *Journal of Non-Crystalline Solids*, Vols. **77&78**, pp. 299-302

[Vaneček-1992] Vaneček, M, Mahan, A. H., Nelson, B. P., Crandall, R. S., "Influence of hydrogen and microstructure on increased stability of amorphous silicon", *Proceedings of the 11th EC Photovoltaic Solar Energy Conference*, (1992), pp. 96-99

[Vaneček-1995] Vaneček, M., Kočka, J., Poruba, A., Fejfar, A., "Direct measurement of the deep defect density in thin amorphous silicon films with the absolute constant photocurrent method", (1995), *Journal of Applied Physics*, Vol. **78**, pp. 6203-6210

[Vaneček-2002] Vaneček. M., Poruba, A., "Fourier-transform photocurrent spectroscopy of microcrystalline silicon for solar cells", (2002) *Applied Physics Letters*, Vol. **80**, pp.719-721

[Vaneček-2010] Vaneček. M., *personal communication*

[Van Wieringen-1956] Van Wieringen, A., Warmoltz, N., "On the permeation of hydrogen and helium in single crystal silicon and germanium at elevated temperatures" (1956), *Physica,* Vol. **22**, pp.849- 865

[Von Löhneysen-1984] Von Löhneysen, H., Schink, H. J., Beyer, W., "Direct experimental evidence for molecular hydrogen in amorphous Si:H", (1984), *Physical Review Letters*, Vol. **52**, pp. 549- 552

[Wanka-1997] Wanka, H. N., Schubert, M. B., "Fast etching of amorphous and microcrystalline silicon by hot-filament generated atomic hydrogen" (1997), *MRS Proceedings,* Vol. **467**, pp. 651- 656

[Wagner-1983] Wagner, H., Beyer, W., "The reinterpretation of the silicon-hydrogen stretch frequencies in amorphous silicon", *Solid-State Communications,* Vol. **48**, pp. 585-587

[Weil-1988] Weil, R., Busso, A., Beyer, W., "Effusion of deuterium from deuterated-fluorinated amorphous silicon under illumination" (1988), *Applied Physics Letters,* Vol. **53**, pp.2477- 2479

[Wickboldt-1997] Wickboldt, P., Pang, D., Paul, W., Chen, J. H., Zhong, F., Chen, C. C., Cohen, J. D., Williamson, D. L., "High performance glow discharge a-Si $_{1-x}$ Ge$_x$ :H of large x", (1997), *Journal of Applied Physics*, Vol. **81**, pp. 6252- 6267

[Williamson-1998] Williamson, D. L., "Structural information on a-Si:H and its alloys from small-angle scattering of X-rays and neutrons", in "*Amorphous Silicon and its Alloys*" (1998), ed. Searle, T., Vol. 19 of EMIS Datareview Series, INSPEC, London

[Wronski-1989] Wronski, C. R., Lee, S., Hicks, M., Kumar, S., "Internal photoemission of holes and the mobility gap of hydrogenated amorphous silicon", (1989), *Physical Review Letters*, Vol. **63**, pp. 1420-1423

[Wyrsch-1991a] Wyrsch, N., Finger F., McMahon T. J., Vanecek M, "How to reach more precise interpretation of sub gap absorption spectra in terms of deep defect density in a-Si: H", (1991), *Journal of Non-Crystalline Solids*, Vol. **137-138**, pp. 347-350

[Wyrsch-1991b] Wyrsch N., Shah. A., "Depth profiles of mobility lifetime products and capture cross-sections in a-Si:H" (1991), *Journal of Non-Crystalline Solids*, Vol. **137-138**, pp. 431-434

[Yang-2009] Yang., J., Guha, S., "Status and future perspective of a-Si:H, a-SiGe:H, and nc-Si: H thin film photovoltaic technology", in "Thin Film Solar Technology" (2009), eds. Delahoy, A. E. and Eldada, L.A., *Proceedings of the SPIE*, Vol. **7409**, pp. 74090C1- 74090C14

[Yelon-1990] Yelon, A., Movaghar, B., "Microscopic explanation of the compensation (Meyer-Neldel) rule", (1990), *Physical Review Letters,* Vol. **65**, pp.618- 620

[Zallen-1983] Zallen, R., "*The Physics of Amorphous Solids*", Wiley-Interscience, John Wiley and Sons, New York, **ISBN** 0471299413, 1998

[Zeman-2006] Zeman, M., "Advanced Amorphous Silicon Solar Cell Technologies" Chapter 5 in "*Thin Film Solar Cells*", (2006). eds. Jef Poortmans and Vladimir Arkhipov, John Wiley and Sons, New York

CHAPTER 3

# BASIC PROPERTIES OF HYDROGENATED MICROCRYSTALLINE SILICON

*Friedhelm Finger*

Thin-film silicon in microcrystalline form (μc-Si:H) is of great importance for the development of thin-film solar cells. The material is still far from being understood even though it has been a matter of intense research in recent years. As a result of its heterogeneous structure, one observes complex relationships between preparation conditions, structure, electrical and optical properties and device performance.

In this chapter a short history of the development of μc-Si:H is presented with some relevant references (Sect. 3.1). Then the structural properties are described (Section 3.2). One cannot emphasize enough that the investigation and understanding of the structural properties are key factors in the research and development of μc-Si:H. In Sections 3.3 to 3.6, optical and electronic properties of μc-Si:H, instability phenomena and developments using μc-Si:H related alloys are described. For details about the preparation and growth process, the reader is referred to Chapter 6 and the references cited there, and for more information on device applications to Chapters 4 and 5. Because disorder still strongly determines the properties of μc-Si:H, it is important to become familiar with the ideas used for describing a-Si:H before trying to understand the properties of μc-Si:H [Street-1991]. The reader should also consult some early review works on the properties of μc-Si:H [Willeke-1991, Schropp-1998, Rath-2003].

## 3.1 HISTORY

Hydrogenated microcrystalline silicon was first described in detail by S. Veprek and co-workers in 1968 [Veprek-1968]. They used a chemical transport technique and a process temperature of 600 °C. The process started out from solid-phase silicon. Using a hydrogen plasma, silicon material is etched from one place in a reactor tube and re-deposited in the microcrystalline form at another place, as long as certain process conditions are maintained. In 1979, the first report about the preparation of μc-Si:H with plasma-enhanced chemical vapor deposition (PECVD) was published

[Usui-1979]. The process was carried out at a lower substrate temperature, and the material was grown from the gas phase using silane strongly diluted with hydrogen. It was soon found that the structure of the resulting layers could easily be varied between the amorphous and microcrystalline phase by appropriate adjustment of the preparation parameters such as discharge power, substrate temperature, process pressure and, most importantly, by modifying the hydrogen dilution of the process gas silane. This compatibility of the μc-Si:H preparation process with the one for amorphous silicon is obviously of great importance for any technological application where combinations of these materials are used. An overview with references to the early work can be found in [Veprek-1983, Matsuda-1983, Willeke-1991]. The material investigated in these early years was typically fairly conductive and *n*-type, probably as a result of impurity doping or inner surface contamination. Furthermore, the deposition rates were very low: in the range of less than 1 Å/s. In the beginning μc-Si:H was mainly investigated and later successfully applied for use as doped layers (contact layers), i.e. for both *n*- and *p*-type layers in single- and multiple-junction *pin*-type solar cells. These doped layers were only used for diode junction formation and as contact layers and not for photogeneration (see Sect. 4.5). During this period, the use of μc-Si:H layers in TFTs was also investigated. Considerable understanding of the PECVD growth processes for μc-Si:H was already developed [Matsuda-1983, Tsai-1988, Veprek-1990]. For the application of μc-Si:H as absorber layers for photogeneration, in thin-film silicon solar cells an important development was the use of higher excitation frequencies (excitation frequencies in the VHF-region) for the PECVD process and its application to μc-Si:H [Oda-1988, Prasad-1990, Finger-1994, Hollingsworth-1994]. With this modified PECVD process, called "VHF-PECVD" [Curtins-1987], a considerable increase in deposition rate could be obtained while still maintaining a high material quality (see also Chapter 6). Now it became possible to deposit material with appropriate thickness for solar cell applications, i.e. with a thickness of 1 to 3 $\mu$m, in a reasonable period of time. Between 1990 and 1992 the first applications of μc-Si:H as absorber layers in solar cells were reported [Wang-1990, Faraji-1992, Flückiger-1992].

A breakthrough in the research and application of μc-Si:H for solar cells came from publications by a group from the IMT Neuchatel [Meier-1994a, Meier-1994b, Meier-1996] on single-junction solar cells with μc-Si:H absorber layers, and on a-Si:H/μc-Si:H tandem junctions which they later baptized as "micromorph" tandem cells. They showed (Fig. 3.1) that their single-junction μc-Si:H cell had a negligible light-induced degradation compared to a-Si solar cells. They also demonstrated that a μc-Si:H cell can work as an effective red-light absorber as a bottom cell in a tandem device, thereby replacing the notoriously unstable amorphous silicon-germanium alloys (a-SiGe:H).

This has stimulated very intense research over the last 15 years on all aspects of μc-Si:H materials and devices, such as deposition processes with higher deposition rates and deposition on large areas, material properties and optimization, and device performance. It is not possible to cover all the important literature and developments on μc-Si:H in this short introduction. So only some milestones are cited as examples:

In 1996 Torres et al. [Torres-1996] reported an improved cell performance upon reduction of the oxygen content in μc-Si:H. The influence of oxygen contamination

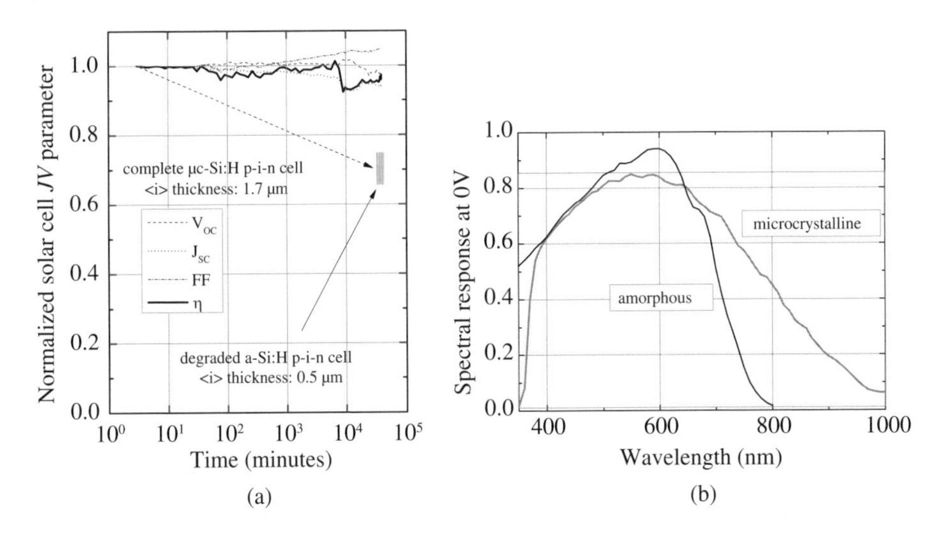


**Fig. 3.1** Comparison of solar cells with μc-Si:H or a-Si:H absorber layer: (a) Behaviour of the solar cell parameters upon exposure to sun light; (b) spectral response of the two types, μc-Si:H or a-Si:H, of solar cells showing the enhanced long (red) wavelength response of the μc-Si:H cell (Data by courtesy of J. Meier et al., IMT Neuchatel; see also [Meier-1994a]).

on the electronic properties and on the structure development during layer growth of μc-Si:H is considered to be of great importance for technological applications and has been the topic of a number of research activities from the very beginning of the work on μc-Si:H [Veprek-1983, Torres-1996, Kamei-1998, Kamei-2001, Finger-2003, Kamei-2004, Kilper-2009].

Investigations on the structural and growth properties and on their influence on material quality or on device performance have been published for example by [Tsai-1988, Collins-1989, Luysberg-1997, Houben-1998, Vallat-Sauvain-2000, Fujiwara-2001, Luysberg-2001, Bailat-2003, Rath-2003, Collins-2003, Droz-2004, Luysberg-2005, Vallat-Sauvain-2006a (with an extensive reference list)]. It was demonstrated that the maximum performance of solar cells with μc-Si:H absorber layers is obtained with material prepared close to the transition to amorphous growth [Vetterl-1999,Vetterl-2000, Droz-2004, Klein-2005]. This was correlated with low defect densities as a result of an effective passivation of the defects on the crystalline columns by an amorphous silicon "skin" [Finger-2004].

Over the following years improvement in solar cell and module efficiencies was pushed forward by several research groups worldwide [Shah-2004, Shah-2006, Mai-2005a, Klein-2005, van den Donker-2007a, Yamamoto-1999, Tawada-2003, Yamamoto-2004, Yang-2005, Yue-2008, Smets-2008, Matsui-2004, Matsui-2006, Gordijn-2006].

Considerable progress beyond the conventional PECVD preparation process was reported, with the exploration of new deposition regimes at considerably higher deposition pressures, both with standard RF excitation as well as with VHF excitation of the plasma [Guo-1998, Kondo-2000, Fukawa-2001, Lambertz-2001, Roschek-2002, Matsui-2002, Kondo-2003, Graf-2003, Rath-2004, Mai-2005b, Smets-2008, van den Donker-2005]. This allows one to prepare material of high quality and cells at rates above 1 nm/s with single-junction efficiencies exceeding 10 % [Mai-2005a]. These deposition regimes with higher deposition pressures and with higher excitation frequencies are at present being intensively studied in connection with the up-scaling of the deposition process to large areas [Niikura-2006, Mashima-2006, Takagi-2006, Strahm-2007a, Meier-2007, Kilper-2008, Rech-2006].

These developments were accompanied by modelling of the growth process of μc-Si:H. Partly competing models exist, with a focus on preferential etching, surface diffusion of radicals, sub-surface reactions involving hydrogen or combinations of these different processes [Matsuda-1983, Tsai-1989, Veprek-1990, Nakamura-1995, Matsuda-1999, Sriraman-2002]. As an alternative to PECVD, Hot-Wire Chemical Vapor Deposition (HWCVD) has also been applied to μc-Si:H [Matsumura-1991, Rath-2003, Matsumura-2008]. This process has the potential for very efficient gas usage and high deposition rates [Cifre-1994, Schropp-2002]. High efficiency solar cells with μc-Si:H absorber layers prepared by HWCVD have already been reported [Klein-2005]. So far, the deposition rates actually used for depositing high-quality material are comparatively low; the goal would therefore be to increase the deposition rate and still obtain excellent solar-cell efficiencies. One of the advantages of the HWCVD process is reflected in the high open circuit voltages of more than 600 mV obtained in the solar cell, possibly as a result of ion-free deposition which reduces damage at the device interfaces [van den Donker-2007a, Mai-2005a]. Such beneficial performance of the HWCVD process for reduced interface damage has also been reported for heterostructure solar cells [Branz-2008] and TFTs [see Matsumura-2008 for a recent summary]. Plasma deposition always leads to a certain amount of ion bombardment on the growing surface. Although the energy of the impinging ions can be reduced through higher excitation frequencies and/or higher deposition pressures, a certain amount of ion-provoked damage always remains; in the case of hot-wire deposition, there are no or very few ions with thermal energy only.

Today μc-Si:H is established as a bottom cell red-light absorber with high potential for improved stable efficiencies. Application as a single-junction device with 10 % efficiency and high stability could even be considered. However, the high current densities in such modules could pose a considerable challenge to the TCO contact materials. The companies Kaneka (Japan) and United Solar (USA) have reported stacked cells with more than 15 % initial efficiencies [Xu-2008, Yamamoto-2006]. "Micromorph" tandems with microcrystalline silicon bottom cells have found their way into production equipment [Meier-2007] and the first commercial stacked modules are already on the market [Tawada-2003, Nagajima-2009].

Today, with the basic μc-Si:H material having been established, the present focus is on further optimization of the deposition process for large-scale production. Furthermore, interesting developments of microcrystalline alloys of Si with C, Ge, O [Huang-2007, Finger-2009, Söderström-2009, Matsui-2009] have been recently reported.

To recapitulate: μc-Si:H has today a key function in the thin-film silicon solar industry and could determine its success. Understanding its properties, its handling and manufacturing will be important for making progress in the near future. The reader will be given some background on the properties of μc-Si:H in this chapter.

## 3.2 STRUCTURAL PROPERTIES OF μc-Si:H

### 3.2.1 Structure

Microcrystalline silicon (μc-Si:H) can easily be prepared with the same deposition process as a-Si:H. The structural composition is the important parameter that determines the distribution and density of defects, as well as the optical and electronic properties.

μc-Si:H is a mixture of crystalline grains, disordered regions and voids in various amounts and dimensions. There is no standard μc-Si:H and a definition just by the grain size is not appropriate. The name "microcrystalline" silicon suggests a predominance of micrometer-size features. This is not the case. Feature sizes vary from a few nanometers to more than a micrometer in one and the same sample. The term "nanocrystalline" is therefore used synonymously.

One can consider three basic configurations (see Fig. 3.2): (a) isolated grains in an amorphous matrix; (b) extended crystalline fibers or crystalline grains forming percolation paths, which would be important for electronic transport; or (c) volume crystalline material with disorder only at the grain boundaries and in the intergrain regions.

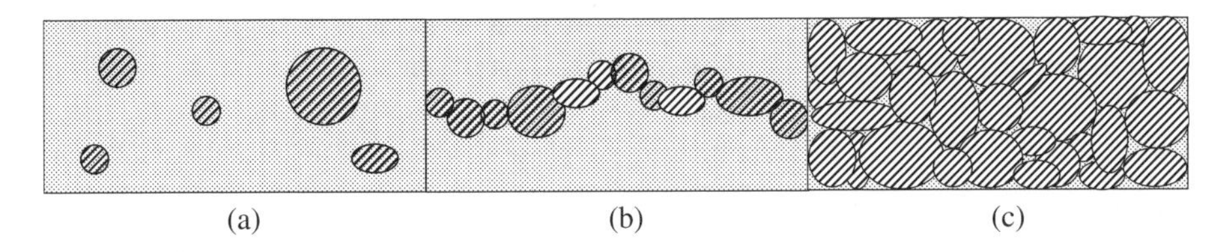


**Fig. 3.2** Schematic of the three principle different configurations of amorphous and crystalline structure in the phase mixture material μc-Si:H. (a) isolated crystalline grains; (b) crystalline regions forming percolation paths; (c) "fully" crystalline with disorder located at the grain boundaries only.

The structural composition depends on the deposition process. By varying the silane/hydrogen mixture in the process gas, one can comfortably grow material all the way from highly crystalline layers to fully amorphous layers. Other deposition parameters such as the process pressure, gas flow, excitation power, excitation frequency, temperature and so on, also influence the resulting material structure. Therefore the gas mixture is, in general, not sufficient to determine the structural composition; this is particularly true when comparing materials from different deposition methods.

The silane/hydrogen gas mixture is quantified as the silane concentration *SC*, i.e. the ratio between the silane gas flow to the process chamber and the total gas flow (silane plus hydrogen)

$$SC = \frac{[\mathrm{SiH_4}]}{[\mathrm{SiH_4}]+[\mathrm{H_2}]} \tag{3.1}$$

where $[SiH_4]$ indicates the flow of silane and $[H_2]$ the flow of hydrogen into the process chamber. Alternatively, the hydrogen dilution *R* defined as

$$R = \frac{[\mathrm{H_2}]}{[\mathrm{SiH_4}]} \tag{3.2}$$

is sometimes used.

To determine the structural properties of μc-Si:H layers, such as crystalline volume fraction, grain sizes and morphology, three experimental techniques are commonly used: Transmission Electron Microscopy (TEM), X-ray Diffraction (XRD) and Raman spectroscopy. Among these three ***Raman spectroscopy*** is the most popular one, because of its easy use. In Raman spectroscopy one investigates the local atom-atom bonding structure by the interaction of light (photons) with the bond vibrations (phonons). The energy and therefore the wavelength of the incoming light is shifted up or down by the energy of a phonon. These shifts are also known as Stokes and anti-Stokes shifts. In a crystalline structure the local bonds have well-defined energies. The Raman resonances have a narrow line width, corresponding to the narrow distribution in bonding energies. In disordered or amorphous structures one has a wide distribution of bonding energies. Consequently, the phonon energies also have an inhomogeneous broadening, resulting in a broadening of the Raman scattering signals. Furthermore the signals are, in general, also shifted in energy. With microcrystalline silicon as a phase mixture one can expect superposition of the signals from the amorphous phase and those from the crystalline phase. Raman scattering is usually measured in the back-scattering configuration and is carried out conveniently with the layers deposited on glass substrates. It can also be used on devices and, by application of different excitation energies with different probe depths or by etching the layer back, allows one to carry out a depth profile of the layer structure [Mai-2005b]. In Figure 3.3 the Raman spectrum of a μc-Si:H layer is shown. The signal contributions are attributed to the amorphous phase at 480 $cm^{-1}$ wavenumbers and to the crystalline phase at 520 and 500 $cm^{-1}$. The unit "wavenumber" $\tilde{\nu}$ is commonly used, both in infrared spectroscopy and in Raman spectroscopy. It is a convenient unit and is defined as the reciprocal of the wavelength $\lambda$. From the relationship between wavelength $\lambda$ and

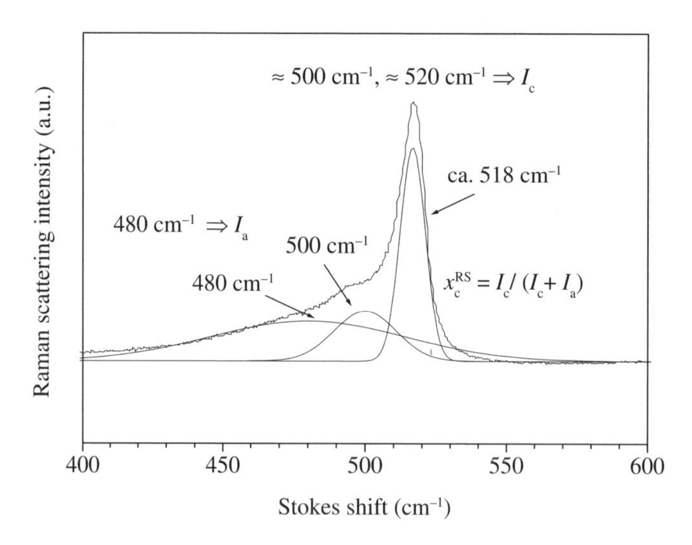


**Fig. 3.3** Measured Raman spectrum of a μc-Si:H sample together with three lines at 480, 500 and 518 $cm^{-1}$, used for a deconvolution of the spectrum. For details see text (By courtesy of Forschungszentrum Jülich).

frequency $\nu$ of electromagnetic radiation $c = \lambda\,\nu$, where $c$ is the light velocity in vacuum, we see that the wavenumber is proportional to the frequency ($\tilde{\nu} = \nu / c$) and therefore proportional to the energy. It is usually measured in $cm^{-1}$. By integrating the different line contributions after numerical deconvolution of the signal, one can calculate the intensity ratio

$$x_C^{RS} = \frac{I_c}{(I_a + I_c)} \tag{3.3}$$

as a semi-quantitative measure of the crystalline volume fraction [Bustarret-1988, Houben-1998, Vallat-Sauvain-2006b, Smit-2003].

Attempts have also been made to derive an absolute measure of the crystalline volume fraction by including in the evaluation procedure the different Raman scattering cross-sections and the optical absorption coefficients of the different phases [Bustarret-1988, Vallat-Sauvain-2006b]. However, there is controversy about this procedure in the literature [Ossadnik-1999]. For all practical purposes, the semi-quantitative value defined above has been established as a very useful tool in optimization processes for the material. With appropriate caution one can refer to the Raman intensity ratio $x_C^{RS}$ as "Raman crystallinity".

In Figure 3.4(a) one can see how the Raman spectra develop when going from amorphous layers (lowest curve in Fig. 3.4a, sample prepared with a high ratio of silane to hydrogen) to layers with high crystalline volume fractions (topmost curves in Figure 3.4(a); samples prepared with a low ratio of silane to hydrogen). Note that for μc-Si:H prepared with PECVD techniques at temperatures of typically 200-300 °C, there always remains a kind of "disorder tail" at low wavenumbers resulting in an asymmetric line, so that the value of the Raman crystallinity never reaches 100 %.

In principle one could also attempt to calculate the grain size from Raman spectroscopy. But such evaluations require various assumptions for parameters, which are not well known, and have therefore so far not been successful. A more straightforward method for evaluating the grain size is ***X-ray diffraction*** (XRD). XRD can be also measured directly on films deposited on substrates. Like Raman it is measured in back-scattering configuration and can, depending on the measurement mode, be rather surface-sensitive, i.e. it will only probe the layer structure close to the surface. A comparison of results on μc-Si:H from different XRD measurement modes is reported for example in [Zhang-2005]. Figure 3.4(b) shows XRD spectra of the same series of samples for which Raman spectra are shown in Figure 3.4(a). The distinct peaks correspond to diffractions from the crystalline planes with the indicated orientation. The grain size is directly correlated with the peak line width and can be calculated from standard diffraction theory. Deviation of the diffraction peak pattern intensities from the one which would be expected from a corresponding polycrystalline powder allows one to determine the possible preferential orientation of the μc-Si:H film grown on the substrate. Evaluation of the diffraction line widths typically yields grain sizes of a few tens of nanometers. The crystalline volume fraction can also be determined from XRD spectra, by deconvolution of the spectra into amorphous and crystalline components [Houben-1998].

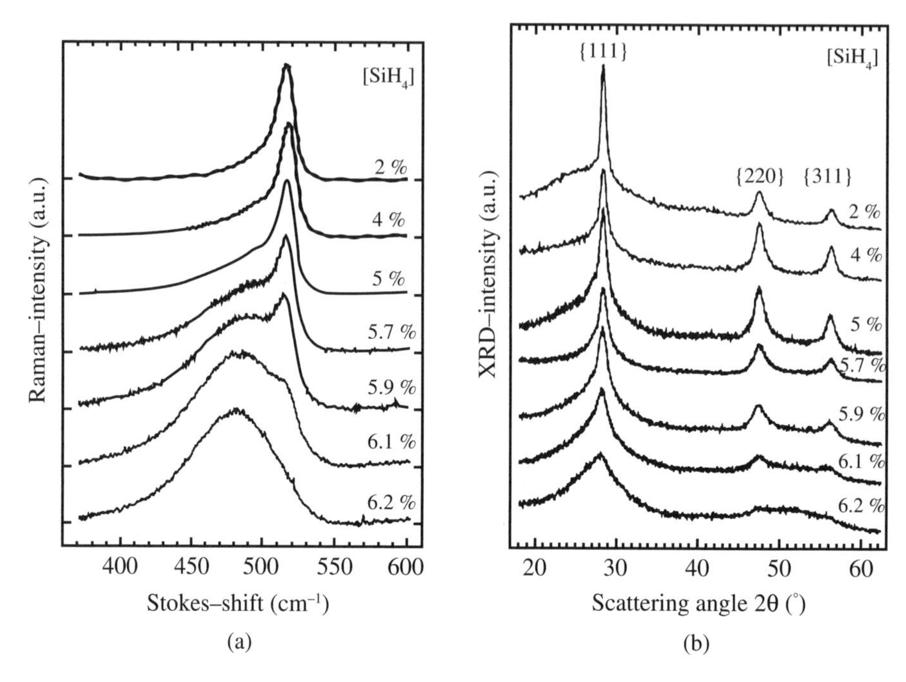


**Fig. 3.4** (a) Raman spectra of a series of silicon films prepared with different silane concentration (as indicated in the figure); (b) XRD spectra with of the same samples with the dominant reflection peaks indicated. All spectra in (a) and (b) are shifted vertically with respect to each other for clarity (By courtesy of L. Houben, Forschungszentrum Jülich).

The most convincing image of μc-Si is derived from ***Transmission Electron Microscopy*** (TEM) on cross-section samples, which gives us information on the actual structure of different μc-Si:H layers and layer stacks in entire devices. Unfortunately, the preparation of such TEM cross-section samples is very elaborate. This means that TEM is not a practical tool for routine investigations. Depending on the TEM measurement mode, i.e. whether one looks at the diffracted electron beam or at the transmitted beam, one distinguishes between dark-field images and bright-field images. In dark-field images crystalline regions appear as white areas, and in bright-field images as dark areas. Figure 3.5(a) shows dark-field images of highly crystalline material. The white regions in the image indicate crystalline columns extending through the entire film from substrate to surface with up to micrometer dimensions and a column base of a few tens of nanometers. It is important to note here that the dark regions in Figure 3.5(a) are not necessarily amorphous regions but crystalline regions of different crystallographic orientation; these regions would then also appear as white regions when the sample is tilted in a different direction in the microscope. High-resolution TEM micrographs (Fig. 3.5b) show perfect crystalline order inside the columns. One also notes frequent twin boundaries - a defect which preserves the tetrahedral coordination - and only very little detectable disordered phase at the grain boundaries. It appears that the crystalline columns are built up from smaller crystalline grains of a few tens of nanometers in size. In low crystalline material the column structure is less

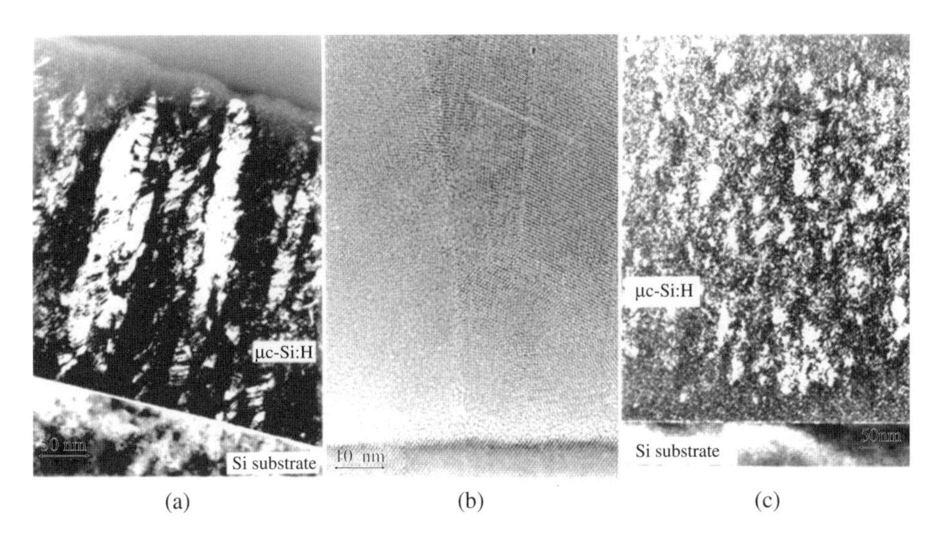


**Fig. 3.5** (a) Dark field cross-section image of a highly crystalline μc-Si:H sample; (b) high resolution TEM micrograph of the sample of the same highly crystalline sample; (c) dark field cross-section image of a μc-Si:H with low crystallinity (By courtesy of M. Luysberg and L. Houben, Forschungszentrum Jülich).

developed and individual features are much smaller (Fig. 3.5c). For such material, high-resolution images show amorphous tissue with isolated grains, similar to the schematic image of Figure 3.2(a).

From bright-field images (not shown [Houben-1998, Luysberg-1997, Vallat-Sauvain-2000]), evidence is found for large crack-like voids in the material preferably along columns and at the interface between the substrate and the μc-Si:H layer. Such voids can be found for all values of the crystalline volume fraction, but are more frequent in highly crystalline material [Houben-1998, Luysberg-1997]. Pronounced crack formation at the film substrate interface is also observed on certain substrate morphologies such as those obtained from textured TCO [Python-2008, Li-2009]. The resulting cracks, susceptible to in-diffusion of impurities, could lead to inferior solar cell performance [Python-2008, Li-2009].

TEM is an ideal tool to demonstrate this very critical substrate-dependence of μc-Si growth. Depending on the substrate and the growth conditions, an extended incubation layer, which can sometimes be completely amorphous, may develop (Fig. 3.6a). On the other hand, a crystalline silicon substrate can promote epitaxial growth at the typical deposition temperature of 200 °C (Fig. 3.6b). This critical substrate-dependence of the structural development is of greatest importance for the optimization of μc-Si:H layers in solar cells, where the μc-Si:H intrinsic absorber layer should grow preferably without any amorphous incubation layer. In fact, the latter generally results in considerable deterioration of the cell performance at the transition from microcrystalline to amorphous growth, with a strong reduction of the fill factor and the short circuit current of the device [Vetterl-2000, Luysberg-2001] (see Section 4.5). Obtaining suitable nucleation of the μc-Si:H layer, directly on the underlying material and thereby avoiding an amorphous incubation layer at the beginning

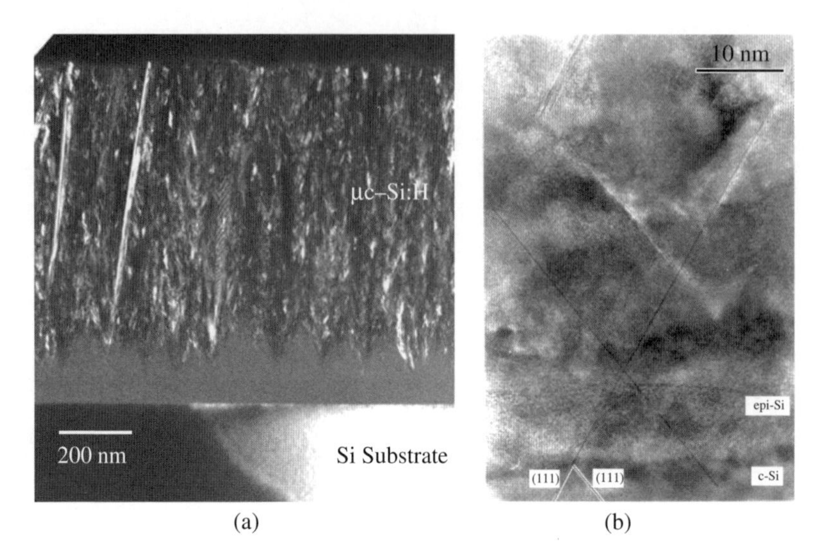


**Fig. 3.6** (a) Dark field cross-section image of a highly crystalline μc-Si:H sample with a pronounced amorphous incubation zone between the glass substrate and the crystalline structure; (b) high resolution image showing epitaxial growth on a pure c-Si wafer substrate (by courtesy of S. Klein and L. Houben, Forschungszentrum Jülich).

of microcrystalline silicon growth, is one of the most critical optimization tasks for the development of μc-Si:H solar cells.

While TEM images give us an useful and valuable visual impression of the structure of μc-Si:H layers, it is difficult to deduce quantitative information about the crystalline volume fraction from them. The amorphous phase, especially for highly crystalline material, will mainly result in loss of contrast. On the other hand, even the smallest crystalline grains can be identified for material which will appear fully amorphous in Raman spectroscopy or XRD. Evaluation of the "average" grain size inside the columns can only be done by visual evaluation and by averaging the TEM images.

We will summarize the information about grain size and crystallinity from the three methods presented here. With all three methods one can identify the variation in crystallinity (Fig. 3.7a). Raman and XRD spectroscopy tend to underestimate the crystallinity for highly crystalline material. For material with low crystallinity these methods may not be sensitive enough. TEM still detects very small traces of crystallinity but a quantitative evaluation of the crystalline volume fraction and the average grain size is difficult and laborious. An average grain size is reliably determined only from XRD, but even here the result depends on which diffraction peak is used for the evaluation (see Fig. 3.7b). It is the size of the coherent regions that are determined by XRD: these coherent regions can be identified with the grains of a few nanometers in size that are embedded within the columns. This would justify calling the material "nanocrystalline". However, it was demonstrated that TEM also shows much larger crystalline regions with single orientations, regions of up to micrometer size. The term "grain size" is therefore not unambiguously defined in μc-Si:H and should be used with care and with reference to the method by which the grain size is determined.

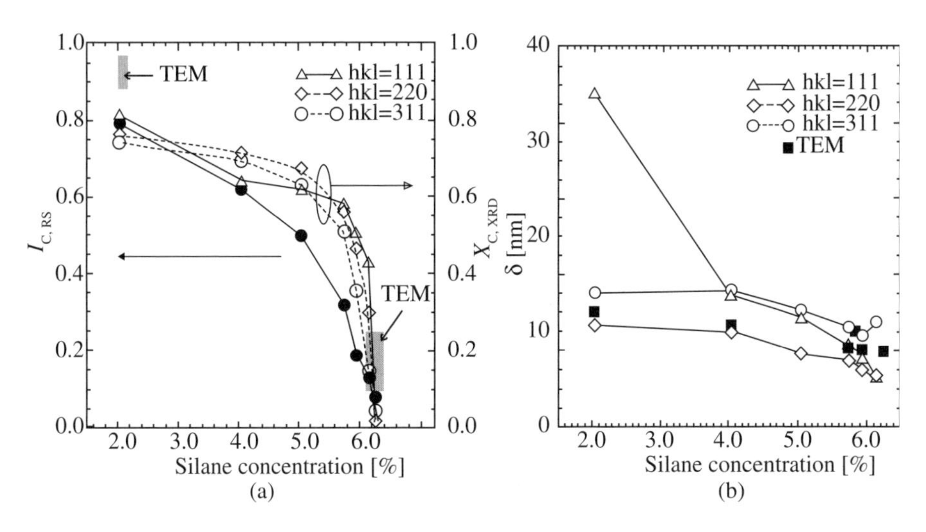


**Fig. 3.7** (a) Crystallinity of μc-Si:H samples determined from three different methods, Raman, XRD and TEM vs. silane concentration; (b) grain sizes of the same samples determined from the XRD spectra and from TEM images (By courtesy of L. Houben, Forschungszentrum Jülich).

### 3.2.2 Defects and gap states

As with amorphous silicon, for the microcrystalline silicon phase one can expect a large number of structural defects, which lead to gap states such as dangling bonds and strained bonds. These states can be active either as carrier traps or as recombination centers (see Chapter 2, Section 2.2). The corresponding defect density is an important measure of quality for μc-Si:H, as it determines, to a large extent, the material performance of devices like solar cells. In order to obtain the electronic density of states, measurement of the defect density, preferably over a whole range of electronic energies, is necessary for the characterization and optimization of μc-Si:H. Therefore, the defects in μc-Si:H and the methods used to evaluate their density are treated here in some detail. The experimental techniques to investigate the defects in μc-Si:H, which we will comment on, are: (i) electron spin resonance (ESR); and (ii) spectroscopic methods based on optical absorption, such as photothermal deflection spectroscopy (PDS) or photocurrent spectroscopy. Of these methods ESR gives direct access to the spin density and possibly the defect density in the material, together with information on the microscopic nature of the prevailing type of defect. It is, however, not preferred for routine measurements because of special requirements for the sample preparation. On the other hand, the methods based on spectroscopic optical absorption are easier to handle with samples on typical glass substrates, but only yield indirect information on the defect density. Also these methods are based on a number of critical assumptions – assumptions that are still being controversially discussed.

Concerning the structural defects in μc-Si:H, the focus will be on the dangling bond defect, which is believed to form electronic states in the mobility gap; these states act in μc-Si:H, just as in a-Si:H, as a recombination center. It is not obvious that the bulk crystalline regions in μc-Si:H would also lead to tail states as a result of strained bonds, at least not to such an extent as is observed in a-Si:H. The crystalline network present in μc-Si:H would not tolerate a wide range of bond energy variation. Therefore tail states would be located preferably at grain or column boundaries. It is not known to what extent they influence transport in μc-Si:H. From photoconductivitiy transient experiments there is evidence of a "multiple-trapping" type transport in μc-Si:H (as in a-Si:H), and this would strongly suggest the existence of tail states [Dylla-2005a, Reynolds-2009]. The possible influence of bandtail states in μc-Si:H will be commented on later in the chapter, when dealing with transport in this material.

An obvious defect to consider is a broken Si-Si bond. In crystalline silicon there is little freedom for the network to relax, i.e. for the atoms to move apart from each other. Therefore, this broken bond would easily reconstruct. But in a disordered network like a-Si:H or μc-Si:H, stable dangling bonds are more likely to occur. Here under-coordinated atoms with unsaturated "dangling" bonds are part of the disordered network at vacancies, inner surfaces, and grain or column boundaries. Furthermore, due to the presence of hydrogen in the material, if one of the orbitals of a broken silicon-silicon bond is stabilized by hydrogen, another orbital (bond) can remain dangling. Evidence for this type of a defect comes from ***Electron Spin Resonance*** (ESR). For thin silicon films one frequently measures ESR on material in the powder form, so as to obtain more sample mass and higher signal intensity. For this the material is grown on metal foil, from where it is removed and collected in quartz tubes. But it can also be measured on thin films on glass substrates, albeit with reduced signal resolution. In general, measurements on both types of samples yield equivalent results; however, especially in the case of μc-Si:H material, the difference in growth kinetics depending on the different substrates has to be taken into account.

Let us look at the situation in amorphous silicon first. (Most of the arguments presented for defects in a-Si:H can be transferred to the situation in μc-Si:H.) In a-Si:H one finds a characteristic ESR signal (Fig. 3.8a).

This signal is described by its intensity, which correlates with the defect density, as well as its line shape and width and its g-values. The g-value is a microscopic quantity characteristic of the defect in its local environment. It is calculated from the resonance condition $h\nu = g\,\mu_B\,B$, where $h$ is the Planck constant, $\nu$ the microwave frequency, $\mu_B$ the Bohr magnetron and $B$ the magnetic field flux. The line shape, g-value and line width of the resonance in a-Si:H can be simulated from the well-known $P_b$ defect [Stutzmann-1989]. The $P_b$ center is the Si dangling bond originating from an oxygen vacancy at the $Si/SiO_2$ interface, which is of great technological relevance for microelectronics and for crystalline silicon solar cells and which has been studied in detail. As described in Chapter 2, the dangling bond orbital has three charge states: (i) positive when it is empty; (ii) neutral when occupied with one electron (this is the paramagnetic state which can be detected by ESR); and (iii) negative when it is occupied by two electrons. The charge states are associated with different electron energies. The energy difference, i.e. the effective correlation energy $U$, is the sum of the Coulomb energy (due to the electrostatic repulsion between the two negatively charged electrons) and a structural relaxation term. In general, $U$ is positive and the doubly occupied state ($E^-$) lies, in the energy diagram, above the neutral paramagnetic state ($E^0$). Occupation of the different charge states in thermodynamic equilibrium

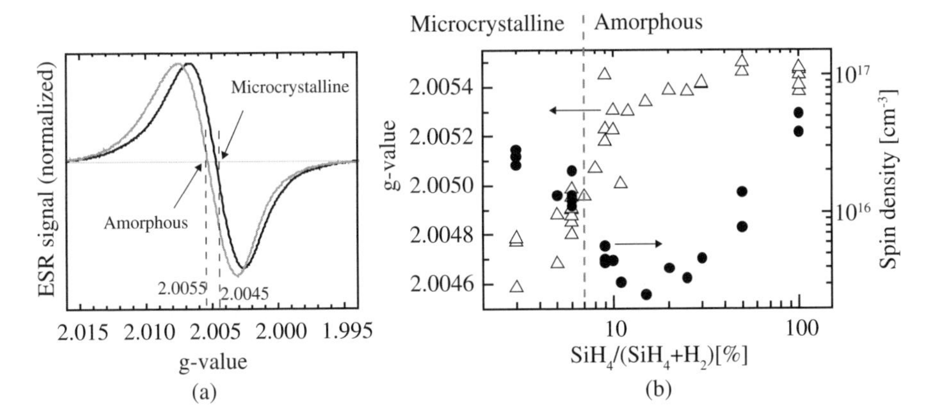


**Fig. 3.8** (a) Electron spin resonance signals of an intrinsic amorphous and a microcrystalline silicon sample (b) g-value and spin density of thin film silicon materials all the way from amorphous to microcrystalline vs. silane concentration during preparation. Note the logarithmic x-axis. The vertical dashed line indicates the transition between material which does or does not show a crystalline peak in the Raman spectra. See text for details (By courtesy of O. Astakhov, Forschungszentrum Jülich).

is determined by occupation statistics, taking into account the correlation energy *U* and the shape and width of the defect distribution [Fritzsche-1980, Schweitzer-1981]. This leads to complicated relationships between the spin density, defect density and the electronic density of states. Although this might sound a little academic, the extent to which one can really detect quantitatively all defects with a method like ESR is nevertheless of considerable importance. In intrinsic a-Si:H with the Fermi level near the midgap, it seems to be correct if one assumes a correlation energy which is of the same magnitude as the width of the dangling bond defect distribution. In such a case a very high percentage of dangling bond states can be singly occupied and paramagnetic.

In μc-Si:H one also has to expect dangling bond defects. In fact, the ESR signal looks very similar (Fig. 3.8a). A complication arises, compared with a-Si:H, from the fact that in μc-Si:H these dangling bond defects could be in different structural environments of the material, namely in the crystalline regions, or on the grains boundaries, or in the amorphous phase. It is not straightforward to distinguish between the different defect locations just by analyzing the results of the different experimental methods presented here. For simplicity one should mainly concentrate on the total measured defect densities. Based on ESR measurements, one finds for μc-Si:H prepared with PECVD or with HWCVD processes, spin densities in the range of $10^{16}$-$10^{17}$ $cm^{-3}$ for deposition temperatures between 200 and 300 °C (Fig. 3.8b). The spin density is highest for material with the highest crystalline volume fractions and it increases considerably to $10^{19}$ $cm^{-3}$ when the deposition temperature is raised to 450 °C [Finger-2004]. The lowest level is found for intermediate values of "crystallinity" i.e. for material close to the transition to amorphous growth. The change in structure from amorphous to microcrystalline is seen in the properties of the paramagnetic states. Both line shape and g-value change (Fig. 3.8b): this is a sign of the change in the microscopic details of the defect, from a defect in a disordered/amorphous environment to a defect in an ordered/crystalline environment. Interestingly, when one looks at the g-value as a function of the gas composition, i.e. of the H-dilution, the g-value already changes in the region where no crystalline phase can be detected by Raman spectroscopy, indicating a structural change has already occurred within the amorphous phase in this "transition zone". Such material, which has very low spin density, is used as a-Si:H with improved stability in single-junction amorphous cells and in stacked "micromorph" solar cells.

The nature of the defects giving rise to ESR signals in the different types of thin-film silicon layers, including the various μc-Si:H material compositions and the details of the ESR signals, are still a matter of research. On the other hand, there is sound evidence from studies of the relationship between doping and the compensation effects of deep defects that in intrinsic μc-Si:H, too, the spin density (as measured by ESR) represents the majority of deep defects [Dylla-2005b].

At present one can, with some confidence but with appropriate caution, use ESR also in μc-Si:H for calibrating the other experimental techniques that are employed to determine the total deep defect density.

Common alternative methods to measure the defect density via the optical absorption are ***photothermal deflection spectroscopy*** (PDS) [Jackson-1982] and photocurrent spectroscopy in different measurement modes such as dual beam spectroscopy (DBS) [Günes-2005], ***constant photocurrent method*** (CPM) [Vanecek-1981], or ***Fourier-transform photocurrent spectroscopy*** (FTPS) [Vanecek-2002] – as already described for a-Si:H in Chapter 2. The advantages of optical absorption measurements over ESR are the ease of use on thin films on substrates or even on entire devices, i.e. in situations where ESR would suffer from low signal resolution or not be applicable at all. The application of both CPM and PDS for the evaluation of the defect density is well established for a-Si:H and methods to identify the contribution of the defects on the absorption spectra have been investigated and described in detail in the literature [Wyrsch-1991, Sauvain-1993, Siebke-1996, Schmidt-1998, Main-2004].

For μc-Si:H, determination of the defect density via optical absorption measurements is basically more difficult. This results from the heterogeneous nature of the material structure, which includes both amorphous and crystalline phases.

In Figure 3.9(a) and (b), the PDS spectra of different μc-Si:H samples are shown. Figure 3.9(a) shows PDS spectra of samples with different "crystallinity", while Figure 3.9(b) shows μc-Si:H layers all having approximately the same "crystallinity" but possessing largely different defect densities. The defects can be linked to absorption contributions below the bandgap. The problem with μc-Si:H is that the bandgap is not well defined and also not well pronounced in the absorption spectra. Nevertheless one can relate the absorption below about 1.0 eV with transition from or into deep defects or tail states. With some arbitrariness one can choose for example the absorption at 0.7 eV and compare it with the spin density measured on the same samples. The result is plotted in Figure 3.9(c). One finds an obvious relationship between defect absorption $\alpha_{def}$ and spin density $N_s$ over three orders of magnitude (Fig. 3.9c). But there is also clear deviation from a simple linear relationship between $N_s$ and $\alpha_{def}$, and furthermore, considerable scatter. This makes it difficult to use PDS for a

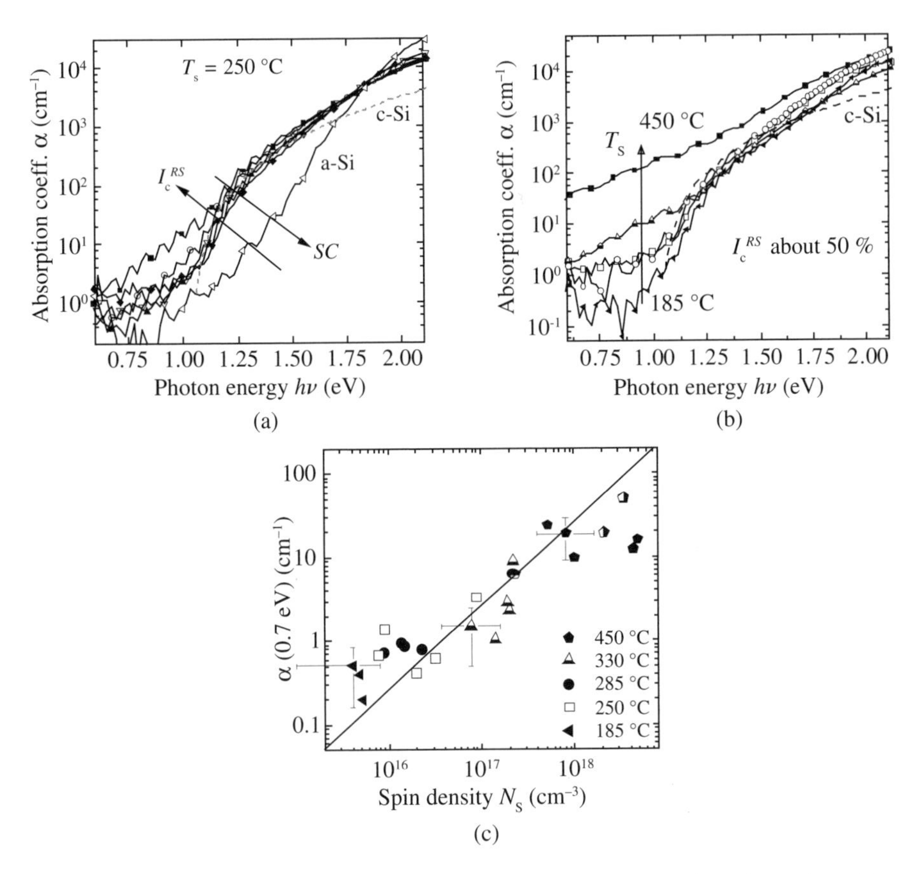


**Fig. 3.9** (a) PDS spectra of μc-Si:H and a-Si:H samples prepared at nearly optimum deposition temperature with different silane concentrations *SC* resulting in different crystallinities $I_c^{RS}$; (b) PDS spectra of μc-Si:H samples prepared at different temperatures; (c) comparison of the value of the defect absorption at 0.7 eV with the spin density for samples from Figure 3.9(a) and (b) (By courtesy of S. Klein et al., Forschungszentrum Jülich).

fine-tuning of the electronic properties of μc-Si:H [Klein-2007]. On the other hand, a distinction between "good" and "really bad" material should be easily possible on the basis of PDS spectra.

In the case of photocurrent spectroscopy, one also has to consider that this technique measures photoexcited carriers in transport paths. It could therefore easily occur that the carrier transport by-passes the more defective regions and that the real defect density is, thus, underestimated. Results from FTPS reveal, in fact, rather low defect densities on the basis of an evaluation (and of a calibration factor) which is borrowed from a-Si:H [Vanecek-2002]. In Figure 3.10(a) the optical absorption coefficient obtained from FTPS is shown for several μc-Si:H samples in comparison with c-Si and silicon on sapphire [Vanecek-2002]. Figure 3.10(b) shows the defect absorption from FTPS vs. "crystallinity" for solar cells before and after light-induced degradation [Vallat-Sauvain-2006a]. Both the variation of the defect absorption with

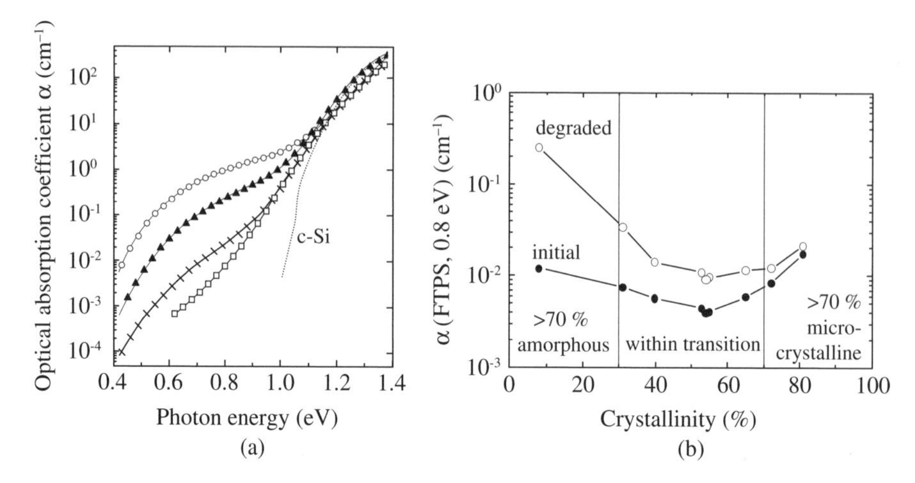


**Fig. 3.10** (a) Optical absorption spectra α(E) obtained from FTPS photocurrent spectroscopy for three microcrystalline silicon samples deposited under different deposition conditions (circles, crosses, and squares), commercial silicon on sapphire thin film (full triangles), and crystalline silicon (dotted line) (b) Optical absorption α(E) at 0.8 eV obtained from FTPS vs. crystallinity for solar cell before and after light induced degradation ((a) by courtesy of M. Vanecek, Institute of Physics, Academy of Sciences, Prague; see also [Vanecek-2002]; (b) by courtesy of F. Sculati-Meillaud et al., IMT Neuchatel; see also [Vallat-Sauvain-2006a]).

crystallinity and the increase upon light soaking is in qualitative agreement with findings from other methods. Using similar calibration constants between the subgap absorption and the defect density as in a-Si:H [Vanecek-2000], one would conclude that the defect densities are in the range of $10^{14}$ $cm^{-3}$, which appears to be unrealistically low. Nevertheless CPM and especially FTPS are used routinely by several laboratories for material optimization of high-quality μc-Si:H layers, which are used as intrinsic layers within *pin*- and *nip*-type solar cells. On the other hand, convincing evidence that photocurrent spectroscopy can be used in μc-Si:H over a wide range of defect densities is so far missing and a comparison of ESR spin density with CPM results would discourage such a use [Astakhov-2009].

The final goal of any defect spectroscopy should be to determine the density of states (*DOS*) within the "gap" of μc-Si:H material and to evaluate its influence on the electronic and optical properties of μc-Si:H layers. Ultimately one desires to evaluate the effect of these "gap states" on the performance of solar cells and other thin-film devices. Most experiments will "see" a spatial and compositional average of the *DOS*. It is, on the other hand, doubtful whether the *DOS* in μc-Si:H layers can be considered as a simple superposition of the *DOS* in crystalline (c-Si) and in amorphous silicon (a-Si:H). As mentioned before, it is not obvious that the bulk crystalline regions in μc-Si:H also develop tail states. Such tail states would be located at the grain or column boundaries. Thereby, features like grain-boundary-related defects and potential offsets (due to the difference in bandgap between the amorphous and the crystalline phases) have to be considered. Convincing models for the *DOS* in μc-Si:H do not exist so far.

### 3.2.3 Hydrogen, defect passivation, impurities and doping

***Hydrogen and defect passivation***

Hydrogen is crucial both for the formation of the microcrystalline structure during the growth process and for the passivation of dangling bond defects in the material. In μc-Si:H material, hydrogen fulfills similar functions as in a-Si:H.

The hydrogen atoms will be located preferably at grain boundaries or in the disordered/amorphous phase since in the pure crystalline silicon phase with its rigid lattice, the solubility of hydrogen is very low [Beyer-1999]. There are no stable Si-H bonds in bulk c-Si material and H will diffuse out from c-Si even at low (room) temperatures. Only after pronounced perturbations to the Si lattice can hydrogen be accommodated in the material at the location of the perturbation.

To investigate hydrogen in μc-Si:H, infrared spectroscopy is a very useful and convenient tool. The infrared spectra are measured in transmission with the films grown on a substrate which is transparent in the relevant infrared wavelength range, such as, for example, a silicon wafer. (A note to the experimentalist: It can be occasionally difficult to grow films with sufficient thickness of typically 500-1000 nm on c-Si wafer or on glass substrate for infrared spectroscopy or for optical and electrical measurements. The films can have internal strain which makes them peel off the substrate. This can be a serious difficulty for the experiment and the data evaluation. Possible solutions are the use of rough substrates or substrate pretreatment for better adhesion.)

With infrared light of specific photon energy, the Si-H bonds can be excited to stretching and bending oscillations. In order to obtain the corresponding infrared spectrum in a rapid and convenient way, one commonly uses a Fourier Transform Infrared (FTIR) spectrometer. FTIR is a well-established measurement tool; it is also used for similar investigations in a-Si:H (see Chapter 2, Section 2.2.2, where the basic functioning of FTIR spectroscopy is briefly explained).

In the case of μc-Si:H material, FTIR will give useful information on:

- structural details regarding the type of silicon-hydrogen bonding configurations (Si-H, $SiH_2$, Si-$H_3$), as found in the different phases of μc-Si:H;
- possible porosity in the material by evaluating the relative amount of H on surfaces or in compact material;
- the total content of bonded hydrogen;
- bonding or in-diffusion of impurities, in particular of oxygen, which can also be an indication of porosity.

Figure 3.11(a) shows typical infrared absorption spectra plotted versus the wavenumber for two distinctly different μc-Si:H samples. For simplicity one can distinguish between bending and stretching vibrations only. In general the bending modes will have lower energies (wavenumbers) than the corresponding stretching modes. For more detailed descriptions, especially of the different bending modes such as the "wagging", "rocking", "twisting" and "scissor" modes, the reader is referred to the literature [Cardona-1983, Street-1991]. One can identify the characteristic absorption modes for Si-H in μc-Si:H, which are all very similar to the ones in a-Si:H: the SiH, $SiH_2$ and $SiH_3$ bending modes at 630 and 800-900 $cm^{-1}$; and the Si-H/Si-$H_2$ stretching modes at 2000/2100 $cm^{-1}$. In addition one sees the strong Si-O modes at 1100 $cm^{-1}$.

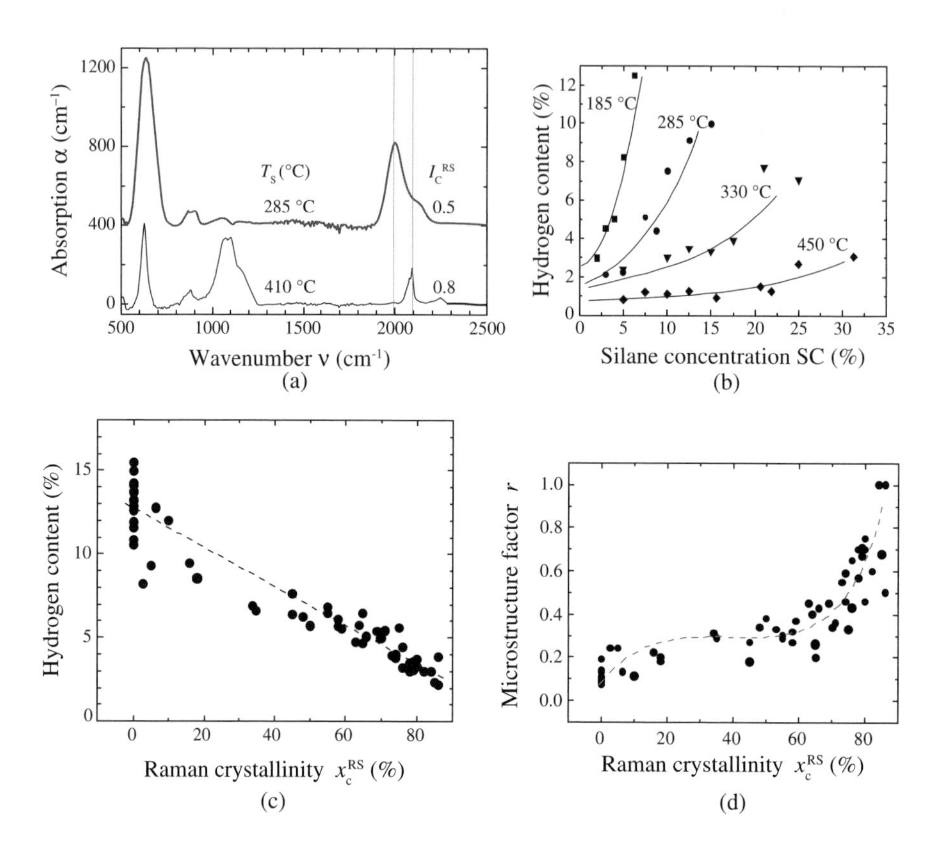


**Fig. 3.11** (a) Infrared absorption spectra of μc-Si:H material prepared at different temperatures with medium and high crystallinity. (b) Total bonded hydrogen content determined from the infrared spectra of μc-Si:H material prepared with different silane concentrations at different temperatures. (c) Total bonded hydrogen content determined from the infrared spectra of μc-Si:H material prepared with HWCVD with a wide range of process conditions as a function of the crystallinity; (d) Microstructure factor *r* as a function of the crystallinity of the same range of samples. The lines are guide to the eye. (By courtesy of S. Klein and J. Lossen, Forschungszentrum Jülich).

The upper trace in Figure 3.11(a) is a typical IR spectrum for a compact μc-Si:H layer with medium crystalline volume fraction of $x_C^{RS} = 0.5$. It shows stretching modes predominantly at 2000 cm$^{-1}$, indicative of H in a compact environment, very little Si-H$_x$ at 800-900 cm$^{-1}$ and no Si-O bonds.

The lower trace in Figure 3.11(a) is typical of highly crystalline material. It exhibits a lower overall hydrogen content, two narrow stretching modes at 2100 cm$^{-1}$ from hydrogen on (inner) crystalline surfaces, which we can identify with the column boundaries (see below), and a strong porosity resulting in the appearance of Si-O modes at 1100 cm$^{-1}$. Because of the low solubility of hydrogen in the crystalline phase, the hydrogen content increases with increasing amorphous volume fraction (Fig. 3.11b and c).

As with a-Si:H, one can define a microstructure factor $r$ (compare Chapter 2) as the ratio of the absorption intensity at 2100 $cm^{-1}$ to that at 2000 $cm^{-1}$. This microstructure factor can be seen as a measure of the porosity of the material; in general, it correlates well with the observation of oxygen in-diffusion. It is a convenient measure for the "device-quality" of the material. But in contrast with the situation in a-Si:H, the absorption modes around 2100 $cm^{-1}$ are, in μc-Si:H, likely to be related to Si-H bonds on crystalline surfaces rather than on internal void surfaces. This is supported by the splitting up of the single mode into a so-called doublet, characteristic for crystalline silicon surfaces, and the much narrower line width of these absorption modes. Figure 3.11(d) shows the microstructure factor as a function of crystallinity. It is seen that, in particular for crystallinities above 60 %, one finds a strong increase in the microstructure parameter as an indication of increasing porosity.

There is also a strong dependence of the hydrogen content on the deposition temperature, similar to that found in a-Si:H (see Fig. 3.11b). At higher substrate temperatures, hydrogen even evolves from the material during the growth process. Such hydrogen out-diffusion at temperatures close to the substrate temperature has been reported for μc-Si:H from hydrogen effusion experiments [Finger-1991]. A reduced effusion temperature is plausible in the case of pronounced porosity or of high crystallinity, where hydrogen can easily move along the crystalline column boundaries or through the void network.

The dependence of the hydrogen content in μc-Si:H on the deposition temperature in μc-Si:H correlates qualitatively with the defect density [Finger-2004]. The defect density increases for deposition temperatures beyond 250 °C. Hydrogen is needed for defect passivation and this can put limits on the deposition temperatures of μc-Si:H from PECVD and HWCVD processes which one would otherwise want to increase, for example to obtain larger grain sizes.

Various reports indicate that a deposition temperature in the range 200-300 °C is beneficial for obtaining μc-Si:H material with low defect density - similar to the situation for "high quality" a-Si:H; this fact underlines the crucial role of hydrogen for "material quality" in both materials [Finger-2004]. On the other hand, it is known for a-Si:H that the out-diffusion of hydrogen during the deposition process can be counteracted by sufficiently high deposition rates and vice versa, thereby allowing combinations of high deposition rates at elevated temperatures above 300 °C [Ganguly-1993, Platz-1998]. Whether similar high temperature/high growth rate conditions are available for μc-Si:H has still to be shown.

One can summarize that infrared spectroscopy is a convenient and very useful tool for investigation of the structural properties of μc-Si:H. The findings on total hydrogen content, on the type of hydrogen bonding, on microstructure and on possible in-diffusion of impurities generally correlate well with the electronic properties and the behavior in devices. Infrared spectroscopy should therefore be considered a standard optimization tool for μc-Si:H process and material development.

### *Effect of oxygen and other impurities*

The incorporation of oxygen plays an important role in the electronic properties of μc-Si:H. Highly crystalline as-deposited μc-Si:H material usually shows a pronounced

$n$-type character with a dark conductivity $\sigma_d$ around $10^{-3}$ S/cm, i.e. a much higher dark conductivity than for high-quality material with reduced crystallinity, which is in the range $10^{-7}$-$10^{-6}$ S/cm (see below). As an explanation for this high $n$-type conductivity, it has been suggested that oxygen forms donor states in μc-Si:H. In view of the application of intrinsic μc-Si:H as absorber layer in solar cells, Torres et al. [Torres-1996] reported that the O concentration and the conductivity in μc-Si:H prepared by PECVD can be reduced by use of a gas purifier during the deposition process, resulting at the same time in much improved solar cell performance. Figure 3.12(a) shows the dark conductivity of μc-Si:H samples vs. the reciprocal temperature for samples with different oxygen impurity concentrations, together with the oxygen content measured by secondary ion mass spectrometry (SIMS). The authors [Torres-1996] conclude that a

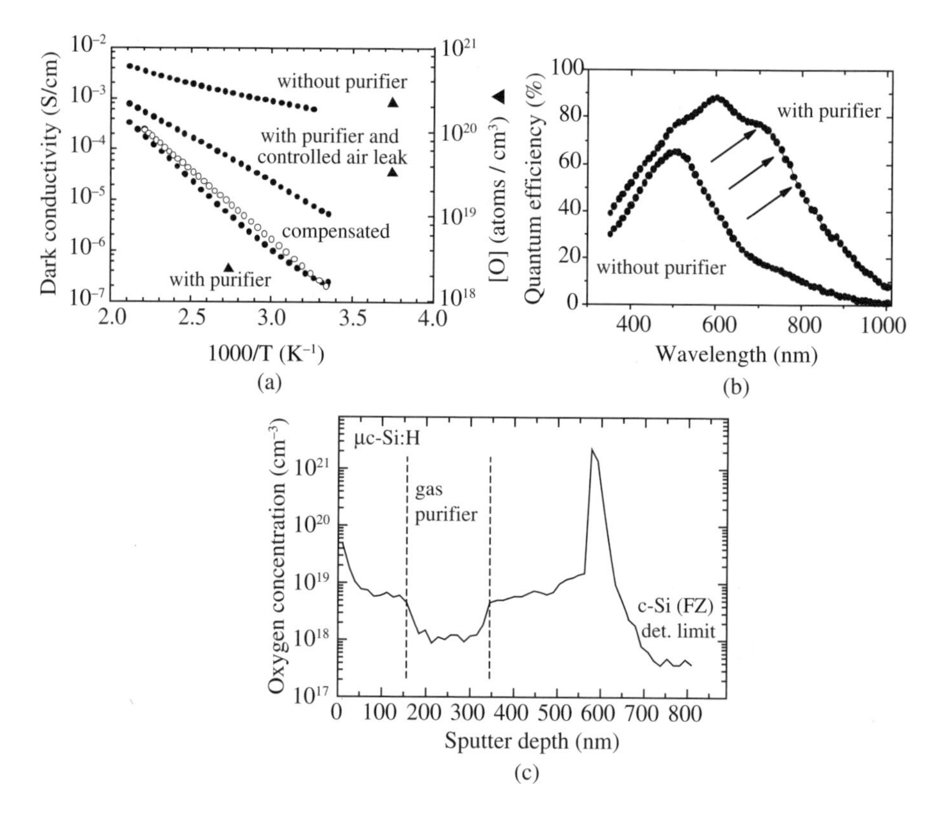


**Fig. 3.12** (a) Dark conductivity vs. 1/T for μc-Si:H with different oxygen contamination levels and for one sample with compensation doping. Oxygen concentrations measured by SIMS are also indicated; (b) Influence of use of a gas purifier on the spectral response of a solar cell with μc-Si:H absorber layer (By courtesy of P. Torres et al., IMT Neuchatel; see also [Torres-1996]); (c) Oxygen concentration profile measured by SIMS of a μc-Si:H film on a c-Si float zone (FZ) wafer showing the influence of switching on and off the gas purifier. Also indicated is the detection limited of the SIMS for oxygen. The peak at around 600 nm sputter depth corresponds to the natural $SiO_x$ layer on the c-Si wafer (By courtesy of O. Vetterl et al., Forschungszentrum Jülich).

reduction of the oxygen concentration to a value of about $2 \times 10^{18}$ cm$^{-3}$ by using a gas purifier yields intrinsic transport behavior similar to material with careful compensation (lower two traces in Fig. 3.12a).

On the other hand, material prepared without a gas purifier shows high conductivity with clear *n*-type behavior. The consequence for solar cells containing these *i*-layers, grown with and without a gas purifier, is seen in Figure 3.12(b). The cells with contaminated material show a strong reduction of the quantum efficiency in the long wavelength range. Figure 3.12(c) shows an SIMS profile with the effect of a gas purifier on the oxygen concentration. The concentration [O] of oxygen atoms is reduced to $10^{18}$ cm$^{-3}$. Oxygen concentrations below $10^{19}$ cm$^{-3}$ are also obtained by employing sufficiently clean gas supply systems and low-leakage deposition systems, and by using high-quality source gases for silane and hydrogen. Additional reduction of the oxygen concentration below $10^{18}$ cm$^{-3}$ is generally not found to reduce further the conductivity or to improve the solar cell performance.

Although this relationship between high oxygen concentration ([O] > $10^{19}$ cm$^{-3}$) and conductivity has been observed, the microscopic doping mechanism is not understood. The influence of the oxygen concentration in the material on structure, on transport properties like carrier density and mobility, and on solar cell performance is a matter of ongoing research [see for example Kamei-2001, Kamei-2004, Kilper-2009]. Kamei et al. [Kamei-2004] discuss the possibility of thermal donors (TD) or oxygen aggregates as donor states. The thermal oxygen donor known from crystalline silicon [Shimura-2004] is formed during annealing above 400 °C, a temperature which is considerably higher than the typical deposition temperatures for μc-Si:H. However, it cannot be excluded that sufficient formation energy for a TD state is supplied during the PECVD growth process.

In addition to the proposed doping effect, oxygen may also lead to modification of the surface potential along the inner surfaces by adsorption or oxidation. The latter effects are observed especially in highly crystalline porous materials (see below) and can have a considerable influence on the conductivity of the material [Veprek-1983, Finger-2003, Smirnov-2006]. Effective non-reversible oxidation occurring even at room temperature can be observed directly in infrared spectroscopy (see Fig. 3.11a) or by electron spin resonance [Veprek-1983, Finger-2003]. Finally, it has been reported that the presence of oxygen in the process gas could also affect the nucleation and growth of μc-Si:H material [Kamei-1998]. The subject is also of great importance for technical applications as it sets requirements for the quality of the deposition and gas supply systems.

What is said for oxygen contamination might be also true for other impurities such as nitrogen, another possible donor. Readers are referred to the literature for details [Kilper-2009]. Furthermore, the columnar and occasionally porous structure of μc-Si:H make the material susceptible to diffusion along these cracks or grain boundaries. Such effects could be very critical in device fabrication, with a possible diffusion of doping atoms (B or P) into the intrinsic layers. Overall one can expect μc-Si:H material to respond more critically to impurity contamination, not least because of its higher doping efficiency, as will be briefly shown in the next section.

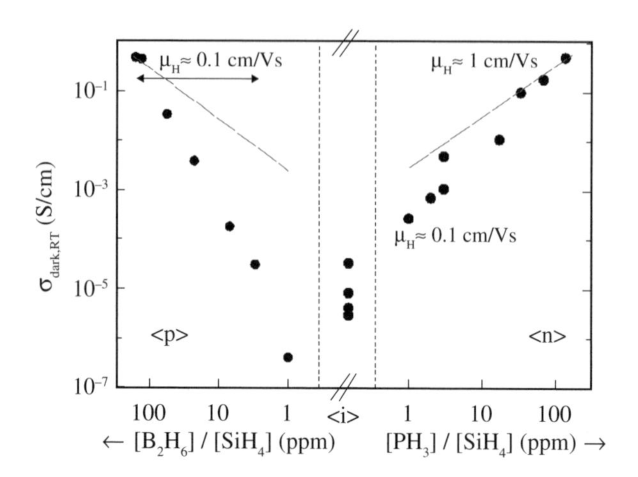


| | a-Si:H | μc-Si:H | c-Si |
|---|---|---|---|
| *n*-type conductivity(S/cm) | $10^{-2}$ | $10^{2}$ | $10^{3}$ |
| *p*-type conductivity(S/cm) | $10^{-2}$ | $10^{2}$ | $10^{3}$ |

**Fig. 3.13** Dark conductivity of μc-Si:H material as a function of the gas phase doping concentrations with $PH_3$ for *n*-type doping and $B_2H_6$ for *p*-type doping. Also indicated are the Hall mobilities measured at the respective doping concentrations. The dashed line corresponds to a linear relationship between doping concentration and conductivitiy (by courtesy of J. Müller et al., Forschungszentrum Jülich).

***Doping***

Microcrystalline silicon can be conveniently doped *n*- or *p*-type by appropriate admixture of doping atoms such as P or B, using gases like $PH_3$, $B_2H_6$ or $B(CH_3)_3$ in the deposition process. It is found that the doping process in μc-Si:H is, not surprisingly, a mixture of the phenomena observed in a-Si:H and in c-Si. For low doping levels, of the order of the defect density in the material, i.e. around $10^{16}$-$10^{17}$ $cm^{-3}$, electrons or holes from the dopant atoms first have to fill up these defect states, just as in a-Si:H. For higher doping levels we observe a 1:1 relationship between the built-in doping atoms and carrier density; i.e. we have a substitutional doping process in the crystalline phase of μc-Si:H similar to that in c-Si [Müller-1999, Dylla-2005b, Finger-2000] (Fig. 3.13). In particular, one does not observe any self-compensation effect of the doping process by thermal equilibrium between four-fold coordinated dopant states and defects. As a result, one can obtain considerably higher conductivities and carrier densities in μc-Si:H, as compared with a-Si:H (see table in Fig. 3.13).

### 3.2.4 Schematic picture for the structure of μc-Si:H

Before the above information is collected in a schematic picture of the μc-Si structure, one can first draw some plausible conclusions from the size, or rather from the "smallness" of the structural features, which will be helpful in obtaining a "working-model" picture for the structure of μc-Si:H. Figure 3.14 shows a schematic view of a crystalline column of size $50 \times 50 \times 200$ cm$^3$, consisting of an agglomerate of perfect grains of 20 nm diameter.

From the atom density of silicon, one calculates that there will be roughly $2 \times 10^5$ atoms in such a grain. When one defect is put on the grain boundary this corresponds to a defect density of $2 \times 10^{17}$ cm$^{-3}$. This means, when using the measured spin density of about $10^{16}$ cm$^{-3}$ (Fig. 3.8), one would have only one defect

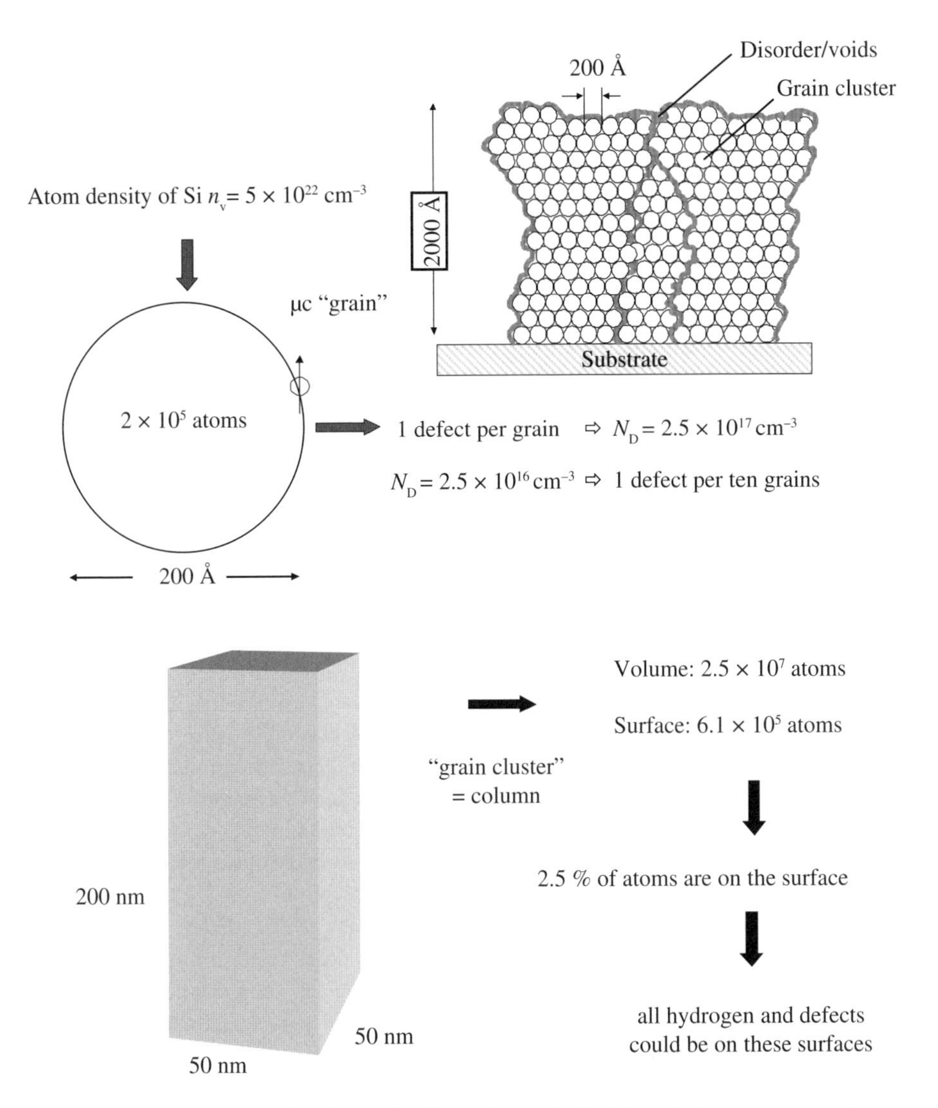


**Fig. 3.14** Schematic view of grains and columns in μc-Si:H with typical dimensions and simple calculations for "grain boundary" defect densities and surface state densitites.

per ten grains. For such small grains, many "grain boundaries" will not contain any defects at all! This idea agrees with the fact that fourfold coordination is preserved between the small size crystallites inside the columns in the case of twin boundaries, which are frequently identified in the TEM images of μc-Si:H. The grain boundaries between the small grains inside the columns can therefore not be seen as "classical grain boundaries" like those that are found in polycrystalline silicon with grain sizes of a few millimeters and more. Instead, one can assume that the column boundaries in μc-Si:H take on the function of the "classical" grain boundaries of polycrystalline silicon with defects and potential barriers; as a consequence, these boundaries have a strong influence on electronic transport and charge carrier trapping and recombination.

For a column with a typical size of $50 \times 50 \times 200$ nm$^3$ one calculates a surface-to-bulk ratio of atoms of 2.5 %. This means that the majority of the measured defects could easily be allocated to the column boundaries. Also the hydrogen could be located there, at least in the case of highly crystalline material where hydrogen contents of 2-4 % are measured. This agrees with the assumption that the column bulk, which is a highly crystalline Si structure, should contain very little hydrogen, similar to the situation in pure crystalline silicon. In the case of material with lower crystallinity, part of the hydrogen is certainly located within the amorphous phase. It has to be stressed that the above considerations should only be used with much care. It will be a challenge for future investigations to investigate chemically the grain and column boundaries.

In Figure 3.15(a) and (b) information about the structure in μc-Si:H is revealed in schematic pictures [Vetterl-2000]. Alternative "artist's impressions" of the structure of μc-Si:H have been published by other research groups [Vallat-Sauvain-2000, Collins-2003]. One should be aware that such schematic representations are necessarily a simplification of the "true" μc-Si:H structure. They are used to condense some of the knowledge one has at present about μc-Si:H and are helpful as a working tool. They should be used with appropriate caution only.

The schematic in Figure 3.15(a) shows a cross-section of a μc-Si:H film deposited on a foreign substrate, such as glass. The film ranges from highly crystalline (left) to amorphous (right). Large crystalline columns containing agglomerates of smaller perfectly crystalline regions are seen in highly crystalline material. Inside the columns twin boundaries dominate. From dimensional considerations and spin density measurements, one can speculate that only very few defects are to be found on these internal grain boundaries. The internal grain boundaries possibly contain tail states. Overall these internal boundaries do not represent a "classical" grain boundary, like those found in polycrystalline silicon. This should have consequences for electronic transport within the columns. The columns are separated by crack-like voids. With decreasing crystalline volume fraction, cracks and column boundaries become covered with a thin skin of amorphous material, thereby efficiently passivating defects in such a "medium-crystallinity" material. In fact, optimum solar cells are found by using intrinsic layers from this intermediate region. Defects could be preferably located on the column boundaries. The same is true for hydrogen which has too low a solubility in the crystalline network to account for the few atomic percent found in μc-Si:H.

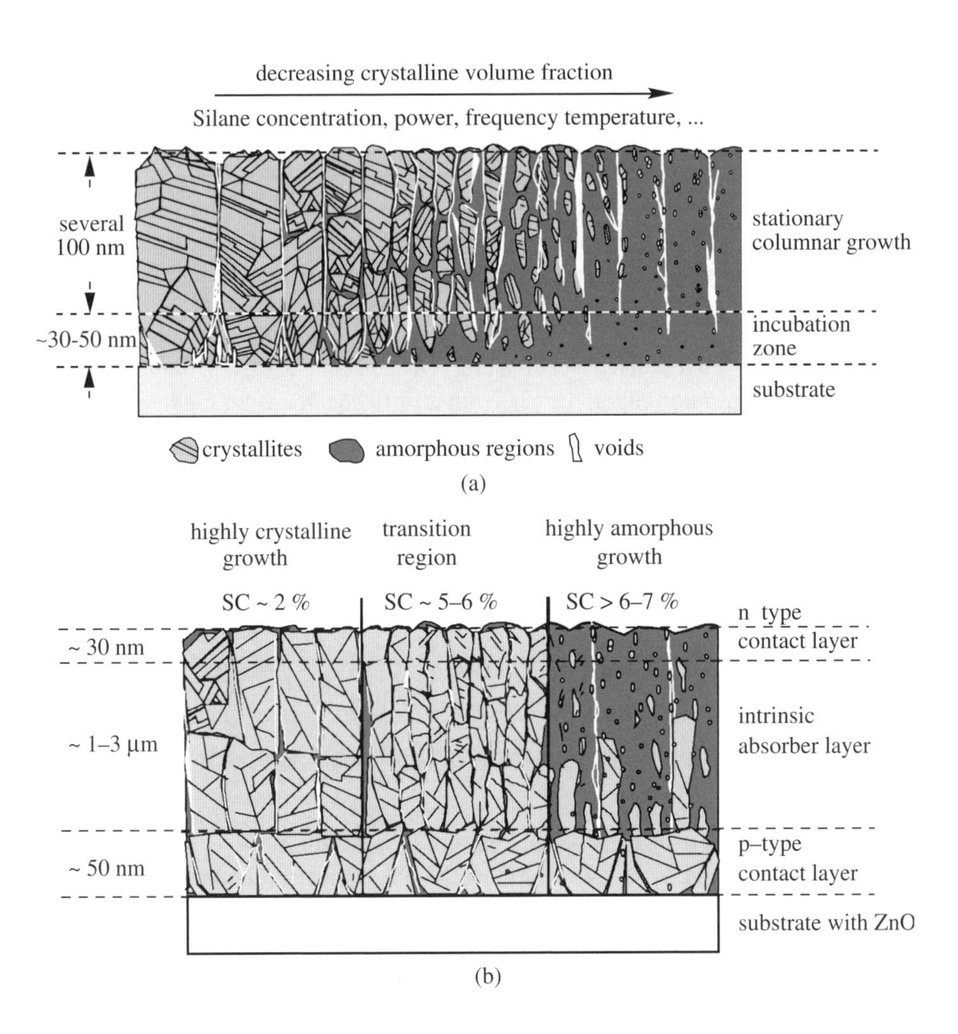


**Fig. 3.15** (a) Schematic view of the structure of μc-Si:H ranging from highly crystalline on the left to amorphous on the right in a cross section view on a foreign substrate like e.g. glass. For details see text (By courtesy of Forschungszentrum Jülich); (b) Schematic cross section view of the structure of μc-Si:H solar cells with the i-layer grown at different silane concentrations on a highly crystalline p-type contact layer. The important local epitaxial growth of the i-layer on the p-layer, thereby avoiding the amorphous incubation layer is indicated. For details see text (By courtesy of C. Scholten et al., Forschungszentrum Jülich).

An amorphous incubation zone, which strongly depends on the substrate and on the growth conditions, may develop. Such an incubation zone will have to be considered for device applications; it generally has an unfavorable effect on solar cells, especially if it is too thick and/or not perforated by microcrystalline columns. This situation is shown in Figure 3.15(b) for μc-Si:H solar cells in the *p-i-n* configuration. The microcrystalline *i*-layer (intrinsic layer) is the absorption layer; it grows on a microcrystalline *p*-layer. The *p*-layer has to serve as a transparent window, it has to be highly conductive, it has to form the *p*/*i*-barrier (for diode formation) and it also has to promote nucleation of the *i*-layer. This makes optimization of the *p*-layer a challenging task in solar cell design [Fujibayashi-2006].

In general, one might have to adjust the process parameters as the film continues to grow, in order to obtain high quality, uniform layers, both for a-Si:H and for μc-Si:H [Collins-2003]. For a-Si:H prepared not from pure silane but with a certain hydrogen dilution $R$ corresponding to growth conditions close to the amorphous-to-microcrystalline transition, one has to prevent microcrystalline growth. Microcrystalline growth would occur after a certain time, i.e. after a certain part of the film has been deposited, if the hydrogen dilution is kept constant. This means one will have to reduce the hydrogen dilution $R$ as the deposition proceeds. On the other hand, for μc-Si:H one has to start the deposition "abruptly" and with growth conditions for highly crystalline growth (in order to avoid the amorphous incubation layer) and then, as the deposition continues, one will again have to reduce the hydrogen dilution $R$ to prevent the formation of too high a crystalline volume fraction with poor defect passivation.

In-situ control with ellipsometry, optical emission spectroscopy or infrared spectroscopy of the exhaust gases has already been successfully applied, in order to adjust accordingly the process parameters [Collins-2003, van den Donker-2007b, Strahm-2007b, Fujiwara-2001]. Similar tools could also be required for process control in production systems.

### 3.2.5 Relationships between structural and other properties of μc-Si:H material

Keeping the schematic view of the structure of μc-Si:H for different crystalline volume fractions in mind, let us now see how the other physical properties of μc-Si:H vary when one modifies the structure by changing the deposition process. Different behavior can be imagined (Fig. 3.16):

(1) a rather abrupt change at a critical deposition condition when the properties change from typically "microcrystalline" to typically "amorphous";

(2) a gradual variation of the properties in the microcrystalline phase as well as in the amorphous phase, with a distinct transition between them; this would mean there is a whole class of microcrystalline materials and also a whole class of amorphous materials;

(3) the properties could simply vary linearly with the crystalline volume fraction, as a superposition of the microcrystalline and the amorphous properties.

To some extent all three possibilities are found. Most experiments will average over the different phases resulting in a macroscopic quantity which may not be representative of the microscopic properties in the individual phases.

In any case it cannot be emphasized enough that when one studies the properties of μc-Si:H it is vital to have information on the exact structural composition of the layers under investigation in order to be able to carry out significant

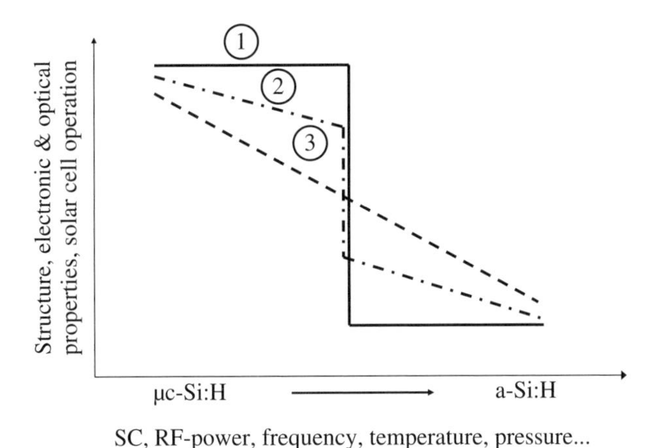


**Fig. 3.16** Illustration of three principal variations of material and solar cell parameters upon variation of the deposition process conditions which lead to a transition from microcrystalline to fully amorphous.

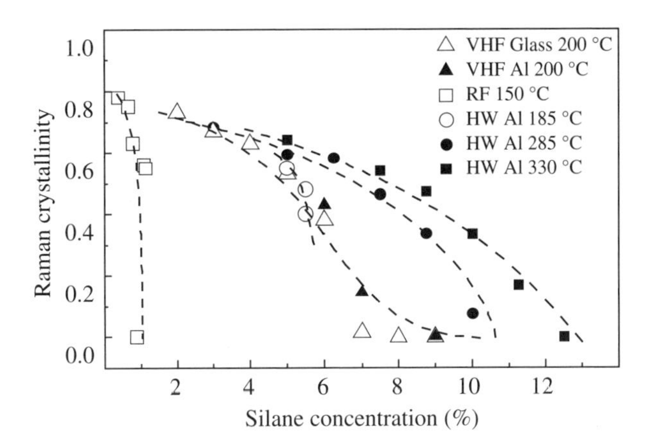


**Fig. 3.17** Raman crystallinity as a function of the silane concentration for different deposition processes (RF-PECVD, VHF-PECVD, HW-CVD), on different substrates (glass and aluminium) and at different substrate temperatures (by courtesy of L. Baia Neto et al., Forschungszentrum Jülich).

evaluation of the data. A characterization that is done only in terms of the variation of a deposition parameter which promotes the structural transition, such as silane concentration *SC*, will, in general, not be sufficient for comparison between different materials.

To demonstrate this with an example, Figure 3.17 shows the Raman intensity ratio versus the silane concentration *SC* for material prepared at different substrate temperatures, with two different processes (HWCVD and PECVD) and, for PECVD, with different plasma excitation frequencies and pressure regimes. One can see that the transition shifts and that the slope at the transition also changes. Optimization or, in general, deposition of μc-Si:H material with a desired structural composition or with specific material properties has to be done in a multi-parameter space. The requirements for μc-Si:H deposition needed to obtain a defined structural composition mean that there is a severe additional constraint compared to a-Si:H. This is a real challenge for developing suitable deposition processes.

## 3.3 OPTICAL PROPERTIES

The optical absorption coefficient versus photon energy is an important quantity for the application of μc-Si:H in solar cells. Optical absorption is commonly measured by combinations of classical transmission and reflection experiments, or by indirect methods such as photocurrent spectroscopy or photothermal deflection spectroscopy, which for certain applications provide a much wider dynamic range for an individual thin film than transmission experiments can (see also the Sect. 3.2.2 for measurement of defects with the related references). Requirements and considerations for the application of the various methods are similar to the ones described for a-Si:H (Chap. 2).

In Figure 3.18(a) the optical absorption spectra of μc-Si:H with different crystalline volume fractions are shown from highly crystalline ($x_C^{RS} = 86\%$) to fully amorphous. The spectra were measured with photothermal deflection spectroscopy (PDS). For $x_C^{RS} = 0.86$ the absorption is very similar to that of crystalline silicon. In particular at photon energies below 1.75 eV, the absorption curve of μc-Si:H resembles that of crystalline silicon, and this is true down to fairly low crystalline volume fractions. This is not surprising if one considers the magnitude of the absorption coefficients in this energy range. Here the absorption contribution from the amorphous phase is very low (1-2 $cm^{-1}$). Even small crystalline volume fractions will show up in the optical absorption in this range of photon energies, and only when the material is fully amorphous will the crystalline contribution disappear.

On the other hand, at photon energies above about 1.8 eV, the absorption is dominated by the amorphous phase. In (Figure 3.18b) the absorption coefficient at 1.4 eV (representative of the crystalline phase) and at 2.1 eV (representative of the amorphous phase) is plotted versus the Raman crystallinity. The optical absorption behaves like curve (3) in Figure 3.16. One finds a good linear relationship between the absorption coefficients at high and low energies and the crystalline volume fraction (Fig. 3.18b).

If one accordingly considers the optical absorption in μc-Si:H as a linear superposition of the absorption in the crystalline and in the amorphous phase of the material, one can then also assume, to a first approximation, that the optical bandgaps within the crystalline and the amorphous phases do not change. But it is difficult to define an effective, average bandgap. Because of the large difference in the absorption coefficients for $h\nu < 1.5$ eV, there is a rather abrupt change in the onset of absorption, and μc-Si:H layers with low crystalline volume fractions will already appear rather like c-Si with respect to their optical absorption.

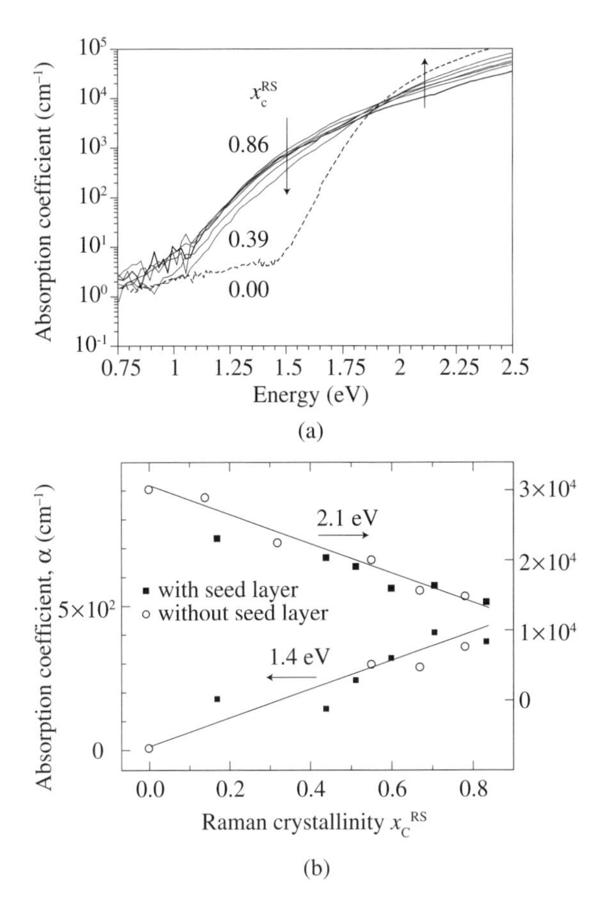


**Fig. 3.18** (a) Optical absorption spectra of μc-Si:H with different crystallinity, as indicated in the figure, and of a-Si:H measured with PDS; (b) Optical absorption coefficients at 1.4 eV and 2.1 eV respectively, determined from the spectra in (a) vs. the Raman crystallinity (by courtesy of C. Ross and R. Carius, Forschungszentrum Jülich).

The highest solar cell efficiencies are obtained with μc-Si:H absorber layers having an intermediate crystalline volume fraction (with "Raman crystallinity" around 50 %). In solar cells with such material one finds higher fill factors, indicative of a more efficient carrier extraction, and higher open circuit voltages $V_{oc}$ when compared to solar cells with highly crystalline material as absorber layers. An important reason for this behavior is the low defect densities obtained by an efficient passivation of grains and column boundaries through the amorphous phase (see Section 3.2). But a decrease of crystalline volume fraction leads to a decrease of the related absorption in the long wavelength range region (energies below 1.5 eV), which is related to the microcrystalline phase, and therefore leads to a reduced current density in the solar cell. The highest current densities are obtained with highly crystalline material; the highest voltages, however, with the highest possible amorphous volume fraction. An important optimization task for solar cell development with μc-Si:H absorber layers is therefore to find material with a medium crystallinity which will give a maximum product $J_{sc} \times V_{oc}$.

## 3.4 ELECTRONIC PROPERTIES AND TRANSPORT

The electronic conductivity in the dark and under illumination will also depend strongly on the structural composition. However, as with some of the other experimental techniques, it will be very difficult to extract microscopic quantities describing electronic transport, such as the mobility or the lifetime for photo-excited carriers, from conventional conductivity measurements, where the sample has to be treated as a "black box" with no or little understanding of the exact transport mechanisms or the detailed transport paths in the heterogeneous phase mixture material μc-Si:H.

One has to assume that the transport behavior in the different structural phases differs considerably and is, furthermore, affected by transport across the boundaries between the phases. A transport model such as that used for disordered but homogeneous amorphous silicon, or that used for polysilicon, where the grains sizes are much larger, will not be applicable to μc-Si:H. So one can expect to be able to measure only "average" or "effective" microscopic quantities for electronic transport in μc-Si:H.

Ignoring for the moment the microscopic details, the conductivity is, in general terms, given by

$$\sigma = q \mu n \tag{3.4}$$

where $n$ is the carrier density, $\mu$ the carrier mobility and $q$ the carrier charge. Adopting this description for our μc-Si:H and looking (as an approximation) only at the electrons, one can write

$$\sigma \approx \sigma_n = e \mu_n^0 n_f \tag{3.5}$$

where $n_f$ is the density of free electrons, $\mu_n^0$ the band mobility and $e$ the electron charge. An analogous equation can be written for holes in $p$-type material. For material with similar electronic conductivity contributions from electrons and holes, the sum of the two quantities would have to be used (see Chapter 2). Both quantities $n_f$ and $\mu_n^0$ certainly vary between the ordered crystalline and the disordered amorphous phase. Let us now look at the carrier density, first under the assumption that the mobility does not vary. As a first approximation one can assume that the carrier density in both phases of undoped μc-Si:H is determined by an activation process like the one found in a-Si:H, i.e.

$$n_f = N_\sigma \exp-(E_{act} / kT) \tag{3.6}$$

where $N_\sigma$ is the density of states at the transport path, $k$ is the Boltzmann constant, $T$ the absolute temperature and $E_{act}$ is an activation energy, which is basically the difference between the Fermi level energy $E_F$ and the electron energy at the transport path $E_\sigma$, i.e. $E_{act} = E_\sigma - E_F$. In crystalline (amorphous) silicon $E_\sigma$ would correspond to the conduction band (mobility) edge $E_c$, or to the valence bandedge $E_V$ for hole transport, and one can probe the position of the Fermi level from conductivity measurements via above equation. In addition, when assuming the Fermi level is positioned near the midgap for material with intrinsic transport behavior, one can even determine the size of the energy (mobility) gap as $E_g = E_c - E_V \approx 2E_{act}$. In practice this is frequently done by measuring of the conductivity whilst varying the Fermi level by doping, and by then assigning the material with the lowest conductivity to the midgap position of the Fermi level.

Let us see to what extent one can apply these simple considerations to our heterogeneous μc-Si:H material. In μc-Si:H one has to assume an additional barrier energy $E_b$, which accounts for possible potential barriers between the grain and column boundaries, i.e. potential barriers that the carriers have to overcome during transport. One then has $E_\sigma = E_c + E_b$ and $n_f = N_\sigma \exp - [(E_c + E_b - E_F)/kT]$. In general, one will only be able to determine the Fermi level position with respect to the bandedge when one has additional information about the barrier energy or when the latter is assumed to be negligible. However one can easily determine relative changes of the Fermi level position by for example doping. Overall one can expect a considerable increase in the carrier density when going from fully amorphous material with a bandgap of 1.7 eV to highly crystalline material with the bandgap of c-Si of 1.1 eV and a related decrease of the activation energy $E_{act}$.

Looking at the variation of the dark conductivity $\sigma_{dark}$ vs. $x_C^{RS}$ one finds that $\sigma_{dark}$ increases typically from $10^{-11}$ S/cm in a-Si:H to $10^{-6}$ S/cm in μc-Si:H; i.e. it increases by five orders of magnitude (Fig. 3.19a). Assuming that this increase is mainly determined by an increase in the density of free carriers (electrons, holes) due to the lower bandgap in the crystalline phase, the linear relationship between ln ($\sigma_{dark}$) and $x_C^{RS}$ seen in Figure 3.19(a) would correspond to a linear variation of the effective activation energy $E_{act}$; i.e. $E_{act}$ is proportional to $x_C^{RS}$. One can then estimate an effective mobility gap $E_G$ from the equation $\sigma = \sigma_0 \exp - (E_{act}/kT)$ using $\sigma_0 = 150$ S/cm [Stuke-1987, Overhof-1989] and assuming a Fermi level position close to the "midgap"; i.e. $E_g = 2 \times E_{act}$. For a-Si:H ($x_C^{RS} = 0$ %) with $\sigma_{dark} = 10^{-11}$ S/cm one obtains $E_{act} = 0.79$ eV and $E_g \approx 1.6$ eV; for highly crystalline μc-Si:H ($x_C^{RS} = 80$ %) with $\sigma_{dark} = 10^{-6}$ S/cm one obtains $E_{act} = 0.49$ eV and $E_g \approx 1.0$ eV. Both these values are close to the "real"

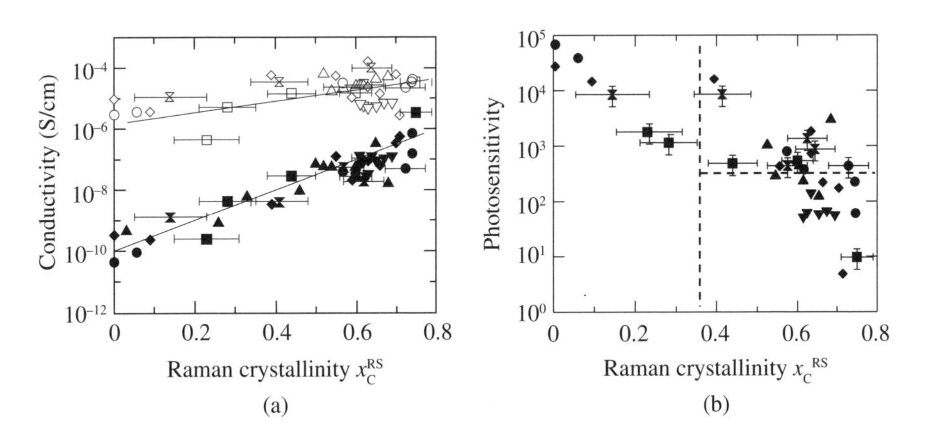


**Fig. 3.19** (a) Dark- and photoconductivity and (b) photosensitivity, the ratio of dark-to-photo conductivity, of μc-Si:H samples as a function of the Raman crystallinity (By courtesy of O. Vetterl et al., Forschungszentrum Jülich).

values determined from other experiments and from theory. So a linear variation of the effective mobility gap seems a fair approximation to describe the change in dark conductivity in μc-Si:H when going from amorphous to highly crystalline layers.

The considerations given above assume that the free carrier mobilities do not vary much as one goes from a-Si:H to highly crystalline μc-Si:H. This is to some extent the case, although exact data on the variation of the band mobilities $\mu_n^0$, $\mu_p^0$ for μc-Si:H layers as a function of the Raman crystallinity are not yet available.

One should in general be careful when using a single effective activation energy for the dark conductivity $\sigma_D$, both in a-Si:H and, especially, in μc-Si:H. Potential fluctuations, in particular in μc-Si:H where they can be provoked by charge in the grain and especially on the column boundaries, would result in a transport path above the mobility edge, such that $E_{act} > E_c - E_F$. In addition, one has to realize that over an extended temperature range the conductivity does not exhibit simple, singly-activated behavior. Plots of $\sigma_{dark}$ vs. $1/T$ often show a pronounced curvature, indicative of a distribution of activation energies [Carius-1999, Müller-1999]. This would correspond to a distribution of grain boundary barriers in μc-Si:H or to the transition or overlap to other transport channels, such as hopping in localized states which could dominate at low temperatures [Carius-1999, Müller-1999].

Also plotted in Figure 3.19(a) is the photoconductivity $\sigma_{photo}$. $\sigma_{photo}$ increases much less than the dark conductivity when going from a-Si:H to μc-Si:H. Therefore the ratio ($\sigma_{photo}$:$\sigma_{dark}$), called the "photosensitivity" decreases a lot from values of $10^6$ in a-Si:H to values close to 1 in highly crystalline μc-Si:H (Fig. 3.19b) [Vetterl-2002]. Photosensitivity is not a good optimization parameter when taken by itself. But it can be stated from empirical observations that intrinsic μc-Si:H layers with a photosensitivity of about 1000 for $x_C^{RS} \approx 50$ % are useful for solar cell applications. These are the samples in the upper right sector in Figure 3.19(b).

The photosensitivity might be much lower in μc-Si:H than in a-Si:H. Nevertheless we find a similar relationship between $\sigma_{dark}$ and $\sigma_{photo}$ in μc-Si:H as in a-Si:H [Beyer-1983, Brüggemann-1998, Astakhov-2009]; this relationship is related to the charge state of defects as a result of the Fermi level position. In Figure 3.20, $\sigma_{photo}$ is plotted vs. $\sigma_{dark}$ for μc-Si:H samples with different "Raman crystallinity", where the Fermi level position is varied by doping, and by defect generation through electron bombardment. With a Fermi level shift towards the conduction band, the dark conductivity increases as a result of the higher carrier density. States in the upper half of the gap become negatively charged. This results in an increase of the majority carrier lifetime (for electrons in this case) and an increase in $\sigma_{photo}$. Such a relationship is well known for a-Si:H. Finding similar dependencies for μc-Si:H underlines the dominant role which defects play, in recombination and trapping, also in μc-Si:H. A key optimization task for μc-Si:H is to increase the carrier lifetime by reducing the density of deep defects.

Indications for significantly different tail state properties in μc-Si:H (as compared to a-Si:H) can be concluded from measurements of the charge carrier drift mobility, as derived from time-of-flight measurements (TOF) [Dylla2005a&b, Reynolds-2006, Reynolds-2009]. For detailed information about the TOF technique, the reader is referred to the literature [Dylla-2005b, Reynolds-2006, Street-1991]. Figure 3.21 shows the hole and electron drift mobilities, measured by time-of-flight experiments, as a function of the "Raman crystallinity". The hole drift mobility increases considerably, by a factor of

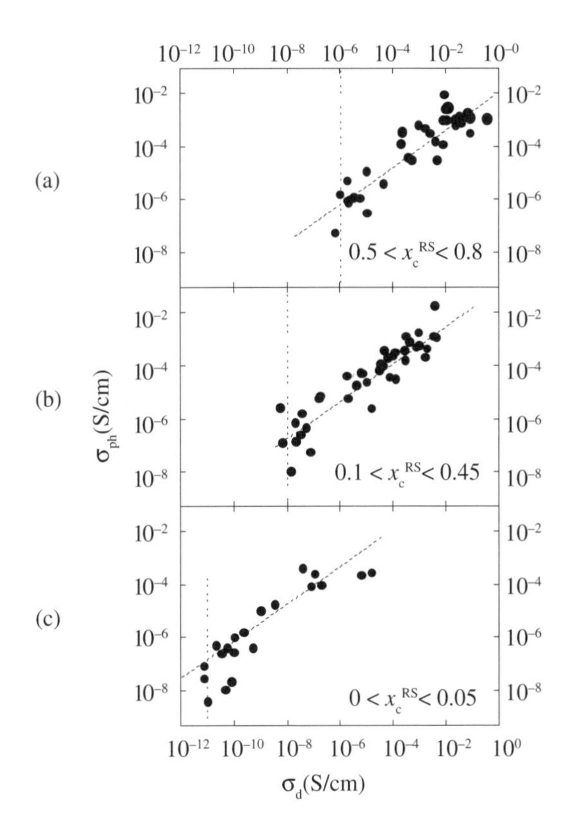


**Fig. 3.20** Photoconductivity vs. dark conductivity of μc-Si:H materials with (a) high, (b) medium and (c) low crystallinity. The Fermi level and therewith the conductivity is shifted by doping (with doping levels in the range of 2-150 ppm $PH_3$) and/or defect creation through electron bombardment. The vertical dashed line indicates the dark conductivity of intrinsic material with the corresponding crystallinity (By courtesy of O. Astakhov et al., Forschungszentrum Jülich).

300, between the amorphous and highly crystalline material. The electron drift mobility increases by a factor of 3 only. Assuming a multiple trapping process (i.e. a transport process where excess charge carriers commute frequently between extended states and traps in the respective tails), one would have to conclude that the valence bandtail in μc-Si:H is considerably steeper than in a-Si:H, whilst the conduction bandtail would have similar slopes in μc-Si:H and in a-Si:H. One comes to the conclusion that the transport properties of holes and electrons are quite similar in μc-Si:H, although they are very different from each other in a-Si:H. It has already been confirmed by experiment [Gross-2001, Huang-2007, Finger-2009] that a solar cell with a μc-Si:H absorber layer should also function with illumination through the *n*-side, where the holes have to travel a longer distance. In this respect the charge carrier properties are more symmetric in μc-Si:H, whilst on the other hand they are still strongly determined by interaction with defects, just as in a-Si:H.

When measuring and evaluating electronic conductivity data of μc-Si:H, it is very important to do these experiments under well-defined conditions, under vacuum and after annealing. In fact, μc-Si:H material shows a considerable influence on the

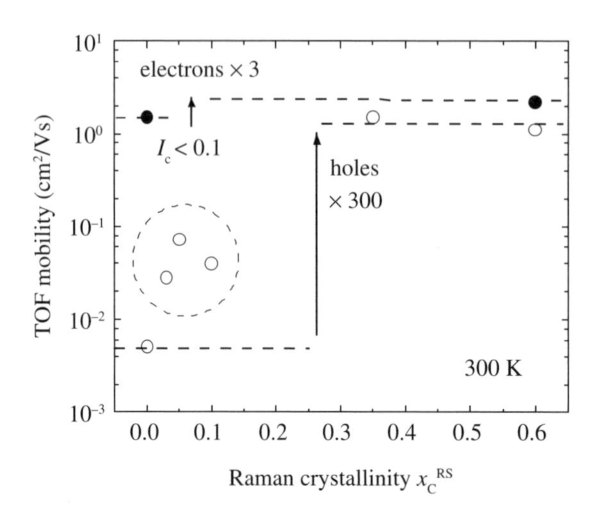


**Fig. 3.21** Drift mobility for electrons and holes in μc-Si:H determined from time-of-flight experiments as a function of the crystallinity (By courtesy of S. Reynolds, Forschungszentrum Jülich).

conductivity of the adsorption of atmospheric gases or even of oxidation. These phenomena are related to the porosity of μc-Si:H layers, and are therefore most pronounced in highly crystalline material. Direct evidence is seen in infrared absorption and in ESR measurements [Veprek-1983, Finger-2003, Smirnov-2006]. One can distinguish here between reversible and non-reversible processes [Smirnov-2006, Finger-2003, Dylla-2005b]. Figure 3.22 gives an example of a reversible process. The conductivity decreases by several orders of magnitude after storage of the material under ambient conditions (in air). This ambient "aging" can be annealed in vacuum or inert gas at temperatures below 200 °C. The aging/annealing cycle can be repeated several times. However, storage or treatment in atmospheric gases for very long times or at elevated temperatures (80 °C) leads to non-reversible effects. Care has therefore to be taken to measure $\sigma_{dark}$ under vacuum and after annealing. Long-time storage, especially for highly crystalline porous material, should be under vacuum or in an inert gas.

The strong scattering in the dark conductivity for a given crystallinity (Fig. 3.19) is an indication of possible variations in impurity doping - variations which are rather typical for thin-film silicon material prepared under a variety of deposition conditions from the gas phase. Furthermore, the effects in μc-Si:H are stronger overall than those observed in a-Si:H. This is a sign of the better doping efficiency in μc-Si:H – it also applies to "unintentional doping" by impurities (see Sect. 3.2.3).

To summarize electronic transport in μc-Si:H one has to confess that the understanding of the various effects is not very mature. The material lies somewhere between a-Si:H as a disordered but homogeneous material and polycrystalline silicon with much larger grains. A straightforward theoretical description does not exist. Simple approaches similar to those used in polycrystalline silicon fail. Also it seems inappropriate to assume a homogeneous disorder.

Nevertheless, it is evident that defects influence the charge carrier properties in a similar strong way as in a-Si:H. This is reflected in the relationships between defects,

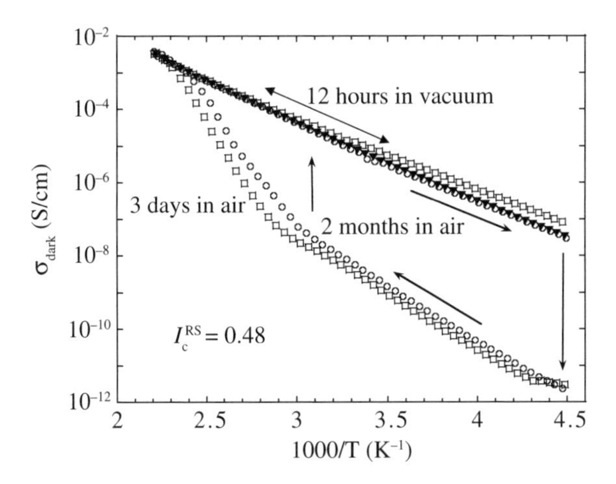


**Fig. 3.22** Dark conductivity vs. the reciprocal temperature for a μc-Si:H sample after storage in vacuum or in air for different times. For details see text (By courtesy of S. Okur et al., Forschungszentrum Jülich).

Fermi level position and photoconductivity. Seeing the conductivity as a compositional average of the electronic transport in the two phases of the material, and adopting the concept of an effective "average mobility gap", could be a useful approach. But one has still to find improved ways for considering the critical influence of the grain and column boundaries on transport, on carrier lifetimes and, finally, on device performance.

## 3.5 METASTABILITY – INSTABILITY

In the transport section we have already introduced metastability and instability phenomena in μc-Si:H. When describing instability in μc-Si:H, it is useful to distinguish between two types of μc-Si:H: a porous type and a compact type. These two types are in many, but not necessarily all, aspects equivalent to material with high "crystallinity" and to material with medium to low "crystallinity", respectively. One finds two main effects:

(1) adsorption and oxidation at (inner) surfaces, related to the porosity of the material;

(2) light-induced degradation (Staebler-Wronski Effect) of the amorphous phase in mixed phase μc-Si:H.

With respect to point (1) above, the effects of adsorption and oxidation on electronic conductivity, infrared spectra, ESR results and solar cells have been investigated and discussed [Veprek-1983, Finger-2003, Smirnov-2006, Sendova-2006, Li-2009]. One observes considerable changes in the dark conductivity (see Fig. 3.19) most probably as a result of surface band bending caused by oxidation/adsorption [Kanschat-1999]. These effects should be considered very carefully when carrying out studies on electronic transport. In devices where μc-Si:H is used as a bottom cell absorber sandwiched between other layers, these effects could be less pronounced [Sendova-2006, Wang-2008]. But one should keep an eye on them and store samples, especially those with high crystallinity, in a safe environment.

With respect to point (2), light-induced degradation depends, not surprisingly, on the amorphous content of the material [Finger-2003, Wang-2008, Meillaud-2005, Vallat-Sauvain-2006a, Yue-2007]. In Figure 3.23 one sees the results of degradation studies of μc-Si:H solar cells with different Raman intensity ratios of the absorber material (i.e. of the intrinsic layer). Highly crystalline material is stable under illumination. Solar cells incorporating *i*-layers with a substantial amorphous phase (i.e. cells which generally have the highest efficiencies because of their high *FF* and their high $V_{OC}$) degrade slightly and, in fact, the degradation depends on the amorphous volume fraction. Overall the degradation is lower as compared with a-Si:H. For a high-efficiency single-junction cell the relative degradation is only about 5 % [Finger-2003, Wang-2008]. In addition, when the μc-Si:H cell is used as a bottom cell in a "micromorph" tandem cell (see Chap. 5), the light intensity and the excitation energy the μc-Si:H *i*-layer "sees" would be much lower than in a single-junction μc-Si:H cell, because the high-energy photons are absorbed by the top cell. One can simulate this situation by using a red filter for the white light excitation. Indeed the cells are stable after red light illumination, even after 1000 h (Fig. 3.23).

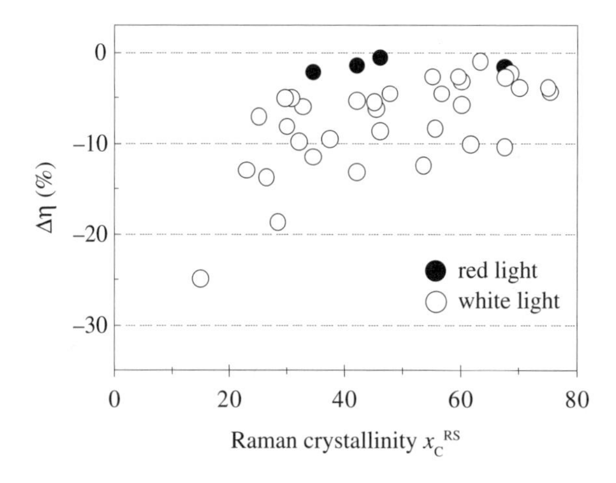


**Fig. 3.23** Relative change in efficiency of solar cells in pin configuration with μc-Si:H absorber layer of different crystallinity upon 1000 h illumination: (open circles) with white light equivalent to AM1.5 and (full circles) red light equivalent to the situation of the μc-Si:H in the bottom cell of a tandem device (By courtesy of Y. Wang et al., Forschungszentrum Jülich).

## 3.6 ALLOYS

As for amorphous silicon, microcrystalline silicon material has the potential of being alloyed with several elements, and alloys of μc-Si:H with Ge, O and C have already been developed and explored. Similar to a-Si:H the alloying is obtained through appropriate admixture of corresponding gases in the deposition process. The main effect of alloying is that the optical gap will be shifted to lower energies (alloying with Ge) or to higher energies (alloying with O, C). The resulting alloy materials have been explored for applications such as infrared absorbers (Ge) [Ganguly-1996, Carius-2001, Matsui-2009], as window layers (C and O) [Konagai-2008, Huang-2007, Chen-2009, Finger-2009], or as intermediate reflectors in stacked cells (O) [Buehlmann-2007, Lambertz-2007, Das-2008].

On the wide bandgap side we have μc-SiC:H, which is the true stochiometric alloy and not the μc-Si:H crystallites in a-SiC:H matrix. Such an alloy would have high potential as a window layer for solar cells. The material is more difficult to prepare. One approach is to use HWCVD with monomethylsilane, a gas which already contains the Si-C bond in a 50-50 mixture. This gives material with excellent crystallinity and a high optical bandgap. In the as-deposited state, this material is very highly conductive *n*-type, which could be due to impurity doping or highly conductive structure phases. Applying such material in *n*-side illuminated solar cells with μc-Si:H absorber layers yields high current values and quantum efficiency due to the good transparency [Huang-2007, Finger-2009]. Figure 3.24 shows some results obtained with this material. Meanwhile, also successful *p*-type doping of μc-SiC:H has been performed [Miyajima-2006, Chen-2010].

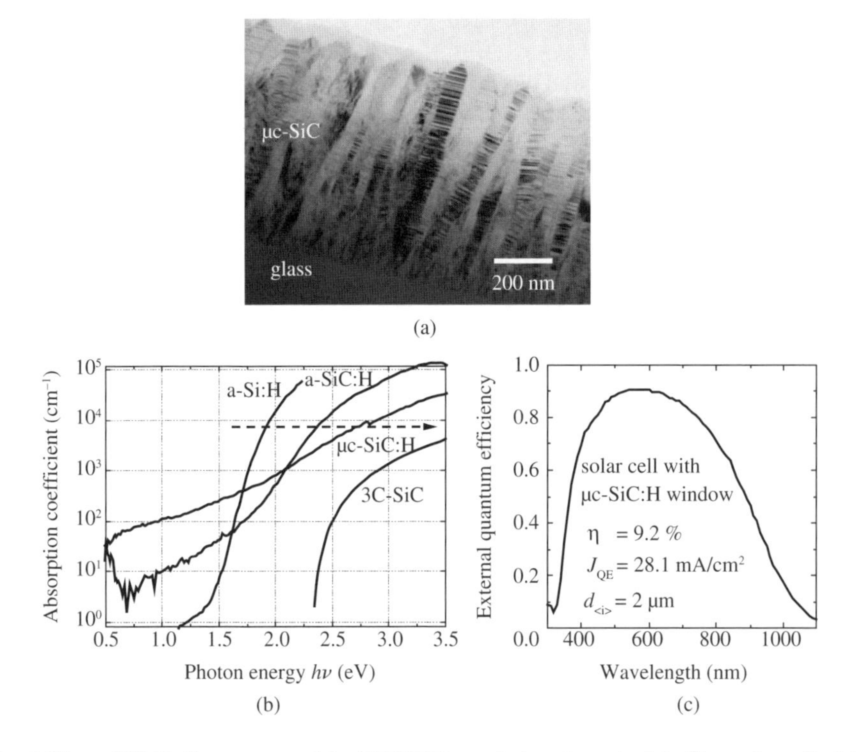


**Fig. 3.24** μc-SiC:H alloys prepared by HWCVD as window layer in thin film solar cells (a) structure in a TEM cross section image, (b) optical absorption coeeficient in comparison with a-Si:H, a-SiC:H and c-SiC, and (c) quantum efficiency of a solar cell with μc-SiC:H window layer and μc-Si:H absorber layer (By courtesy of R. Carius et al., Forschungszentrum Jülich).

For microcrystalline SiO and SiGe, very promising results have already been published and one expects in future to see much progress in the application of these alloys. The reader is referred to the literature for recent results [Buehlmann-2007, Das-2008, Matsui-2009].

## 3.7 SUMMARY

This section on the properties of μc-Si:H concludes with some important statements on the material and with guiding principles for material optimization:

- microcrystalline silicon has a complex structure of crystalline, disordered and void regions and this structure will depend on layer thickness and on the substrate;
- the structure can be modified by preparation conditions and may also depend on the type of substrate and the sample thickness;
- the structure determines the electronic and optical properties of the material;
- the properties of μc-Si:H are a "mixture" between a-Si:H and c-Si properties;
- as in a-Si:H, hydrogen is essential for defect passivation;
- the material has to be prepared at low temperatures to prevent hydrogen out-diffusion;
- material with very good electronic properties is available and applications are already manifold;
- the material shows metastable effects, which depend on the structural composition;
- as in a-Si:H, the material can be alloyed to yield higher or lower bandgap materials.

To conclude, we provide a guideline for material optimization:

(1) Make silane concentration (SC) series to cover the Raman crystalinity range $x_c^{RS}$ from 0 to 0.8 on glass and c-Si. Preferably use temperatures around 200 °C.

(2) Measure Raman or XRD to obtain the "crystalline content" (e.g. the Raman crystallinity $x_c^{RS}$). Aim for an "optimum phase mixture " at around $x_c^{RS} = 0.6$. [Mai-2005b]

(3) Measure dark conductivity (after annealing in vacuum for 30 min; 20° below deposition temperature) and photoconductivity (AM1.5 or 60 mW/cm$^2$ halogen lamp). $\sigma_{photo}/\sigma_{dark}$ should be 300-1000 for $x_c^{RS} = 0.6$ [Vetterl-2002].

(4) Measure infrared absorption. There should be no Si-O modes and no or little 2100 mode.

(5) Measure PDS or some other optical absorption (below bandgap) method. $\alpha_{0.8\,eV}$ should be around 1-2 cm$^{-1}$, or lower [Klein-2007, Günes-2005, Vanecek-2002].

(6) Alternatively measure ESR: Spin density should be close to or below $10^{16}$ cm$^{-3}$ for $x_c^{RS} = 0.6$ [Finger-2004].

(7) Measure oxygen content (SIMS etc). The oxygen content should be below $1 \times 10^{19}$ cm$^{-3}$; to obtain this, a gas purifier has often to be used [Torres-1996].

# REFERENCES

[Astakhov-2009] Astakhov, O., Carius, R., Finger, F., Petrusenko, Y., Borysenko, V., Barankov, D., "Relationship between defect density and charge carrier transport in amorphous and microcrystalline silicon" (2009), *Phys. Rev. B*, Vol. **79**, p. 104205.

[Bailat-2003] Bailat, J., Vallat-Sauvain, E., Feitknecht, L., Droz, C., Shah, A., "Microstructure and open circuit voltage of n-i-p microcrystalline silicon solar cells" (2003), *J. Appl. Phys.* Vol. **93**, p. 5727-5732.

[Beyer-1983] Beyer, W., Hoheisel, B., "Photoconductivity and dark conductivity of hydrogenated amorphous silicon" (1983), *Solid State Commun.* Vol. **47**, p. 573.

[Beyer-1999] Beyer, W., "Hydrogen Phenomena in Hydrogenated amorphous silicon" (1999), *Semiconductors and Semimetals* Vol. **61**, p. 165-239.

[Brüggemann-1998] Brüggemann, R., Main, C., "Fermi-level effect on steady-state and transient photoconductivity in microcrystalline silicon" (1998), *Phys. Rev. B* Vol. **57**, p. R10580.

[Branz-2008] Branz, H. M., Teplin, C. W., Young, D. L., Page, M. R., Iwaniczko, E., Roybal, L., Bauer, R., Mahan, A. H., Xu, Y., Stradins, P., Wang, T., Wang, Q., "Recent advances in hot-wire CVD R&D at NREL: From 18% silicon heterojunction cells to silicon epitaxy at glass-compatible temperatures" (2008), *Thin Solid Films* Vol. **516**, p. 743.

[Buehlmann-2007] Buehlmann, P., Bailat, J., Domine, D., Billet, A., Meillaud, F., Feltrin, A., Ballif, C., "In-situ silicon oxide based intermediate reflector for thin-film silicon micromorph solar cells" (2007), *Appl. Phy. Lett.* Vol. **91**, p. 14305.

[Bustarret-1988] Bustarret, E., Hachicha, M.A., Brunel, M., "Experimental determination of the nanocrystalline volume fraction in silicon from Raman spectroscopy" (1988), *Appl. Phys. Lett.* Vol. **52**, p. 1675.

[Carius-1999] Carius, R., Finger, F., Backhausen, U., Luysberg, M., Hapke, P., Otte, M., Overhof, H., "Electronic properties of microcrystalline silicon" (1997), *Materials Research Society Symposia Proceedings* Vol. **467**, p. 283.

[Carius-2001] Carius, R., Krause, M., Finger, F., Voigt, F., and Stiebig, H. "Structural and electronic properties of microcrystalline silicon-germanium alloys" (2001), in *Materials for Information Technology in the New Millenium*, eds. J. M. Marshall, A. G. Petrov, A. Vavrek, D. Nesheva, D. Dimova-Malinovska, and J. M. Maud, Bath (U.K), p. 18.

[Cardona-1983] Cardona, M., "Vibrational Spectra of Hydrogen in Silicon and Germanium", (1983), *Phys. Stat. Sol. (b)* Vol. **118**, p. 463.

[Chen-2009] Chen, T., Huang, Y., Yang, D., Carius, R., and Finger, F., "Microcrystalline silicon-carbon alloys as anti-reflection window layers in high efficiency thin film silicon solar cells" (2008), *Physica Status Solidi RRL*, Vol. **2**, 160.

[Chen-2010] Chen, T., Schmalen, A., Wolff, J., Yang, D., Carius, R., Finger, F., "Aluminum Doped Silicon-Carbon Alloys Prepared by Hot Wire Chemical Vapor Deposition", (2010) *Phys. Stat. Sol. C* (in press).

[Cifre-1994] Cifre, J., Bertomeu, J., Puigdollers, J., Polo, M., Andreu, J., and Lloret, A., "Polycrystalline silicon films obtained by hot-wire chemical vapour deposition" (1994), *Appl. Phys. A: Solids Surf. A* Vol. **59**, p. 645.

[Collins-1989] Collins, R. W., and Yang, B. Y., "In situ ellipsometry of thin film deposition: Implications for amorphous and microcrystalline Si growth" (1989), *J. Vac. Sci. Technol. B* Vol. **7**, p. 1155.

[Collins-2003] Collins, R. W., Ferlauto, A. S., Ferreira, G. M., Chen, C., Koh, J., Koval, R.J., Lee, Y., Pearce, J.M., and Wronski, C.R., "Evolution of microstructure and phase in amorphous, protocrystalline, and microcrystalline silicon studied by real time spectroscopic ellipsometry" (2003), *Sol. Energ. Mat. Sol. C.* Vol. **78**, pp. 143-180.

[Curtins-1987] Curtins, H., Wyrsch, N., and Shah, A.V., "High rate deposition of amorphous hydrogenated silicon: effect of plasma excitation frequency" (1987), *Electron. Lett.* Vol. **23**, p. 228.

[Das-2008] Das, C., Lambertz, A., Huepkes, J., Reetz, W., Finger, F. "A constructive combination of antireflection and intermediate-reflector layers for a-Si/μc-Si thin film solar cells" (2008), *Applied Physics Letters*, Vol. **92**, 053509.

[Droz-2004] Droz, C., Vallat-Sauvain, E., Bailat, J., Feitknecht, L., Meier, J., Shah, A., "Relationship between Raman crystallinity and open-circuit voltage in microcrystalline silicon solar cells" (2004), *Solar Energy Materials & Solar Cells* Vol. **81**, pp. 61–71.

[Dylla-2005a] Dylla, T., Finger, F., Schiff, E. A., "Hole drift-mobility measurements in microcrystalline silicon" (2005) *Appl. Phys. Lett.* Vol. **87**, 032103.

[Dylla-2005b] Dylla, T., "Electron Spin Resonance and Transient Photocurrent Measurements on Microcrystalline Silicon" (2005) in *Schriften des Forschungszentrums Jülich, Energietechnik*, Vol. 43 (Forschungszentrum Jülich GmbH, Jülich.

[Faraji-1992] Faraji, M., Gokhale, S., Choudari, S. M., Takwale, M. G., Ghaisas, S.V., "High mobility hydrogenated and oxygenated silicon as a photo-sensitive material in photovoltaic application" (1992), *Appl. Phys. Lett.* Vol. **60**, p. 3289.

[Finger-1991] Finger, F., Prasad, K., Dubail, S., Shah, A., Tang, X.-M., Weber, J., Beyer, W., "Influences of doping on the structural properties of microcrystalline silicon prepared with the VHF-GD technique at low deposition temperatures" (1991), *Mat. Res. Soc. Symp. Proc.* Vol. **219**, p. 469.

[Finger-1994] Finger, F., Hapke, P., Luysberg, M., Carius, R., Wagner, H., Scheib. M., "Improvement of grain size and deposition rate of microcrystalline silicon by use of very high frequency glow discharge" (1994), *Appl. Phys. Lett.* Vol. **65**, p. 2588

[Finger-2000] Finger, F., Müller, J., Malten, C., Carius, R., Wagner, H., "Electronic properties of microcrystalline silicon investigated by electron spin resonance and transport measurements" (2000), *J. Non-Cryst. Solids* Vol. **266-269**, p. 511.

[Finger-2003] Finger, F., Carius, R., Dylla, T., Klein, S., Okur, S., and Günes, M., "Stability of microcrystalline silicon for thin film solar cell applications" (2003), *IEE Proc.-Circuits Devices Syst.* Vol. **150**, p. 300.

[Finger-2004] Finger, F., Baia Neto, L., Carius, R., Dylla, T., Klein, S., "Paramagnetic defects in undoped microcrystallin silicon" (2004) *Phys. Stat. Sol. (c)* Vol. **1**, pp. 1248-1254.

[Finger-2009] Finger, F., Astakhov, O., Bronger, T., Carius, R., Chen, T., Dasgupta, A., Gordijn, A., Houben, L., Huang, Y., Klein, S., Luysberg, M., Wang, H., and Xiao, L. "Microcrystalline silicon carbide alloys prepared with HWCVD as highly transparent and conductive window layers for thin film solar cells" (2009), *Thin Solid Films*, Vol. **517**, p. 3507.

[Flückiger-1992] Flückiger, R., Meier, J., Keppner, H., Kroll, U., Greim, O., Morris, M., Pohl, J., Hapke, P., and Carius, R., "Microcrystalline silicon prepared with the very high frequency glow discharge technique for p-i-n solar cell applications" (1992), in *Proc. 11th EC PV Solar Energy Conference*, p. 617.

[Fujibayashi-2006] Fujibayashi, T., Kondo, M., "Roles of microcrystalline silicon p layer as seed, window, and doping layers for microcrystalline silicon p-i-n solar cells" (2006), *J. Appl. Phys.* Vol. **99**, 043703.

[Fujiwara-2001] Fujiwara, H., Kondo, M., Matsuda, A., "Real-time spectroscopic ellipsometry studies of the nucleation and grain growth processes in microcristalline silicon thin films" (2001), *Phys. Rev. B* Vol. **63**, 115306.

[Fukawa-2001] Fukawa, M., Suzuki, S., Guo, L., Kondo, M., Matsuda, A., "High rate growth of microcrystalline silicon using a high-pressure depletion method with VHF plasma" (2001), *Sol. Energ. Mat. Sol. C.* Vol. **66**, pp. 217-223.

[Fritzsche-1980] Fritzsche, H., "Effect of compensation and correlation on near the metal non-metal transition" (1980), *Phil. Mag. B* Vol. **42**, pp. 835-844.

[Ganguly-1993] Ganguly, G., Matsuda, A., "Defect formation during growth of hydrogenated amorphous silicon" (1993), *Phys. Rev. B* Vol. **47**, p. 3661.

[Ganguly-1996] Ganguly, G., Ikeda, T., Nishimiya, T., Saitoh, K., Kondo, M., Matsuda, A., "Hydrogenated microcrystalline silicon germanium: A bottom cell material for amorphous silicon-based tandem solar cells" (1996) *Appl. Phys. Lett.* Vol. **69**, p. 4224.

[Gordijn-2006] Gordijn, A., Rath, J. K., Schropp, R. E. I., "High-efficiency μc-Si solar cells made by very high-frequency plasma-enhanced chemical vapor deposition" (2006) *Progress in Photovoltaics* Vol. **14**, pp. 305-311.

[Graf-2003] Graf, U., Meier, J., Kroll, U., Bailat, J., Droz, C., Vallat-Sauvain, E., Shah, A., "High rate growth of microcrystalline silicon by VHF-GD at high pressure" (2003), *Thin Solid Films* Vol. **427**, p. 37.

[Gross-2001] Gross, A., Vetterl, O., Lambertz, A., Finger, F., Wagner, H., and Dasgupta, A., "N-Side illuminated Microcrystalline Silicon Solar Cells" (2001), *Appl. Phys. Lett.* Vol. **79**, p. 2841.

[Guo-1998] Guo, L., Kondo, M., Fukawa, M., Saitoh, K., Matsuda, A., "High Rate Deposition of Microcrystalline Silicon Using Conventional Plasma Enhanced Chemical Vapour Deposition" (1998), *Jpn. J. Appl. Phys.* Vol. **37**, p. L1116.

[Günes-2005] Günes, M., Göktas, O., Okur, S., Isik, N., Carius, R., Klomfass, J., and Finger, F., "Sub-bandgap absorption spectroscopy and minority carrier transport properties of hydrogenated microcrystalline silicon thin films" (2005), *J. Optoelectronics and Advanced Materials* Vol. **7**, p. 161.

[Hollingsworth-1994] Hoolingswirth, R. E., Bhat, P.K., "Doped microcrystalline silicon grown by high frequency plasmas" (1994), *Appl. Phys. Lett.* Vol. **64**, p. 616.

[Houben-1998] Houben, L., Luysberg, M., Hapke, P., Carius, R., Finger, F., Wagner, H., "Structural properties of microcrystalline silicon in the transition from highly crystalline to amorphous growth" (1998) *Philos. Mag. A* Vol. **77**, p. 1447.

[Huang-2007] Huang, Y., Dasgupta, A., Gordijn, A., Finger, F., and Carius, R., "Highly transparent microcrystalline silicon carbide grown with hot wire chemical vapor deposition as window layers in n-i-p microcrystalline silicon solar cells" (2007), *Applied Physics Letters*, Vol. **90**, 203502.

[Jackson-1982] Jackson, W. B., and Amer, N. M., "Direct measurement of gap-state absorption in hydrogenated amorphous silicon by photothermal deflection spectroscopy" (1982), *Phys. Rev. B* Vol. **25**, p. 5559.

[Kamei-1998] Kamei, T., Kondo, M., Matsuda, A., "A significant reduction of impurity content in hydrogenated microcrystalline silicon films for increased grain size and reduced defect density" (1998), *Jpn. J. Appl. Phys.* Vol. **37**, pp. L265-L268.

[Kamei-2001] Kamei, T., Wada, T., Matsuda, A., "Where oxygen donors live in microcrystaline silicon" (1991), *Mat. Res. Soc. Symp. Proc.* Vol. **664**, p. A10.1.1.

[Kamei-2004] Kamei, T., Wada, T., Matsuda, A., "Oxygen impurity doping into ultrapure hydrogenated microcrystalline Si films" (2004), *J. Appl. Phys.* Vol. **96**, pp. 2087-2090.

[Kanschat-1999] Kanschat, P., Lips, K., Brüggemann, R., Hierzenberger, A., Sieber, I., Fuhs, W., "Paramagnetic Defects in Undoped Microcrystalline Silicon Deposited by the Hot-Wire Technique" (1999), *Mater. Res. Soc. Symp. Proc* Vol. **507**, p. 793.

[Kilper-2008] Kilper, T., van den Donker, M.N., Carius, R., Rech, B., Bräuer, G., Repmann, T., "Process control of high rate microcrystalline silicon based solar cell deposition by optical emission spectroscopy" (2008), *Thin Solid Films* Vol. **516**, pp. 4633-4638.

[Kilper-2009] Kilper, T., Beyer, W., Bräuer, G., Bronger, T., Carius, R., van den Donker, M. N., Hrunski, D., Lambertz, A., Merdzhanova, T., Mück, A., Rech, B., Reetz, W., Schmitz, R., Zastrow, U., and Gordijn, A., "Oxygen and nitrogen impurities in microcrystalline silicon deposited under optimized conditions: Influence on material properties and solar cell performance" (2009), *J. Appl Phys.* Vol. **105**, 074509.

[Klein-2005] Klein, S., Finger, F., Carius, R., and Stutzmann, M., "Deposition of microcrystalline silicon prepared by hot-wire chemical-vapor deposition: The influence of the deposition parameters on the material properties and solar cell performance" (2005), *Journal of Applied Physics*, Vol. **98**, 024905.

[Klein-2007] Klein, S., Finger, F., Carius, R., Dylla, T., and Klomfass, J., "Relationship between the optical absorption and the density of deep gap states in microcrystalline silicon" (2007), *J. Appl. Phys.* Vol. **102**, 103501.

[Konagai-2008] Konagai, M., "Deposition of new microcrystalline materials, μc-SiC, μc-GeC by HWCVD and solar cell applications" (2008), *Thin Solad Films* Vol. **516**, p. 490.

[Kondo-2000] Kondo, M., Fukawa, M., Guo, L., Matsuda, A, "High rate growth of microcrystalline silicon at low temperature" (2000), *J. Non-Cryst. Solids* Vol. **266-269**, p. 84.

[Kondo-2003] Kondo, M., Suzuki, S., Nasuno, Y., Tanda M., and Matsuda, A., "Recent developments in the high growth rate technique of device-grade microcrystalline silicon thin film" (2003), *Plasma Sources Sci. Technol.* Vol. **12**, p. S111.

[Lambertz-2001] Lambertz, O. Vetterl and F. Finger, "High deposition rate for μc-Si:H absorber layers using VHF PECVD at elevated discharge power and deposition pressure" (2001), *17th European Photovoltaic Solar Energy Conference and Exhibition*, B. McNelis et al. (WIP-Renewable Energies, Munich), p. 2977.

[Lambertz-2007] Lambertz, A., Dasgupta, A., Reetz, W., Gordijn, A., Carius, R., Finger, F. "Microcrystalline Silicon Oxide as Intermediate Reflector for Thin Film Silicon Solar Cells" (2007). *Proc. 22nd (European Photovoltaic Solar Energy Conference)* Milan/Italy, p. 1839.

[Li-2009] Li, H., van der Werf, K. H.M., Rath, J. K., Schropp, R. E.I., "Hot wire CVD deposition of nanocrystalline silicon solar cells on rough substrates" (2009), *Thin Solid Films* Vol. **517**, pp. 3476–3480.

[Luysberg-1997] Luysberg, M., Hapke, P., Carius, R., and Finger, F., "Structure and growth of microcrystalline silicon: Investigation by TEM and Raman spectroscopy of films grown at different plasma excitation frequencies" (1997), *Philos. Mag. A* Vol. **75**, p. 31.

[Luysberg-2001] Luysberg, M., Scholten, C., Houben, L., Carius, R., Finger F., and Vetterl, O., "Structural properties of microcrystalline Si Solar cells" (2001), *Mat. Res. Soc. Symp.* Vol. **664**, p. A15.2.

[Luysberg-2005] Luysberg M., and Houben, L., "Structure of microcrystalline solar cell materials: What can we learn from electron microscopy?" (2005), *Mater. Res. Soc. Symp. Proc.* Vol. **862**, p. A24.1.

[Mai-2005a] Mai, Y., Klein, S., Carius, R., Stiebig, H., Geng, X., Finger, F., "Open circuit voltage improvement of high-depositon-rate microcrystalline silicon solar cells by hot wire interface layers" (2005), *Applied Physics Letters*, Vol. **87**, 073503.

[Mai-2005b] Mai, Y., Klein, S., Carius, R., Wolff, J., Lambertz, A., Finger, F., Geng, X. "Microcrystalline silicon solar cells deposited at high rates" (2005), *Journal of Applied Physics*, Vol. **97**, pp. 114913.

[Main-2004] Main, C., Reynolds, S., Zrinscak, I., and Merazga, A., "Comparison of AC and DC constant photocurrent methods for determination of defect densities", (2004), *J. Non-Cryst. Solids* Vol. **338**, p. 228.

[Mashima-2006] Mashima, H., Yamakoshi, H., Kawamura, K., Takeuchi, Y., Noda, M., Yonekura, Y., Takatsuka, H., Uchino, S., Kawai, Y., "Large area VHF plasma production using a ladder-shaped electrode" (2006), *Thin Solid Films* Vol. **506–507**, pp. 512–516.

[Matsuda-1983] Matsuda, A., "Formation kinetics and control of microcrystallite in μc-Si:H from glow discharge plasma" (1983), *J. Non-Cryst. Solids*, Vol. **59&60**, p. 767.

[Matsuda-1999] Matsuda, A., "Growth mechanism of microcrystalline silicon obtained from reactive plasmas" (1991), *Thin Solid Films* Vol. **337**, pp. 1-6.

[Matsui-2002] Matsui, T., Tsukiji, M., Saika, H., Toyama, T., Okamoto, H., "Correlation between Microstructure and Photovoltaic Performance of Polycrystalline Silicon Thin Film Solar Cells" (2002), *Jpn. J. Appl. Phys.* Vol. **41**, pp. 20-27.

[Matsui-2004] Matsui, T., Matsuda, A., and Kondo, M., "High-Rate Plasma Process for Microcrystalline Silicon: Over 9% Efficiency Single Junction Solar Cells" (2004), *Mat. Res. Soc. Symp. Proc.* Vol. **808**, p. A8.11.

[Matsui-2006] Matsui, T., Matsuda, A., and Kondo, M., "High-rate microcrystalline silicon deposition for p–i–n junction solar cells" (2006), *Sol. Energy Mater. Sol. Cells* Vol. **90**, p. 3199.

[Matsui-2009] Matsui, T., Chang, C.W., Takada, T., Isomura, M., Fujiwara, H., Kondo, M., "Thin film solar cells based on microcrystalline silicon–germanium narrow-gap absorbers" (2009), *Solar Energy Materials & Solar Cells* Vol. **93**, pp.1100–1102.

[Matsumura-1991] Matsumura, H., "Formation of Polysilicon Films by Catalytic Chemical Vapor Deposition (cat-CVD) Method" (1991), *Jpn. J. Appl. Phys.* Vol. **30**, pp. L1522-L1524.

[Matsumura-2008] Matsumura, H., Ohdaira, K., "New application of Cat-CVD technology and recent status of industrial implementation" (2009), *Thin Solid Films* Vol. **517**, pp. 3420–3423.

[Meier-1994a] Meier, J., Flückiger, R., Keppner, H., and Shah, A., "Complete microcrystalline p-i-n solar cell-Crystalline or amorphous cell behavior?" (1994) *Appl. Phys. Lett.* Vol. **65**, p. 860.

[Meier-1994b] Meier, J., Dubail, S., Flückiger, R., Fischer. D., Keppner, H., Shah, A., "Intrinsic microcrystalline silicon (μc-Si:H) – A promising new thin film solar cell material" (1994), *Proc. 1st World Conference on Photovoltaic Energy Conversion*, IEEE, p. 409.

[Meier-1996] Meier, J., Torres, P., Platz, R., Dubail, S., Kroll, U., Anna Selvan, J. A.,Pellaton Vaucher, N., Hof, Ch., Keppner, H., Shah, A., Ufert, K. D., Giannoules, P., Koehler, J., "On the way towards high effciency thin film silicon solar cells by the "micromorph" concept" (1996) *Mater. Res. Soc. Symp. Proc.*, p. 3.

[Meier-2007] Meier, J., Kroll, U., Benagli, S., Roschek, T., Huegli, A., Spitznagel, J., Kluth, O., Borello, D., Mohr, M., Zimin, D., Monteduro, G., Springer, J., Ellert, C., Androutsopoulos, G., Buechel, G., Zindel, A., Baumgartner, F., Koch-Ospelt, D., "Recent Progress in Up-scaling of Amorphous and Micromorph Thin Film Silicon Solar Cells to 1.4 m2 Modules" (2007), *Mater. Res. Soc. Symp. Proc.* Vol. **989**, pp. A24-01.

[Meillaud-2005] Meillaud, F., Vallat-Sauvain, E., Niquille, X., Dubey, M., Bailat, J., Shah, A., Ballif, C., "Light-induced degradation of thin film amorphous and microcrystalline silicon solar cells" (2005) *Proc. 31th IEEE Photovoltaic Specialist Conference*, Lake Buena Vista, FL, USA, pp. 1412-1415.

[Miyajima-2006] Miyajima, S., Yamada, A., Konagai, M., "Aluminum-doped hydrogenated microcrystalline cubic silicon carbide films deposited by hot wire CVD", (2006), *Thin Solid Films*, Vol. **501**, p. 186.

[Müller-1999] Müller, J., Finger, F., Carius, R., Wagner, H. "Electron spin resonance investigation of electronic states in hydrogenated microcrystalline silicon" (1999), *Physical Review B*, Vol. **60**, p. 11666.

[Niikura-2006] Niikura, C., Kondo, M., Matsuda, A., "High rate growth of device-grade microcrystalline silicon films at 8 nm/s" (2006), *Solar Energy Materials & Solar Cells* Vol. **90**, pp. 3223–3231.

[Nakamura-1995] Nakamura, K., Yoshino, K., Takeoka, S., Shimizu, I., "Roles of Atomic Hydrogen in Chemical Annealing" (1995), *Jpn. J. Appl. Phys.* Vol. **34** pp. 442-449.

[Nagajima-2009] Nakajima, A., Gotoh, M., Sawada, T., Fukuda, S., Yoshimi, M., Yamamoto, K., Nomura, T., "Development of thin-film Si HYBRID solar module" (2009), *Solar Energy Materials and Solar Cells*, Vol. **93**, pp. 1163-1166.

[Oda-1988] Oda, S., and Noda, J., "Preparation of a-Si:H films by VHF plasma CVD" (1988) *Mat. Res. Soc. Sym. Proc.* Vol. **118**, p. 117.

[Ossadnik-1999] Ossadnik, Ch., Veprek, S., and Gregora, I., "Applicability of Raman scattering for the characterization of nanocrystalline silicon" (1999) *Thin Solid Films* Vol. **337**, pp.148.

[Overhof-1989] Overhof, H., and Thomas, P., "Electronic Transport in Hydrogenated Amorphous Semiconductors" (1989) Springer-Verlag, New York.

[Platz-1998] Platz, R., Hof, C., Fischer, D., Meier, J., Shah A., "High-Ts amorphous top cells for increased top cell currents in micromorph tandem cells" (1998) *Solar Energy Materials and Solar Cells*, Vol. **53**, pp. 1-13.

[Prasad-1990] Prasad, K., Finger, F., Curtins, H., Shah, A., and Bauman, J., "Preparation and characterization of highly conductive (100S/cm) phosphourus doped μc-Si:H films deposited using the VHF-GD technique" (1990), *Mat. Res. Soc. Symp. Proc.* Vol. **164**, p. 27.

[Python-2008] Python, M., Vallat-Sauvain, E., Bailat, J., Dominé, D., Fesquet, L., Shah, A., Ballif, C., "Relation between substrate surface morphology and microcrystalline silicon solar cell performance" (2008), *J. Non-Crys. Solids*, Vol. **354**, pp. 2258-2262.

[Rath-2003] Rath, J.K. "Low temperature polycrystalline silicon: a review on deposition, physical properties ad solar cell applications" (2003), *Solar Energy Materials and Solar Cells*, Vol. **76**, pp. 431-487.

[Rath-2004] Rath, J.K., Franken, R.H.J., Gordijn, A., Schropp, R.E.I., Goedheer, W.J., "Growth mechanism of microcrystalline silicon at high pressure conditions" (2004) *J. Non-Cryst. Solids,* Vol. **338-340**, p. 56.

[Rech-2006] Rech, B., Repmann, T., van den Donker, M. N., Berginski, M., Kilper, T., Hüpkes, J., Calnan, S., Stiebig, H., Wieder, S., "Challenges in microcrystalline silicon based solar cell technology" (2006) *Thin Solid Films,* Vol. **511-512**, pp. 548-555.

[Reynolds-2006] Reynolds, S., "Time-resolved Photoconductivity as a Probe of Carrier Transport in Microcrystalline Silicon" (2006), *Mater. Res. Soc. Symp. Proc.* Vol. **910**, p. 0910-A01.

[Reynolds-2009] Reynolds, S., Carius, R., Finger, F., Smirnov, V., "Correlation of structural and optoelectronic properties of thin film silicon prepared at the transition from microcrystalline to amorphous growth" (2009), *Thins Solid Films* Vol. **517**, pp. 6392-6395.

[Roschek-2002] Roschek, T., Repmann, T., Müller, J., Rech, B., and Wagner, H., „Comprehensive study of microcrystalline silicon solar cells deposited at high rate using 13.56 MHz plasma-enhanced chemical vapor deposition" (2002), *J. Vac. Sci. Technol. A* Vol. **20**, p. 492.

[Sauvain-1993] Sauvain, E., Mettler, A., Wyrsch, N., and Shah, A., "Subbandgap absorption spectra of slightly doped a-Si:H measured with constant photocurrent method (CPM) and photothermal deflection spectroscopy (PDS)" (1993), *Solid State Comm.* Vol. **85**, p. 219

[Schmidt-1998] Schmidt, J.A., and Rubinelli, F.A., "Limitations of the constant photocurrent method: A comprehensive experimental and modeling study" (1998), *J. Appl. Phys.* Vol. **83**, p. 339.

[Schropp-1998] Schropp, R.E.I., Zeman, M. "Amorphous and Microcrystalline Silicon Solar Cells: Modeling, Materials and Device Technology" (1998) Kluver Academic Publishers, Boston, MA, ISBN 0-7923-8317-6.

[Schropp-2002] Schropp, R. E. I., Xu, Y., Iwaniczko, E., Zaharias, G. A., and Mahan, A. H., "Microcrystalline Silicon for solar cells at high deposition rates by hot wire CVD" (2002), *Mat. Res. Soc. Symp. Proc.* Vol. **715**, p. A26.3.

[Schweitzer-1981] Schweitzer, L., Grünewald, M., Dersch, H., "Correlation effects and the density of states in amorphous silicon" (1981) *Sol. State. Comm.* Vol. **39**, pp. 355-358.

[Sendova-2006] Sendova-Vasileva, M., Klein, S., and Finger, F. "Instability phenomena in μc-Si:H solar cells prepared by hot-wire CVD" (2006), *Thin Solid Films*, Vol. **501**, p. 252.

[Shah-2004] Shah, A. V., Schade, H., Vanecek, M., Meier, J., Vallat-Sauvain, E., Wyrsch, N., Kroll, U., Droz, C., and Bailat, J., "Thin-film Silicon Solar Cell Technology" (2004), *Prog. Photovolt: Res. Appl.* Vol. **12**, pp. 113–142.

[Shah-2006] Shah, A., Meier, J., Buechel, A., Kroll, U., Steinhauser, J., Meillaud, F., Schade, H., Dominé, D., "Towards very low-cost mass production of thin-film silicon photovoltaic (PV) solar modules on glass" (2006), *Thin Solid Films* Vol. **502**, p. 292-299.

[Shimura-2004] *Semiconductors and Semimetals 42,* Oxygen in Silicon ed. F. Shimura, Academic Press, San Diego, CA (1994).

[Siebke-1996] Siebke, F., Stiebig, H., Abo-Arais, A., and Wagner, H., "Charged and neutral defect states in a-Si:H determined from improved analysis of the constant photocurrent method" (1996), *Solar Energy Materials and Solar Cells,* Vol. **41**, p. 529.

[Smets-2008] Smets, A. H. M., Matsui, T., Kondo, M., "High-rate deposition of microcrystalline silicon p-i-n solar cells in the high pressure depletion regime" (2008), *J. Appl. Phys.* Vol. **104**, 034508.

[Smirnov-2006] Smirnov, V., Reynolds, S., Finger, F., Carius, R., Main, C., "Metastable effects in silicon thin films: Atmospheric adsorption and light induced degradation" (2006), *J. Non-Cryst. Solids* Vol. **352**, p. 1075.

[Smit-2003] Smit, C., van Swaaij, R. A. C. M. M., Donker, H., Petit, A. M. H. N., Kessels, W. M. M., and van de Sanden, M. C. M., "Determining the material structure of microcrystalline silicon from Raman spectra" (2003), *J. Appl. Phys.* Vol. **94**, p. 3582.

[Söderström-2009] Söderström, T., Haug, F.-J., Niquille, X., Terrazzoni, V., and Ballif, C., "Asymmetric intermediate reflector for tandem micromorph thin film silicon solar cells" (2009), *Appl. Phys. Lett.* Vol. **94**, 063501.

[Stutzmann-1989] Stutzmann, M., Biegelsen, D.K., "Microscopic nature of coordination defects in amorphous silicon" (1989), *Phys. Rev. B* Vol. **40**, p. 9834.

[Stuke-1987] Stuke, J., "Problems in the understanding of electronic properties of amorphous silicon" (1987), *J. Non-Cryst. Solids* Vol. **97-98**, p. 1.

[Strahm-2007a] Strahm, B., Howling, A.A., Sansonnens, L., Hollenstein, Ch., Kroll, U., Meier, J., Ellert, Ch., Feitknecht, L., Ballif, C., "Microcrystalline silicon deposited at high rate on large areas from pure silane with efficient gas utilization" (2007), *Sol. Energy Mat. Sol. Cells*, Vol. **91**, pp. 495-502.

[Strahm-2007b] Strahm, B., "Investigation of radio-frequency capacitively-coupled large area industrial reactor: Cost-effective production of thin film microcrystalline silicon solar cells" (2007) Thesis, EPFL (Lausanne).

[Street-1991] Street, R. A., (1991) *Hydrogenated Amorphous Silicon*, Cambridge University Press, Cambridge.

[Sriraman-2002] Sriraman, S., Agarwal, S., Aydil, E. S., Maroudas, D., "Mechanism of hydrogen-induced crystallization of amorphous silicon" (2002) *Nature* Vol. **418**, pp. 62-65.

[Takagi-2006] Takagi, T., Ueda, M., Ito, N., Watabe, Y., Kondo, M., "Microcrystalline Silicon Solar Cells Fabricated using Array-Antenna-Type Very High Frequency Plasma-Enhanced Chemical Vapor Deposition System" (2006), *Jpn. J. Appl. Phys.* Vol. **45**, pp. 4003–4005.

[Tawada-2003] Tawada, Y., Yamagishi, H., Yamamoto, K., "Mass productions of thin film silicon PV modules" (2003), *Solar Energy Materials & Solar Cells* Vol. **78**, pp. 647–662.

[Tsai-1988] Tsai, C. C. (1988). "Plasma deposition of amorphous and crystalline silicon: The effect of hydrogen on the growth, structure and electronic properties." *Amorphous silicon and related materials*, (1988), H. Fritzsche, ed., World Scientific Publishing Company, pp. 123-147.

[Tsai-1989] Tsai, C. C., Anderson, G.B., Thompson, R., Wacker, B., "Control of silicon network structure in plasma deposition" (1989), *J. Non-Crystl. Solids* Vol. **114**, p. 151.

[Torres-1996] Torres, P., Meier, J., Flückiger, R., Anna Selvan, J.A., Keppner, H., Shah, A., Littlewood, S.D., Kelly, I.E., Giannoules, P., "Device grade microcrystalline silicon owing to reduced oxygen contamination" (1996), *Appl. Phys. Lett* Vol. **69**, p. 1373.

[Usui-1979] Usui, S., Kikuchi, M., "Properties of heavily doped gd-Si with low resistivity" (1979), *Journal of Non-Crystalline Solids*, Vol. **34**, pp. 1-11.

[van den Donker-2005] van den Donker, M., Rech, B., Finger, F., Kessels, W. M. M., and van de Sanden, M. C. M., "Highly Efficient Microcrystalline Silicon Solar Cells Deposited from a Pure SiH4 Flow" (2005), *Applied Physics Letters*, Vol. **87**, 263503.

[van den Donker-2007a] van den Donker, M., Klein, S., Rech, B., Finger, F., Kessels, W. M. M., and van de Sanden, M. C. M "Microcrystalline silicon solar cells with an open-circuit voltage above 600 mV" (2007), *Applied Physics Letters*, Vol. **90**, 183504..

[van den Donker-2007b] van den Donker, M., „Plasma Deposition of Microcrystalline Silicon Solar Cells: Looking Beyond the Glass" (2007), in *Schriften des Forschungszentrums Jülich, Energietechnik*, Vol. **57** (Forschungszentrum Jülich GmbH, Jülich).

[Vallat-Sauvain-2000] Vallat-Sauvain, E., Kroll, U., Meier, J., Shah, A., Pohl, J., "Evolution of the microstructure in microcrystalline silicon prepated by very high frequency glow-discharge using hydrogen dilution" (2000), *J. Appl. Phys* Vol. **87**, pp. 3137-3142.

[Vallat-Sauvain-2006a] Vallat-Sauvain, E., Shah, A., Bailat, J., "Advances in microcrystalline silicon solar cell technology" in "*Thin-Film Solar Cells*" (2006), ed. J. Portmans and V. Arkhipov, Wiley&Sons, Chihcester, UK.

[Vallat-Sauvain-2006b] Vallat-Sauvain, E., Droz, C., Meillaud, F., Bailat, J., Shah, A., Ballif, C., "Determination of Raman emission cross-section ratio in microcrystalline silicon" (2006), *Journal of Non-Crystalline Solids*, Vol. **352**, pp. 1200-1203.

[Vanecek-1981] Vanecek, M., Kocka, J., Stuchlik, J., Triska, A., "Direct Measurement of the gap states and band tail absorption by constant photocurrent method in amorphous silicon" (1981), *Sol. State. Comm.* Vol. **39**, pp.1199-1202.

[Vanecek-2000] Vanecek, M., Poruba, A., Remes, Z., Rosa, J., Kamba, S., Vorlícek, V., Meier, J., Shah, A., "Electron spin resonance and optical characterization of defects in microcrystalline silicon" (2000), *J. Non-Cryst. Solids,* Vol. **266-269**, pp. 519.

[Vanecek-2002] Vanecek, M., Poruba, A., "Fourier-transform photocurrent spectroscopy of microcrystalline silicon for solar cells" (2002), *Appl. Phys. Lett.* Vol. **80**, p. 719

[Veprek-1968] Veprek, S., Marecek, V., "The preparation of thin layers of Ge and Si by chemical hydrogen plasma transport" (1986), *Solid State Electronics*, Vol. **11**, pp. 683-684.

[Veprek-1983] Veprek, S., Iqpal, Z., Kühne, R. O., Capezzuto, P., Sarott, F.-A., and Gimzewski, J. K., "Properties of microcrystalline silicon: IV. Electrical conductivity, electron spin resonance and the effect of gas adsorption" (1983), *J. Phys. C: Solid State Phys*. Vol. **16**, pp.. 6241-6262.

[Veprek-1990] Veprek, S. "Chemistry and solid state physics of microcrystalline silicon" (1990), *Materials Research Society Symposium Proceedings*, Vol. **164**, pp. 39-49.

[Vetterl-1999] Vetterl, O., Hapke, P., Kluth, O., Lambertz, A., Wieder, S., Rech, B., Finger, F., Wagner, H., "Intrinsic microcrystalline silicon for solar cells" (1999), *Solid State Phenomena,* Vols. **67-68**, pp. 101-106.

[Vetterl-2000] Vetterl, O., Finger, F., Carius, R., Hapke, P., Houben, L., Kluth, O., Lambertz, A., Mück, A., Rech, B., and Wagner, H. "Intrinsic microcrystalline silicon: A new material for photovoltaics" (2000), *Sol. Energ. Mat. Sol. C.* Vol. **62**, p. 97.

[Vetterl-2002] Vetterl, O., Gross, A., Jana, T., Ray, S., Lambertz, A., Carius, R., and Finger, F., "Changes in electric and optical properties of intrinsic microcrystalline silicon upon variation of the structural composition" (2002), *Journal of Non-Crystalline Solids*, Vol. **299-302**, p. 772.

[Wang-1990] Wang C., and Lucovsky, G., "Intrinsic microcrystalline silicon deposited by remote PECVD: a new thin-_lm photovoltaic material" (1990), in *Proceedings of the 21st IEEE Photovoltaic Specialists Conference*, p.1614.

[Wang-2008] Wang, Y., Geng, X., Stiebig, H., and Finger, F., "Stability of Microcrystalline Silicon Solar Cells with HWCVD Buffer Layer" (2008), *Thin Solid Films*, Vol. **516**, p. 733.

[Willeke-1991] Willeke, G., "Physics and Electronic Properties of Microcrystalline Semiconductors" (1991), in Kanicki, J., Ed., "*Amorphous and Microcrystaline Devices-Materials and Device Physics*", Artech House, Norwood, MA, ISBN 0-89006-379-6, pp. 55-88.

[Wyrsch-1991] Wyrsch, N., Finger, F., McMahon, T. J. and Vanecek, M., "How to reach more precise interpretation of subgap absorption spectra in terms of deep defect density in a-Si:H" (1991), *J. Non-Cryst. Solids* Vol. **137**, p. 347.

[Xu-2008] Xu, X., Yan, B., Beglau, D., Li, Y., DeMaggio, G., Yue, G., Banerjee, A., Yang, J., Guha, S., Hugger, P.G., and Cohen, J. D., "Study of Large Area a- Si:H and nc-Si:H Based Multijunction Solar Cells and Materials" (2008), *Mater. Res. Soc. Symp. Proc.* Vol. **1066,** A14-02.

[Yamamoto-1999] Yamamoto, K., Yoshimi, M., Tawada, Y., Okamoto, Y., Nakajima, A., Igari, S., "Thin film poly-Si solar cells on glass substrate fabricated at low temperature" (1999), *Appl. Phys. A* Vol. **69**, pp. 179-185.

[Yamamoto-2004] Yamamoto, K., Nakajima, A., Yoshimi, M., Sawada, T., Fukuda, S., Suezaki, T., Ichikawa, M., Koi, Y., Goto, M., Meguro, T., Matsuda, T., Kondo, M., Sasaki, T., Tawada Y., "A high efficiency thin film silicon solar cell and module" (2004), *Solar Energy* Vol. **77**, pp. 939-949.

[Yamamoto-2006] Yamamoto, K., Nakajima, A., Yoshimi, M., Sawada, T., Fukuda, S., Suezaki, T., Ichikawa, M., Koi, Y., Goto, M., Meguro, T., Matsuda, T., Kondo, M., Sasaki, T., Tawada Y., "High efficiency thin film silicon hybrid cell and module with newly developed innovative interlayer" (2006), *IEEE 4th World Conf. PV Energy Conversion*, Hawaii, USA, p. 1489.

[Yang-2005] Yang, J., Yan, B., Guha, S., "Amorphous and nanocrystalline silicon-based multi-junction solar cells" (2005), *Thin Solid Films* Vol. **487**, pp. 162-169.

[Yue-2007] Yue, G., Yan, B., Ganguly, G., Yang, J., Guha, S., "Metastability in hydrogenated nanocrystalline silicon solar cells" (2007), *J. Mater. Res.* Vol. **22**, p. 1128.

[Yue-2008] Yue, G., Yan, B., Teplin, C., Yang, J., Guha, S., "Optimization and characterization of i/p buffer layer in hydrogenated nanocrystalline silicon solar cells" (2008), *Journal of Non-Crystalline Solids* Vol. **354**, pp. 2440-2444.

[Zhang-2005] Zhang, X. D., Zhao, Y., Gao, Y. T., Zhu, F., Wei, C. C., Sun, J., Geng, X. H., Xiong, S. Z., "Analysis of structural properties of microcrystalline silicon thin films for solar cells using x-ray diffraction" (2005), *20th European Photovoltaic Solar Energy Conference*, Barcelona, Spain.

CHAPTER 4

# THEORY OF SOLAR CELL DEVICES (SEMI-CONDUCTOR DIODES)

## PART I: INTRODUCTION AND "*pin*-TYPE" DIODES

*Arvind Shah and Corinne Droz*

## 4.1 CONVERSION OF LIGHT INTO ELECTRICAL CARRIERS BY A SEMI-CONDUCTOR DIODE

A solar cell is a semiconductor device that produces electrical current (and then sends it through an external load) when exposed to light (photovoltaic effect). The operation of a solar cell consists of two main steps:

(1) Creation of electron-hole pairs, i.e. generation of free electrons and holes through the absorption of photons.

(2) Separation of the free electrons and holes in order to actually produce electricity.

The efficiency $\eta$ of a solar cell is the ratio of the electrical output power to the incident light power.

### 4.1.1 First step: generation of electron-hole pairs

***Absorption and energy conversion of a photon***

When light illuminates a solar cell, photons are absorbed by the semiconductor material and pairs of free electrons and holes are created (see Fig. 4.1). However, in order to be absorbed, the photon must have an energy $E_{ph} = h\nu$ (where $h$ is the Planck's constant and $\nu$ the frequency of light) higher or at least equal to the bandgap energy $E_g$ of the semiconductor. The bandgap energy is the difference in energy levels between the lowest energy level of the conduction band ($E_C$) and the highest energy level of the valence band ($E_V$) (Fig. 4.2). For a given semiconductor, $E_g$ is a fixed material constant that only slightly depends on temperature. Figure 4.3 gives the values of $E_g$ for various amorphous and crystalline semiconductors at $T = 25$ °C.

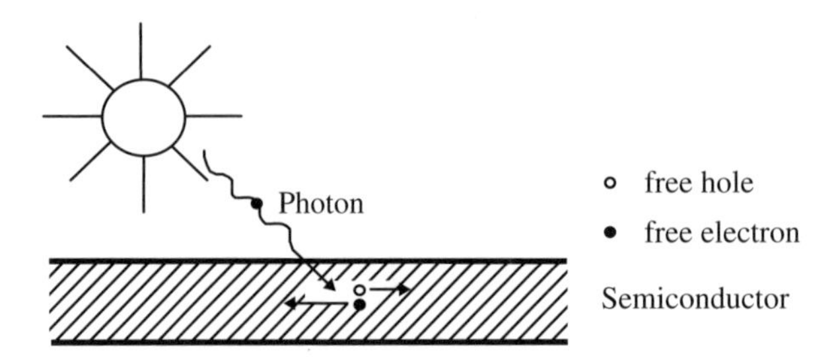


**Fig. 4.1** Creation of an electron-hole pair through absorption of a photon of energy $E_{ph} = h\nu$.

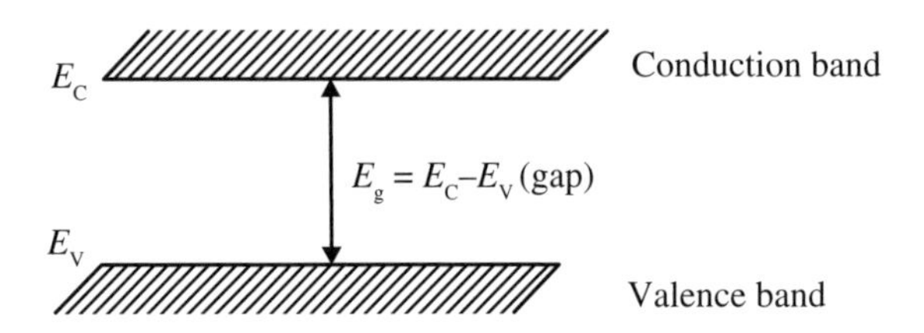


**Fig. 4.2** Bandgap energy $E_g$ of a semiconductor.

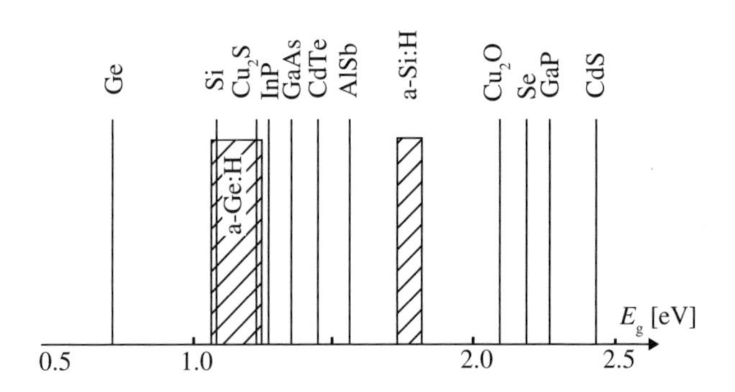


**Fig. 4.3** Bandgap energy $E_g$ of various semiconductors at temperature $T = 25$ °C.

Depending on the energy of the photon and on the bandgap energy of the solar cell material, three cases can occur:

(1) $E_{ph} = E_g$: in this case, the photon can be absorbed and will then generate a single electron-hole pair (Fig. 4.4), without any loss of energy.

(2) $E_{ph} > E_g$: in this case, the photon can be easily absorbed and will then, in general, create a single electron-hole pair. The excess energy $E_{ph}–E_g$ is rapidly transformed into heat (thermalization; Fig. 4.5). In some rare cases (which have absolutely no bearing on present-day solar cells), a second electron-hole pair can be successively created by impact ionization, provided that $E_{ph} \gg E_g$.

(3) $E_{ph} < E_g$: the energy of the photon is not high enough to be absorbed. The photon will either be reflected or absorbed elsewhere, and its energy is lost.

Note that in some rare cases where a sufficiently high density of states exists within the gap (due to impurities or defects in crystallinity), a photon with $E_{ph} < E_g$ can still be absorbed (Fig. 4.6). However, in general, and for present-day solar cells, this mechanism will not contribute to the effective electric current produced by the cell.

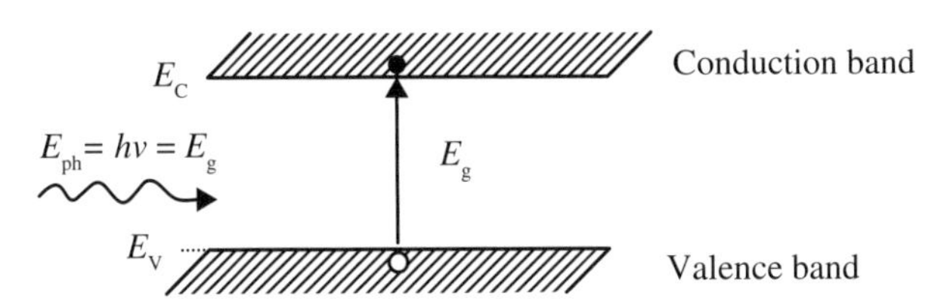


**Fig. 4.4** Absorption of a photon when $E_{ph} = E_g$.

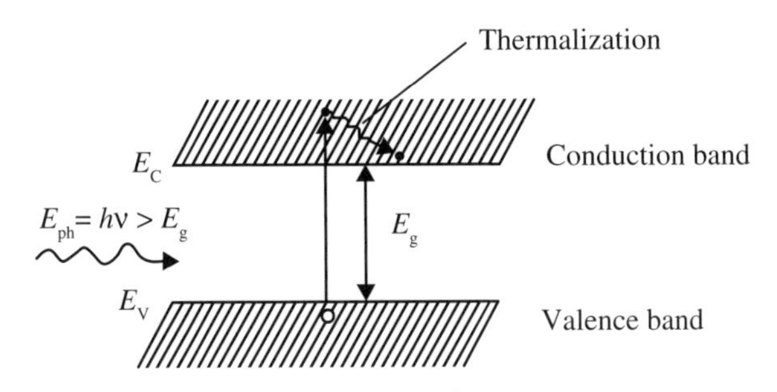


**Fig. 4.5** Absorption of a photon when $E_{ph} > E_g$; thermalization (conversion of excess energy $E_{ph}–E_g$ into heat) takes place very rapidly.

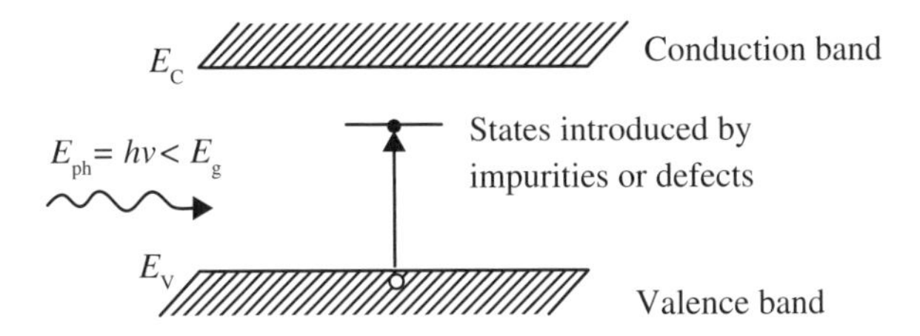


**Fig. 4.6** Possible absorption of a photon when $E_{ph} < E_g$, thanks to states within the gap.

### *Direct and indirect bandgaps*

The terms "direct" and "indirect" refer here to the interaction between incoming light and the semiconductor itself. The semiconductor is considered to have a direct gap when the absorption of an incoming photon is directly possible without the interaction of a phonon within the semiconductor. In this case, the maximum of the valence band corresponds to the minimum of the conduction band. GaAs, CdTe and $CuInSe_2$ are typical examples of semiconductors with a direct gap. The semiconductor is considered to have an indirect gap when the absorption of an incoming photon is only possible with the interaction of a phonon within the semiconductor. In this second case, the maximum of the valence band does *not* correspond to the minimum of the conduction band. Because a phonon must now be emitted (or absorbed) for a photon to be absorbed (see Fig. 4.7), the absorption

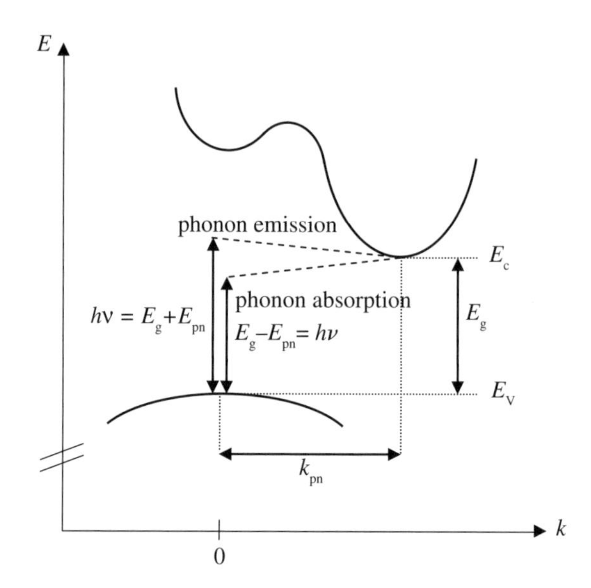


**Fig. 4.7** Diagram $E(k)$ of energy $E$ versus momentum, for a semiconductor with indirect bandgap $E_g$. The absorption of a photon requires here the emission or absorption of a phonon, so as to obey the rule of momentum conservation. Whereas a photon has practically only an energy component ($E = h\nu$), a phonon possesses both energy ($E_{pn}$) and momentum ($k_{pn}$); $E_{pn}$ has been drawn in the figure about ten times larger than it really is.

probability is significantly reduced, as compared to the case of a semiconductor with a direct bandgap. Si, Ge and C are typical examples of semiconductors with indirect gaps. The latter, thus, have absorption coefficients that are relatively low (see Fig. 4.8, as well as [Shah-2006]). Therefore, if silicon is used as a solar cell material, we basically have to use relatively thick silicon wafers; this is in contrast with the case for GaAs, CdTe and $CuInSe_2$, where thin photoabsorbing layers are already sufficient. For this reason, sophisticated light-trapping techniques have been developed for silicon solar cells; especially for thin-film silicon solar cells (see Sects. 4.5.18, 5.3.1 and 6.4.5).

### *Spectrum of the incoming light*

The quantity of light absorbed by a semiconductor depends on the bandgap energy (as previously discussed), but also on the spectrum of the incoming light (i.e. the energy distribution of the incident light as a function of wavelength). Figure 4.10 shows the spectrum of light emitted by the sun (which can be considered as a black body at 6000 K[1]), as well as the spectrum of sunlight received at the earth, both

[1]A more exact value for the surface temperature of the sun is 5778 K or 5505 °C. In this book we give approximate, rounded values, which can be easily memorized, for many of the key parameters. In Figure 4.10 we have therefore plotted the radiation spectrum of a black body at 6000 K. Had we plotted the radiation spectrum of a black body at a lower temperature, then the difference between that curve and the AM0 curve would have been slightly different. There would, however, still be a residual difference due to surface effects of the sun's surface and also due to interstellar absorption.

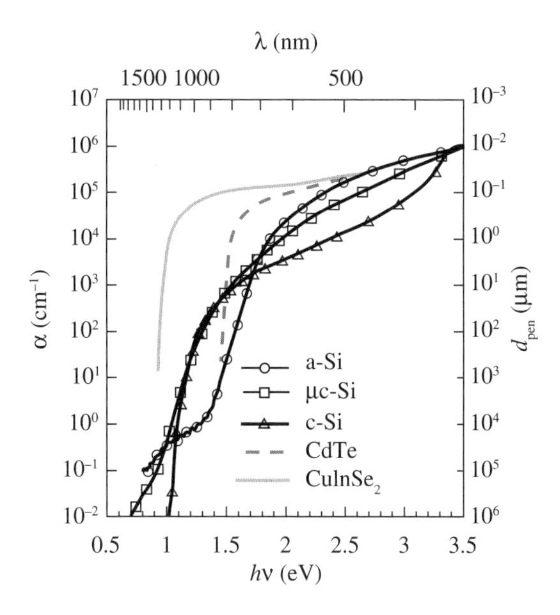


**Fig. 4.8** Absorption coefficient α and penetration depth $d_{pen}$ (of monochromatic light) as a function of wavelength λ and of photon energy $h\nu$, for various semiconductor materials commonly used in solar cells (a-Si: amorphous silicon, μc-Si: microcrystalline Si, c-Si: crystalline Si) [Shah-2006].

outside the atmosphere and at ground level. The difference between the last two is due to the path through the earth's atmosphere, which attenuates the solar spectrum (Rayleigh scattering, scattering by aerosols, absorption by the constituent gases of the atmosphere: UV absorption by ozone, infrared absorption by water vapor, etc.). This attenuation effect is characterized by the "Air Mass" coefficient AM = $1/\cos\alpha$, where α is the angle between the solar ray and a vertical line (Fig. 4.9). Thus, AM0 corresponds to the solar spectrum outside the atmosphere and AM1 to the solar spectrum on the earth's surface when the sun is directly (vertically) overhead. During mornings and evenings, when the incidence of sunlight is almost horizontal, we may have "Air Mass" coefficients AM3 or even higher. AM1.5 (i.e. $\alpha \approx 45°$)[2] is considered to be the "standard" value for testing and specifications in terrestrial applications, as it corresponds to a "reasonable average value" of the air mass through which the incoming sunlight passes. Note that not only the spectrum is defined by this coefficient, but also the intensity of the incoming sunlight. However, the intensity is generally normalized to 1000 W/m$^2$ in order to calibrate the efficiency of photovoltaic systems.

A semiconductor with a lower value of bandgap energy will be able to absorb a wider range of the solar spectrum (i.e. more photons) compared to one with a larger bandgap. However, in the first case, a substantial part of the incident energy will be lost by thermalization (i.e. by the energy difference between the energy of each

[2]A more precise value for the angle of incidence is $\alpha = 48.2°$. Again with a view to give rounded values, which can be easily memorized, we have written $\alpha \approx 45°$. One often specifies "AM 1.5g", where the letter "g" denotes the global incident sunlight (as opposed to direct incident sunlight). The AM 1.5g spectrum includes diffuse light from the sky and diffuse reflected light from the ground.

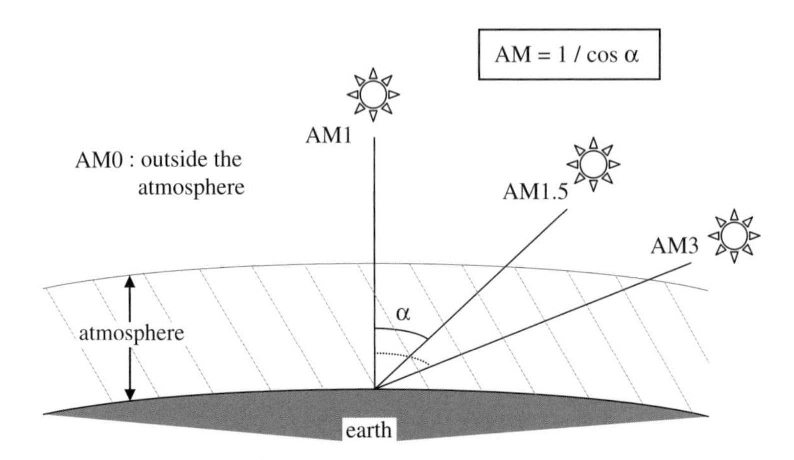


**Fig. 4.9** Diagram illustrating the definition of AM0, AM1, AM1.5 and AM3. α is the angle of incidence of sunlight, defined as the angle between a vertical line and the incident sunrays.

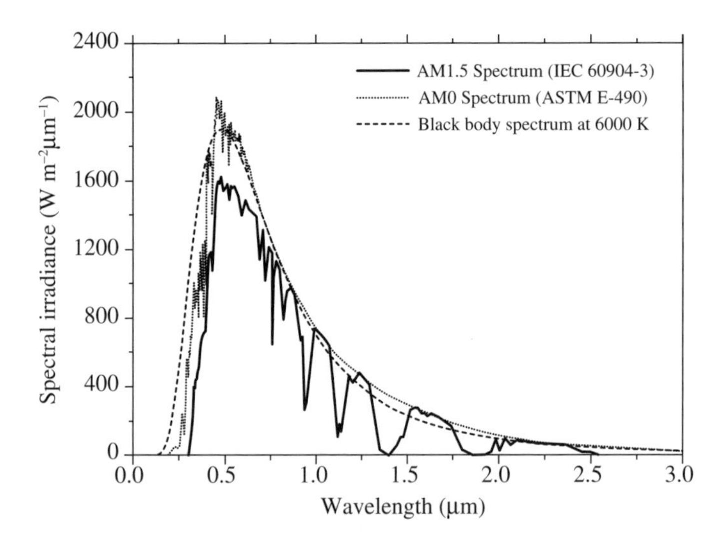


**Fig. 4.10** Spectral distribution of the solar spectrum received on the earth's surface (AM1.5) and outside the atmosphere (AM0); given here are the standardized spectra according to standards IEC 60904-3 (for AM1.5) and ASTM E-490 (for AM0); the AM0 spectrum is compared with the radiation of a blackbody at 6000 K. Courtesy of Adinath Funde, Photovoltaic Materials Laboratory, Department of Physics, University of Pune.

photon and the bandgap energy). On the other hand, a semiconductor with a higher value of bandgap energy will only be able to absorb a relatively narrow range of the solar spectrum (i.e. a relatively small amount of high-energy photons only), but less energy will be lost through thermalization. One can thus intuitively understand that to maximize the spectral conversion efficiency, one must choose an optimum value for the bandgap that is somewhere in an intermediate range, corresponding approximately to the bandgap of crystalline silicon of $E_g = 1.12$ eV.

The actual latent energy that each electron-hole pair (generated by one photon) holds is the bandgap energy $E_g$. Assuming that $\phi$ photons are absorbed per time unit, $\phi$ electrons-hole pairs are thus generated, having a corresponding total latent energy of $\phi \cdot E_g$ per time unit. The *spectral conversion efficiency* $\eta_S$ can thus be defined as:

$$\eta_S = \frac{\phi \cdot E_g}{P_{sun}}, \tag{4.1}$$

where $P_{sun}$ is the "power of the sun", i.e. the total energy within the solar spectrum per unit of time.

The previous expression (Eq. 4.1) can be rewritten as:

$$\eta_S = \frac{(\phi \cdot q) \cdot (E_g / q)}{P_{sun}}, \text{ in units:} \left[ \frac{\left( \frac{1}{s} \cdot C \right) \cdot (eV/C)}{W} \right] \equiv \left[ \frac{(A) \cdot (V)}{W} \right], \tag{4.2}$$

where $q$ is the elementary charge.

In this form, the first term in the numerator $(\phi \cdot q)$ corresponds to an electrical current, whereas the second term $(E_g/q)$ corresponds to an electrical potential (voltage).

Figure 4.11 shows the theoretical upper limit obtained for the spectral conversion efficiency $\eta_S$ – i.e. for the first step of the conversion of light into electrical carriers – as a function of the bandgap energy of the semiconductor, considering an AM1.5 illumination. Some typical examples are also given in Table 4.1. Maximal values for $\eta_S$ are almost 50 % and are obtained for gaps in the range between 0.8 eV and 1.5 eV. However, this limit does not apply, not even as a theoretical upper limit, to the entire

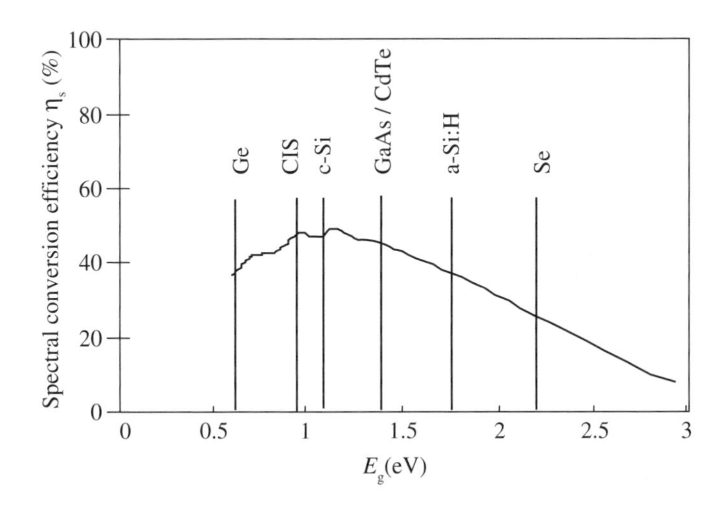


**Fig. 4.11** Maximal limit for the spectral conversion efficiency $\eta_S$ ('first step') as a function of the bandgap energy $E_g$ for AM1.5 illumination; values of bandgap energy for different semiconductor materials commonly used in solar cells are also shown.

**Table 4.1** Comparison of the spectral conversion efficiency $\eta_S$ of three semiconductors: Ge, Si and Se.

| Material | Bandgap $E_g$(eV) | Absorption | Voltage $E_g/q$ | Current $\phi \cdot q$ | Spectral conv. eff. $\eta_S$ |
|---|---|---|---|---|---|
| Ge | 0.66 (small) | almost all photons are absorbed | small | large | average (~40 %) |
| Si | 1.12 (average) | most infrared photons not absorbed | average | average | quite large (~48 %) |
| Se | 2.15 (large) | green, red + infrared photons not absorbed | large | very small | small (~27 %) |

solar cell, as there will be losses during the second step (separation of electrons and holes).

Note that these values correspond to the case of a solar cell with a single junction; here we have only considered one single gap. When two (or more) semiconductor materials with different gaps are superposed, the limit for the spectral conversion efficiency $\eta_S$ can indeed be increased. (This case will be treated in Chapter 5.)

### 4.1.2 Second step: separation of electrons and holes

The first step within the photovoltaic conversion process, which we have just described in Section 4.1.1, is the generation of pairs of electrons and holes within the semiconductor. Now, as the *second* step, these electron-hole pairs have to be separated: the electrons have to be sent to one side and the holes to the other side of the solar cell device. Such a separation is only possible by an electric field. Electrons and holes have opposite electric charge and therefore move in opposite directions within an electric field. To create an electric field, solar cells are formed as *diodes:* usually as *pn*-type diodes but, in the case of thin-film silicon solar cells, as *pin*-type diodes. Within a diode, an internal electric field is always formed. As schematically shown in Figure 4.12, the internal electric field is limited to the depletion layer in the *pn*-type diode, whereas it extends over the whole of the *i*-layer (intrinsic layer) in the *pin*-type diode. We will discuss these differences as well as their consequences later in Section 4.5.1. The generation of the internal electric field is governed by the so-called "built-in voltage $V_{bi}$". $V_{bi}$ is always smaller than $(E_g/q)$. For those readers who are familiar with band diagrams, we have sketched the evaluation of $V_{bi}$ in Figure 4.13. For the other readers let us just say that the value of $V_{bi}$ for *all forms* of *silicon* solar cells is roughly 1 V, provided that the *p*- and *n*-type layers within the solar cell have been "properly" doped.

This second step of charge separation comes at the price of additional energy losses: about one-third to one-half of the remaining energy has now to be sacrificed. The charge separation process is evidently more efficient if the internal electric field is stronger, i.e. if $V_{bi}$ is higher. This means that, in the optimal case, the charge process is more efficient if the bandgap energy is higher. Therefore the optimum

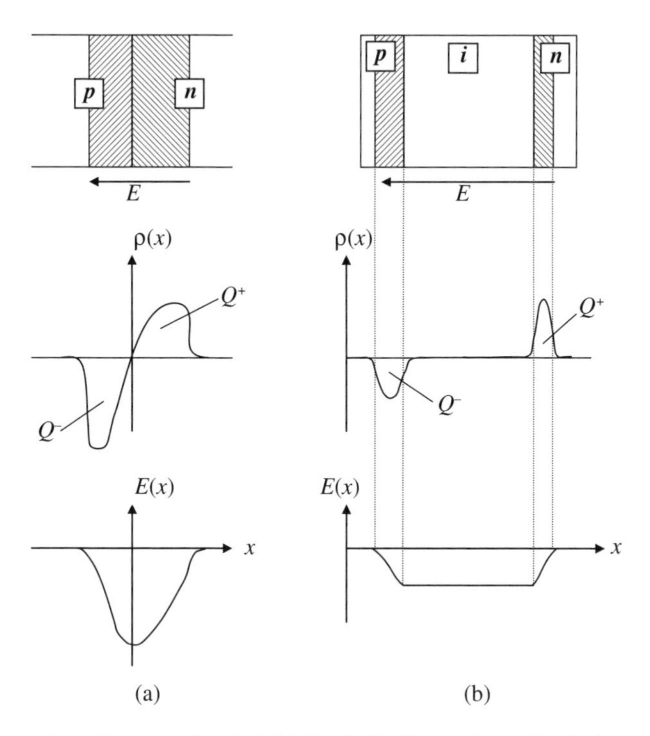


**Fig. 4.12** Formation of internal electric field $E$ and of built-in voltage $V_{bi}$: (a) in a *pn*-type diode; (b) in a *pin*-type diode. Shaded areas denote the depletion zones; $\rho(x)$ is the space charge density, given by $n_A-$ (in the *p*-depletion zone) and by $n_D+$ (in the *n*-depletion zone) (see Eq. 4.5 for difinitions), $Q^-$ is the total negative charge within the depletion zone of the *p*-layer; $Q^+$ is the total positive charge within the depletion zone of the *n*-layer, note that $Q^- = Q^+$; $E(x)$ is the electric field strength as a function of the position $x$; the negative sign of $E(x)$ signifies that the electric field is pointing from right to left, i.e. from the *n*-layer to the *p*-layer; the built-in voltage $V_{bi} = \int E(x)\mathrm{d}x$ is not directly shown in the figure, but corresponds to the area under the $E(x)$ curve.

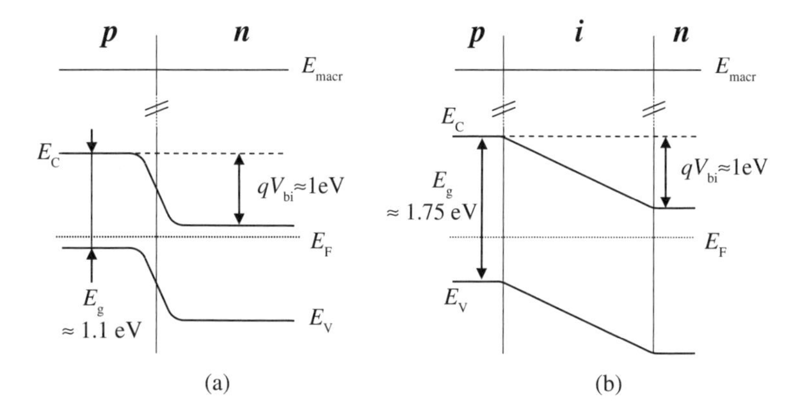


**Fig. 4.13** Band diagram showing schematically the evaluation of $V_{bi}$: (a) in a c-Si *pn*-type diode; (b) in an a-Si *pin*-type diode. $E_{macr}$ is the "macroscopic energy level" outside the solar cell, also called "local vacuum energy level".

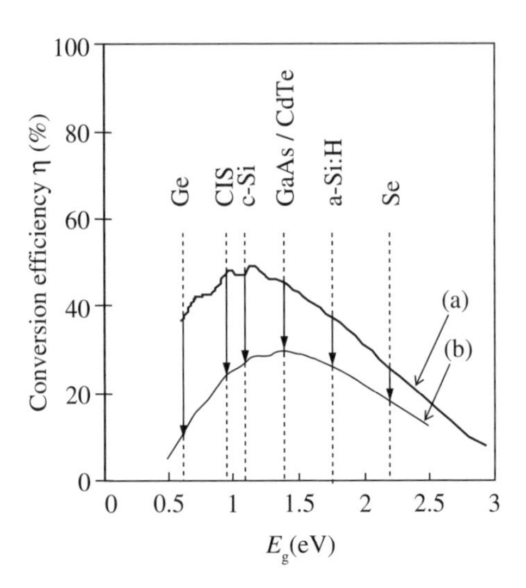


**Fig. 4.14** Comparison between (a) the maximum limit values for the spectral conversion efficiency $\eta_S$ ('first step') and (b) those for the overall efficiency $\eta$ ('first and second steps') of solar cells, as a function of the bandgap energy $E_g$, for AM1.5 illumination; the arrows indicate efficiency losses due to the separation of holes and electrons, i.e. due to the second step.

bandgap energy for maximizing the energy conversion efficiency of the whole solar cell, including the second step of charge separation, is shifted to higher values when compared to the optimum for the first step alone. This is shown in Figure 4.14. The optimum bandgap for the whole energy conversion process will be around 1.5 eV, as opposed to 1.1 eV for the first step alone. We will come back to these considerations in Section 4.4.

We will now examine in more detail the mechanics of charge separation. To do this we will first study carefully the *pn*-diode.

## 4.2 THE "*pn*-TYPE" OR "CLASSICAL" DIODE: DARK CHARACTERISTICS

A solar cell is basically a diode. The simplest case is an ideal *pn*-junction, i.e. a "junction" formed by a *p*-type region next to an *n*-type region. The *n*-type region corresponds to the region of semiconductor material doped with donor impurities – leading to an increased density of electrons – and the *p*-type region corresponds to the same semiconductor material doped with acceptor impurities – leading to a decreased density of electrons (or, as one also says, an increased density of holes). When no light is applied (dark condition), the relation between the current density $J_{dark}$ flowing through the device and the voltage $V$ applied to its boundaries is given by the following equation:

$$J_{dark} = J_0\left[\exp\left(\frac{qV}{kT}\right) - 1\right], \tag{4.3}$$

where: $q = 1.6022 \times 10^{-19}$ [C] is the elementary charge,
$k = 1.3807 \times 10^{-23}$ [J/K] is the Boltzmann constant,
$T$ [K] is the absolute temperature,
$J_0$ is the reverse saturation current density of the diode (see Eq. 4.5 below).

A typical $J(V)$ characteristic for an ideal *pn*-diode is shown in Figure 4.15. The above exponential expression (Eq. 4.3) is obtained from the so-called "diffusion model" of the diode, assuming an ideal *pn*-junction with an abrupt transition between the two doped regions (Fig. 4.16). The "diode model" or "diffusion model of the diode" is presented in most textbooks on semiconductor devices (see e.g. [Sze-2006]).

In the "diffusion model", first the diffusion currents of the minority carriers on both sides of the "depletion zone" are determined as a function of the applied voltage $V$. They are then added together to give the total current density $J_{dark}$. In order

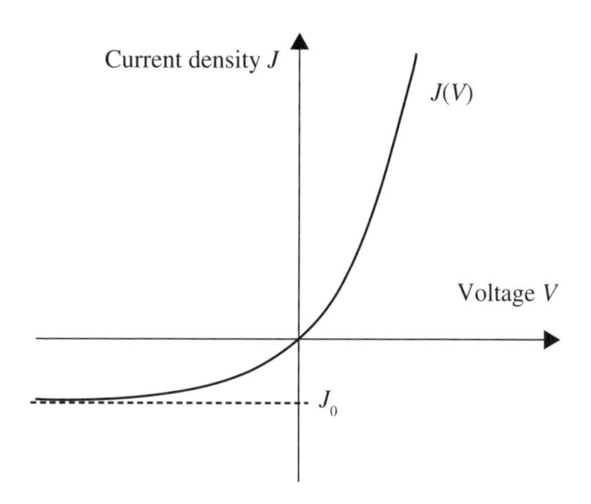


**Fig. 4.15** Typical $J(V)$ characteristic for a diode. $J_0$ is the reverse saturation current density.

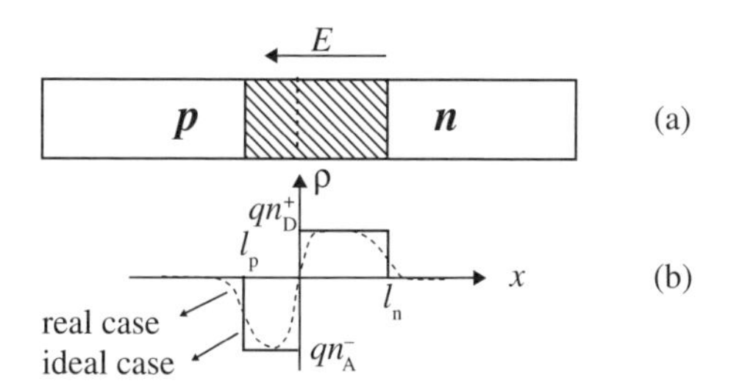


**Fig. 4.16** (a) Ideal *pn*-junction with an abrupt transition between the two doped regions resulting in a very thin depletion zone (hatched zone). (b) Space-charge distribution $\rho(x)$; $l_n$ and $l_p$ are the widths of the respective depletion zones; $n_A^-$, $n_D^+$ are the ionized acceptor and donor densities, respectively.

to carry out the calculation of $J_{dark}(V)$ for this model, the following approximations are generally used:

(1) Assumption of "total depletion". The device is assumed to be divided into two kinds of regions: two quasi-neutral regions ($n$ and $p$ bulk regions) where the space-charge density is assumed to be zero, and in between a depletion zone where the concentration of free carriers [i.e. the electron density $n(x)$ and hole density $p(x)$] is assumed to be negligible, so that the charge density here is assumed to be given only by the density of ionized dopants.

(2) For the depletion zone, one assumes that the drift current density $J_{drift}$ and diffusion current density $J_{diff}$ are very large, and have opposite signs, and magnitudes that are approximately equal:

$$J_{drift_{n,p}} \approx -J_{diff_{n,p}} \tag{4.4}$$

(3) In the bulk regions, the minority carrier concentration is much lower than the majority carrier concentration (i.e. $n \gg p$ in the $n$ region and $p \gg n$ in the $p$ region).

(4) In the bulk regions, the minority carriers are mainly transported by diffusion (minority carrier transport by drift is considered to be negligible in these regions).

(5) The depletion zone is very thin, so that the recombination-generation phenomena in this zone are considered to be negligible.

Note that for the case of a *pin*-type diode (see Sect. 4.5), the three last assumptions are definitely not satisfied Since an intrinsic *i*-layer where $n \approx p$ is added inbetween the *p*- and *n*-layers.

Based on the "diffusion model" and on the five assumptions/approximations listed above, one finds the following theoretical expression for the reverse saturation current density $J_0$:[3]

$$J_0 = \left[ \frac{qD_n n_i^2}{L_n n_A^-} + \frac{qD_p n_i^2}{L_p n_D^+} \right], \tag{4.5}$$

where: $n_i$ is the intrinsic carrier density;
$n_A^-$, $n_D^+$ are the densities of the ionized acceptor and donor atoms;
$D_p$, $D_n$ are the diffusion constants of minority carriers in the $p$ and $n$ regions;
$L_p$, $L_n$ are the diffusion lengths of minority carriers in the $p$ and $n$ regions, respectively.

[3]According to recent analysis [Miazza-2006], it would appear that the reverse saturation current density $J_0$ is, for an ideal diode and for typical sets of material/diode parameters, not defined by the "diffusion model", but rather by recombination/generation in the depletion zone. However, the "diffusion model" does render correctly the $J(V)$ characteristics for the forward-biased diode.

The reverse saturation current density $J_0$ is a key parameter for solar cell analysis.

The above expression (Eq. 4.5) is, however, of no direct use for solar cell analysis, as it contains material/diode parameters that are not accessible, except to the manufacturer of the diode. Furthermore, the bandgap energy does not explicitly appear in Equation 4.5.

Martin Green [Green-1982] has proposed, on the other hand, a simple empirical relation that can be used as a reasonable estimate for the minimal value of the reverse saturation current density:

$$J_0 = 1.5 \times 10^8 \cdot \exp\left(\frac{-E_g}{kT}\right) \quad \text{in} \left[\frac{\text{mA}}{\text{cm}^2}\right] \tag{4.6}$$

This relation has the advantage of linking $J_0$ to the bandgap energy $E_g$; it has the further advantage of not being based on the "diffusion model", but directly on measured data. We will use it in the following Section 4.3 to obtain an estimate for the maximum values of the main solar cell parameters $V_{oc}$ and $FF$.

We will thereby write:

$$J_0 = J_{00} \cdot \exp(-E_g/kT) \tag{4.7}$$

For Green's simple empirical relationship, one then has:

$$J_{00} = J_{00}^{\text{Green}} = 1.5 \times 10^8 \ [\text{mA/cm}^2] \tag{4.8}$$

The fifth approximation mentioned above can be abandoned if one takes into account, when solving the equations, the generation $G$ and recombination $R$ in the depletion zone.[4] This leads to a modified expression for the dark current density: one has to add a second term, with a different pre-factor $J_{02}$. The pre-factor $J_{02}$ also basically depends on material/diode parameters (life times, dopant concentrations, etc.):

$$J_{\text{dark}} = J_{01}\left[\exp\left(\frac{qV}{kT}\right) - 1\right] + J_{02}\left[\exp\left(\frac{qV}{2kT}\right) - 1\right] \tag{4.9}$$

The complete expression for the dark current density is commonly rewritten in a simpler, albeit slightly incorrect way, by "combining" the two additive (!) exponential functions given in Equation 4.9 into a *single* exponential function, Equation 4.10; thereby, one introduces the so-called "ideality factor $n$" of the diode.

$$J_{\text{dark}} = J_0\left[\exp\left(\frac{qV}{nkT}\right) - 1\right] \tag{4.10}$$

[4]The inclusion of thermal generation within the depletion zone appears to be a necessary condition for correctly evaluating the diode characteristics under reverse bias and, thus, assessing the reverse saturation current density $J_0$, see Footnote 3. At any rate in actual solar cells, many other factors, such as surface recombination, decisively influence $J_0$. There exist a large amount of scientific publications on the minimal values for recombination in "real" solar cells of different types – especially for solar cells based on silicon wafers; see e.g. [Green-1995].

The value of $n$ lies basically within the range between 1 and 2. For crystalline silicon, $n = 1$ when the diffusion current is preponderant (ideal case, with $J_{01} >> J_{02}$), whereas $n = 2$ when the generation/recombination current dominates ($J_{01} << J_{02}$). Note, that for *pin*-type diodes made with amorphous or microcrystalline silicon, experimentally measured values for $n$ are, in general, situated in the range between 1.4 and 1.8. Note also that an ideal *pin*-type diode is generally assumed to have a value $n = 2$.

## 4.3 THE "*pn*-type" OR "CLASSICAL" DIODE: PROPERTIES UNDER ILLUMINATION

### 4.3.1 Photo-generation and superposition principle (ideal case)

A solar cell is always "just" a diode. In the dark, a solar cell behaves *exactly* like a diode, in all respects. Its $J(V)$ curve is "ideally" given by the expression (see Sect. 4.2 above):

$$J = J_{\text{dark}} = J_0 \left[ \exp\left( \frac{qV}{nkT} \right) - 1 \right] \tag{4.11}$$

As soon as light hits the solar cell, a photo-generated current density $J_{\text{ph}}$ will be superimposed on the dark current density $J_{\text{dark}}$. Equation 4.11 then becomes:

$$J_{\text{illum}} = J_{\text{ph}} - J_{\text{dark}} = J_{\text{ph}} - J_0 \left[ \exp\left( \frac{qV}{nkT} \right) - 1 \right] \tag{4.12}$$

There is now a negative sign in front of $J_{\text{dark}}$, because the photo-generated current and the diode forward current flow in opposite directions. The photo-generated current is composed of holes flowing to the $p$-side and electrons flowing to the $n$-side, whereas in the forward operation of a (dark) diode, the flow of carriers is just the opposite. This situation is schematically sketched in Figure 4.17, both for the *pn*-diode, discussed in the present Section, as well as for the *pin*-diode, which will be discussed later in Section 4.5 to 4.9.

Equation 4.12 implicitly assumes that the photo-generated current and the dark current of the diode can simply be additively superimposed, without any additional term. This is only true for the *pn*-diode; the *pin*-diode will require an additional term. Furthermore, even for the simple (and "classical") case of the ideal *pn*-diode, it is in no way a trivial assumption. Indeed, photo-generation completely changes the carrier profiles within the diode; because of this, it becomes in fact actually necessary to solve the drift-diffusion equations for the semi-conductor diode once again for the case of photo-generation. However, because the relationships used for deriving the $J(V)$ curve of the dark diode are basically linear, such a superposition *can indeed, at least intuitively, be justified* within a quite general framework [Green-1982, Möller-1991]. Additionally, the superposition principle is, also, well demonstrated by a host of experimental results on actual solar cells.

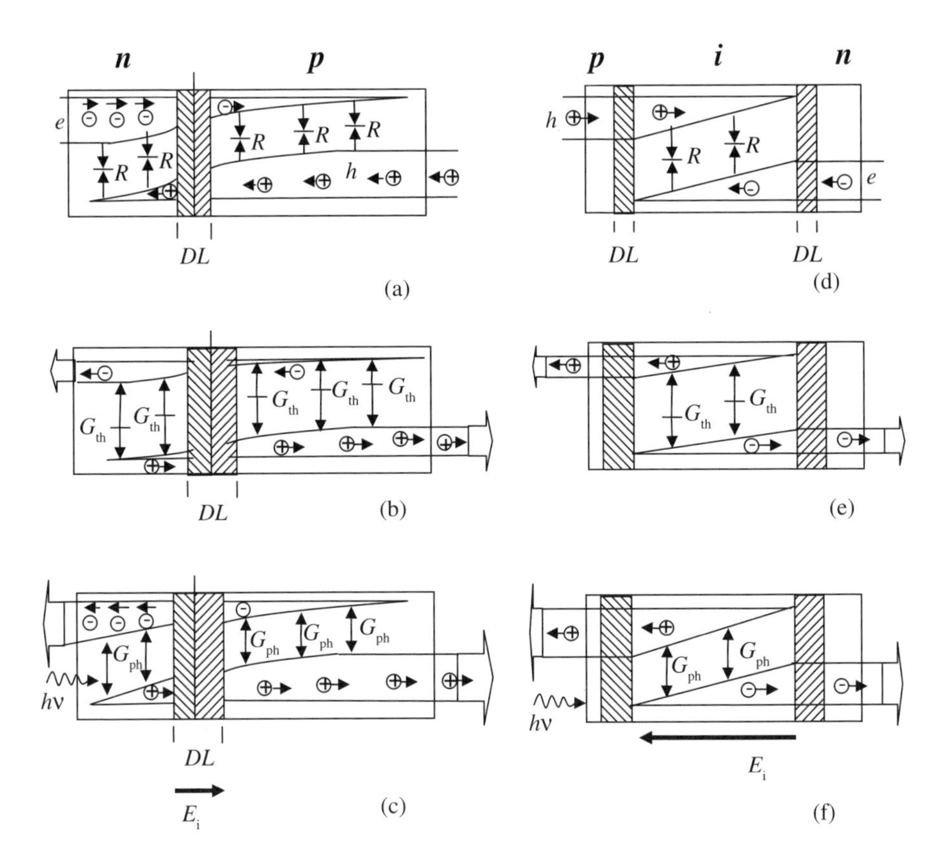


**Fig. 4.17** Flow of electrons and holes; recombination and generation in diodes and solar cells: (a), (b), (c) *pn*-diode, typical for wafer-based c-Si solar cells; (d), (e), (f) *pin*-diode, typical for thin-film silicon solar cells. (a), (d): dark diode, forward-biased; (b), (e): dark diode, reverse-biased; (c), (f): illuminated diode (solar cell) under reverse bias; the forward biased conditions for illuminated diodes (solar cells) are given by the superposition principle (for *pin*-diodes see Sect. 4.9.1). Abbreviations: *e*: electrons; *h*: holes; *DL*: depletion layer; *R*: Recombination; $G_{th}$: thermal generation; $G_{ph}$: photo-induced generation; $h\nu$: incoming light (photons), $E_i$: internal electric field.

The superposition principle is therefore commonly accepted for all solar cells of the *pn*-type.

An extended form of the superposition principle can be applied to solar cells of the *pin*-type, and will be introduced in Section 4.9.

The *actual carrier profiles* within an illuminated *pn*-type solar cell are shown schematically in Figure 4.18, which is based on [Green-1982].

One may note in Figure 4.18 how the minority carriers, especially the holes within the *n*-type region, are transported by diffusion to the depletion layer (actual *pn*-junction) where they are "whisked away" by drift (i.e. by the internal electric field within the depletion region) to the other side of the junction, i.e. to the *p*-type region.

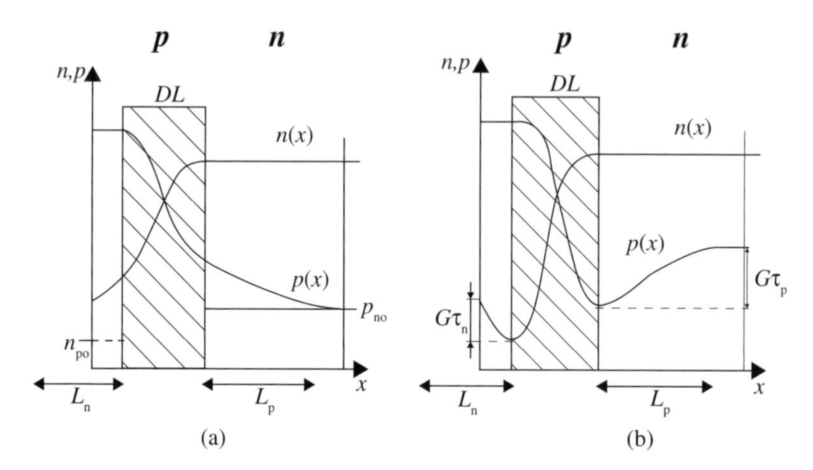


**Fig. 4.18** Carrier profiles (electron $n(x)$ and hole $p(x)$ concentrations) within a *pn*-type solar cell in the dark with forward bias (a), and under illumination in short-circuit condition (b). Adapted from [Green-1982].

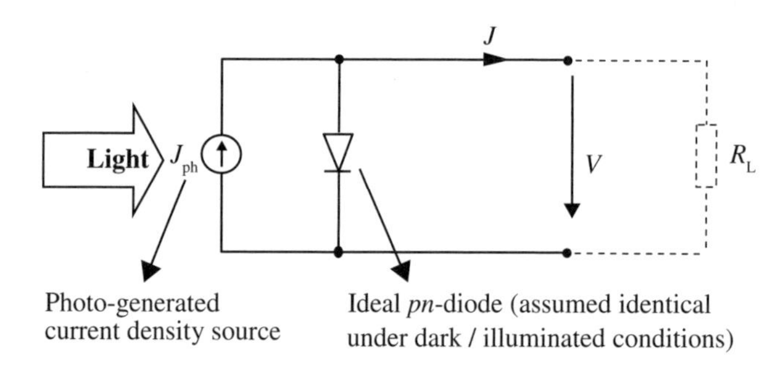


**Fig. 4.19** Simplest equivalent circuit, for an "ideal" solar cell; an external load resistance $R_L$ has also been drawn.

Thus, electrons travel from left to right in the above structure. The opposite is true for holes, which travel from right to left. Therefore one can say that within a *pn*-type solar cell, (minority) carrier transport is by diffusion, and carrier separation is (as in *all* solar cells) by drift. The limiting parameter for carrier transport and collection is here given by the minority carrier diffusion length (in this case, by the diffusion length of the holes).

Equation 4.12 leads to the solar cell equivalent circuit shown in Figure 4.19; this equivalent circuit contains just a diode and a current density source for the photo-generated current. The corresponding $J(V)$ characteristics are shown in Figure 4.20.

### 4.3.2 Limitations of a "real" diode (under illumination)

In reality, other circuit elements have to be added to the basic solar cell equivalent circuit of Figure 4.19. These are the series resistance $R_s$ and the parallel resistance $R_p$.

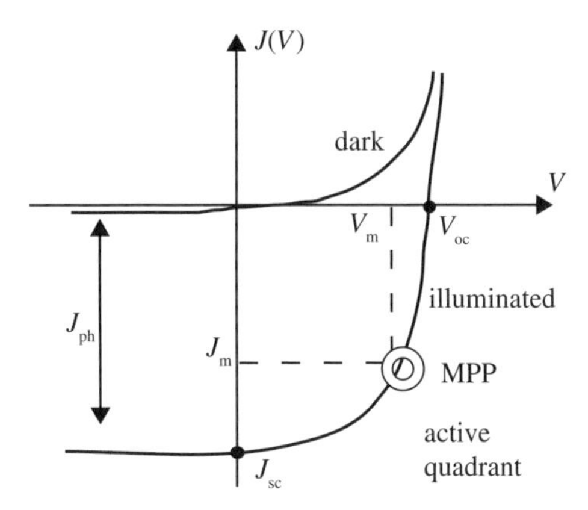


**Fig. 4.20** $J(V)$ characteristics of an "ideal" solar cell (*pn*-diode) under dark and illuminated conditions. According to the superposition principle, the curve for illuminated conditions is exactly the same as the one for dark conditions, but is simply shifted down by the value $J_{ph.}$ The point "MPP" is the "Maximum Power Point", i.e. the operating point at which the solar cell gives maximum power to the attached load (see Sect. 4.3.3, below).

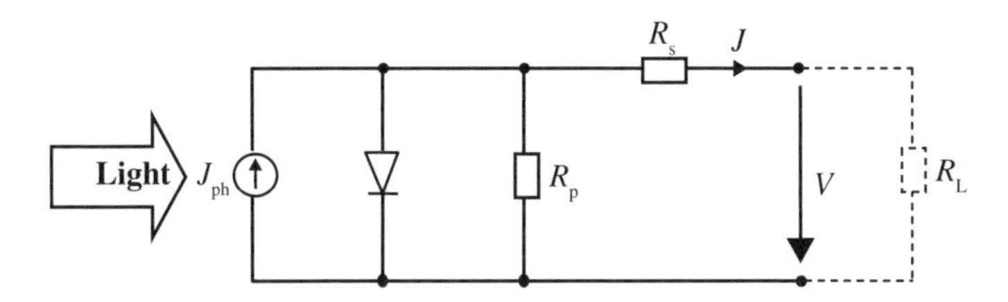


**Fig. 4.21** Universally used equivalent circuit for a solar cell.

We thereby obtain the equivalent circuit of Figure 4.21, which is the one universally used to represent a solar cell within electrical systems.

The series resistance $R_s$ represents thereby ohmic resistance in the connection wires up to the solar cell and also other ohmic resistances within the solar cell, between the external connection point and the actual active device.

The parallel resistance $R_p$ is employed not only to designate actual ohmic shunts, but also, and mainly, to symbolize the recombination losses within the solar cell. Physically, this is not a satisfactory way of representing the recombination losses: the parallel resistance $R_p$ does render the dependence of the recombination losses on the solar cell voltage $V$ more or less correctly, but it does not at all show the dependence of the recombination losses on the photocurrent $J_{ph}$. Still, if one is designing an electrical system design around the solar cell, then this representation is truly a very convenient way. It is an acceptable representation, as long as the light intensity, i.e. the value of the photo-generated current density $J_{ph}$, does not vary significantly. If we have a situation where the variation of the light intensity is important, we need another representation: Indeed the higher the light intensity, the higher the value of $J_{ph}$ becomes, but also the higher are the (minority) carrier

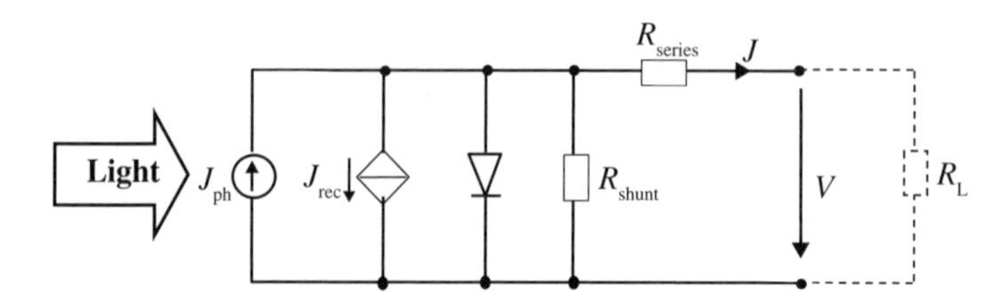


**Fig. 4.22** "Physically acceptable" equivalent circuit for a "real" solar cell, including series resistance $R_{series}$, shunt resistance $R_{shunt}$, as well as a recombination current density sink $J_{rec}$. The value of the latter is proportional to $J_{ph}$, but also depends on the external voltage. In [Merten-1998] precise expressions are given for the dependence of $J_{rec}$ on $V$, derived, however, for the case of thin-film silicon solar cells of the *pin*-type (see Sect. 4.9.1). So far, this equivalent circuit has *not* been used for *pn*-type solar cells.

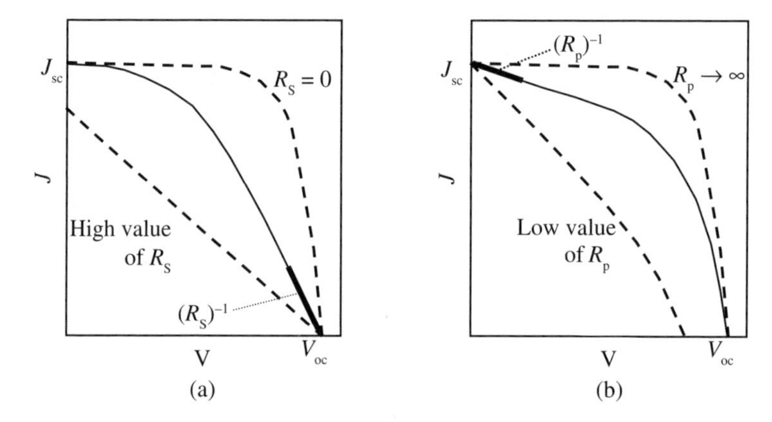


**Fig. 4.23** Effect of (a) the series resistance $R_s$ and of (b) the parallel resistance $R_p$, as given in the "universal" circuit of Fig. 4.22 on the $J(V)$ characteristics of a solar cell. Values of $(R_s)$–1 and $(R_p)$–1 are given by the slopes of the $J(V)$ characteristics, at the short-circuit point ($V = 0$) and at the open-circuit point ($J = 0$), respectively.

concentrations within the solar cell (see Fig. 4.18); the latter basically increase proportionally to $J_{ph}$. Also at higher light intensity, the recombination losses increase (in absolute terms). This cannot be represented by a constant parallel resistance $R_p$, but rather leads us intuitively to the equivalent circuit of Figure 4.22. The equivalent circuit of Figure 4.22 contains a "recombination current density sink $J_{rec}$", i.e. a circuit element representing the fact that recombination losses are proportional to $J_{ph}$. It was introduced by Merten and Shah [Merten-1998] for use in thin-film *pin*-type solar cells. It would be interesting to investigate experimentally how far it can also be used for *pn*-type solar cells.

The equivalent circuit of Figure 4.22 models quite well the physics behind the functioning of the solar cell (at least for *pin*-type solar cells), but it is far too unwieldy to be used in electrical system design. For this reason, the simpler equivalent circuit of Figure 4.21 still continues to be universally used.

The effect of the parallel resistance $R_p$ and of the series resistance $R_s$ on the $J(V)$ characteristics is shown schematically in Figure 4.23.

### 4.3.3 Maximum power point (MPP) and fill factor (*FF*) of a solar cell

In order for the solar cell to deliver maximum power to a load, it has to be operated at a point known as the "Maximum Power Point (MPP)". The power delivered by the solar cell to a load is proportional to the product *I* times *V*, if *I* and *V* are the current and the voltage at the external contacts of the solar cell. They are linked to each other by the load resistance $R_L$, according to Ohm's law: $V = R_L \times I$. In practice the load resistance $R_L$ has to be chosen in such a way as to maximize the product $(I \times V)$. This is done by electronic circuits called Maximum Power Point Trackers. Or, in a simpler manner, it is often done just in a crude way by selecting a fixed load resistance that more or less keeps the solar module near its MPP for "usual" illumination conditions.

At this stage, we are not concerned with the external circuit, but just wish to compute the maximum power that the solar cell could ideally and hypothetically deliver to a well-adjusted load. For this the solar cell itself has to be operated at the "Maximum Power Point (MPP)", as shown schematically in Figure 4.24. Note that in accordance with the other parts of this section, we use in Figure 4.24 "current densities *J*" and not simply currents *I*. The relationship between the two parameters is simply given by the solar cell area *A*, with: $I = A \times J$.

The fill factor *FF* is defined as the ratio of the maximum output power (density) of the solar cell to the product of the short-circuit current (density) $J_{sc}$ times the open-circuit voltage $V_{oc}$:

$$FF = \frac{J_m V_m}{J_{sc} V_{oc}} \tag{4.13}$$

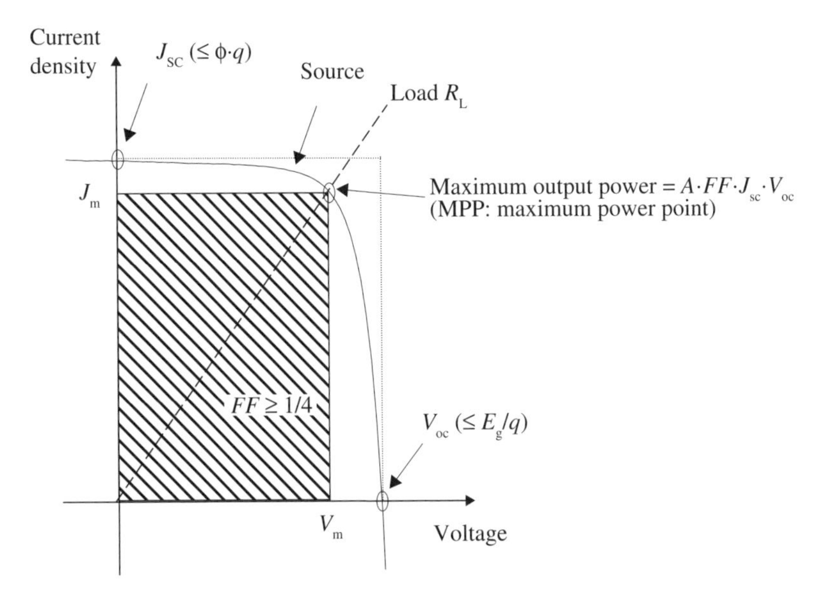


**Fig. 4.24** Nonlinear source characteristic of a typical solar cell (schematically drawn) and indication of the maximum power point MPP. This point defines the fill factor *FF*, according to Equation (4.13) hereunder.

where $J_m$, $V_m$ are the current (density) and voltage, respectively, corresponding to the maximum power point MPP. For a solar cell of good quality, one obtains *FF* values between 0.7 and 0.85 (i.e. between 70 % and 85 %).

It is interesting to note that, if the solar cell would constitute a linear source, the fill factor would be as low as 25 % (see Fig. 4.25). In this case, the maximum power point MPP is given by a voltage that is simply ½ $V_{oc}$ and a current (density) that is also simply half $J_{sc}$.

The solar cell's output characteristic is, however, strongly non-linear; it is in fact basically given by an exponential function, with the expression $(qV/nkT)$ in its exponent; this allows for fill factor values that are well above 50 %. One can indeed easily visualize that the value of the fill factor depends on the ratio $\tilde{v}_{oc} = (qV_{oc})/(nkT)$.

The higher this ratio, the more the $J(V)$ curve of the solar cell will be strongly curved and approach a rectangle; the lower this ratio, the more the $J(V)$ curve of the solar cell will look like a straight line. Figure 4.26(a) and (b) schematically show how this looks for $n = 1$, $T = 300$ K (approx. 25 °C) and for four hypothetical cases of ideal solar cells with bandgap energies $E_g$ between 0.7 V and 2.15 V.

The fill factor *FF* is a very sensitive indicator of solar cell quality; this will be discussed in more detail, and specifically for *pin*-type thin-film silicon solar cells, in Section 4.9.5.

### 4.3.4 Basic solar cell parameters $J_{sc}$, $V_{oc}$, *FF*

***Short-circuit current density $J_{sc}$***

In the short-circuit condition, i.e. when no voltage is applied ($V = 0$), Equation 4.12 gives in the case of an ideal illuminated diode:

$$J_{sc} = J_{ph} \tag{4.14}$$

In concrete terms, the electrical current delivered by the solar cell directly depends on the quantity of solar photons absorbed by the semiconductor and on the elementary charge $q$. Assuming that losses are neglected, the theoretical maximum value for the short-circuit density is then given by:

$$J_{sc}^{max} = \phi \cdot q \tag{4.15}$$

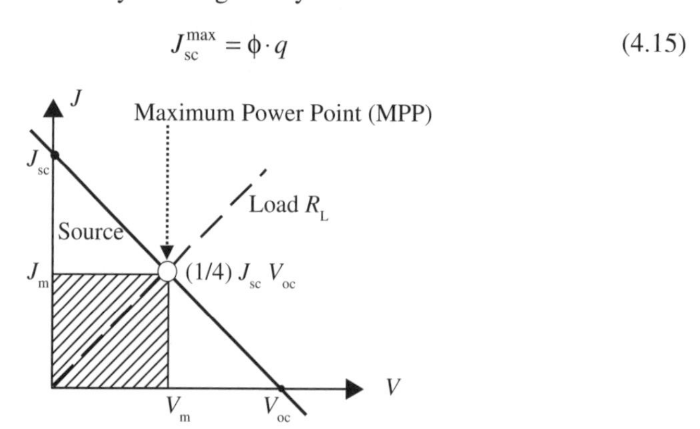


**Fig. 4.25** Hypothetical case of solar cell with linear source characteristic and $FF = 25$ %.

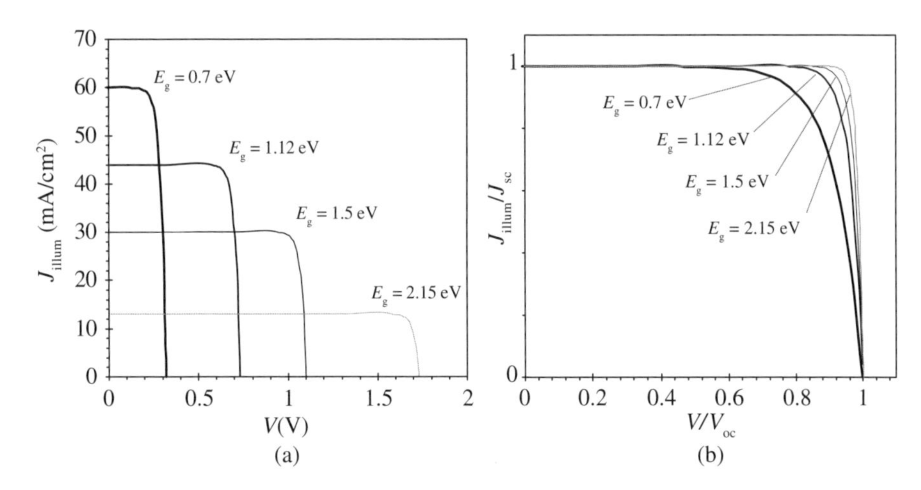


**Fig. 4.26** Theoretical and ideal solar cell $J(V)$ curves: (a) in absolute units and (b) in relative units. These curves are drawn for ideal, hypothetical solar cells with diode quality factor $n = 1$ and $J_{00} = 1.5 \times 10^8$ mA/cm$^2$ (see Eqs. 4.6 to 4.8); 4 different bandgap energies $E_g$ have been assumed: $E_g = 0.7$ eV (corresponding to Ge), 1.12 eV (corresponding to c-Si), 1.5 eV (corresponding to GaAs) and 2.15 eV (corresponding to Se). One clearly sees here how with an increase in $V_{oc}$ (i.e. an increase in $E_g$), the $J(V)$ curve becomes more curved and approaches a rectangle, and the value of the *FF* correspondingly increases.

where $\phi$ is the number of electron-hole pairs generated per unit time and depends on the bandgap energy of the semiconductor and on the spectral distribution of the light (see also Sect. 4.1.1).

Figure 4.27(a) shows the curve for the theoretical maximum value of $J_{sc}$ as a function of the semiconductor bandgap, considering AM1.5 illumination, as well as some experimental laboratory values obtained for very good solar cells made of different semiconductors [Green-2010]. The theoretical physical upper limit is based on an "ideal" *pn*-diode where (A) all the useful light is absorbed (i.e. a *pn*-diode with sufficient thickness), and (B) the collection is perfect (i.e. no recombination takes place).

Also given in Figure 4.27(a) is, as a straight line, a simple approximation function for the maximum value of the short-circuit current density:

$$J_{sc} \approx 80 \text{ mA/cm}^2 - 34 \text{ mA/(cm}^2\text{eV)} \times E_g; \; E_g \text{ in eV} \tag{4.16}$$

Figure 4.27(a) shows that the short-circuit current density $J_{sc}$ increases as the bandgap energy $E_g$ decreases: a fact that can be very easily understood as more photons of the solar spectrum have sufficient energy to create electron-hole pairs when the gap decreases.

Note that $J_{sc}$ is only very slightly affected by a change in temperature. $E_g$ decreases when the temperature increases, so that the absorption of light is facilitated. However, this effect is quite small and the increase of $J_{sc}$ with temperature is almost negligible.

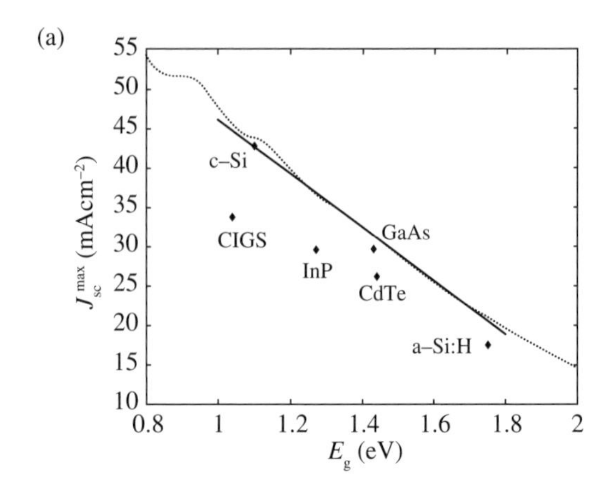


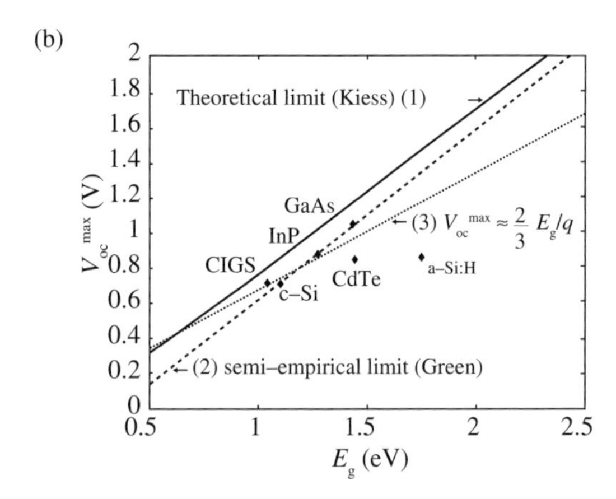


**Fig. 4.27** "Theoretical" limit values and experimentally obtained values, as a function of the bandgap energy $E_g$, for single-junction solar cells under 1 sun-illumination (AM 1.5 spectrum) and at 25 °C: (a) for the short-circuit current density $J_{sc}$; (b) for the open-circuit voltage $V_{oc}$; (c) for the fill factor $FF$; and (d) for the energy conversion efficiency η. Courtesy of M. Stückelberger, IMT PV Lab, Neuchâtel.

### *Open-circuit voltage $V_{oc}$*

The open-circuit voltage $V_{oc}$ can be found by setting $J_{illum} = 0$ in Equation 4.12, rendering:

$$J_{ph}/J_0 = \left[\exp\left(\frac{qV_{oc}}{nkT}\right) - 1\right] \tag{4.17}$$

This leads to:

$$V_{oc} = \frac{nkT}{q}\ln\left(\frac{J_{ph}}{J_0} + 1\right) \approx \frac{nkT}{q}\ln\left(\frac{J_{ph}}{J_0}\right) \tag{4.18}$$

We can immediately note that $V_{oc}$ depends on the value of the photo-generated current density, i.e. on the illumination level. By decreasing the incoming light by a

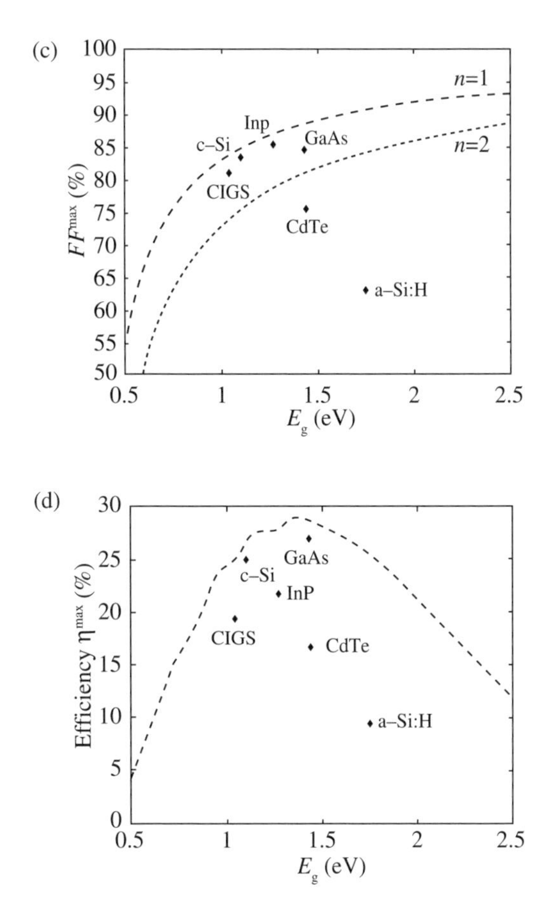


**Fig. 4.27** (Continued)

factor of 10, i.e. by a factor of approximately $e^2$, $V_{oc}$ will decrease by roughly $n$ times 52 mV, because $kT/q$ is approximately 26 mV at room temperature.

$V_{oc}$ also depends on temperature; this is on one hand given by the pre-factor $(nkT/q)$, but it is also given by the quantity $J_0$, which has a very strong temperature dependence. As a net result, $V_{oc}$ always decreases with increasing temperature $T$. In a similar manner, $V_{oc}$ always decreases if the diode quality factor $n$ increases (indicating a diode with a *lower* quality!).

Experimentally, one of the key quantities determining $V_{oc}$ is the diode reverse saturation current $J_0$. This parameter will be discussed in more detail in Section 4.5.13, in the context of *pin*-type solar cells.

At First, we will just look at the practical (semi-empirical) limit value for $J_0$ given by Martin Green [Green-1982], which was already given in Section 4.2:

$$J_0 = 1.5 \times 10^8 \cdot \exp\left(\frac{-E_g}{kT}\right) \qquad \text{in} \left[\frac{\text{mA}}{\text{cm}^2}\right]$$

where $E_g$ is the bandgap energy.

This leads to the following expression for $V_{oc}$, if one sets at the same time $n = 1$:

$$V_{oc} = \frac{kT}{q}\ln\left(\frac{J_{ph}}{J_0}+1\right) \approx \frac{kT}{q}\ln\left(\frac{J_{ph}}{J_{00}}\right)+\frac{E_g}{q}, \tag{4.19}$$

where one has, for the practical limit case given by [Green-1982] (see Equations 4.6, 4.7 and 4.8) :

$$J_{00} = J_{00}^{\text{Green}} = 1.5 \times 10^8 \,[\text{mA/cm}^2]$$

Note that $\ln(J_{ph}/J_{00})$ is always negative, so that $V_{oc} < E_g/q$.

On the other hand, there exists a *fundamental thermodynamic limit* for $V_{oc}$, as given for example by Schockley and Queisser [Schockley-1961] and by Kiess and Rehwald [Kiess-1995]. The latter derive the following expression for $V_{oc}$:

$$V_{oc} = E_g/q + (kT/q)\ln\left\{(h^3c^2/2\pi kT)(N_{\text{incident}}/E_g^2)\right\}, \tag{4.20}$$

where $h$ is Planck's constant, $c$ the velocity of light and $N_{incident}$ the number of photons incident per unit area and unit time with energies $h\nu > E_g$. $N_{incident}$ corresponds to our quantity $\phi$ (see above). Fanny Sculati-Meillaud shows in her Ph.D. thesis [Sculati-Meillaud-2006] that the theory of Schockley and Queisser [Schockley-1961] leads to practically the same limit value as the developments by Kiess and Rehwald [Kiess-1995].

In Figure 4.27(b) we show: (1) the fundamental, theoretical limit for $V_{oc}$ (Eq. 4.20), (2) the semi-empirical limit (Equation 4.19) according to Martin Green's formula for $J_0$, and (3) a simple rule of thumb $V_{oc} \approx 2/3\ (E_g/q)$, which is very easy to memorize and is therefore a useful approximation in the range of bandgaps, which normally concerns us in silicon solar cell work.

We will later on (see Sect. 4.5.13) look at the limit values and the experimentally obtained values for the open-circuit voltage $V_{oc}$, in the case of the *pin*-type solar cell. Here, $J_0$ is always higher and $V_{oc}$ always lower than for a *pn*-type solar cells (because of recombination/thermal generation phenomena within the *i*-layer, and because $n \rightarrow 2$).

### *Fill factor FF*

The limit values for the fill factor *FF* are found by taking the exponential function given as Equation 4.12, and by computing the product of voltage × current, i.e. by evaluating

$$P_{ideal} = (J_{illum} \times V),$$

where $P_{ideal}$ denotes the limit value for the output power density that the solar cell can deliver in the ideal case, corresponding to Figures 4.19 and 4.24. We then mathematically maximize $P_{ideal}$ as a function of $V$, to find the maximum power point MPP (in this ideal case) and compute the quantity

$$FF = P_{max}/(V_{oc} \times J_{sc}) = (V_m \times J_m)/(V_{oc} \times J_{sc}).$$

This gives us the expression [Green-1982, Green-1983]:

$$FF = \frac{\tilde{v}_{oc} - \ln(\tilde{v}_{oc} + 0.72\text{V})}{\tilde{v}_{oc} + 1} \tag{4.21}$$

with:

$$\tilde{v}_{oc} = \frac{V_{oc}}{nkT/q}$$

Inserting the semi-empirical limit value for $V_{oc}$, from the preceding paragraph (curve (2) in Fig. 4.27(b)), we obtain the curves[5] shown in Figure 4.27(c) for the fill factor $FF$, for $n = 1$ and $n = 2$.

An approximation for the $FF$ as a function of the bandgap $E_g$ (in eV) is $FF = 1 - (n \times 15\,\%/E_g); n = 1, 2$. This approximation follows, for bandgaps between 1 eV and 2 eV, the curves in Figure 4.27(c) with a difference of less than 2 to 3 %.

We will later (see Sect. 4.9.5) look, in a closer way, at the limit values and the experimentally obtained values for the fill factor $FF$, in the case of the *pin*-type solar cell; they correspond to the case $n = 2$ in Figure 4.27(c).

## 4.4 LIMITS ON SOLAR CELL EFFICIENCY

### 4.4.1 Limits at standard test conditions (STC)

To calculate the *overall energy conversion efficiency* $\eta$ of the solar cell, we must now divide the electrical output power $P_{out}$ (at the maximum power point) by the input (solar) power $P_{in} = P_{sun}$. One has:

$$P_{out} = V_m \times J_m = J_{sc} \times V_{oc} \times FF \tag{4.22}$$

[5]The curves for $V_{oc}$ in Figure 4.27(b) have been computed for the case $n = 1$; in fact, curve (2) in Figure 4.27(b) was obtained from Equation 4.19, by specifying $n = 1$; in order to obtain the corresponding curves for $V_{oc}$ for the case $n = 2$, we need to know more about the recombination within the depletion layer or within the intrinsic layer of the solar cell; this will be investigated for *pin*-type solar cells in Section 4.9 and depends on the defect densities prevailing in the intrinsic layer: no general conclusion can be made. We have therefore here simply combined Equation 4.19 for $V_{oc}$ with $n = 1$ and Equation 4.21 with $n = 2$, in order to obtain a coarse but general approximation for the fill factor in solar cells, where $n \to 2$, as shown in Figure 4.27(c).

where $J_{sc}$, $V_{oc}$ and $FF$ have been calculated above. On the other hand, $P_{in} = P_{sun}$ is the "power of the sun", i.e. the total energy within the solar spectrum per unit of time. It was already used in Section 4.1.1 to calculate the *spectral conversion efficiency* $\eta_S$ (Eq. 4.2).

Using the previous result for $\eta_S$ we can now write:

$$\eta = \eta_S \times (J_{sc} \times V_{oc} \times FF)/(\phi\, E_g) \tag{4.23}$$

For the AM 1.5 solar input spectrum, the value $J_{sc}$ is given in Figure 4.27(a) as a function of the semiconductor bandgap energy $E_g$. Thereby, one had assumed that *all* incoming photons with a quantum energy $h\nu$ higher than the bandgap energy $E_g$ usefully contribute to the current $J_{sc}$, whereas *no* photons with a quantum energy lower than $E_g$ contribute to the current $J_{sc}$. This means that one had assumed that the solar cell absorbs all incoming light with wavelengths shorter than the absorption edge of the semiconductor, and that collection within the solar cell is ideal.

From Equations 4.19, 4.21 and 4.22 and from Figure 4.27(a), one can now compute, numerically, the semi-empirical limit efficiency for single-junction solar cells. Thereby, $V_{oc}$ is determined by the diode reverse saturation current $J_0$ according to Martin Green's formula (Eq. 4.6). The result is represented in Figure 4.27(d) as a function of the semiconductor bandgap energy.

From Figure 4.27(d) we can see that the optimum bandgap for a single-junction solar cell is around 1.5 eV. This corresponds to the bandgap of GaAs. This is one of the reasons why GaAs is used to obtain solar cells with particularly high efficiencies. Crystalline silicon has a slightly lower bandgap of 1.1 eV, but this decrease in bandgap just accounts for a drop in efficiency of about 2.5 %, from 29 % to 26.5 %. Indeed the maximum in Figure 4.27(d) is a quite broad maximum. Surprisingly enough, experimental values as high as 25.0 % have so far been obtained for high-quality single-junction (c-Si) laboratory cells [Green-2010]. This shows that crystalline silicon (c-Si) technology is now a very mature technology.

In practice, commercial cells and modules made of c-Si have efficiencies in the range of 12 to 20 %, depending upon the sophistication of their fabrication process, as well as on the detailed design of the solar cell itself. For more details on this topic, the reader is referred to the books by Martin Green: [Green-1982, Green-1987, Green-1995].

It is interesting to note that the maximum of the curve has been shifted to higher bandgap values if one looks at Figure 4.27(d) and compares this figure to Figure 4.11. The maximum in Figure 4.11 was around 1.1 eV, i.e. the highest conversion efficiency for the *first step* of photovoltaic energy conversion (generation of electron-hole pairs) is obtained for semiconductors with a bandgap of 1.1 eV. The maximum is now shifted to 1.5 eV for the overall conversion process. This means that the *second step* of photovoltaic energy conversion (separation of electron-hole pairs) is more efficient for higher bandgaps. This is not at all surprising: The higher the bandgap, the higher is the internal electric field within the diode, and it is precisely this field that separates holes from electrons. In terms of the solar cell parameters $J_{sc}$, $V_{oc}$ and $FF$, it is mainly the $FF$ which increases with the bandgap (as shown in Fig. 4.27(c)), whereas the product $V_{oc} \times J_{sc}$ remains more or less constant.

Is it possible to increase the solar cell conversion efficiency over the limit values shown in Figure 4.27(d)? Yes, and this can be done in a very straightforward manner, by one of two strategies:

- *Use concentration of sunlight*: By this method $V_{oc}$, and indirectly also *FF*, will be increased as can be seen from Equation (4.18): if $J_{ph}$ is increased, then the ratio $J_{ph}/J_0$ also increases and $V_{oc}$ increases according to a logarithmic function. In practice, single-junction silicon solar cells have been reported to obtain 27.6 % efficiency under approximately 100 suns (100-times concentration of sunlight) [Green-2010].
- *Use multijunction cells*, e.g. by combining silicon with a material with higher bandgap; the best results have been obtained here not with silicon, but with various alloys of gallium, partly combined with germanium as low-gap material. The record efficiencies so far measured are for triple-junction cells: 35.8 % for 1 sun and 41.6 % for 364 suns [Green-2010].

At this stage it is useful to discuss how the cell efficiency $\eta$ varies if the operating conditions of the cell vary. We may thereby consider all the following possible variations:

- variation in light intensity (already partially considered);
- variation in the operating temperature (the standard value is 25 °C, i.e. almost 300 K; typically solar cells/modules operate at much higher temperature values, up to about 80 °C, i.e. up to about 350 K);
- variation in the spectrum of the incoming light: the standard spectrum under which all solar cell test measurements are carried out is the AM 1.5 spectrum; as the sun descends towards the horizon in the evening, or when the sun rises in the morning, we have a spectrum AM *k*, where *k* increases as the sun's light reaches us from a point that is further away from the zenith and nearer to the horizon; typical spectra are AM 3 and AM 4 spectra; they are red-shifted with respect to the standard AM 1.5 spectrum; on the other hand, on a clear winter day with blue sky but without direct sunlight, the spectrum is blue-shifted.

### 4.4.2 Variation in light intensity

Here, as long as we only consider the *ideal* solar cell, $\eta$ is clearly reduced as the light intensity is decreased; this occurs because of the corresponding decrease in $V_{oc}$ according to Equation 4.18. Because of the decrease in $V_{oc}$, there also will be, correspondingly, a slight decrease in *FF*. For the ideal *pn*-diode (diode ideality factor $n = 1$) and for $T = 300$ K, we have $nkT/q = 26$ mV, which means that for every factor 10 decrease in light intensity, $V_{oc}$ will be reduced by ln(10) = 2.303 times 26 mV, i.e. by about 60 mV. (In the case of the "ideal" pin-diode ($n = 2$), the decrease of $V_{oc}$ per decade of decrease in light intensity is double, i.e. 120 mV.)

In the actual, *real* solar cell, we have to look into the effects of $R_{shunt}$ and $R_{series}$, as given in Figure 4.22: The effect of $R_{shunt}$ is to cause a pronounced fall-off of $\eta$ towards very low light intensities. The effect of $R_{series}$ is to induce a fall-off of $\eta$ towards high light intensities. The net result is a curve like those given in Figure 4.28. The curves in Figure 4.28 are typical measurement results for silicon solar cells designed for non-concentrating solar applications, and having a maximum of $\eta$ at about 1 sun

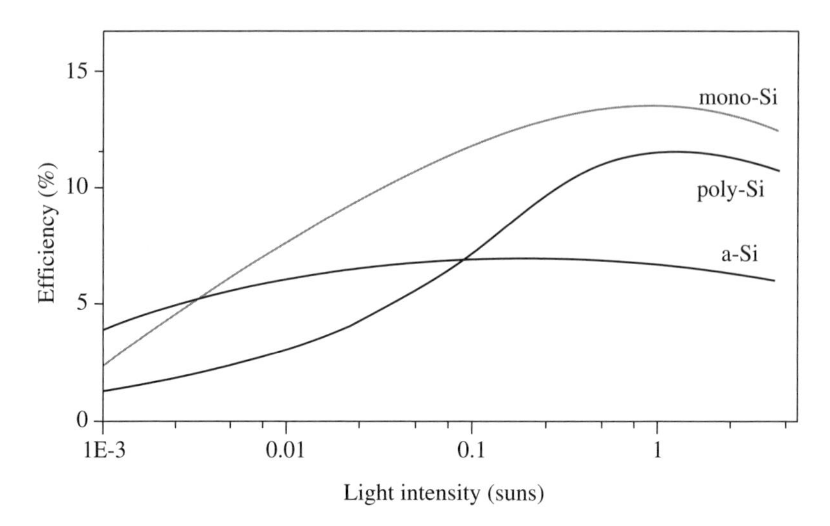


**Fig. 4.28** Effect of light intensity on solar cell efficiency for three typical silicon solar cells (monocrystalline silicon, polycrystalline silicon and amorphous silicon). Note that for cells made of the same material, the behavior at low levels of light intensity can vary a lot, depending on the fabrication process and the cell design.

(or 100 mW/cm$^2$); concentrator cells, on the other hand, are so designed as to have their maximum at a much higher light level, e.g. at 100 suns (or even more). Needless to say, they are designed with very low values of $R_{series}$.

A word of caution is warranted here: $R_{shunt}$ is a solar cell parameter that strongly depends on the fabrication process and the detailed structure of the solar cell. There is, thus, a large variation in $R_{shunt}$ for solar cells made from the same material and even for solar cells from the same manufacturing lot. The solar cells shown in Figure 4.26 are typically solar cells selected for having an excellent low-light performance. One will, on the other hand, easily find amorphous silicon solar cells, which have a much lower efficiency already at 0.01 suns.

## 4.4.3 Variation in operating temperature

This effect is of very great importance for the practical application of solar cells and solar modules. The operating temperature can, in fact, easily be as high as 80 °C or more. In general, solar cell efficiency decreases with an increase in operating temperature. At 80 °C, the efficiency can be about 20 to 30 % lower than at Standard Test Conditions (STC, 25 °C).

The solar cell parameter that varies most with temperature is $V_{oc}$, whereas $FF$ and $J_{sc}$ hardly change.

If we take Equation 4.19 to compute $\partial V_{oc}/\partial T$, we find

$$\frac{\partial V_{oc}}{\partial T} = \frac{k}{q} \ln \frac{J_{ph}}{J_{00}} = \frac{k}{q} \left\{ \ln \frac{J_{ph}}{J_{00}^{Green}} - \ln \frac{J_{00}^{Green}}{J_{00}} \right\} \tag{4.24}$$

Note that $k/q = 0.086$ mV/K and that $J_{00}$ is a "quality parameter" defining the reverse saturation current of the diode; in the semi-empirical optimal case $J_{00} = J_{00}^{Green}$.

The second term within the brackets in Equation (4.24) is, thus, a correction factor, which does not depend on $E_g$, but depends on the quality of the solar cell; in the optimal case it becomes zero; for "real" solar cells, it may add between 2 and 12 times {0.86 (mV/°C)} to the magnitude of $\partial V_{oc}/\partial T$; the first term within the brackets can be rewritten, by setting $J_{ph} = J_{sc}$ and by using Equation 4.16 to express $J_{sc}$:

$$\frac{\partial V_{oc}}{\partial T} \approx \frac{k}{q}\ln\left\{\frac{80\,\text{mA/cm}^2 - 34\,\text{mA/cm}^2 \times E_g}{J_{00}^{\text{Green}}}\right\} + \text{correction factor}$$

The above expression can be evaluated numerically; one sees that in the ideal case (correction factor zero, i.e. minimum recombination) $\partial V_{oc}/\partial T$ is between –1.5 and –1.6 mV/°C and hardly depends on $E_g$; in the "real" case (higher recombination in actual solar cells) it may become (depending on the value of the "quality parameter" $J_{00}$ ) as low as –2.5 mV/°C.

We can now divide the result by $V_{oc} \approx 2/3\ (E_g/q)$, so as to obtain the *relative* variation of $V_{oc}$ i.e. to obtain $(\partial V_{oc}/\partial T)/V_{oc}$. As very coarse approximation we set $(\partial\eta/\partial T)/\eta \approx (\partial V_{oc}/\partial T)/V_{oc}$.

Taking the lower limit (–2.5 mV/°C) for $\partial V_{oc}/\partial T$, we obtain, for the relative variation of the efficiency $\eta$ with temperature, the approximate relationship (valid for temperatures around 25°C):

$$(\partial\eta/\partial T)/\eta \approx -0.4\ \%/^\circ\text{C} \times (1\text{eV}/E_g);\ E_g \text{ in eV} \qquad (4.25)$$

Equation 4.25 is just a very simple "rule of thumb", which enables us to roughly assess the order of magnitude of the temperature coefficient of the efficiency of solar cells and modules.

The following Table 4.2 gives a brief overview of the relative temperature coefficients (*TC*), in [%/°C], for different types of solar cells. The solar cells are classified according to the semiconductor material out of which they are made. The values from literature are mainly from the work of Wilhelm Durisch; a few other, selected authors are also cited. The values from the datasheets are purely indicative: a full survey of the *TC* of commercially available solar modules would be far beyond the scope of the present book. In Figure 4.29, the measured relative decrease in conversion efficiency, as a function of operating temperature, is shown for eight different modules.

Let us conclude this Section on the temperature coefficient of solar cells/modules with three remarks:

(1) The experimental verification of the temperature coefficient of photovoltaic modules is relatively difficult, because it is, in general, not easy to evaluate the actual temperature of the semiconductor junctions, within the PV module, especially under outdoor conditions.

(2) In reality, not only $V_{oc}$ changes with temperature, but also $J_{sc}$ and *FF*; furthermore, $V_{oc}$ depends on recombination processes, which are quite complex in actual solar cells (see e.g. [Green-2003] and references cited therein). Therefore, the temperature behavior of a solar module should be directly measured for each type of commercial module.

**Table 4.2** Relative temperature coefficients $TC = (\partial\eta/\partial T)/\eta$ of photovoltaic solar cells and modules, in [%/°C], for cells/modules made of different semiconductor materials: Values from literature and from datasheets, compared with "theoretical" values from our Equation 4. 25.

| Material | $E_g$(eV) | *TC* a/c to Eq. (4.25) | *TC* literature | Author(s) | *TC* datasheets |
|---|---|---|---|---|---|
| c-Si "standard" cells | 1.1 | –0.36 | –0.40 | [Durisch-2008] | –0.38 to –0.48 |
| High-η c-Si (HIT) | 1.1 | –0.36 | –0.25 | [Maruyama-2006] | –0.3 |
| μc-Si:H | 1.1 | –0.36 | –0.4 | [Meier-1998] | – |
| a-Si:H, 2-junction | 1.75 | –0.23 | –0.23 | [King-2000] | –0.2 |
| a-Si,Ge:H/a-Si:H triple-junction | 1.3/ 1.5/1.75 | – | –0.14 | [Durisch-2007] | –0.21 |
| μc-Si:H/a-Si:H tandem | 1.1/ 1.75 | – | –0.26 | [Meier-1997] | –0.24 to –0.33 |
| CdTe | 1.44 | –0.28 | –0.16 | [Durisch-2010] | –0.25 |
| CIGS | ~1.2 | ~ –0.33 | –0.27 to –0.43 | [Durisch-2007] | –0.36 |

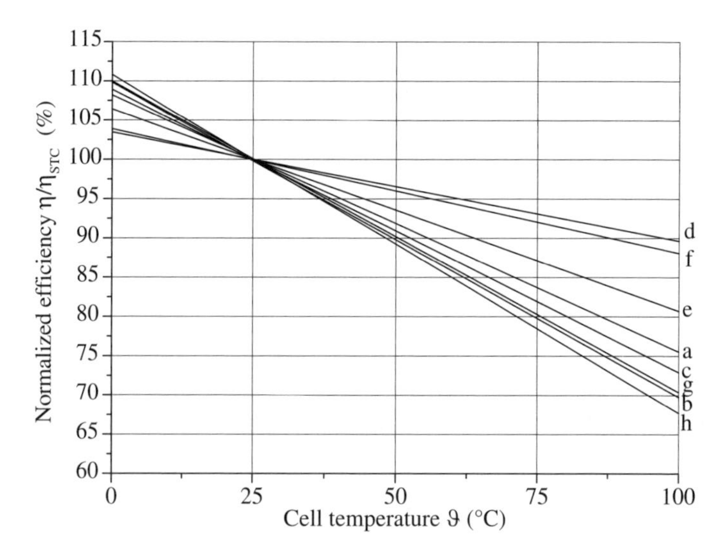


**Fig. 4.29** Representation of the effect of operating temperature $T$ on the normalized value of the efficiency $\eta_{rel} = \eta(T)/\eta(25\ °C)$ of typical solar modules, for various cell technologies: (a)-(c) wafer-based crystalline silicon (c-Si) modules; (e)-(h) thin-film modules; (a) high efficiency c-Si; (b) monocrystalline c-Si; (c) polycrystalline c-Si; (d) amorphous silicon; (e) "micromorph" tandem; (f) CdTe; (g), (h) CIGS. Courtesy of Wilhelm Durisch, PSI, Villigen, Switzerland.

(3) There exists indeed quite a spread for the values of *TC*, for different cells/modules, within the same semiconductor material. Thus, high efficiency c-Si cells/modules have, in general, much lower values of *TC* than "standard" c-Si

cells/modules; this is specifically the case for "HIT" cells (Heterojunction with Intrinsic Thin layer), see e.g. [Maruyama-2006].

### 4.4.4 Variation in the spectrum of the incoming light

The variation in the spectrum has a direct influence on $J_{sc}$. The exact value of $J_{sc}$ has to be calculated as a convolution between the spectrum of the incoming light and the spectral response of the solar cell. In the coarse approximation used so far (ideal solar cell), and described in Section 4.3.4, the spectral response of the cell is simply considered to be zero for $h\nu < E_g$ and one for $h\nu > E_g$.

The following are some simple rules, based on intuitive reasoning and on experience:

- If the bandgap energy $E_g$ of the semiconductor is lower than the optimum value of 1.5 eV (as is the case for crystalline silicon, c-Si and microcrystalline silicon, μc-Si:H), then a red shift in the spectrum of the incoming light will lead to an increase of efficiency and a blue shift to a decrease in efficiency.
- If the bandgap energy $E_g$ of the semiconductor is higher than the optimum value of 1.5 eV (as is the case for a-Si:H), then as a red shift in the incoming spectrum of the incoming light will lead to a decrease of efficiency and a blue shift to an increase in efficiency.
- We may note that under "average" climatic conditions, we have, on a sunny day, an AM1.5-like spectrum at noon, whereas we have a red shift of the spectrum (towards what is called an AM3- or AM4-like spectrum) during morning and evening. Snow and the vicinity of large water surfaces (lakes, large rivers, the sea) shift on the other hand, the spectrum towards the blue.
- Finally, we should note that single-junction cells are, in principle, less sensitive to variations in the spectrum of the incoming light than tandem and multi-junction cells (see also Sect. 5.3.2 in Chap. 5).

Examples in the literature [Durisch-2007] show that several single-junction c-Si modules and one single-junction CIGS module have hardly any efficiency variation as the air mass is increased from AM 1 to AM 4, whereas a triple-junction thin-film silicon module shows over 20 % relative loss in efficiency; [Durisch-2008] reports a 10 % loss for a single-junction CdTe module and [Durisch-2000] a 15 % loss for a "micromorph" tandem module. The practical importance of these losses in efficiency due to red shifts should not be over-estimated, as a large part of the incoming *energy* of the sunlight is very often obtained under spectral conditions, which are near to AM 1.5. On the other hand, there is to the author's knowledge so far no systematic comparative data for various solar cell technologies and cell structures as a result of a blue-shift in the spectrum of the incoming light.

There are many investigations that have been carried out on how the efficiency of solar modules varies when the spectrum (and the intensity) of the incoming sunlight changes during the course of a day. These investigations have usually been conducted for outdoor conditions, at a particular test site (see e.g. [Durisch-2006] and references cited therein, as well as [Nann-1992]). In practice it is necessary to monitor the behavior of the PV modules at the site where they are effectively being deployed.

# THEORY OF SOLAR CELL DEVICES (SEMI-CONDUCTOR DIODES)

## PART II: “*pin*-TYPE” SOLAR CELLS

*Arvind Shah*

## 4.5 INTRODUCTION TO “*pin*-TYPE” SOLAR CELLS

### 4.5.1 Basic structure and properties

In Sections 4.1 to 4.4 of this chapter, we have introduced the basic functioning of wafer-based crystalline silicon solar cells. We shall now study both amorphous and microcrystalline versions of thin-film silicon solar cells. When entering the “world” of thin-film silicon solar cells, we must deal with two different aspects, as compared to the crystalline counterparts:

(I) The absorption coefficients of amorphous, and to some extent also of microcrystalline, silicon are higher (see Fig. 4.8). Therefore, the penetration depth of the incident irradiance is correspondingly smaller, and thinner cell material suffices to fully absorb the incoming light. The ratio between the light penetration depth and cell thickness may be further enhanced by light-trapping techniques, which are particularly relevant for microcrystalline silicon (to be discussed in Sect. 4.11.2).

(II) The transport properties of photo-generated carriers in both amorphous, and microcrystalline silicon layers are far inferior compared to those of crystalline silicon. As a result, the excess carriers recombine within very short distances from their origin, within distances much shorter than the cell thickness. The question is: how can we design and fabricate a thin-film silicon solar cell where the collection length for the photo-generated carriers is comparable or even higher than the penetration depth of sunlight?

The principle of any solar cell is the separation of electron-hole pairs by the action of an internal electric field. The location of this electric field is governed by the requirement that all photo-generated carriers can travel sufficiently far to reach that field.

In crystalline silicon solar cells this travel distance is determined by the carrier diffusion length (see below), which is of the same order of magnitude as the cell layer thicknesses. The electric field, concentrated within the depletion layers of the *p*- and *n*-layers, i.e. within a very thin zone at the *p/n* junction, is within reach for all the photo-generated carriers which have to be separated. Thus, crystalline silicon solar cells work as *pn*-diodes.

Due to the extremely short travel distances of photo-generated carriers in amorphous and microcrystalline silicon materials, the electric field needs to be present at the origin of photo-generation, in order to assist carrier travel and "immediately" separate electrons and holes and, thus, to avoid their recombination. The field-assisted travel distance is given by the drift length (see below). For optimum utilization of the incident irradiance, the electric field must extend throughout the entire cell thickness. This is accomplished by inserting an *i*-layer between the *p*- and *n*-layers, whereby a *pin*-diode is formed. With the relatively low light-penetration depths and, hence, with the low values of *i*-layer thicknesses (see aspect (I) above), the electric field strength $E(x)$, determined by the built-in voltage (see below) and the *i*-layer thickness, reaches a sufficient magnitude for efficient carrier separation. This means that it is possible, thanks to the *pin*-structure, to use materials such as amorphous and microcrystalline silicon, which would give only very poor results if the *pn*-structure were used.

The main difference between *pn*-diodes and *pin*-diodes is the extension of the internal electric field in the diode; this is illustrated in Figure 4.30.

The overwhelming majority of semiconductor devices use doped regions (*p*- and *n*-type regions) and *not* intrinsic regions (*i*-regions). The *pin*-diodes employed as solar cells for the case of *thin-film* silicon (amorphous and microcrystalline silicon) constitute one of the very few exceptions to this general rule.

In the case of the *pn*-diode, collection is governed by the minority carrier diffusion length $L_{\text{diff}}$, as was described above. In the case of the *pin*-diode, collection is governed by the drift length $L_{\text{drift}}$ of both electrons and holes within the intrinsic layer. It can be shown, that, under "reasonable conditions" (which are in general fulfilled), the drift length $L_{\text{drift}}$ has a value that is about 10 times larger than the minority carrier diffusion length $L_{\text{diff}}$; see [Shah-1995], Equation (10), with $\kappa = d_i/L_{\text{drift}} \approx 0.5$, where $d_i$ is the thickness of the intrinsic layer.

One therefore has the following situation: if material quality (as expressed by the diffusion length) is excellent, a *pn*-diode should be used as solar cell structure; if material quality is "mediocre", a *pin*-diode should be used; if material quality is poor, no reasonable solar cell can be fabricated. It so happens that the best *intrinsic* amorphous and microcrystalline layers fall precisely into the category of layers with "mediocre" quality.

In amorphous silicon there is yet another reason to use *pin*-type diodes instead of *pn*-diodes: it turns out to be basically impossible to deposit doped amorphous layers with reasonable quality. In fact, as we increase the doping level of the layer, we invariably increase the defect density, i.e. the density of recombination centers (see Chap. 2, Sect. 2.5). Therefore, the doped layers within an amorphous silicon solar cell are only used to form a difference in potential and to create an internal electrical field, and do not actively contribute to the photo-generated current. All light that is absorbed within these layers is "lost": as soon as the hole-electron pairs are generated here, they almost immediately disappear through recombination. Because of

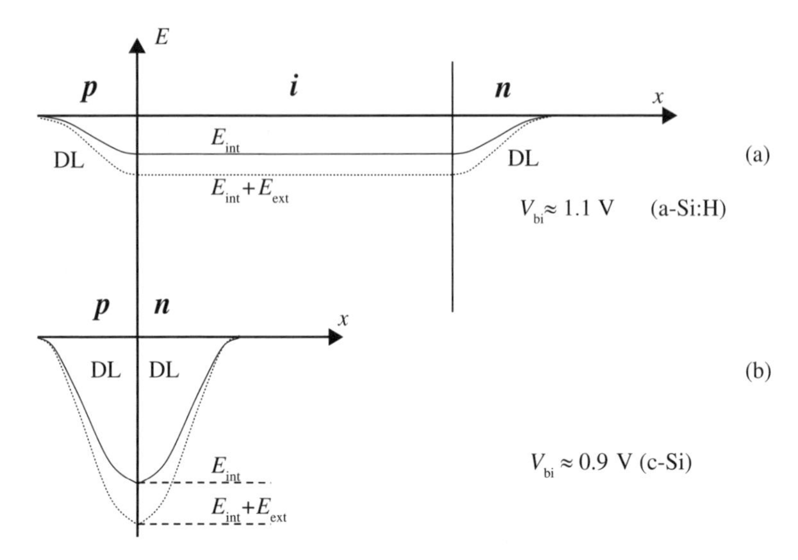


**Fig. 4.30** Internal electric field $E_{int}(x)$, for (a) *pin*- and (b) *pn*-type diodes; note that for zero external (applied) voltage, the integral $\int E_{int}(x)dx$ is, for both types of diodes, equal to the built-in voltage $V_{bi}$, a parameter that is approximately equal to 1 V for all forms of silicon. $E_{int}$ is the internal electric field for zero applied voltage, $E_{ext}$ is the additional electric field due to an externally applied voltage $V$; $E = E_{int} + E_{ext}$ is the total electric field. One can therefore write: $V = \int -E_{ext}(x)dx$: if the applied voltage is negative, it gives a reverse bias to the diode and the magnitude of the electric field is augmented (as shown in the Figure); if the applied voltage is positive, it gives a forward bias to the diode and the magnitude of the electric field is reduced. DL is the depletion layer.

this, the doped layers in an amorphous silicon solar cell are usually kept very thin. This is especially true of the *p*-type layer, which is a "window layer" through which the light is usually made to enter into the cell. It has to be just sufficiently thick to enable the internal electric field to be formed (see Sect. 4.5.2). In practice, the *p*-type layer in amorphous silicon is additionally alloyed with carbon, so as to increase its bandgap and further reduce its absorption coefficient (see Chap. 2, Sect. 2.7 and Chap. 6, Sect. 6.3). Figure 4.31 shows the typical structure of a *pin*-type amorphous silicon solar cell.

How far this reasoning applies also to microcrystalline silicon solar cells is less clear, because too little is known about microcrystalline silicon doped layers. In practice, however, one tends to keep the doped layers very thin, also within microcrystalline silicon solar cells.

Our goal is now to understand more about the functioning of thin-film silicon *pin*-type solar cells: Which are the factors that limit cell efficiency? How can the cell design be optimized? What does light-induced degradation (which is, alas, an unavoidable facet of amorphous silicon) do to the functioning of the solar cell? Which is the main material parameter that limits collection? (In *pn*-type solar cells it was the minority carrier diffusion length; here it will be the drift lengths of both carriers.)

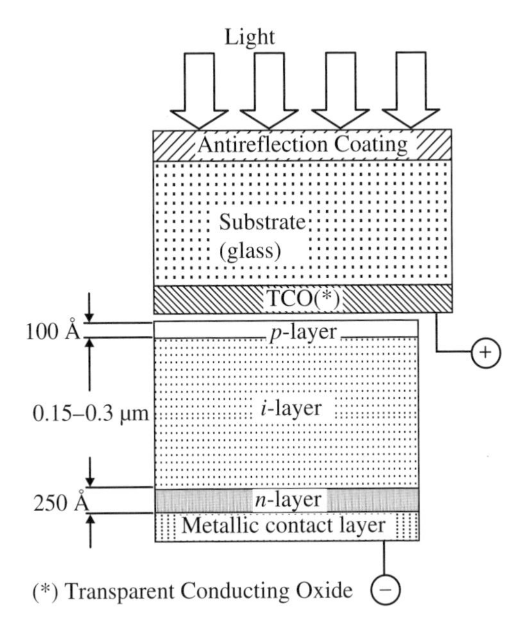


**Fig. 4.31** Typical structure of *pin*-type amorphous silicon solar cell on glass substrate; micro-crystalline silicon solar cells have a similar structure; however in the latter case, the *i*-layer is much thicker, i.e. 1 to 2 μm thick.

As the internal electric field, which extends throughout the photoactive intrinsic layer, is a key feature of *pin*-type solar cells, we will first try to understand how such a field is formed.

## 4.5.2 Formation of the internal electric field

There are two ways to understand the formation of the internal electric field within the *i*-layer:

(1) We can evaluate the *space charge regions* that are formed and apply Poisson's equation;

(2) We can look at the *band diagrams*.

We will present both methods here, as they are somewhat complementary.

Thereby, we shall assume, for the sake of simplicity, that the externally applied voltage is (at first) zero.

Figure 4.32 illustrates the formation of *space-charge regions* at the edges of the *p*- and *n*-type doped regions. Free carriers ($p_f$ and $n_f$), as well as carriers trapped in band tail states ($p_t$ and $n_t$) are depleted from this region, leaving behind them the ionized dopant impurities (negatively charged acceptors $n_A^-$, and positively charged donors $n_D^+$, respectively). This creates a space charge region, with space charge density ρ. If the applied external voltage $V$ is 0, the electric field at the far edges of the *p*- and *n*-layers, adjacent to the contacts, will be zero. Therefore, an electric field $E_{int}$ is formed in the

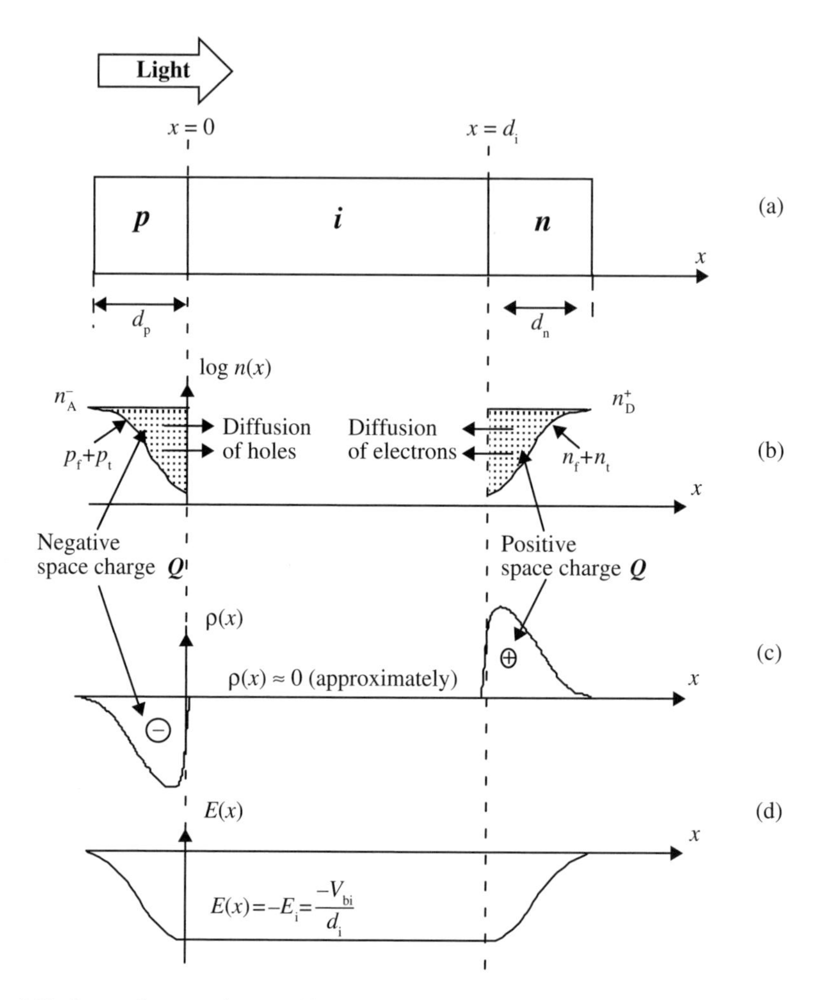


**Fig. 4.32** Space charge regions and internal electric field $E(x)$ within a *pin*-type solar cell, for the case of zero external applied voltage $V = 0$. The electric field $E(x)$ shown here has a negative value $(-V_{bi}/d_i = E_i)$, because it is directed from the *n*-layer to the *p*-layer and points in the direction opposite to the *x*-axis. $n_D^+$ and $n_A^-$ are the densities of the ionized dopant atoms (of the positively ionized donor atoms and of the negatively ionized acceptor atoms, respectively). Holes and electrons do not actually continue to diffuse (as indicated in the Figure) become, in reality, the diffusion current is compensated by a drift current in opposite direction. Holes and electrons can be *imagined* to have diffused, i.e. moved away from the depletion regions, as soon as the latter are made adjacent to an intrinsic layer, with much lower carrier density.[6]

intrinsic layer. This internal electric field will have a negative sign, because it will be directed from right to left, i.e. from the *n*-layer to the *p*-layer. As long as the space charge density within the *i*-layer can be considered to be almost zero, the resulting internal electric field $E_{int}$ will be approximately constant and equal to $-E_i$. The magnitude of $E_i$

[6]Both free and trapped carries move away, as there is proportionality between free and trapped carriers, see Section 4.5.4 hereunder and especially Figure 4.38.

is approximately $V_{bi}/d_i$, i.e. it is given by the built-in voltage $V_{bi}$ divided by the *i*-layer thickness $d_i$; these quantities are shown in the next figure, Figure 4.33.

On the other hand, the present figure, Figure 4.32, allows us to visualize that the thickness $d_p$ and $d_n$ of the two doped layers cannot be reduced below certain minimal values, otherwise sufficient space charge cannot be formed. This issue is particularly important for the *p*-layer, as the *p*-layer is the "window layer", through which light enters into the cell and we therefore want to keep it as thin as possible (see also Chap. 6, Sect. 6.3.1). Now, we need for a cell of unit cross-sectional area, according to Poisson's equation, a total areal charge $Q = \int \rho(x)\, dx = E_{int} \times \varepsilon$, at the *p*/*i* interface, in order to build-up the internal electric field $E_{int}$. (Note that in this Section the charge $Q$ will always be expressed as "charge for unit cross-sectional area $A$"; the total charge is then $Q \times A$.) This means that if we consider the space charge density in the depletion region to be equal to $n_A^-$, then the minimum *p*-layer thickness $d_{p.min}$ will be

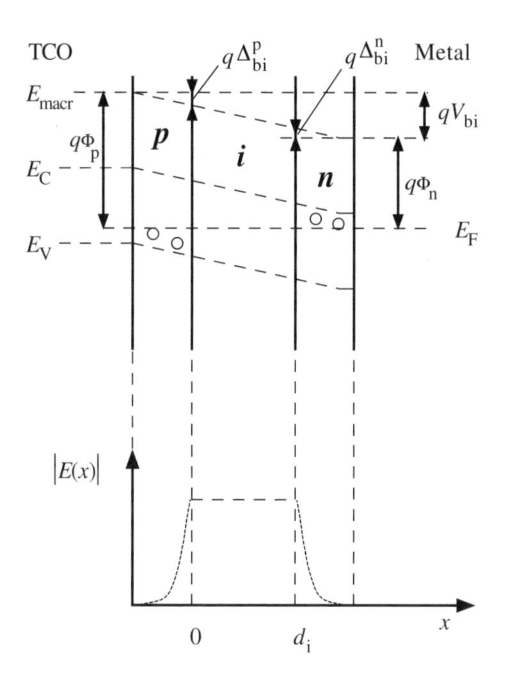


**Fig. 4.33** Band diagram for the ideal *pin*-type solar cell at zero applied voltage. $E_c$ is the energy level of the bottom of the conduction band; $E_v$ is the energy level of the top of the valence band. $E_{macr}$ is the macroscopic energy level outside the solar cell, also called "local vacuum energy level". The difference between the level of $E_{macr}$ on the *p*-side and the level of $E_{macr}$ on the *n*-side gives us the built in electrostatic potential $V_{bi}$. The Fermi level $E_F$ is constant throughout the semiconductor. The energy difference $q \times \Phi$ between the Fermi level $E_F$ and the macroscopic potential or local vacuum level is called the "work function" or "extraction energy". In semiconductors it depends on doping. It takes on a value $q \times \Phi_p$ on the *p*-side of the semiconductor and a value $q \times \Phi_n$ on the n-side, due to doping. The built in potential $V_{bi}$ can, thus, be directly related to the relative positions of the Fermi level on the *p*- and *n*-sides of the pin-solar cell. The slope of the energy level $E_{macr}$, i.e. $\partial E_{macr}/\partial x$, directly gives us the quantity $q \times E_{int}(x)$, i.e. it is directly proportional to the internal electric field $E_{int}(x)$.

$\{E_{int} \times \varepsilon \times (qn_A^-)^{-1}\} = \{V_{bi} \times \varepsilon\ (d_i \times q \times n_A^-)^{-1}\}$, where $\varepsilon = \varepsilon_0 \times \varepsilon_r$ is here the dielectric constant of silicon[7], with $q = 1.6 \times 10^{-19}$ Coulomb, $\varepsilon_0 = 8.9 \times 10^{-14}$ Farad/cm and $\varepsilon_r \approx 10$. If we take, for amorphous silicon solar cells, $V_{bi} = 1$ V and, as very rough estimate[8], $n_A^- = 3 \times 10^{17}$ cm$^{-3}$, we will obtain as limit value for the *p*-layer thickness $d_{p.min} \approx 6$ nm, if $d_i$ is 0.3 μm, as in typical amorphous silicon solar cells. If we reduce the *i*-layer thickness, the minimal *p*-layer thickness must be increased in order to provide sufficient areal charge at at the *p*/*i* interface to support a stronger electric field with the given built-in voltage $V_{bi}$. For example, for an *i*-layer thickness of 150 nm, instead of 300 nm, the minimal *p*-layer thickness $d_{p.min}$ will have to be increased from 6 to 12 nm. The important points to note here are:

(1) There exists a minimal *p*-layer thickness $d_{p.min}$ below which the internal electric field and, hence, also the $V_{oc}$-value will suffer;

(2) The value for a given situation is best determined experimentally, by a procedure of trial and error (make a thickness series for $d_p$ and measure $V_{oc}$);

(3) If we decrease the *i*-layer thickness, the value of $d_{p.min}$ will increase proportionally.

In the case of microcrystalline silicon solar cells the situation is much more complicated, as (a) the layers are not homogeneous in growth direction, (b) the doped layers (especially the *p*-type layers) used have large amorphous volume fractions, and (c) too little is known about doping of these "mixed phase" (amorphous/microcrystalline) layers.

Let us now look at the second method, i.e. proceed by drawing the *band diagram* for the *pin*-diode under zero applied voltage. This is shown in Figure 4.33. What is given, within this Figure, by the solar cell fabrication process itself, are the relative positions of the Fermi level in the doped regions. This directly gives us $V_{bi}$. Under the assumption that the quantities $\Delta V_{bi}$ in the two doped regions are negligibly small (see Fig. 4.33), we immediately obtain that $|E_i|$ is approximately equal to $V_{bi}/d_i$.

We have now established the value of the internal electric field within the *i*-layer of a *pin*-solar cell, under "ideal" conditions and for zero external voltage. The electric field is inversely proportional to the thickness $d_i$ of the *i*-layer. Thin solar cells have a strong electric field. This is why amorphous silicon solar cells, where collection is particularly

[7]The relative dielectric constant $\varepsilon_r$ depends on the doping level, on the density of the layer etc.; in this Chapter we are not concerned with calculating exact values for the electrical field, but rather with introducing the *concepts and mechanisms* of field formation and field deformation. We have therefore taken an average, rounded value $\varepsilon_r \approx 10$, throughout the Chapter.

[8]The density of ionized dopant atoms in amorphous silicon is much lower than the total density of dopant atoms, because of the formation of additional dangling bonds through doping, see Chapter 2, Section 2.6. On the other hand, the density of free carriers is here much lower than the density of ionized dopant atoms, because a large fraction of the carriers, which become available during the ionization process, are trapped in the bandtails.

critical (especially in the "degraded" state, after light-induced degradation), are kept very thin ($d_i = 0.15$ to $0.3$ μm). If a forward bias voltage develops (or is applied), as is the case for actual solar cell operation, the electric field is correspondingly reduced.

The electric field is, furthermore, deformed and reduced by space charge within the *i*-layer. Space charge within the *i*-layer can be due to one of the four following sources:

(A) ionized impurities, like oxygen;
(B) free charge carriers $p_f$ and $n_f$;
(C) trapped charge carriers $p_t$ and $n_t$;
(D) ionized defects i.e. charged dangling bonds.

Factor A can be avoided, by appropriate measures, during solar cell deposition. It is especially important to avoid (or, at least, reduce) oxygen contamination within the intrinsic layers of microcrystalline silicon solar cells. This can be achieved by the use of a gas purifier (see [Torres-1996], [Kroll-1995]).

Factor B can be neglected, because the densities of free carriers are usually much lower than the densities of trapped carriers and of charged dangling bonds.

Factors C and D are, thus, in practice, the most important factors that influence field deformation. They depend, however, entirely on the densities of the free carriers $p_f$ and $n_f$; for this reason we shall now look in more detail at the concentration profiles of the free carriers $p_f(x)$ and $n_f(x)$ within the intrinsic layer (*i*-layer) of a *pin*-solar cell under illumination.

### 4.5.3 Carrier profiles in the intrinsic layer: free carriers $p_f$ and $n_f$

It is a tedious task to solve the nonlinear differential equations that govern carrier transport and carrier concentrations within the *i*-layer of a *pin* solar cell. Both types of charge carriers, holes and electrons, have similar concentrations and, therefore, no simple linear approximations are available here, as was the case for the doped regions of *pn*-type solar cells (see the five approximations in Sect. 4.2). Of course, the solution can be found numerically, but that method generally provides little insight, unless one already has an intelligent guess of what to look out for.

We will therefore take another approach here: we will consider, first, only a *very simple case*. From this simple case we can obtain a very rough, intuitive idea of how the carrier profiles look like, in actual practical situations. This simple case corresponds to the *pin*-type solar cell in short-circuit (SC) or reverse-bias (RB) operation. It is based on the following three hypotheses:

(a) uniformly absorbed light;
(b) no recombination within *i*-layer, in SC- and RB-operation;
(c) carrier transport within *i*-layer only by drift (not by diffusion).

We will start out with *hypothesis (a)*: we will assume using light (e.g. red light) that is uniformly absorbed within the whole *i*-layer. We will also immediately introduce *hypothesis (b)*: for a first estimation of the carrier profiles within the *i*-layer, we will assume that all photo-generated carriers can be collected, i.e. that recombination is negligibly small.

We, thus, obtain the situation shown in Figures 4.34 and 4.35: pairs of holes and electrons are uniformly photo-generated throughout the whole *i*-layer; the holes drift under the influence of the electric field towards the *p*-region, and the current density of holes $|J_p|$ increases linearly until it reaches its maximum value at the *p*/*i* interface; the electrons drift under the influence of the electric field towards the *n*-region, and the current density of electrons $|J_n|$ increases linearly until it reaches its maximum value at the *n*/*i* interface. One, therefore, obtains linear current density profiles $J_p(x)$ and $J_n(x)$, as shown schematically in Figure 4.35.

We will now introduce a further, decisive assumption: *hypothesis (c)*. This hypothesis postulates that carrier transport within the *i*-layer is mainly by drift and that the contribution of diffusion can be neglected. This is generally true, for *most* of the *i*-layer, as long as the *i*-layer thickness $d_i$ is not too large, and as long as the external voltage $V$ is not too positive, i.e. approximately up to the maximum power point

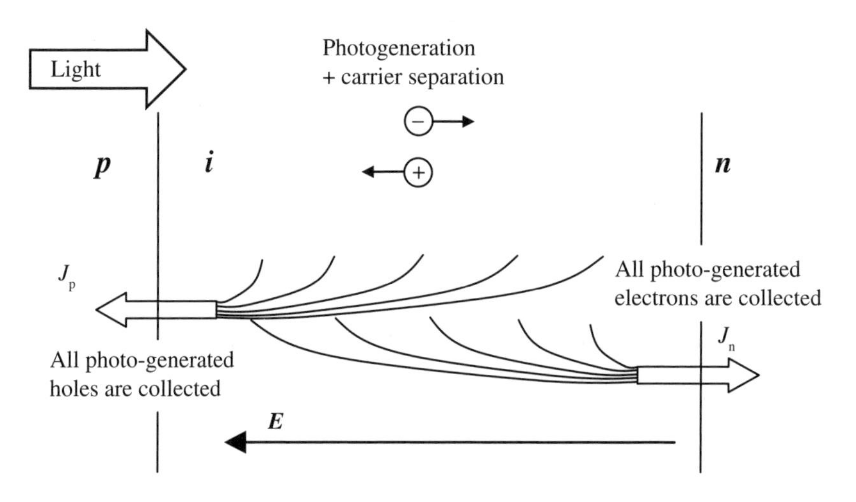


**Fig. 4.34** Current collection within the *i*-layer for uniformly absorbed light.

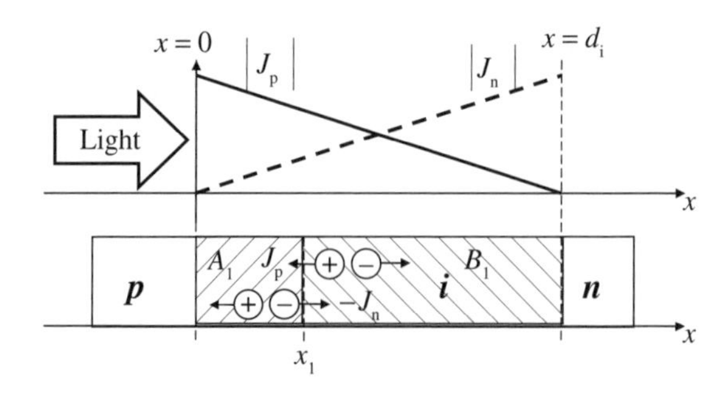


**Fig. 4.35** Current density profiles for holes: $J_p(x)$; and for electrons: $J_n(x)$. At point $x = x_1$ the hole current density $J_p(x_1)$ is proportioned to the surface $B_1$ and the electron current density $J_n(x_1)$ is proportional to the surface $A_1$.

(MPP). In fact, one can show that the *average* drift current within the *i*-layer is much larger than the average diffusion current within the *i*-layer, provided

$$\frac{kT}{q} << V_{\text{bi}} - V$$

In reality diffusion currents do play a role, but only locally, especially near the *p*/*i* and *n*/*i* interfaces.

If only drift is present, and if the electric field is uniform, then we will have proportionality between the current densities $J_{\text{p.drift}}(x)$ or $J_{\text{n.drift}}(x)$ and the free carrier concentrations $p_{\text{f}}(x)$ or $n_{\text{f}}(x)$, respectively.

$$J_{\text{p}}(x) = J_{\text{p.drift}}(x) = (q\mu_{\text{p}}^{0}\, E) \times p_{\text{f}}(x),$$

and similarly for electrons:

$$J_{\text{n}}(x) = J_{\text{n.drift}}(x) = -\,(q\mu_{\text{n}}^{0}\, E) \times n_{\text{f}}(x);$$

here, $\mu_{\text{n}}^{0}$ and $\mu_{\text{p}}^{0}$ are the band mobilities, i.e. the mobilities of free electrons and free holes, respectively. One generally assumes that $\mu_{\text{n}}^{0} \geq \mu_{\text{p}}^{0}$. The corresponding free carrier profiles are shown schematically in Figure 4.36:

- If one sets: $\mu_{\text{n}}^{0} \approx 3 \times \mu_{\text{p}}^{0}$ (as is often done) one would obtain, within the *i*-layer, free carrier profiles $p_{\text{f}}(x)$ and $n_{\text{f.3}}(x)$ {uninterrupted line for the electron density}; in this case the density of holes would be approximately three times larger than the density of electrons, because it would take about three times more holes to "carry" the same electric current.
- If one sets: $\mu_{\text{n}}^{0} \approx \mu_{\text{p}}^{0}$ (as we will do, for simplicity's sake, in our numerical computations) one would obtain, within the *i*-layer, free carrier profiles $p_{\text{f}}(x)$ and $n_{\text{f.1}}(x)$ {dashed line for the electron density}; now, the density of holes would be approximately equal to the density of electrons, because it would take about the same quantity of holes and electrons to "carry" the same electric current.

At this point it is useful to drop one of our three simplifying hypotheses, namely hypothesis (a), where we had postulated that we would be using light that is uniformly absorbed within the whole *i*-layer.

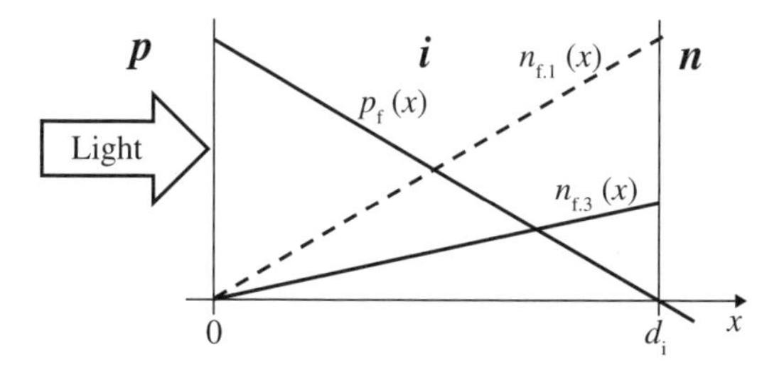


**Fig. 4.36** Free carrier profiles $p_{\text{f}}(x)$ and $\text{n}_{\text{f}}(x)$ for *pin*- and *nip*-type solar cells.

Hypothesis (a) led us to a situation where photo-generation was uniform throughout the *i*-layer (Fig. 4.37(a)). Uniform photo-generation is, however, only observed for light with very long wavelengths, for which the penetration depth is much larger than the *i*-layer thickness. This is the case for red light, in amorphous silicon solar cells, and for near-infrared light, in microcrystalline silicon solar cells. It is certainly not the case for sunlight with for example an AM 1.5 spectrum. Indeed, a large part of the incoming sunlight is in the green and blue regions of the spectrum, and has penetration depths that are far smaller than the *i*-layer thickness; this part of the light is absorbed near the *p*/*i*-interface, within the front part of the *i*-layer.

Furthermore, because of light scattering (see Sect. 4.11.2), light does not enter into the silicon layers vertically, but rather within a whole range of scattering angles (see Fig. 4.59) – this factor further accentuates the absorption of a large part of the light in the front half of the i-layer. Finally, in actual solar cells there are also back reflectors, which will tend to have the opposite effect, i.e. to increase the absorption of light within the whole *i*-layer.

The Ljubljana group has extensively investigated these effects, with the help of an optical simulation program called SunShine [Krc-2003a]. Results can be found in the thesis of Janez Krc [Krc-2003b]: He had simulated there the case of an amorphous silicon solar cell with an *i*-layer thickness of $d_i = 450$ nm, and with realistic assumptions for the roughness of the front TCO layer, as well as for the back reflector. The authors obtained from Janez Krc [Krc-2009], simulation results also for an *i*-layer thickness $d_i = 300$ nm, which are presented here: Figure 4.37 shows schematically how the absorption/photo-generation profiles, the current density profiles and the profiles of free carriers (free holes $p_f$ and free electrons $n_f$) change in a typical a-Si:H solar cell, if one goes from a situation with uniformly the absorbed light (a) to the more realistic situation (b) of an AM 1.5 solar spectrum entering into the cell plus light trapping within the cell.

For microcrystalline silicon solar cells of the *pin*- and *nip*-type, the mechanisms of photo-generation and carrier transport by drift can be considered to be very similar, at least as long as one is at or near the short-circuit operating point. The plot of absorption coefficient versus light wavelength is, however, very different. Absorption is much lower for blue and green light, but extends right up to the near infrared. Because the absorption coefficient is very low, in the case of microcrystalline silicon, μc-Si:H solar cells will have *i*-layer thicknesses $d_i$ that are in the order of 1 to 2 μm, i.e. much higher than typical thicknesses of a-Si:H solar cells.

### 4.5.4 Trapped charge carriers $p_t$ and $n_t$ in bandtails

Now we must remember that one of our main goals here, in this part of our development, is to evaluate the space charge contributions B, C, and D, as already listed in the preceding section:

(B) Free charge carriers $p_f$ and $n_f$,
(C) Trapped charge carriers $p_t$ and $n_t$,
(D) Ionized defects, i.e. charged dangling bonds.

*Free carriers* are much less numerous than the other species and their contribution to space charge can certainly be neglected. *Trapped carriers* are a more important

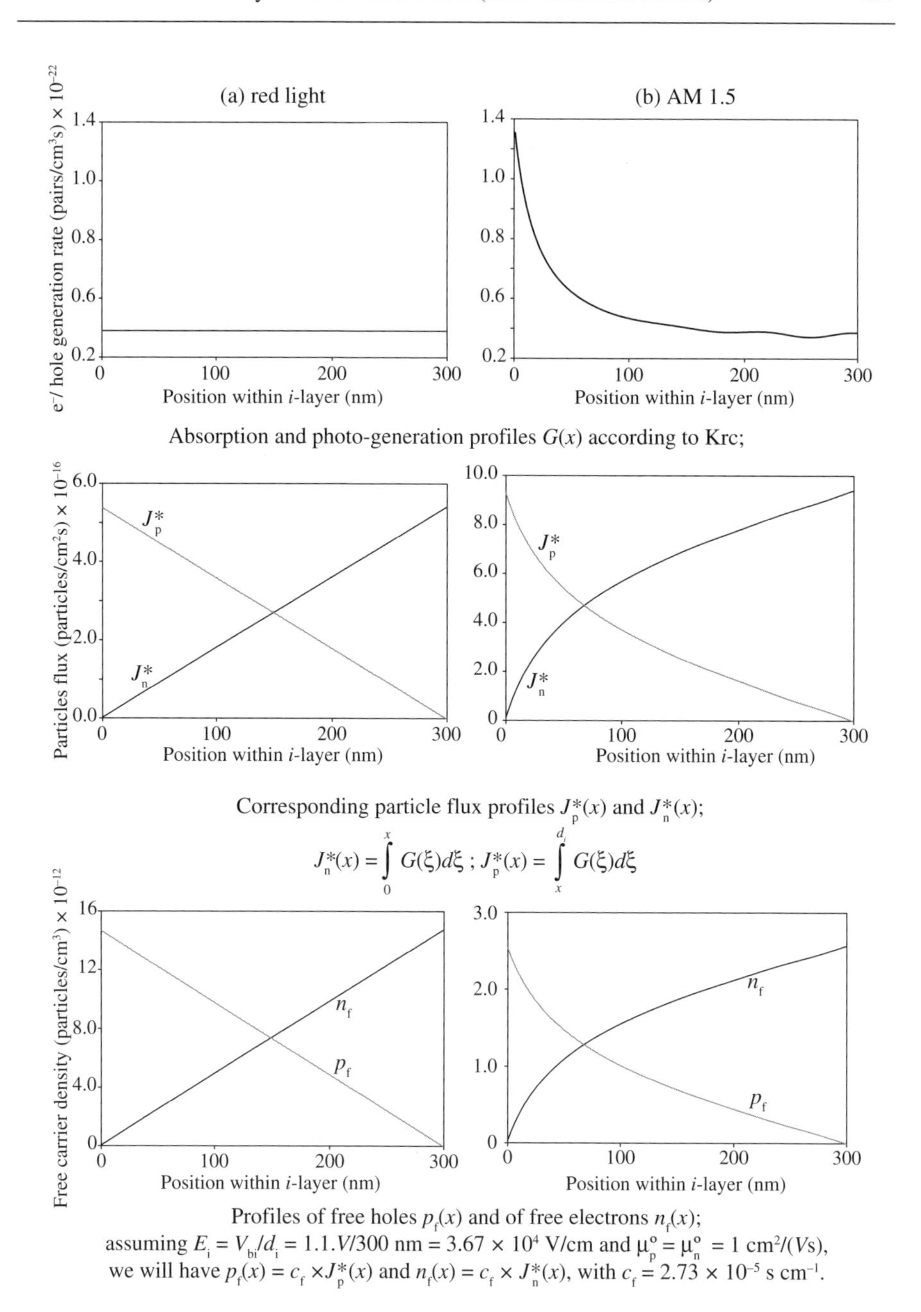


$$J^*_n(x) = \int_0^x G(\xi)d\xi \; ; \; J^*_p(x) = \int_x^{d_i} G(\xi)d\xi$$

assuming $E_i = V_{bi}/d_i = 1.1.V/300$ nm $= 3.67 \times 10^4$ V/cm and $\mu_p^o = \mu_n^o = 1$ cm²/(Vs), we will have $p_f(x) = c_f \times J^*_p(x)$ and $n_f(x) = c_f \times J^*_n(x)$, with $c_f = 2.73 \times 10^{-5}$ s cm⁻¹.

**Fig. 4.37** Absorption/photo-generation profiles, particle flux profiles and profiles of free carriers, within the *i*-layer of an amorphous silicon *pin*-solar cell, for two cases of incoming light: (a) uniformly absorbed light; (b) AM 1.5 spectrum plus light trapping. Case (b) has been calculated with the help of Janez Krc [Krc-2009] using his optical simulator "SunShine" as well as the assumptions described in his thesis [Krc-2003b], adapted here to the case where i-layer thickness is 300 nm.

contribution to space charge; it is indeed these trapped carriers, which we will look into, in the present Section 4.5.4.

Trapped carriers are, in equilibrium with free carriers, because of the continuous interaction (by capture and release) between bandtail states and band states, as schematically represented in Figure 4.38.

It can be shown (see [Rose-1963]) that the density of trapped carriers is more or less proportional to that of free carriers. The following approximate relationships hold:

$$p_{\mathrm{f}} = \Theta_{\mathrm{p}}\left(p_{\mathrm{f}} + p_{\mathrm{t}}\right) \approx \Theta_{\mathrm{p}} p_{\mathrm{t}} \quad \text{with } \Theta_{\mathrm{p}} = \frac{p_{\mathrm{f}}}{p_{\mathrm{f}} + p_{\mathrm{t}}} \approx 10^{-3} \ldots 10^{-2} \tag{4.26a}$$

$$n_{\mathrm{f}} = \Theta_{\mathrm{n}}\left(n_{\mathrm{f}} + n_{\mathrm{t}}\right) \approx \Theta_{\mathrm{n}} n_{\mathrm{t}} \quad \text{with } \Theta_{\mathrm{n}} = \frac{n_{\mathrm{f}}}{n_{\mathrm{f}} + n_{\mathrm{t}}} \approx 10^{-1} \tag{4.26b}$$

$\Theta_{\mathrm{p,n}}$ are the "trapping factors", which were introduced by Al Rose [Rose-1963].

There is no exact, fixed proportionality between free carriers and trapped carriers. Rather there are general rules, to the effect, that (1) as the density of free carriers increases, the density of trapped carriers also increases in roughly the same manner and (2) trapping is a much more pronounced effect for holes than for electrons. Thus, Equations 4.26(a) and (b) are no strict mathematical relationships, but just guidelines to assess the approximate density of trapped carriers, once the density of free carriers is known. Nevertheless, they will allow us to determine, for amorphous silicon solar cells, – albeit in a very approximate manner – the electric charge trapped in the bandtails, and with that, by using Poisson's equation, the deformation of the internal electric field due to this charge. This will be done in the Section 4.6.

**Note:** In the case of microcrystalline silicon, trapping in bandtails is, as far as one can judge today, much less pronounced than in amorphous silicon. Therefore one may safely assume that the deformation of the internal electric field is a negligible effect within μc-Si:H solar cells.

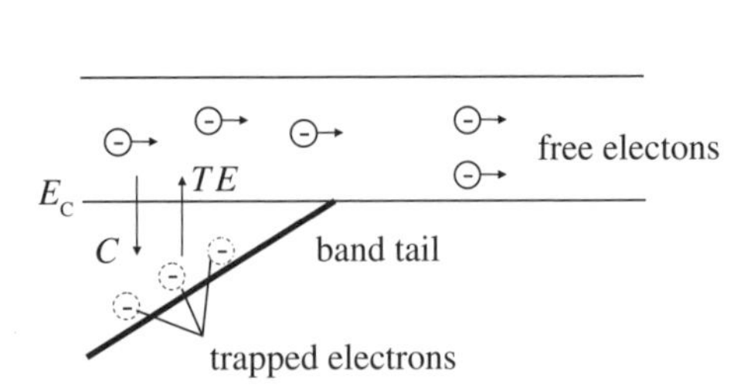


**Fig. 4.38** Band diagram sketch, showing schematically the exchange between (localized) conduction band tail states with trapped electrons $n_t$ and (extended, mobile) conduction band states with free electrons $n_f$. In the case of a steady-state situation, an equilibrium exists between the two processes, i.e. between the capture (*C*) of free electrons (trapping) and the release of trapped electrons (by thermal excitation *TE*). A very similar exchange takes place on the side of the valence band, and of its band tail, i.e. between trapped holes and free holes.

## 4.6 EFFECT OF TRAPPED CHARGE IN VALENCE AND CONDUCTION BANDTAILS ON ELECTRIC FIELD AND CARRIER TRANSPORT

### 4.6.1 Deformation of electric field in *i*-layer by trapped carriers: Concept

We will now proceed to evaluate the space charge due to the valence and conduction bandtails and the deformation of the internal electric field $E(x)$ in the *i*-layer, resulting from this space charge.

To obtain a very general idea of how this is done, we shall make, at first the following assumptions:

(I) $\mu_n^0 \approx 3 \times \mu_p^0$

(II) $\Theta_p << \Theta_n$

This leads us to the situation sketched in Figure 4.39.

Figure 4.39 shows, schematically, for the case of "red light", i.e. for uniformly absorbed light: (a) the densities of particles fluxes $J_p$ and $J_n$, for holes and electrons, respectively; (b) the mechanism of photo-generation and particle flow; (c) the profiles of free holes $p_f(x)$ and of free electrons $n_f(x)$; (d) the profiles of carriers trapped in the band tails, i.e. the profiles of trapped holes $p_t(x)$ and of trapped electrons $n_t(x)$; (e) the profile of the resulting space charge $\rho(x) = q(p_t - n_t)$; and (f) the resulting internal electric field $E(x)$within the *i*-layer, with the corresponding field deformation $\Delta E(x) = (1/\varepsilon_r\varepsilon_0)\int \rho(x)\mathrm{d}x$, where $\varepsilon_r$ is the dielectric constant or "relative permittivity" (of the thin intrinsic silicon layer) and $\varepsilon_0$ is the vacuum permittivity.

One may, based on the concepts sketched in Figure 4.39, make the following remarks:

- In amorphous silicon solar cells, there is usually quite a large amount of charge stored in the valence bandtail as trapped holes $p_t$. The charge stored in the conduction bandtail as trapped electrons $n_t$ is much less pronounced, the ratio between $\Theta_n$ and $\Theta_p$ being often as high as a factor 100. This means that $L_1 \gg L_2$ in Figure 4.39(e); in fact, the width $L_2$ of the zone where there are more trapped electrons than trapped holes is generally negligibly small with respect to the *i*-layer thicknesses $d_i$.
- The field deformation leads to a reduction of the electric field near the *i/n*-interface. Thus, field deformation due to trapped carriers can be a factor that limits drift-assisted charge collection, especially in amorphous silicon solar cells: If one keeps amorphous silicon solar cells sufficiently thin ($d_i < 300$ nm), one may avoid a reduction in fill factor and efficiency due to this effect.
- The densities of trapped holes and electrons are proportional to those of free holes and free electrons, respectively; they are therefore, also proportional to

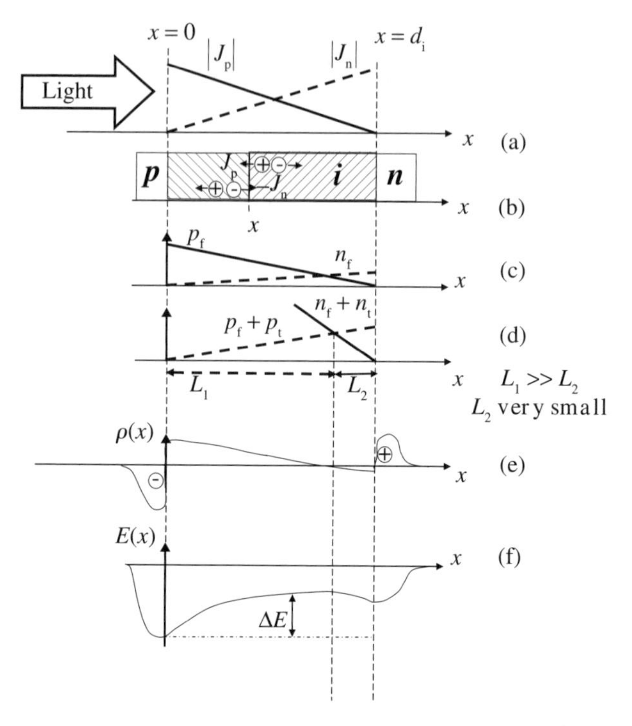


**Fig. 4.39** Schematic representations of space charge and electric field in the whole *pin*-solar cell, showing the deformation of the electric field due to trapped carriers (mainly trapped holes) in the intrinsic layer as well as the resulting field reduction $\Delta E$ near the *i*/*n*-interface (see text).

the illumination level. Field deformation due to trapped charge is therefore of particular importance at high illumination levels. On the other hand, bandtails are not affected by light-induced degradation (Staebler-Wronski effect) [Wyrsch-1995]. This means that this specific contribution to electric field deformation and electric field reduction is not influenced by light-induced degradation.

In device-quality microcrystalline layers, bandtails are much less pronounced than in device-quality amorphous silicon layers; this is especially true of the valence bandtail. The effect described here is therefore not important at all for "good quality" microcrystalline silicon solar cells.

### 4.6.2 Deformation of electric field in i-layer by trapped carriers: numerical simulations for amorphous silicon

So far, we have just looked at the effect of field deformation by trapped carriers in a qualitative manner. We shall now look at some quantitative, numerical results: Figure 4.40 shows the profiles of carriers trapped in the bandtails, i.e. the profiles of

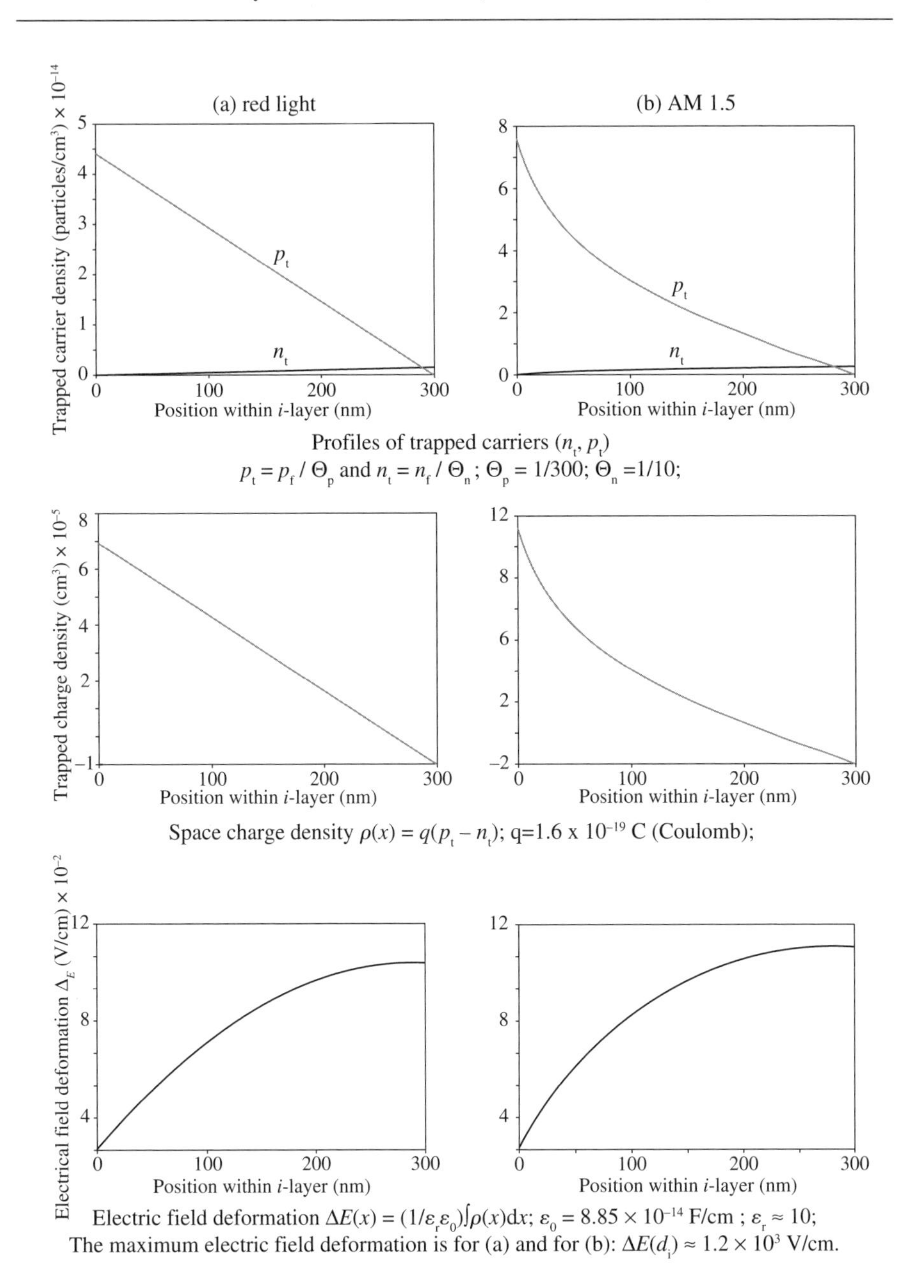


**Fig. 4.40** Comparison of the two "standard" cases of input light: (a) "red light" (uniformly absorbed light) and (b) AM 1.5 spectrum plus light trapping (usual situation in solar cell). Case (b) has been computed by Janez Krc using the optical simulator SunShine, for a *pin*-type solar cell with 300 nm *i*-layer thickness, deposited on a glass + rough $SnO_2$ substrate, with an Al back reflector; simulation details are given in [Krc-2003b]. Given here are the profiles of trapped carriers ($n_t$, $p_t$), the space charge density $\rho(x) = q(p_t - n_t)$ due to these trapped carriers, and the resulting electric field deformation $\Delta E(x) = (1/\varepsilon_r\varepsilon_0)\int \rho(x)dx$ according to Poisson's equation. Calculations and figure courtesy of Michael Stückelberger, IMT PV Lab Neuchâtel.

trapped holes $p_t(x)$ and of trapped electrons $n_t(x)$: (a) for the case of "red light", i.e. for uniformly absorbed light, and (b) for AM 1.5 light spectrum entering into the cell plus light trapping within the cell (according to Janez Krc's optical simulation, see above). Figure 4.40 also represents the additional space charge $\rho(x) = q[p_t(x) - n_t(x)]$ due to these carriers and, according to Poisson's equation, the deformation of the electric field $\Delta E(x) = (1/\varepsilon_r\varepsilon_0)\int \rho(x)dx$, as provoked by the trapped carriers.

We have thereby made the following, additional assumptions:

(III) $\mu_n^0 \approx \mu_p^0$

(IV) $\Theta_n = 30 \times \Theta_p$

On the other hand, the difference between $\Theta_n$ and $\Theta_p$ is very often often as much as a factor of 100. Thus, the trapping of holes is, in amorphous silicon layers, in most cases even more pronounced than the effect shown in Figure 4.40. This means: within the *i*-layer of amorphous silicon solar cells, the space charge due to trapped carriers is essentially a positive charge extending over the whole *i*-layer, except for a very narrow region towards the *i/n*-interface. Furthermore, as can be visualized by comparing cases (a) and (b) in Figure 4.40, there is, as far as this effect is concerned, not much difference between the two different cases of illumination.

### 4.6.3 Mobilities in amorphous and microcrystalline silicon

There are four different kinds of mobilities used to characterize transport in amorphous silicon (and, by extension, also in microcrystalline silicon):

(1) Band mobilities $\mu_p^0$, $\mu_n^0$ that describe the transport of free holes and free electrons (used for the description of devices, like solar cells, in steady-state operation).

(2) Effective mobilities $\mu_p^{eff}$, $\mu_n^{eff}$ that describe the relationship between free and trapped carriers and their transport (used for the description of devices, like solar cells, in steady-state operation, when one has to look also at the space charge constituted by the trapped carriers).

(3) Field-Effect Transistor (FET) mobilities $\mu_p^{FET}$, $\mu_n^{FET}$ that describe the operation of thin-film field-effect transistors (used for evaluating the characteristics of FETs; the latter are majority carrier devices : in a *n*-channel FET, for example, transport is by free electrons ($n_f$), but the applied voltage acts on the total charge constituted by free electrons ($n_f$), trapped electrons ($n_t$), and negatively charged dangling bonds ($N_{D^-}$): $\mu_n^{FET}$ describes the relationship between the transport of free electrons and the total charge $Q_{n_f+n_t+N_{D^-}}$).

(4) Drift mobilities $\mu_p^{drift}$, $\mu_n^{drift}$ that describe the behaviour in transient processes, i.e. in drift processes. These mobility values are effective when a whole sheet of charge drifts through the semiconductor and during this movement all the

traps have to be filled and emptied again. They are not of importance for steady-state situations, such as are prevalent in solar cells. The values of the drift mobilities depend strongly on the prevalent experimental conditions.

One has, under certain conditions, the approximate relationships:

$$\mu_p^{eff} \approx \mu_p^{FET} \approx \mu_p^{drift} = \Theta_p \times \mu_p^0,$$

$$\mu_n^{eff} \approx \mu_n^{FET} \approx \mu_n^{drift} = \Theta_n \times \mu_n^0,$$

where $\Theta_p$ and $\Theta_n$ are the trapping factors according to Rose (see above).

For amorphous silicon (a-Si:H) one finds the approximate values in Table 4.3 (all mobility values are in [$cm^2/(V \cdot s)$] (see [Spear-1988], as well as [Schiff-2004] and [Schiff-2009a,b] and references cited therein).

For microcrystalline silicon (μc-Si:H) one only has very coarse approximations, see Table 4.4; these values are simply "orders of magnitude", for "best device-quality layers" (i.e. layers for solar cells[9] with 50 % to 60 % Raman crystallinity); (see [Dylla-2006], as well as [Schiff-2004] and [Schiff-2009a,b] and references cited therein):

**Table 4.3** Range of values for band- and effective-mobilities for a-Si:H intrinsic layers.

| | $\mu_p^0, \mu_n^0$ | $\Theta_p, \Theta_n$ | $\mu_p^{eff}, \mu_n^{eff}$ |
|---|---|---|---|
| Holes | 0.2 to 2 | 0.001 to 0.01 | $10^{-4}$ to $10^{-2}$ |
| Electrons | 1 to 10 | 0.1 to 0.3 | 0.1 to 1 |

**Table 4.4** Range of values for band- and effective-mobilities for μc-Si:H intrinsic layers.

| | $\mu_p^0, \mu_n^0$ | $\Theta_p, \Theta_n$ | $\mu_p^{eff}, \mu_n^{eff}$ |
|---|---|---|---|
| Holes | 0.3 to 3 | 0.1 ? | around 0.1 |
| Electrons | 1 to 10 | 0.1 to 0.3 | 0.3 to 3 |

## 4.7 DANGLING BONDS AND THEIR ROLE IN FIELD DEFORMATION

### 4.7.1 Dangling bond charge states

*Ionized defects (charged dangling bonds)* have a charge state that is "amphoteric", i.e. it can take on both charge signs, and be either positively or negatively charged. They

[9]Microcrystalline layers for thin-film transistors have much higher crystallinity values, and, thus, also much higher mobilities [Lee-2006]. They have, however, also much higher values of defect densities and are not at all suitable for the intrinsic layers of solar cells, because of the resulting high recombination.

become charged by capturing free holes or free electrons; their charge state is, thus, governed by the *ratio* $p_f(x)$: $n_f(x)$.

Intuitively one can visualize the following behavior:

- if the density of free holes is much larger than the density of free electrons, the dangling bonds will be mainly positively charged, because they can easily capture a free hole but have to wait a long time before capturing one of the scarce free electrons;
- on the other hand, if the density of free electrons is much larger than the density of free holes, the dangling bonds will be mainly negatively charged;
- and, finally, if the densities of both carriers are in the same range, the dangling bonds will be mainly neutral.

To *calculate* the occupation functions for the three charge states one has to look in detail at the capture processes shown in Figure 4.41. The procedure has already been briefly described in Chapter 2, Section 2.4.4; it is similar to that used for "classical" Schockley-Read-Hall recombination theory (see e.g. [Sze-2006]).

Just as in the "classical" case, we may also assume for amorphous silicon that:

- the capture rates for electrons, i.e. the rates $r_a$ and $r_c$ are proportional to the density of free electrons $n_f$ that can be captured;
- the capture rates for holes, i.e. the rates $r_b$ and $r_d$ are proportional to the density of free holes $p_f$ that can be captured;
- free electrons $n_f$ have a mobility $\mu_n^0$ that is about 1 to 3 times higher than the mobility $\mu_p^0$ of free holes $p_f$; thus, they have a thermal velocity $v_{th.n}$ that is 1 to 3 times higher than the thermal velocity $v_{th.p}$ of free holes, and are therefore also about 1 to 3 times more likely to be captured for otherwise comparable parameters. The $\mu\tau$-products are thus the same for electrons and holes. This was experimentally confirmed by [Beck-1996].

We have now to take into account the *amphoteric nature* of the defect states in amorphous silicon. This is a situation, which is not taken into account in "classical"

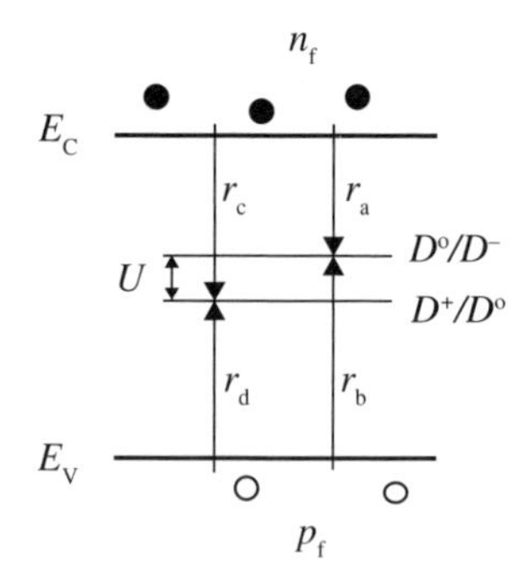


**Fig. 4.41** Capture of free holes $p_f$ and free electrons $n_f$ on dangling bond states; thermal generation is not shown as it can certainly be neglected under normal circumstances of a sufficiently illuminated solar cell. $U$ is the so-called "correlation energy" or "Hubbard energy" (see Chap 2, Sect. 2.2.2)

Schockley-Read-Hall recombination theory, where the defect states are *not* considered to be "amphoteric". In fact, according to Chapter 2, Section 2.2.2, we may have a dangling bond without any electron, which is positively charged ($D^+$), or a dangling bond with a single electron, which is neutral ($D^0$), or finally a dangling bond with two electrons, which is negatively charged ($D^-$). Because of the "amphoteric" nature of the dangling bonds, we will now have to distinguish between Coulomb-assisted (charge-assisted) capture processes (which will have relatively large capture cross-sections) and other capture processes (with much smaller capture cross-sections). Thus, we can list the following capture processes:

- $r_a$ is the capture rate of a free electron by a neutral dangling bond $D^0$; it is governed by a relatively small capture cross-section $\sigma_n^0$ for capture that is not assisted by Coulomb attraction; it is proportional to the density of dangling bonds that are neutral $N_{db} \times f^0$, where $N_{db}$ is the total density of dangling bonds and $f^0$ is the fraction of dangling bonds that are neutral ($D^0$).
- $r_b$ is the capture rate of a free hole by a negatively charged dangling bond $D^-$; it is governed by a relatively large capture cross-section $\sigma_p^-$ for capture that is assisted by Coulomb attraction; it is proportional to the density of dangling bonds that are negatively charged $N_{db} \times f^-$ where $N_{db}$ is the total density of dangling bonds and $f^-$ is the fraction of dangling bonds that are negatively charged ($D^-$).
- $r_c$ is the capture rate of a free electron by a positively charged dangling bond $D^+$; it is governed by a relatively large capture cross-section $\sigma_n^+$ for capture that is assisted by Coulomb attraction; it is proportional to the density of dangling bonds that are positively charged $N_{db} \times f^+$, where $N_{db}$ is the total density of dangling bonds and $f^+$ is the fraction of dangling bonds that are positively charged ($D^+$).
- $r_d$ is the capture rate of a free hole by a neutral dangling bond $D^0$; it is governed by a relatively small capture cross-section $\sigma_p^0$ for capture that is not assisted by Coulomb attraction; it is proportional to the density of dangling bonds that are neutral $N_{db} \times f^0$, where $N_{db}$ is the total density of dangling bonds and $f^0$ is the fraction of dangling bonds that are neutral ($D^0$).

Assuming $\sigma_n^+ = \sigma_p^- = \sigma_{n,p}^{\pm}$ and $\sigma_n^0 = \sigma_p^0 = \sigma_{n,p}^0$ and writing $\sigma_{n,p}^{\pm} = \varsigma \times \sigma_{n,p}^0$, we find, through straightforward calculations, based on the above guidelines, the following values for the dangling bond occupation functions [Hubin-1992]:

$$f^+ = \varsigma \times p_f^2 / (\varsigma \times p_f^2 + p_f n_f + \varsigma \times n_f^2) \quad (4.27a)$$

$$f^- = \varsigma \times n_f^2 / (\varsigma \times p_f^2 + p_f n_f + \varsigma \times n_f^2) \quad (4.27b)$$

$$f^0 = 1 - f^+ - f^- \quad (4.27c)$$

These occupation functions are represented in Figure 4.42.

From this we can conclude that: charged dangling bonds ($D^+$, $D^-$) do not occur very often; indeed provided there is a sufficiently high density of free carriers of the opposite polarity ($n_f$, or $p_f$, respectively) these free carriers will be easily captured by

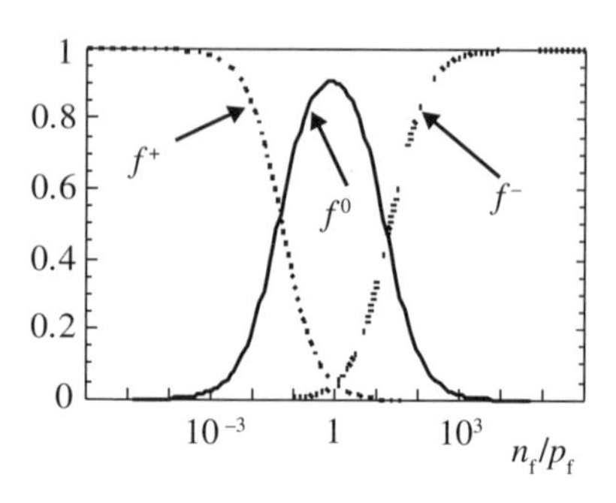


**Fig. 4.42** Occupation functions $f^+$, $f^0$ and $f^-$ for Dangling Bond States $D^+$, $D^0$ and $D^-$, respectively, according to Equations 4.27 a,b,c; adapted from [Hubin-1992], for the assumptions given above, and for $\sigma^{\pm}_{n,p} = \varsigma \times \sigma^{0}_{n,p}$ with $\varsigma = 100$.

the charged dangling bonds, which then become neutral dangling bonds ($D^0$). This happens because Coulomb-assisted capture (capture aided by the attraction of two opposite charges) is much more likely than "neutral" capture (capture by an uncharged dangling bond) : $\sigma^{\pm}_{n,p} \gg \sigma^{0}_{n,p}$, i.e. $\varsigma \gg 1$.

For the following, we will assume that the capture cross-section ratio $\sigma^{\pm}_{n,p}$: $\sigma^{0}_{n,p} = \varsigma$ is approximately equal to 10 and that $\mu^0_p = \mu^0_n$. Compared to the wide range of values given in the literature, $\varsigma = 10$ is somewhat at the lower end; one would nowadays probably be somewhat more justified to assume $\varsigma \approx 30$ [Wyrsch-2010]. However by setting $\varsigma = 10$, as will be done hereunder, the resulting effects are easier to distinguish in the following Figures.

One then finds that:

- neutral dangling bonds ($D^0$) are likely to occur when the free carrier ratio $\xi = p_f(x) : n_f(x)$ lies in the range 0.1 to 10;
- positive dangling bonds ($D^+$) are likely to occur when the free carrier ratio $\xi = p_f(x) : n_f(x)$ lies in the range over 10;
- negative dangling bonds ($D^-$) are likely to occur when the free carrier ratio $\xi = p_f(x) : n_f(x)$ lies in the range below 0.1, i.e. provided the ratio $n_f(x) : p_f(x)$ lies in the range over 10.

## 4.7.2 Field deformation by charged dangling bonds within the *i*-layer: Concept

From the considerations in the preceding section, it becomes clear that in a *pin*-type amorphous silicon solar cell,

- the very "first" part of the *i*-layer, adjoining the *p*-layer, will have positively charged dangling bonds (because in this region $p_f \gg n_f$);
- the "last" part of the *i*-layer, adjoining the *n*-layer, will have negatively charged dangling bonds (because in this region $n_f \gg p_f$);
- the whole central part of the *i*-layer has predominantly neutral dangling bonds.

(The terms “first” and “last” refer here to the passage of light. In the case of *pin*-type solar cells, these terms also refer to the deposition sequence, whereas in the case of *nip*-type solar cells, the deposition sequence is just the opposite.)

For the practical case of a *pin*-type amorphous silicon solar cell (solar cell deposited on a transparent substrate, i.e. on glass + TCO), with AM 1.5 input light and with light trapping, the zone with positively charged dangling bonds will, in fact, be extremely thin and of no practical importance, whereas the zone with negatively charged dangling bonds in the “last” part of the *i*-layer, adjoining the *n*-layer, is quite substantial, and will have the decisive effect in deforming the internal electric field.

This statement will be substantiated by a numerical simulation (Sect. 4.7.3). For the present, just let us have a look at the general concept: The situation is shown schematically in Figure 4.43.

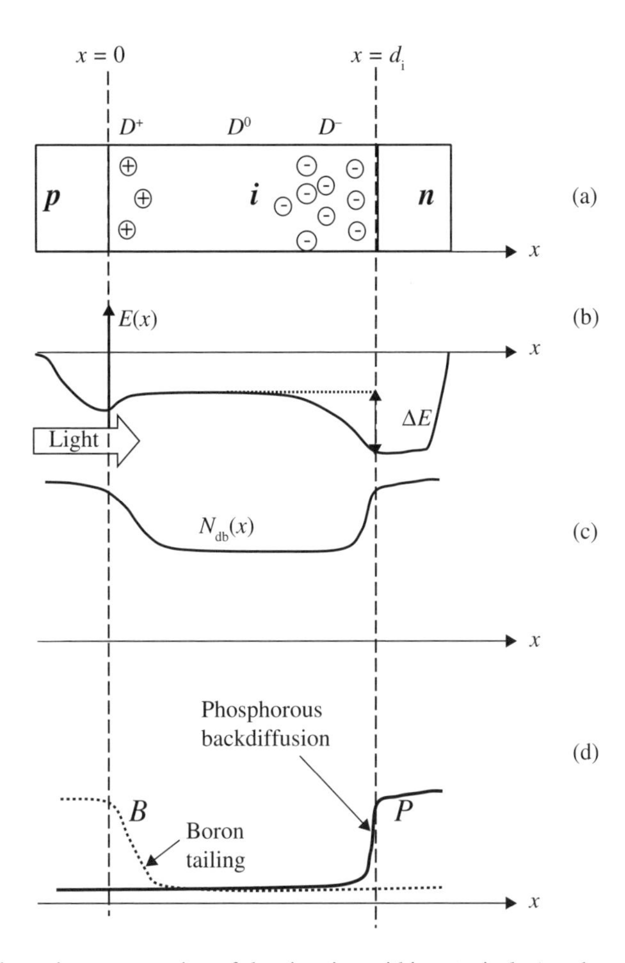


**Fig. 4.43** Schematic representation of the situation within a typical *pin*-solar cell: (a) charge state of dangling bonds; (b) indication of the resulting deformation of the electric field $E(x)$ with field reduction $\Delta E$; (c) generally presumed profile of dangling bond density $N_{db}(x)$; (d) concentration profiles of dopant atoms boron (B) and phosphorous (P).

### 4.7.3 Field deformation by charged dangling bonds within the *i*-layer: numerical simulation for an amorphous silicon solar cell with $d_i = 300$ nm

We can now look at quantitative values for the ratio $p_f(x) : n_f(x)$ of the free carrier profiles; the profiles for $p_f(x)$ and $n_f(x)$ were evaluated in Section 4.5.3 above, for a *pin*-type solar cell with an *i*-layer thickness $d_i = 300$ nm and for the band mobility values $\mu_p^0 = \mu_n^0 = 1\ \text{cm}^2/(\text{V}\cdot\text{s})$.

The ratio of $p_f(x)$ to $n_f(x)$ is shown in the top row of Figure 4.44: in column (a) for red light and in column (b) for AM 1.5. In the second (middle) row, the zones with charged dangling bonds are depicted; we have thereby assumed that $D \rightarrow D^+$, if $p_f/n_f \geq 10$ and that $D \rightarrow \text{D}^-$, if $p_f/n_f \leq 0.1$; this means that we take the capture cross-section ratio $\sigma^{\pm}_{\text{n,p}} : \sigma^{0}_{\text{n,p}} = \varsigma$ to be approximately equal to 10.

We immediately see that in the main part of the intrinsic layer the dangling bonds will be neutral. We also note from Figure 4.44: It is only in a very thin zone at the edge of the *i*-layer, towards the *p*/*i*-interface, that there will be positively charged dangling bonds, and in a relatively thin zone at the other edge of the *i*-layer, towards the *n*/*i*-interface, that there will be negatively charged dangling bonds.

For red light, the two zones with charged dangling bonds will have approximately equal thicknesses. On the other hand, for AM 1.5 illumination, the zone with positively charged dangling bonds towards the *p*/*i*-interface will be extremely thin, whereas the zone with negatively charged dangling bonds towards the *n*/*i*-interface will be quite substantial.

The resulting profile of the electric field $E(x)$ and the field reduction $\Delta E$ is shown in the last (bottom) row of Figure 4.44. The curves refer to short-circuit conditions, when no external voltage is present.

### 4.7.4 Field deformation within the *i*-layer: summary of situation for different *i*-layer thicknesses

The field reduction $\Delta E$ due to negatively charged dangling bonds is given by the expression $\Delta E = (1/\varepsilon_r\varepsilon_0)qN_{\text{db}} \times d^-$, where $d^-$ is the width of the negatively charged zone within the *i*-layer. The width $d^-$ depends solely on the capture cross-section ratio $\sigma^{\pm}_{\text{n,p}}/\sigma^{0}_{\text{n,p}} = \varsigma$. If we assume a higher value for $\varsigma$, we will find a thinner region which is negatively charged, and, thus, a lower value of $d^-$ and also a lower value of $\Delta E$.

As was stated above, we are thereby somewhat underestimating $\varsigma$, and therefore also somewhat overestimating $\Delta E$. However, if one considers the interface effects, as described in Section 4.8.1, these effects almost surely lead to a significant increase of the dangling bond density $N_{\text{db}}(x)$ near the *p*/*i*- and *i*/*n*-interfaces (represented in Fig. 4.43c), and this will cause a corresponding increase in $\Delta E$. Thus, the comments given in Table 4.5 hereunder and based upon the data of Figures 4.40 and 4.44, and on analogous calculations for $d_i = 180$ µm and $d_i = 500$ µm are probably quite realistic.

The field reduction $\Delta E$ evaluated here is proportional to the dangling bond density; it increases with light-induced degradation and is of substantial importance

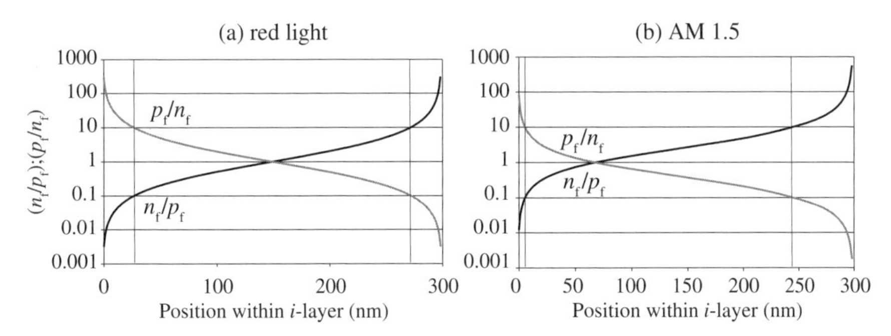


Ratios $\xi = p_f(x)/n_f(x)$ and $\xi^1 = n_f(x)/n_f(x)$ between the densities of free carriers;

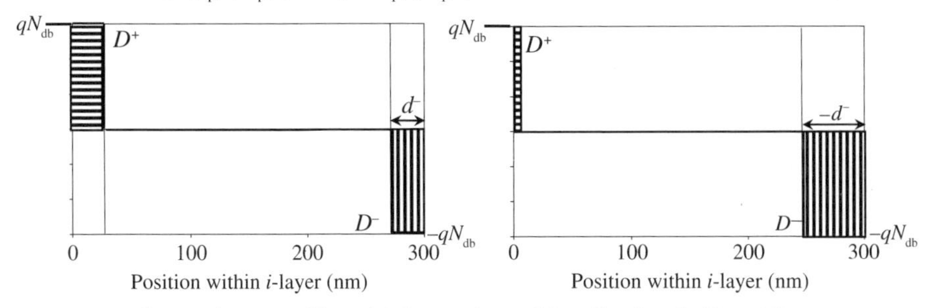


Space charge profiles $\rho(x)$ due to charged dangling bonds $D^+$ and $D^-$ assuming $D \rightarrow D^+$, if $\xi > 10$ and $D \rightarrow D^-$, if $\xi < 0.1$; as well as: $N_{db} = 10^{16}/\text{cm}^3$;

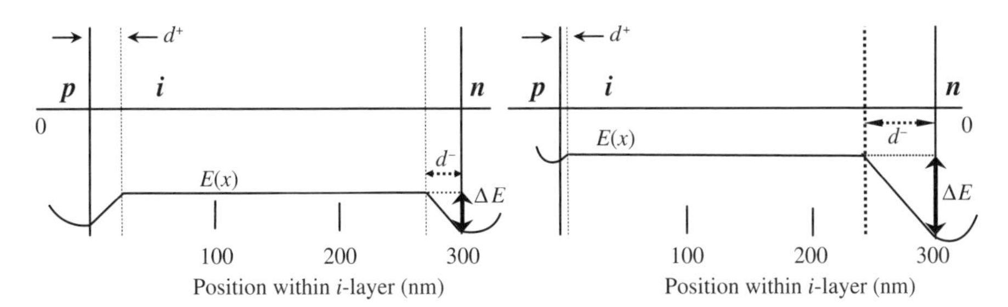


Electric field profile $E(x)$; $\Delta E = (1/\varepsilon_r\varepsilon_0)qN_{db}d^-$; $\varepsilon_0 = 8.85\times10^{-14}$ F/cm, $\varepsilon_r = 10$; for the parameters given here, one finds, in the case of the AM 1.5 spectrum: $\Delta E \approx \{8.85\times10^{-13})\}\times(1.6\times10^{-19})\times(10^{+16})\times(0.6\times10^{-5}) \approx 1\times10^4 V/\text{cm}$.

**Fig. 4.44** Profiles of the ratio ($p_f$ / $n_f$) of free carriers within the intrinsic layer, and, schematically, of the space charge $\rho(x) = q \times N_D+$ and $\rho(x) = -q^- N_D-$, due to charged dangling bonds $D^+$, $D^-$, respectively; and of the resulting electric field $E(x)$with the field reduction $\Delta E^-$ according to Poisson's equation; $p_f$ and $n_f$ are taken from the data displayed in Figure 4.40. The two "standard" cases of input light are again compared: (a) "red light" (uniformly absorbed light) and (b) AM 1.5 spectrum plus light trapping (usual situation in solar cell). Calculations and figure: Courtesy of Michael Stückelberger, IMT PV Lab Neuchâtel.

in stabilized (degraded) amorphous silicon solar cells. If the electric field becomes strongly deformed, then carrier drift will be strongly reduced and the collection of photo-generated carriers will no longer be assured: charge collection will "break

**Table 4.5** Estimated values of electric field deformation due to trapped holes in valence bandtail and due to charged dangling bonds, for *pin*- and *nip*-type amorphous silicon solar cells, with "standard" light trapping conditions, operating under AM 1.5 light, and with *i*-layer thicknesses of 500 nm, 300 nm and 180 nm, from [Stückelberger-2010].

| Effect | *i*-layer thickness (nm) | Hypotheses | Thickness of charged zone (nm) | Field deformation (V/cm) | In % of non-deformed field | Comments |
|---|---|---|---|---|---|---|
| Deformation due to valence bandtail | 500 | $N_{db} = 10^{16}$/cm3 Boundaries at $n_f/p_f$ = 10; 0.1 | ≈300 | $3.8 \times 10^3$ | 20 % | Substantially deformed field |
| | 300 | | ≈150 | $1.2 \times 10^3$ | 4 % | Slightly deformed field |
| | 180 | | ≈100 | $5.2 \times 10^2$ | 1 % | No noticeable field deformation |
| Deformation due to negatively charged dangling bonds near *i*/*n*-interface | 500 | $\mu_p^0 = 1$ | ≈100 | $1.8 \times 10^4$ | 90 % | Field totally distorted |
| | 300 | $\mu_p^{eff} = 0.0033$ | ≈60 | $1.0 \times 10^4$ | 30 % | Substantially deformed field |
| | 180 | | ≈30 | $5.4 \times 10^3$ | 10 % | Slightly deformed field |
| Comments | | | | Non-deformed field: 500 nm: $2.0 \times 10^4$ 300 nm: $3.5 \times 10^4$ 180 nm: $6.0 \times 10^4$ | | |

down". This is a further reason why amorphous silicon solar cells have to be kept very thin ($d_i < 300$ nm).

For the practical case of a *nip*-type amorphous silicon solar cell (solar cell deposited on an opaque substrate, like stainless steel), with AM 1.5 input light and with light trapping, one finds again that the zone with negatively charged dangling bonds, adjoining the *n*-layer, is quite substantial, and that it is this zone that will have a decisive effect in deforming the internal electric field. This zone corresponds now to that part of the *i*-layer that is deposited at first, just after deposition of the *n*-layer. It will usually have some P-contamination (phosphorus contamination). Now, P is a *n*-type dopant; the P-atoms will, almost all, be positively ionized, creating thereby a positive space charge, which, to a certain extent, counteracts the negative space charge due to the dangling bonds. This may indeed be one of the reasons why *nip*-type solar cells have somewhat superior $V_{oc}$ values, as compared to *pin*-type solar cells.

Because microcrystalline silicon layers have only a very mild form of light-induced degradation, and because in "device-quality" microcrystalline silicon layers, the dangling bond density is (as far as can be judged from absorption measurements) certainly quite a bit lower than in amorphous silicon layers, the effect is less pronounced for microcrystalline silicon solar cells. The latter can, thus, have thicker *i*-layers; $d_i$-values up to a few μm are possible in microcrystalline silicon solar cells without carrier collection breaking down.

Table 4.5 gives a summary of the estimated deformation/reduction of the internal electric field, at short-circuit conditions, due to trapped holes in valence bandtail states, and due to charged dangling bonds, for the case of amorphous silicon *pin*- and *nip*-type solar cells.

**Note:** For a given solar cell and under fixed illumination, the electric field depends on the following:

- *for the deformation due to the valence bandtail*: *only on the value of the effective mobility* $\mu_p^{eff}$ of the holes; this parameter is assumed to be 1/300 cm²/Vs = 0.00333 cm²/Vs, for the above estimations; if $\mu_p^{eff}$ is larger, than the field deformation will be smaller, because there will be fewer trapped holes for a given drift current density.
- *for the deformation due to negatively charged dangling bonds*: only on the capture cross-section ratio $\sigma_{n,p}^{\pm} : \sigma_{n,p}^{0} = \varsigma$; this parameter is assumed to be 10 for the above estimations; if $\varsigma$ is larger, than the field deformation will be smaller, because the region with negatively charged dangling bonds will be thinner.

## 4.8 RECOMBINATION AND COLLECTION

### 4.8.1 *p/i* and *i/n* interfaces

During the fabrication of a *pin*-type solar cell by plasma deposition, the part of the *i*-layer that is deposited first is especially critical, because it tends to be contaminated by the dopant (boron) in the underlying *p*-type layer (so-called boron tailing). In fact, the plasma will "attack" or "etch" the underlying *p*-type layer and set free a substantial quantity of boron atoms, which interfere with the deposition of the first part of the *i*-layer. If the solar cell is deposited in a single-chamber system, without any special precautions, the whole deposition chamber will at first be contaminated by the boron gas used for the underlying *p*-type layer, and the phenomenon of boron tailing will be even stronger.

During subsequent deposition of the *n*-type layer, the directly underlying part of the *i*-layer, i.e. the part that is deposited last, is also somewhat affected by the subsequent plasma that now contains phosphorus (so-called *phosphorus back-diffusion*).

Figure 4.45 illustrates (as a follow-up on Figure 4.44(d) the effects of boron tailing and phosphorus back-diffusion on impurity concentration levels within the *i*-layer of a *pin* solar cell. To obtain satisfactory solar cell performance, phosphorus back-diffusion and especially boron tailing would have to be considerably lower than what is shown in this Figure. The level of oxygen contamination would also have to be reduced (see Chap. 6, Sect. 6.1.3).

In *nip*-type solar cells, one will have phosphorus tailing in the first part of the *i*-layer and boron back-diffusion in the last part.

Now impurities, such as P, B, O and N, will introduce additional defects, i.e. additional dangling bonds. These defects act as recombination centers and will deteriorate the performance of the solar cell.

In microcrystalline silicon solar cells these effects are even more critical, for three reasons:

- the presence of atomic hydrogen in the deposition plasma;
- the initial growth of the crystallites is very critical;
- the influence of impurities on microcrystalline silicon is far stronger than their effect on amorphous silicon (see e.g. [Kilper-2009]).

In amorphous silicon, theoretical (thermodynamic) considerations based on the so-called "defect pool model" predict that, even if there are no structural or chemical

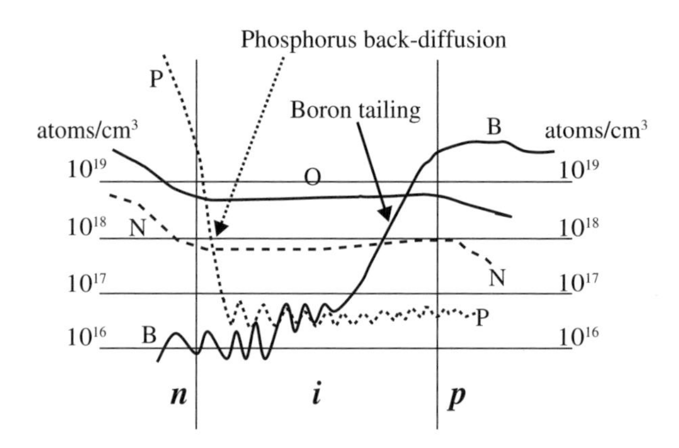


**Fig. 4.45** SIMS (secondary ion mass spectroscopy) profile, as typically obtained for *pin*-type solar cells, showing: boron tailing, phosphorous back-diffusion and a relatively high level of oxygen contamination. The solar cell on which these SIMS measurements were made did NOT have satisfactory performance.

changes, the defect density will increase as soon as the Fermi level is shifted away from the midgap (see e.g. [Powell-1992], [Winer-1990]). This again would mean that both ends of the *i*-layer would have an increased defect density.

For all these reasons one considers that the *p*/*i*- and *i*/*n*-interfaces are regions with higher defect densities and with additional recombination (compare with Fig. 4.44(c)). This is especially true of that interface which is deposited first, i.e. the *p*/*i*-interface in the *pin*-solar cell (deposited on a glass superstrate) and the *i*/*n*-interface in the *nip*-solar cell (deposited on an opaque substrate).

The interface defects play a particularly important role in limiting the open circuit voltage $V_{oc}$ of the cell, because it is during open-circuit conditions that the free carrier densities of both free carriers are particularly large at both ends of the intrinsic layer.

### 4.8.2 Recombination

Recombination is a pair process; it takes a pair consisting of an electron and of a hole to recombine. Therefore, we can consider that recombination takes place mainly in the center of the intrinsic layer, at least for operation conditions near the short-circuit and maximum power points. As we approach the open-circuit point on the $J(V)$ curve of the solar cell, transport by drift breaks down and both types of carriers diffuse more easily towards both ends of the intrinsic layer; thus, interface defects (see Sect. 4.8.1) now begin to play a larger role.

Excluding open-circuit conditions and looking, therefore, at the centre of the intrinsic layer, we can see that the recombination centres will be constituted, to a large extent, by neutral dangling bonds ($D^0$ state, see Figs. 4.41 and 4.42). The recombination function for neutral dangling bonds is given by

$$R = n_f/\tau_n^0 + p_f/\tau_p^0, \tag{4.28}$$

where $\tau_n^0 = (v_{th.n}\sigma_n^0 N_{db})^{-1}$ and $\tau_p^0 = (v_{th.p}\sigma_p^0 N_{db})^{-1}$ are capture times, for capture by neutral dangling bonds, of electrons and holes, respectively [Hubin-1992].

If the interfaces are very defective, a significant amount of additional recombination may take place at one or both interfaces, but especially at the *p*/*i*-interface. At the latter location the dangling bonds will be mainly positively charged, because of the large current flow of holes towards the *p*-layer; here, Equation 4.28 should be replaced by

$$R = n_f/\tau_n^+, \tag{4.29a}$$

where $\tau_n^+ = (v_{th.n}\sigma_n^+ N_{db})^{-1}$ is the capture time, for capture of electrons by positively charged dangling bonds.

A similar reasoning can be applied to the *i*/*n*-interface; here, recombination will be governed by the expression

$$R = p_f/\tau_p^-, \tag{4.29b}$$

where $\tau_p^- = (v_{th.p}\sigma_p^- N_{db})^{-1}$ is the capture time, for capture of holes by negatively charged dangling bonds. Note that Equations 4.28, 4.29(a) and 4.29(b) are all linear functions of the free carrier densities $n_f$ and $p_f$.

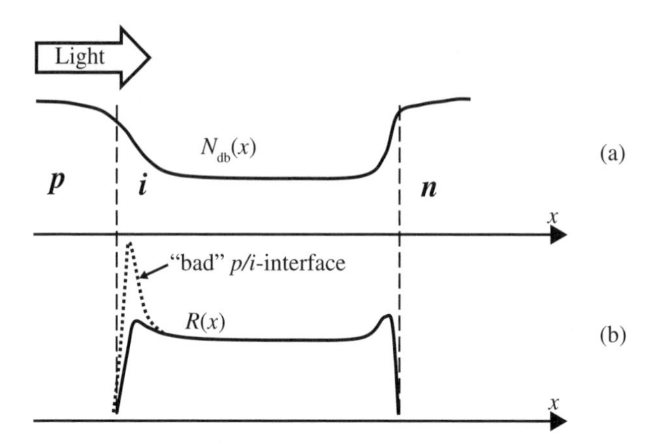


**Fig. 4.46** Schematic representation of the dangling bond density $N_{db}(x)$ and of the recombination $R(x)$ in the intrinsic layer of a *pin*-type solar cell; if the cell has a problematic *p*/*i*-interface. e.g. due to boron contamination (boron tailing), there will be strong, supplementary recombination as indicated by the dotted line.

Figure 4.46 shows schematically dangling bond density $N_{db}(x)$ and recombination function $R(x)$ for the intrinsic layer of a *pin*-type solar cell: we indicate here that the major part of the recombination takes place in the "bulk" of the *i*-layer, but that there can be additional, enhanced recombination at the interfaces, especially at the critical *p*/*i*-interface.

## 4.8.3 Collection and drift lengths

Transport of carriers within the intrinsic layer of the *pin*-type solar cell is mainly by drift, as already mentioned above. Transport by diffusion can be neglected here, at least as far as the centre of the intrinsic layer is concerned. Therefore, carrier collection is governed by the *drift lengths* of free electrons and free holes, i.e. by

$$l_{\text{drift.n}} = \mu_n^0 \tau_n^0 \times E$$

and

$$l_{\text{drift.p}} = \mu_p^0 \tau_p^0 \times E$$

Here, $\mu_n^0$ and $\mu_p^0$ are the band mobilities, i.e. the mobilities of free electrons and free holes, respectively. It has been shown, for amorphous silicon layers, that $\mu_n^0 \tau_n^0 \approx \mu_p^0 \tau_p^0 \approx \mu^0 \tau^0$ [Beck-1996]. This assumption probably also holds for microcrystalline silicon layers, as indeed similar defects limit transport and govern recombination in both types of materials [Droz-2000]. This means that both drift lengths $l_{\text{drift.n}}$, $l_{\text{drift.p}}$ are approximately equal:

$$l_{\text{drift}} \approx l_{\text{drift.n}} \approx l_{\text{drift.p}} \approx \mu^0 \tau^0 \times E, \tag{4.30}$$

where the product $\mu^0 \tau^0$ is a quality parameter of the intrinsic layer.

For the solar cell to function properly, the drift lengths must be, at least for short circuit conditions and for the maximum power point, much larger than the thickness $d_i$ of the intrinsic layer. The drift lengths depend on the quality of the intrinsic layer, i.e. on the $\mu^0\tau^0$-product. If the dangling bond density $N_{db}$ is increased, then $\tau^0$ as well as the whole $\mu^0\tau^0$-product will decrease proportionally. The drift lengths depend also on the electric field $E$ in the center of the intrinsic layer.

There is no straightforward way to measure the drift lengths prevailing in the intrinsic layers of amorphous and microcrystalline silicon solar cells. The best one can do is to evaluate the dangling bond density $N_{db}$, within the intrinsic layer of the completed *pin*-solar cell, by performing Fourier Transform Photocurrent Spectroscopy (FTPS) measurements [Vaneček-2002]. These are, in fact, simply absorption measurements: one evaluates, with FTPS, the absorption provoked by midgap defects (dangling bonds) in the material. As far as the other parameters intervening in the $\mu^0\tau^0$-product are concerned (i.e. band mobilities, capture cross-sections, etc.), one may assume that they do not vary significantly between different layers of the same material (amorphous or microcrystalline silicon).

It is even more difficult to evaluate the electric field within the *i*-layer (intrinsic layer). All one can do here is to keep the *i*-layer as thin as possible and avoid contamination by ionized impurity atoms, oxygen, boron and phosphorus atoms. Of course, both charged dangling bonds and charged carriers trapped in bandtail states will also contribute to deform the electric field and reduce its value in the centre of the *i*-layer, as described above. A substantial reduction or deformation of the electric field usually shows up in the EQE measurements of the solar cell as a decrease in the EQE (external quantum efficiency) curve (or alternatively, in the spectral response (SR) curve) towards long wavelengths. (The EQE/SR measurement techniques will be described in Sect. 4.11.1.) If such a decrease is perceived, even though the FTPS result indicates a low dangling bond density, then the problem must be attributed to a shortcoming of the electric field. This is indicated in Figure 4.47.

Such spectral response measurements are done under bias light conditions comparable to AM 1.5 light and with zero applied voltage $V = 0$; they evaluate the response of the solar cell to an additional modulated light signal superimposed on the bias light. A reduction of the spectral response curve in the longer wavelengths, for a solar cell with identical optical properties, means that the light absorbed at the far end of the *i*-layer, towards the *i*/*n*-interface, generates carriers that partly recombine before reaching the other end of the *i*-layer.

## 4.9 ELECTRICAL DESCRIPTION OF THE PIN-SOLAR CELL

### 4.9.1 Equivalent circuit and extended "superposition principle"

Let us look at Figure 4.48, which shows schematically the flow of electrons and holes in a *pin*-diode for 3 cases:

(a) dark conditions and forward bias;
(b) dark conditions and reverse bias;
(c) illuminated conditions.

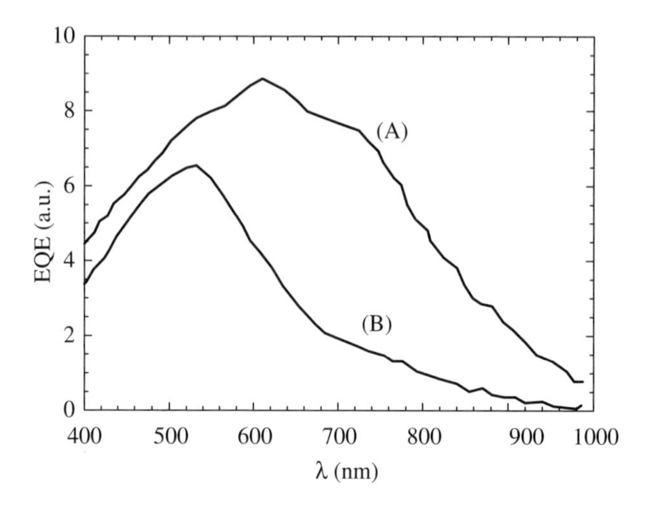


**Fig. 4.47** Curves of the external quantum efficiency (EQE) for the example of two microcrystalline silicon solar cells ($d_i = 2.8$ μm), which are otherwise identical: (A) with an *i*-layer deposited under oxygen-free conditions; (B) with electric field deformation/reduction in the *i*-layer, due to the presence of oxygen contamination in the *i*-layer reproduced with permission from [Torres-1996].

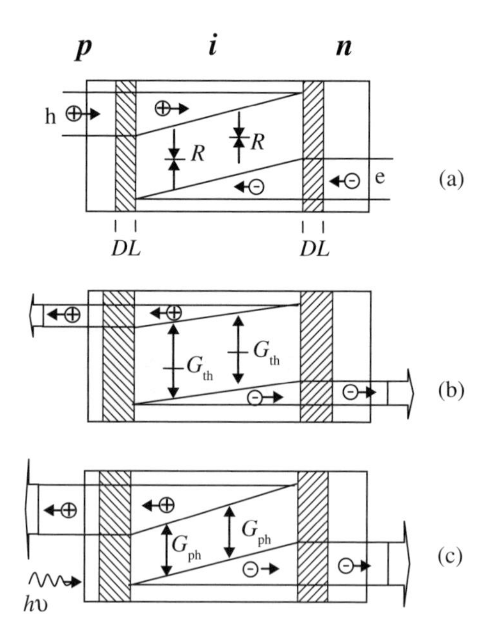


**Fig. 4.48** Schematic representation of the flow of electrons and holes and of recombination/generation phenomena in a *pin*-type diode: (a) forward-biased diode, in dark conditions; (b) reverse-biased diode in dark conditions; (c) illuminated diode with photo-generation; recombination and generation take place in connection with mid-gap defects, i.e. dangling bonds, which are considered to be predominantly in the neutral state. *DL* is the depletion layer; $G_{th}$ stands for thermal generation and $G_{ph}$ for photo-generation.

Figure 4.48 corresponds to Figure 4.17 for the *pn*-diode. At the left (*p*-side) of the diode, the flow of electric current is entirely carried by holes; on the right (*n*-side) of the diode, the flow of electric current is entirely carried by electrons. This means that current continuity can physically only be maintained by recombination/generation phenomena, which in the *pin*-diode almost exclusively take place in the *i*-layer, and are moderated, for dark conditions, to a great extent by neutral dangling bonds ($D^0$).

*In the dark*, the *pin*-diode is generally considered to be governed by an exponential function

$$J_{\text{dark}} = J_0 \left[ \exp\left( \frac{qV}{nkT} \right) - 1 \right], \tag{4.31}$$

with a "diode ideality factor" $n = 2$.

There exists, to the author's knowledge, no direct and general derivation of Equation 4.31 for the case of the *pin*-diode. It can, however, be intuitively justified by looking, as a first step, at the *pn*-diode and considering there the case where recombination in the depletion layer becomes dominant. In this case, textbooks on semiconductor devices generally make certain assumptions, notably:

(a) that recombination takes place mainly in the centre of the depletion region of the *pn*-type solar cell; and
(b) that the prevailing recombination center is a midgap defect of the type assumed by classical Shockley-Read-Hall recombination theory;

and come to the conclusion that the diode ideality factor then becomes $n = 2$ (see e.g. [Sze-2006], or [Gray-2003], p. 90).

Afterwards, in a second step, we look at the *i*-layer within the *pin*-diode and, as a very coarse approximation, assimilate it to the depletion layer of a *pn*-diode. We can imagine taking the *pn*-diode and pulling it apart, so that at the place where we previously had the frontier between *p*- and *n*-zones, a new undoped zone (*i*-layer) is formed. We will now obtain a sort of "extended depletion layer (DL)" (see also Fig. 4.17), consisting of the original depletion layers and the newly created *i*-layer. In this way it is probably easy to accept the idea that a *pin*-diode corresponds to a *pn*-diode with additional recombination in the depletion layer.

One now can intuitively understand that above Equation 4.31, with $n = 2$, should be valid for the *pin*-diode, at least as long as

- interface defects are negligible;
- the recombination centres behave similarly to those in crystalline silicon;
- recombination/thermal generation takes place mainly in the center of the *i*-layer.

In reality, the situation is much more complex in a-Si:H and μc-Si:H solar cells.

[Deng-2005] have made a comprehensive study, from both experimental and theoretical viewpoints, of the diode ideality factor *n* for amorphous silicon solar cells. They come to the conclusion that a voltage-dependent ideality factor $n(V)$, with values of *n* varying between 1.2 and 1.9, has to be introduced. Such a complex model does not, however, give much insight into basic mechanisms governing the dark characteristics $J_{dark}(V)$ of amorphous silicon solar cells.

In practice, for amorphous and microcrystalline silicon solar cells one uses Equation 4.31 with a diode ideality factor $n$ having a value between 1.5 and 1.8; a higher value (approaching 2) is seen as indication that bulk recombination within the *i*-layer is dominant, whereas a lower value (around 1.5 to 1.6) is often taken as indication that recombination at the *p*/*i*- or *n*/*i* interfaces are dominant.

Now, let us add photo-generation to the picture. Experimentally one finds for illuminated *pin*- and *nip*-type solar cells the situation shown in Figure 4.49, where the various $J(V)$ curves all (more or less) meet at a single intersection point P; this behavior is generally seen both with amorphous silicon solar cells [Kusian-1989] and with microcrystalline silicon solar cells [Sculati-Meillaud-2006]. It is therefore evident that the superposition principle, as introduced by Equation 4.12 in Section 4.3, for *pn*-type solar cells, is not at all valid for *pin*- and *nip*-type solar cells.

*Theoretically*, we shall now proceed by looking at recombination and not at transport: Photo-generation will increase the densities of both carriers in the *i*-layer. Thus, photo-generation will lead to additional recombination, mainly in the center of the *i*-layer, according to Equations 4.28. It will also lead to additional recombination at the *p*/*i*- and *i*/*n*-interfaces according to Equations 4.29a and 4.29b. Because Equations 4.28 and 4.29a,b govern recombination both in the dark and in the illuminated case, and because Equations 4.28 and 4.29a,b are all linear, we can superpose the dark current density $J_{dark}$ according to Equation 4.31 and the photo-generated current, as long as we consider recombination/generation in the *i*-layer and not transport. This leads to the following expression for the illuminated *pin*- or *nip*-type solar cell:

$$J_{illum} = J_{ph} - J_{dark} - J_{rec}, \tag{4.32}$$

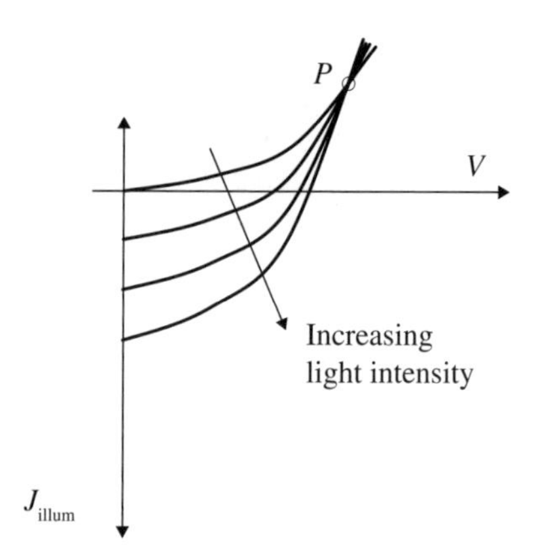


**Fig. 4.49** Schematic illustration of the various $J(V)$ curves, for different illumination levels intersecting at a single point $P_0$. This behavior has so far been observed on all thin-film silicon solar cells with a *pin* (or *nip*) structure.

where $J_{ph}$ is the photogenerated current density and $J_{rec}$ is the recombination current density:

$$J_{rec} = q\int_0^{d_i} R(x)\mathrm{d}x \tag{4.33}$$

Here, $R(x)$ denotes supplementary recombination, due only to the photo-generated free carriers.

For the case where interface defects are important, additional recombination via interface defects will be governed by Equation 4.28 for the *p*/*i*-interface, and by Equation 4.29 for the *n*/*i*-interface. These are again linear functions of the free carrier densities. So, here again, the superposition principle can be applied when considering recombination. This leads us to the equivalent circuit of Figure 4.50.[10]

In Figure 4.50(a), $J_{rec}$ constitutes a recombination current sink; its value is proportional to $J_{ph}$. For the simplest case where the electric field is uniform (constant) with $E(x) = E_0$, throughout the whole *i*-layer, $J_{rec}$ reduces to the simple form

$$J_{rec} = J_{ph}\frac{d_i}{l_{drift}}, \tag{4.34}$$

with $l_{drift} = \mu^0\tau^0 E_0$ and $E_0 = (V_{bi} - V)/d_i$.

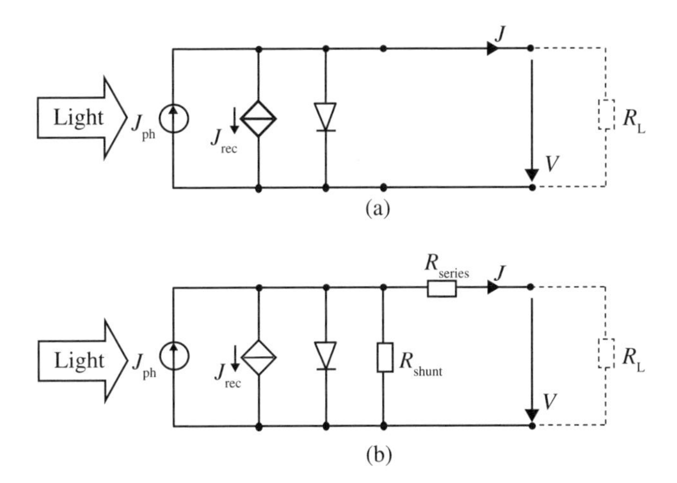


**Fig. 4.50** Equivalent circuits for *pin*- and *nip*-type thin-film silicon solar cells: (a) Basic equivalent circuit, based on the superposition of a diode D without illumination, a photo-generated current source $J_{ph}$, as well as recombination through a current sink $J_{rec}$ [Merten-1998]; (b) Complete equivalent circuit including the effect of recombination and also of series and shunt resistances, $R_{series}$ and $R_{shunt}$, respectively. The diode D is characteried by its ideality factor *n* and by its reverse saturation current density $J_0$.

[10]Note that in Chapter 6, Section 6.6.4, the traditional (universal) equivalent circuit for the solar cell (Fig. 4.21) is used, and the parallel resistance $R_p$ is divided into two parts: "photo-shunts" (corresponding to $J_{rec}$) and "dark shunts" (called simply "shunts" in this Chapter).

In reality, the electric field will be deformed by space charge within the *i*-layer, especially for the case of thick *i*-layers; the electric field will also be significantly reduced as the applied external voltage $V$ is increased towards the open circuit voltage $V_{oc}$. Nevertheless, Equation 4.34 can be used as a first approximation and will describe reasonably well the behavior of a high-quality amorphous or microcrystalline solar cell from the short-circuit point up to the maximum power point (MPP). For the other cases (lower-quality solar cells, all cells approaching $V_{oc}$), it will be sufficient to state that $l_{drift}$ will be lower than the value given above and that the recombination current $J_{rec}$ will be correspondingly higher, but that it will still be proportional to $J_{ph}$.

The equivalent circuit of Figure 4.50(a) can be used to explain the $J(V)$ curves shown in Figure 4.49 with the intersection point P.

(It is, in reality, not permissible to use the equivalent circuit of Figure 4.50(a), together with Equation 4.34, for voltages higher than the voltage $V_{MPP}$ at the maximum power point; still it does allow one to visualize the existence of the single intersection point P. There seems to be a yet unknown reason why this model works so well, even for voltages higher than $V_{MPP}$.)

### 4.9.2 Shunts[10]

The equivalent circuit of Figure 4.50(a) may now be completed, by adding series resistance $R_{series}$ and shunt resistance $R_{shunt}$, as already described in Section 4.3. In thin-film silicon solar cells $R_{series}$ represents contact resistance due to metal paths, due to the TCO-layer (see Chap. 6, Sect. 6.5) and also due to insufficiently doped *p*- and *n*-layers, whereas $R_{shunt}$ represents ohmic shunts through the diode, due to pinholes and especially in microcrystalline silicon solar cells shunts due to cracks within the *i*-layer, as caused by growth problems. One arrives, thus, at the equivalent circuit of Figure 4.50(b), which is identical with the one shown in Figure 4.22 (Sect. 4.3).

Figure 4.51 illustrates schematically the formation of shunts due to dust particles sitting on the underlying layer before and during plasma deposition of the thin-film silicon layer. Shunts can also be created during structuring of the solar cell, e.g. by laser patterning (see Chap. 6. Sect. 6.5.1). These shunts can result in actual short circuits and, thus, completely disable the whole solar cell; more frequently they

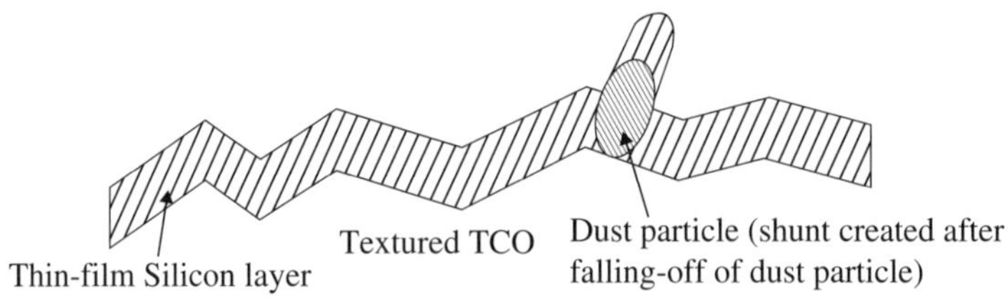

Textured TCO

Dust particle (shunt created after falling-off of dust particle)

Thin-film Silicon layer

**Fig. 4.51** Schematic representation of the formation of pinholes and shunts due to dust particles sitting on the underlying layer before and during thin-film silicon deposition.

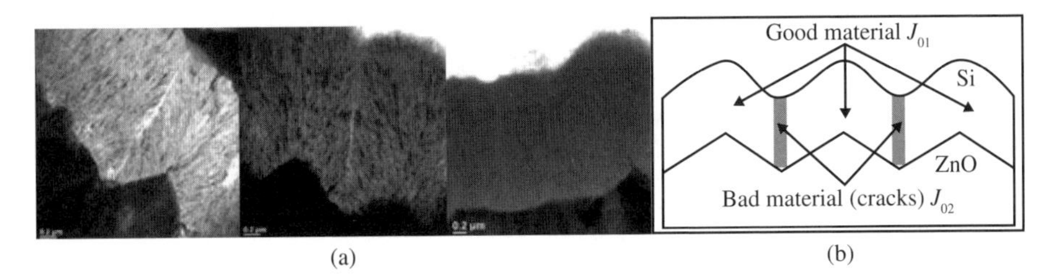


**Fig. 4.52** (a) Electron micrograph showing the growth of a microcrystalline silicon solar cell on a rough substrate and the formation of cracks starting out in the valleys of the substrate layer. Here the substrate layer is a ZnO layer grown by low-pressure chemical vapor deposition (LP-CVD) [Python-2008]. (b) Schematic drawing.

simply lead to a very low value of $R_{shunt}$ and will, thus, reduce the fill factor $FF$ and the efficiency $\eta$ of the whole solar cell.

Figure 4.52 is an electron micrograph of a microcrystalline silicon solar cell, illustrating the appearance of cracks during the microcrystalline growth process. Such cracks often start off in the valleys of an underlying rough substrate (as used for light scattering and light trapping, see Sect. 4.11.2); they can lead to low values of $R_{shunt}$, but can also result in high values of the diode's reverse saturation current density $J_0$. The latter effect can be explained if one considers that diodes of very bad quality may be formed within the regions surrounding the cracks, for example because of the diffusion of contaminants through the cracks.

### 4.9.3 Variable illumination measurements (VIM)

The individual elements of the equivalent circuit of Figure 4.50 can be evaluated, for a given solar cell, by measuring the $J(V)$-curve of the solar cell under variable levels of input light, i.e. by so-called *variable-illumination measurements* (VIM). It is intuitively clear that at very low levels of light, the behavior of the solar cell, as represented by the equivalent circuit of Figure 4.50(b), will be governed by $R_{shunt}$. At high illumination levels it will, on the other hand, be governed by $R_{series}$.

It is of specific interest to measure the slope of the $J(V)$-curve at the short-circuit point, i.e. the short-circuit differential resistance $R_{sc} = \partial V/\partial J|_{V=0}$, for different illumination levels. It can easily be shown (see e.g. [Merten-1998, Sculati-Meillaud-2006, Meillaud-2006, Shah-2010a]) that $R_{sc}$ has three distinct regions (see Fig. 4.53):

(a) A low-intensity region, where $R_{sc} \to R_{shunt}$.
(b) A medium intensity region, where $R_{sc} \approx \mu^0\tau^0(V_{bi}^2/d_i^2)(1/J_{ph})$; here, the term $\mu^0\tau^0(V_{bi}^2/d_i^2)$ will be called "collection voltage $V_{coll}$"; it is the slope of the $R_{sc}(1/J_{ph})$-curve in Figure 4.53; it is also a measure of the collection efficiency within the $i$-layer of the solar cell; it can be used also when the electric field within the $i$-layer is non-uniform. It can also be written as $V_{coll} = l_{drift}(V=0)\times(V_{bi}/d_i)$, i.e. it is proportional to the drift length at zero applied voltage.

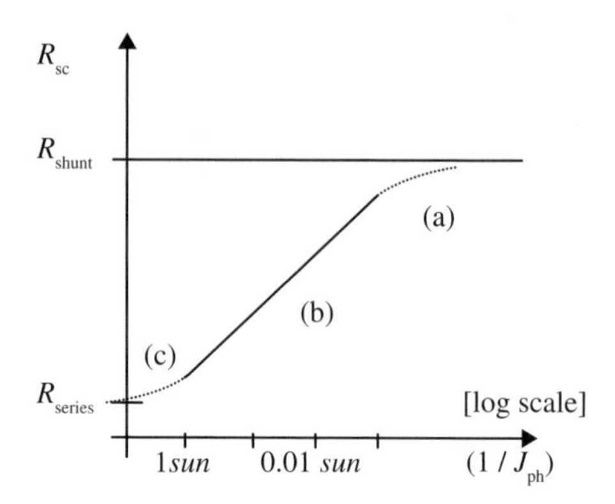


**Fig. 4.53** General behavior of the short-circuit differential resistance $R_{sc} = \partial V / \partial J\big|_{V=0}$, as a function of the illumination level, i.e. as a function of the photo-generated current $J_{ph}$. Because of the logarithmic scale used here for both co-ordinates, the slope of the curve in region b) in this Figure is unity, and the collection voltage $V_{coll}$ is given by the position of the curve. Note that the light level increases as we go from right to left, because we have represented here $\log(1/J_{ph})$ as the $x$-axis.

(c) A high intensity region, where $R_{sc}$ $R_{series}$; it is, however, much more convenient to evaluate $R_{series}$ from the asymptotic value of the open circuit differential resistance $R_{oc} = \partial V / \partial J\big|_{J=0}$, which we obtain as we approach high light intensities.

Variable-illumination measurements (VIM) are a very convenient method for discriminating between three different deficiencies commonly found in thin-film silicon solar cells, i.e.

(a) ohmic shunts (if one finds $R_{shunt} < 10$ kΩcm²);
(b) unsatisfactory collection efficiency, due for example to insufficient material quality and/or contamination problems (if one finds $V_{coll} < 20$ Volts in amorphous silicon solar cells and $V_{coll} < 50$ Volts in microcrystalline silicon solar cells);
(c) high series resistance, due e.g. to poor TCO-layers (if $R_{series}$ is too high; typically if $R_{series} > 10$ Ωcm²).

These three deficiencies (a), (b), (c) have a direct effect in reducing the fill factor *FF* of the cell. We will discuss their diagnosis in more detail in Section 4.9.5.

The measurement of the *J*(*V*)-curve in the dark, without illumination, is a further important step in assessing solar cell quality for amorphous and microcrystalline silicon solar cells. This assessment is especially important for microcrystalline silicon solar cells: if the reverse saturation current density $J_0$ in the dark is too high (higher than $10^{-7}$ A/cm², for microcrystalline silicon solar cells), this is probably due to cracks that have occurred during the microcrystalline silicon growth process (see Sect. 4.9.2)

### 4.9.4 Reverse saturation current $J_0$ and open circuit voltage $V_{oc}$

According to Section 4.3.4, and based on the diode equation, the open-circuit voltage $V_{oc}$ can be basically written as:

$$V_{oc} = \frac{nkT}{q}\ln\left(\frac{J_{ph}}{J_0}+1\right) \approx \frac{nkT}{q}\ln\left(\frac{J_{ph}}{J_{00}}\right)+\frac{E_g}{q}, \tag{4.35}$$

if we define $J_{00}$ such that $J_0 = J_{00} \exp(-E_g/nkT)$; thereby $n = 2$ for "ideal" *pin*-type solar cells. $J_0$ and $J_{00}$ have now different values than in the case of *pn*-diodes. Thus, at least in the ideal case, Equation 4.35, gives us for *pin*-type solar cells a link between $V_{oc}$ and $J_0$: The higher the reverse saturation current $J_0$ of the diode in the dark is, the lower will be the open circuit voltage $V_{oc}$.

Because of additional recombination/generation within the *i*-layer, $J_0$ will always be higher, under otherwise comparable circumstances, in a *pin*-type diode than in a "classical" *pn*-type diode. This also leads to somewhat lower limit values of $V_{oc}$, as can be physically and intuitively easily grasped. Because of the change in the diode ideality factor $n$, this conclusion does not come in a straightforward way from the mathematical expressions, but needs to be substantiated by experimental measurements for "device-quality" layers and by numerical calculations, such as given in [Sculati-Meillaud-2006].

Experimentally one finds, in thin-film silicon *pin*-type solar cells, that in "state of the art" solar cells" $J_0$ is around $10^{-12}$ A/cm$^2$ (see e.g. [Tchakarov-2003]) for amorphous silicon and around $10^{-8}$ to $10^{-9}$ A/cm$^2$ [Python-2008], for microcrystalline silicon. The latter $J_0$-value (as measured for microcrystalline silicon) corresponds [Sculati-Meillaud-2006] to a $J_{00}$-value that is around 50 A/cm$^2$ and a $V_{oc}$-limit that is about 50 mV lower than the semi-empirical limit for *pn*-type wafer-based solar cells, which was shown in Section 4.3. The $J_0$-value measured for amorphous silicon corresponds, on the other hand, to a $J_{00}$-value that is around 1500 A/cm$^2$; this will give us a $V_{oc}$-limit that is about 200 mV lower than the semi-empirical limit for *pn*-type wafer-based solar cells.

In practice, state-of-the art microcrystalline silicon solar cells have $V_{oc}$-values that are in the range of 550 to 600 mV [van den Donker-2007] and are, thus, very near to what can be expected from the semi-empirical limit value, as derived above. The best experimental amorphous silicon solar cells, however, have $V_{oc}$-values that are in the range 900 mV to 1 V and are, therefore, much lower than the limit value. There are various reasons for that, i.e.:

- the relatively high reverse saturation current prevailing in actual amorphous silicon solar cells;
- the difficulty in displacing the Fermi-level by doping in amorphous silicon layers, because of the presence of bandtails.

In spite of these reasons, the relatively low $V_{oc}$-values obtained in amorphous silicon solar cells (despite a band gap higher than 1.7 eV) are not fully understood; they do constitute, however, a major limitation for applications. For microcrystalline silicon solar cells, the situation is, in the ideal case, better, and one is able, as already stated, to reach $V_{oc}$-values that are near the theoretical limit value, provided the cell is properly fabricated. However, if the microcrystalline *i*-layer has cracks, or if in any other way the diode is of "insufficient" quality, $J_0$ will be higher and may attain a value as high as$10^{-6}$ A/cm$^2$, leading then to a low value of $V_{oc}$.

Additionally, in both cases $V_{oc}$ will be further reduced by interface recombination (see Sect. 4.9.1 above).

One can therefore understand that $V_{oc}$ is indeed a critical parameter in thin-film silicon solar cells. It is only in a well-designed and well-fabricated cell that a reasonable value of $V_{oc}$ will be reached. Still, not all solar cell defects show up as low values of $V_{oc}$. As an example: a high value of series resistance (due to unsatisfactory contact layers) will not show up as a low value of $V_{oc}$.

### 4.9.5 Fill factor in *pin*-type thin-film silicon solar cells

The measured value of the fill factor is a very sensitive experimental indication of the quality of a solar cell and it should be used as a complementary quantity, in addition to $V_{oc}$, to characterize solar cell quality.

Amorphous silicon solar cells have, in practice, values of the fill factor, which are, in the best case, between 75 % (in the initial state) and 65 % (in the degraded state). This again is lower than what can be expected from theory: This relatively low value of the fill factor in the initial state is certainly (to a large extent) due to the "practical" limitations on the value of $V_{oc}$ in amorphous silicon (see above). The difference in the fill factors between the initial and degraded states can, on the other hand, be explained by the effect of light-induced degradation on collection.

Microcrystalline silicon solar cells, on the other hand, can attain fill factors that are near their theoretical limit value of 75 to 78 % as defined by the bandgap and by the $V_{oc}$-value.

If the fill factor *FF* in a thin-film silicon solar cell is lower than the values just stated, then the cell has a specific problem, which can be diagnosed. For example by VIM measurements and by other diagnostic methods.

The most common problems encountered are:

(a) collection problems due to insufficient *i*-layer quality or to problems with the *p*/*i*- and *i*/*n*-interfaces; this will show up in a low measured value of the collection voltage $V_{coll}$, and also in the spectral response measurement of the cell;
(b) high series resistance due to unsatisfactory contact layers;
(c) low parallel resistance due to cracks and structuring;
(d) high diode reverse current due to leakage paths and bad quality diodes which are formed within the regions surrounding cracks.

Problems (b) and (c) show up in VIM-measurements, and problem (d) can be identified by dark $J(V)$-measurements. The following Table 4.6 indicates qualitatively, based on calculations given (for microcrystalline silicon, in [Sculati-Meillaud-2006, Meillaud-2006], and, for amorphous silicon in [Shah-2010a]) the effect of various faults (a) to (d) on the fill factor *FF*.

**Table 4.6** Effect of various shortcomings (faults) within a single-junction thin-film silicon solar cell on the fill factor *FF.*

| Fault | Measured parameter | *FF* reduction (amorphous silicon cells) | *FF* reduction (microcrystalline silicon cells) |
|---|---|---|---|
| Collection problem | $V_{coll}$ [V] | $(V_{bi}/V_{coll}) \times 90$ % | $(V_{bi}/V_{coll}) \times 90$ % |
| High series resistance | $R_{series}$ [$\Omega$cm$^2$] | $1.5 \times R_{series}$ | $3 \times R_{series}$ |
| Low parallel resistance | $R_{shunt}$ [k$\Omega$cm$^2$] | $4 \times (1/R_{shunt})$ | $2 \times (1/R_{shunt})$ |
| High diode reverse current | $J_0$ [A/cm$^2$] | So far not treated... | So far not treated in literature |

### 4.9.6 Limits for the short-circuit current $J_{sc}$ in *pin*-type thin-film silicon solar cells

The short-circuit current $J_{sc}$ in thin-film silicon solar cells is mainly limited by four factors

(a) A significant amount of light is reflected back before entering the photoactive *i*-layer, e.g. it may be reflected by the glass cover, or at the TCO-silicon interface, etc.
(b) The solar cell is too thin to absorb a sufficiently large part of the solar spectrum and the light trapping scheme used (see Sect. 4.11.2) does not increase the equivalent optical thickness to a sufficient extent.
(c) A significant amount of the light may be absorbed within layers that do not contribute to the photo-generated current $J_{ph}$, i.e. within transparent contact layers, doped layers, reflector layers etc., rather than within the photoactive *i*-layer. Actually, light trapping and light scattering will increase these losses.
(d) Part of the photo-generated current $J_{ph}$ is lost due to recombination within the *i*-layer, even at short-circuit conditions. This will occur, if the *i*-layer is too thick or if its quality is too poor.

At present, the best amorphous silicon solar cells have values of the short-circuit current density $J_{sc}$ that are around 18 mA/cm$^2$, which is not very far away from the limit value that can be expected with a material having a gap of 1.75 eV (This limit value is about 21 mA/cm$^2$ and depends on the exact value of the gap, see Fig. 4.27(a). On the other hand, the best microcrystalline silicon solar cells have values of the short-circuit current density $J_{sc}$ that are around 25 to 30 mA/cm$^2$, which is, in its turn, rather far away from the limit value that can be expected with a material having a gap of 1.1 eV (This limit value is about 44 mA/cm$^2$, see Fig. 4.27(a)).

It is indeed particularly difficult to design a microcrystalline silicon solar cellthat absorbs a sufficient quantity of light and this because of two main reasons:

- the low absorption coefficient of microcrystalline silicon (as shown in Fig. 4.8);

- The microcrystalline silicon solar cells have a relatively low bandgap value and are therefore destined to absorb light in the near-infrared region, where transparent conductive oxides (TCO) used as contact layers have increased absorption due to free carrier absorption effects, and where metal layers used as reflecting layers also often show increased absorption losses.

At present there is a considerable research effort under way to improve the optics of thin-film silicon solar cells and especially of microcrystalline silicon solar cells, and one may expect the $J_{sc}$-values to be gradually increased over the next few years.

## 4.10 LIGHT-INDUCED DEGRADATION OR "STAEBLER-WRONSKI EFFECT" IN THIN-FILM SILICON SOLAR CELLS

As already described in Section 2.2.3, amorphous silicon layers suffer from light-induced degradation: under the influence of light exposure amorphous silicon layers show a pronounced increase in dangling bond density by about 1 to 2 orders of magnitude, before they finally more or less stabilize, after some 1000 hours of light exposure. This phenomenon is known as the "Staebler-Wronski effect".

Now, dangling bonds act as recombination centers, so that an increased dangling bond density results in reduced drift lengths. It also means, for the *pin*-solar cell, an increased space charge within the *i*-layer, due to charged dangling bonds, especially due to negatively charged dangling bonds near the *n*/*i*-interface. This leads to a deformation/reduction in the internal electric field and again to a decreased drift length. It is therefore evident that amorphous silicon solar cells will also be affected by light-induced degradation and that their conversion efficiency will be reduced, especially in the case of thick cells. Figure 4.54 shows the typical behaviour of a pin-

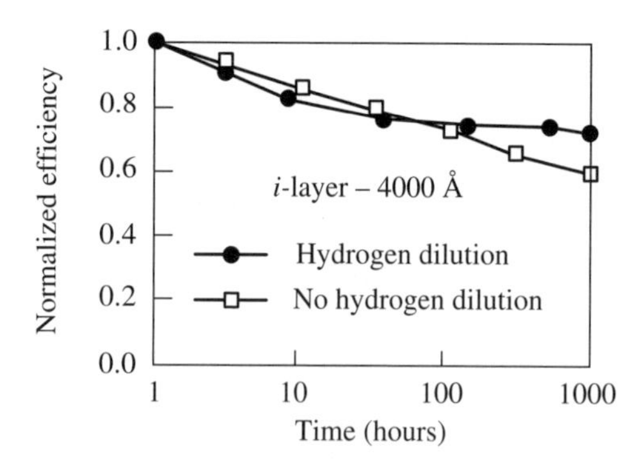


**Fig. 4.54** Typical behavior of two *pin*-type cells with 400 nm thick *i*-layers, with and without hydrogen dilution during plasma deposition. Clearly visible are the reduction and (at least in the case of hydrogen dilution) subsequent stabilisation of normalized efficiency during light-soaking. Figure taken with permission from [Wronski-1996].

type solar cell with a 400 nm thick *i*-layer, during light-induced degradation under "standard conditions" (illumination of 100 mW/cm$^2$, spectrum corresponding to AM 1.5, and cell temperature at 50 °C, cell at the maximum-power point).

In amorphous silicon solar cells, light-induced degradation mainly affects the fill factor and, usually to a lesser extent, the short-circuit current $J_{sc}$; it has, in general, a very limted effect on the open circuit voltage $V_{oc}$.

Physically, light-induced degradation is caused by the recombination of carriers within the cell; it is therefore much stronger when the cell is open-circuited, i.e. when the internal field is strongly reduced, and is still substantial when the cell is operated with a load resistance close to the maximum power point. Under short-circuit conditions, the internal field is larger, and hence the recombination and resulting degradation are weaker.

Furthermore, light-induced degradation is a reversible effect; cells can be annealed by heating them for a few hours, to around 150 °C, where they attain their initial efficiency again. In a very simple model, one can view the actual process always as equilibrium between a degradation component (i.e. the creation of additional dangling bonds) and an annealing component (i.e. the suppression or "healing" of dangling bonds). For this reason degradation is more severe (and stabilized efficiency values reached are lower) when the operating temperature is lower.

In practice, for present state-of-the-art amorphous silicon solar cells one finds a degradation of around 10 % to 30 % for cells with *i*-layer thicknesses between 200 nm and 400 nm; cells with thicker *i*-layers have higher relative degradation, however, *i*-layer thickness is not the only factor involved. Unfortunately, the short-circuit current density $J_{sc}$ decreases with decreasing *i*-layer thickness, so that a compromise has to be found. With present methods of light trapping the optimal *i*-layer thickness is somewhere around 200 to 250 nm.

In spite of a huge research effort, so far no method has been found to avoid the Staebler-Wronski effect (note that the effect can be somewhat reduced by using a moderate amount of hydrogen dilution during plasma deposition of the amorphous layer; see Chap. 6, Sect. 6.1.5).

Microcrystalline silicon solar cells only show a very mild form of light-induced degradation. The extent of degradation observed depends on the crystalline volume fraction: hydrogenated microcrystalline silicon is a "mixture" of an amorphous matrix and of crystallites. The best cells so far fabricated have a crystalline volume fraction of around 60 to 70 %. These cells show a relative efficiency degradation of 5 % or less, even when their *i*-layer is 2 μm thick [Meillaud-2008, Meillaud-2005, Sculati-Meillaud-2006].

The use of tandem and multi-junction solar cells is one way to reduce the effect of light -induced degradation: the top cell in a tandem can be kept very thin and, therefore shows little light-induced degradation; the bottom cell is not exposed to the full AM 1.5 spectrum and therefore shows once again less light-induced degradation. In fact, for a "micromorph" (microcrystalline/amorphous) tandem structure, no light-induced degradation whatsoever has been observed in the microcrystalline bottom cell.

# 4.11 SPECTRAL RESPONSE, LIGHT TRAPPING AND EFFICIENCY LIMITS

## 4.11.1 Spectral response (SR) and external quantum efficiency (EQE) measurements

One of the most important measurement methods, which is of particular use for evaluating thin-film silicon solar cells and for carrying out a diagnosis, i.e. localizing an area with problems, is the *spectral response* measurement method. In addition to the term *spectral response* (SR) the term external quantum efficiency (EQE) is also used, as the SR and EQE curves are related to each other (see below).

Figure 4.55 illustrates the *principle of the spectral response (SR) measurement*: A monochromator provides light with a tuneable wavelength, which is chopped at a frequency $f_{chopper}$ and illuminates the solar cell under examination. The chopped light has an amplitude corresponding to a relatively low light intensity. The solar cell is, at the same time, also illuminated by a superimposed white light bias, with an intensity and a spectrum approximately equal to that of AM 1.5 sunlight. The role of the white light source is to ensure that the occupation of dangling bond and bandtail states remains more or less the same as during the actual operation of the solar cell under STC (standard test conditions). The chopped monochromatic light constitutes the probe beam; the solar cell is operated under short-circuit conditions and a phase-lock detector synchronized with the monochromatic light source measures, as a function of light frequency $\nu_{probe}$ or, in other words, as a function of the light wavelength $\lambda_{probe}$, how large the corresponding response is.

We now evaluate, for every wavelength $\lambda_{probe}$, the energy of the incident probe beam (in mW) and of the corresponding current response (in mA); their ratio, calibrated in mA/mW is the spectral response (SR) curve.

On the other hand, the *external quantum efficiency* (EQE) is the ratio between collected electron-hole pairs and incoming photons. EQE is a ratio of two particle

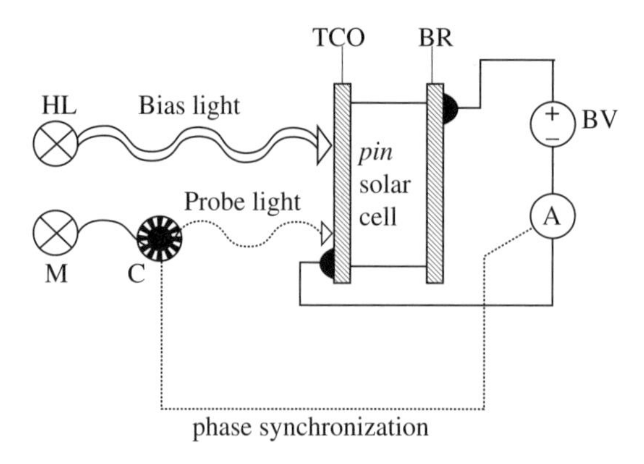


**Fig. 4.55** Measurement set-up for spectral response (SR) and external quantum efficiency (EQE).

flows and is a number without physical dimensions. If the EQE curve reaches (in the limit case) 100 %, at a certain wavelength $\lambda_{probe}$ or light frequency $\nu_{probe}$, this would mean that every single photon of this particular frequency/wavelength hitting the solar cell is converted by photo-generation into an electron-hole pair, and that all the charge carriers resulting from this conversion process are usefully collected as current contribution by the solar cell. If this curve reaches only a value of 50 %, then this means that half of the contribution is lost, either through "parasitic" absorption of the photons themselves, through part of the photons not being absorbed at all within the solar cell, or finally through the recombination of electron-hole pairs.

The spectral response SR($\lambda$) is related to EQE($\lambda$) via the relation $\mathrm{SR}(\lambda) = q \times \mathrm{EQE}(\lambda)/E(\lambda)$, where $q$ is the charge of the electron and $E(\lambda)$ the energy of the photon of wavelength $\lambda$.

On the other hand, *the internal quantum efficiency* (IQE) curve looks at the ratio between the collected charge carriers and the photo-generated electric charge carriers. The IQE measurement evaluates only recombination losses, whereas the EQE curve evaluates both recombination losses and optical losses. The IQE curve is more difficult to obtain than the EQE curve. In many cases, especially by laboratories working on crystalline silicon wafer-based solar cells, the IQE curve is approximated by the curve $\mathrm{IQE}^* = \mathrm{EQE}/(1-R)$, $R$ being the total reflection of the cell, as measured via integrating sphere. The IQE* curve evaluates losses in the whole cell, including those due to the TCO. The IQE and IQE* curves will not be discussed further in this book.

Figure 4.56 shows typical EQE curves for amorphous and microcrystalline single-junction *pin*-type solar cells. The cells were deposited on glass substrates covered with ZnO layers deposited by low-pressure chemical vapor deposition (LP-CVD). Such ZnO layers have a very pronounced surface texturing ([Faÿ-2007], [Faÿ-2008]), which scatters the incoming light (see also Fig. 4.57(a)) and allows the cells to actively absorb and convert light up to relatively long wavelengths: Note how the

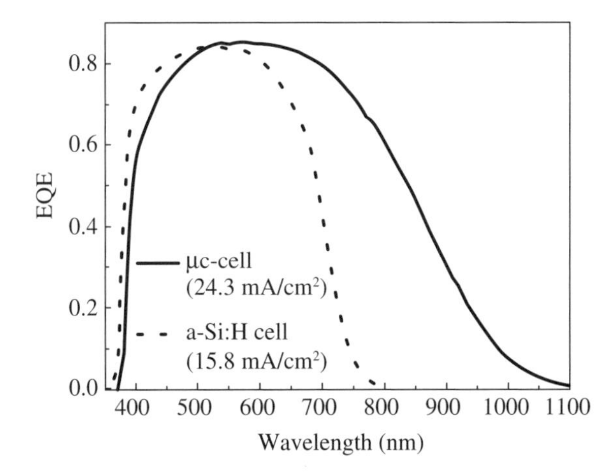


**Fig. 4.56** Typical external quantum efficiency (EQE) curves, for single-junction *pin*-type solar cells: solid line for an amorphous silicon solar cell; dashed line for a microcrystalline silicon solar cell. Courtesy of Julien Bailat and Peter Cuony, IMT PV Lab, Neuchâtel.

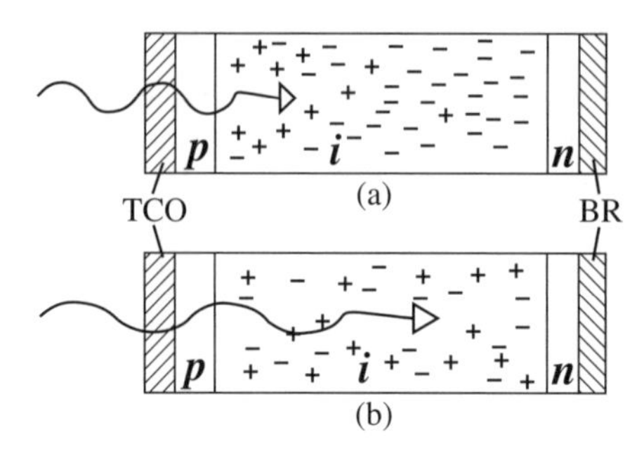


**Fig. 4.57** Photo-generation, flow of carriers and recombination within a *pin* solar cell: (a) in the case of blue light; (b) in the case of red light. Abbreviations: TCO: transparent conductive oxide; BR: back reflector.

microcrystalline cell has a response that extends well into the near-infrared spectral range. On the other hand, one sees that the amorphous cell has an enhanced response in the region of blue light. This is due to the use of an amorphous silicon-carbon alloy material for the *p*-layer.

SR (or EQE) curves are usually measured, on one hand for zero V electrical bias over the solar cell (i.e. under short-circuit conditions, as mentioned above), and, additionally, under reverse bias conditions, often at a voltage of –2 V, for single-junction individual solar cells. The reverse bias strengthens the electric field within the *i*-layer, so as to basically avoid almost all recombination. Thus, the difference SR(–2 V) – SR(0 V) [or, alternatively the difference EQE(–2 V) – EQE(0 V)] is of particular significance: a large difference is an indication of large recombination losses, which will become especially important when the solar cell is forward-biased, as is the case for usual operating conditions. A small or negligible difference indicates that the recombination losses are not that important.

*One can now assign different regions of the solar cell to different parts of the SR- or EQE-curve.* To do this, one has to look at the penetration depths for light of different wavelengths, as shown in Figure 4.57, for amorphous and microcrystalline silicon layers.

As an example, blue light with a wavelength of 420 nm will have a penetration depth of 15 nm in amorphous silicon and of 50 nm in microcrystalline silicon. (The penetration depth being defined as the depth after which only $1/e = 1/2.718\ldots$ of the original light intensity persists.) This is schematically shown in Figure 4.57(a). Now, if we take a probe light of this wavelength and notice a substantial difference between SR(–2V) and SR(0V), then we can assert that there is substantial recombination in the region into which this light penetrates, In fact, for recombination to occur (in the chopped, phase-locked-mode), there have to be both holes and electrons, modulated with that particular chopping frequency *f*, present in the corresponding region (one has to remember that recombination is a pair process and requires the presence of both holes and electrons.)

Figure 4.57(b) shows the situation occurring with light of longer wavelength, as an example, with red light with 680 nm wavelength. (Here, the penetration depth is about 1200 nm, for flat layers in both cases, i.e. for both amorphous silicon and microcrystalline silicon layers. In this case the recombination may take place anywhere in the whole region under penetration from the light. If, however, the difference between SR(−2V) and SR(0V) *only* shows up for red light, then we conclude that recombination only shows up for red light, and we can assign the recombination to the far end of the solar cell, i.e. the end that is opposite to the side at which light enters into the solar cell.

Spectral Response (SR), or External Quantum Efficiency (EQE) measurements, thus, constitute a method for selectively looking at different regions of the solar cell (depth-wise). The geometrical resolution is, however, very good at the front end of the solar cell (i.e. at that side where light enters the solar cell, which is the *p*-side for both *pin-* and *nip-type* solar cells) but very poor on the far end. To obtain a good resolution on both sides, it is necessary to provide the possibility for light to enter from both sides, i.e. to perform "bifacial" spectral response (see [Fischer-1993]).

Figure 4.58 shows EQE curves obtained for microcrystalline silicon solar cells with various types of shortcomings.

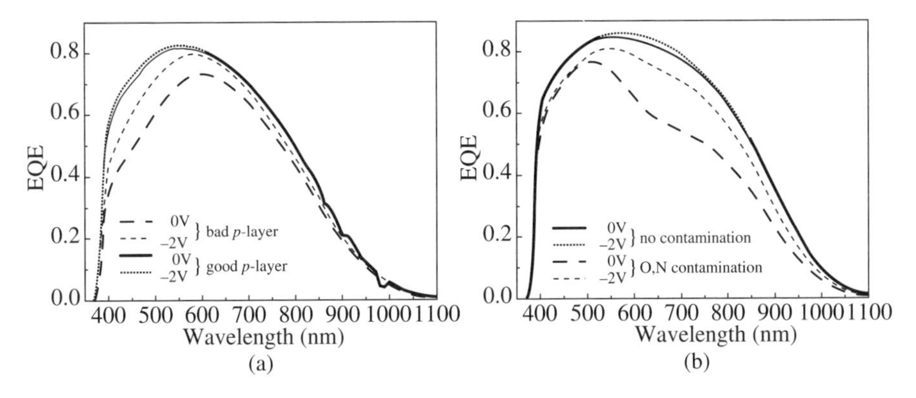


**Fig. 4.58** External quantum efficiency (EQE) curves obtained for various *pin*-type microcrystalline silicon solar cells, with typical shortcomings/modifications: (a) cell with different *p*-layers; (b) cell with/without oxygen and nitrogen contamination. Courtesy of Didier Dominé and Gaetano Parascondolo, IMT PV Lab, Neuchâtel.

## 4.11.2 Light trapping in thin-film silicon solar cells

Light trapping is an essential part of all thin-film silicon solar cells. In amorphous silicon solar cells it is necessary to keep the intrinsic layer very thin in order to obtain a strong electric field within the *i*-layer and, thus, to reduce the light-induced degradation effect. Such thin amorphous silicon cells, which may have *i*-layer thicknesses in the range of 180 to 300 nm would absorb only a relatively small part of the light in the range of wavelengths below 700 nm, unless the light is scattered and trapped within the cell. In microcrystalline silicon solar cells, the absorption coefficient has a very low value in the near-infrared and red, and light trapping is necessary in order to keep the *i*-layer thickness below 2 μm; cells with *i*-layer thicknesses above 2μm would, on the other hand, lead to unduly long deposition times (typically over 1

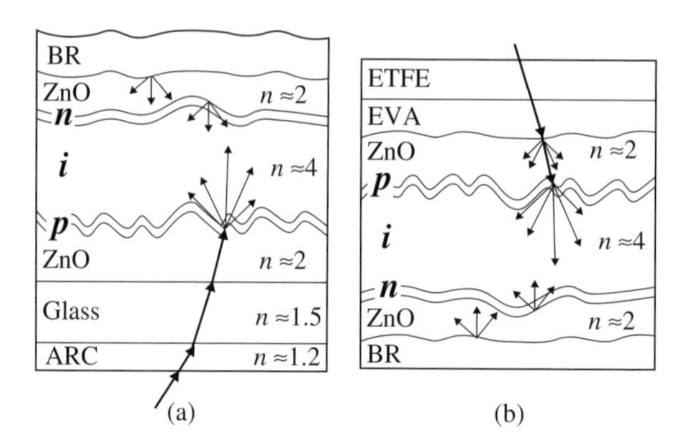


**Fig. 4.59** Principle of light trapping in a single-junction thin-film silicon solar cells: (a) *pin*-type solar cell; (b) *nip*-type solar cell. Abbreviations used: BR: back reflector; ZnO: as grown textured zinc oxide layer, produced by LP-CVD (low-pressure chemical vapor deposition); ARC: antireflective coating; ETFE: ethylene tetra-fluor-ethylene polymer foil; EVA: ethylene-vinyl acetate polymer foil.

hour) and therefore become uneconomical. Figure 4.59 shows the principle of light-trapping in single-junction *pin*- and *nip*-type thin-film silicon solar cells. There are in Figure 4.59 two main places, where the incoming light is scattered and/or reflected: *First*, at the interface between the ZnO layer (optical refractive index $n \approx 2$) and the thin-film silicon layer (optical refractive index $n \approx 4$). Here, due to the roughness (texture) of the interface and due to the difference in optical refractive indices, the light gets diffracted in all directions, i.e. it gets scattered. The *second* place is at/near the Back Reflector (BR), with the silicon/ZnO and ZnO/BR interfaces. Here, the light is reflected and scattered. The reflected light then goes partly back to the first place, i.e. to the interface between ZnO layer and thin-film silicon layer, where it reflected and scattered once again. Thus, multiple reflections, associated each time with additional scattering, are achieved.

In current *pin*-type solar cells, light scattering is achieved mainly through the roughness of the “top” TCO layer, through which the light enters the cell, and through reflection by the aluminum or silver back contact. In commercial amorphous silicon modules, the top TCO layer was, until very recently, composed of $SnO_2$ and was almost always fabricated at the glass factory which supplied the glass substrate. At present, ZnO layers are being increasingly used, both for commercial modules and for cells and modules fabricated in R & D laboratories. Section 6.4 describes in more detail the TCO layers used for thin-film silicon modules; a recent book edited by Ellmer, Klein and Rech [Ellmer-2008] deals, on the other hand, extensively with the various research and production issues raised by ZnO layers.

Additionally, one has also investigated [Zhang-2010] the possibility of using rough textured glass substrates in order to obtain (or at least, to enhance) light scattering: rough glass can add additional texture for more efficient scattering at longer wavelengths (around 1 μm); this is necessary for microcrystalline silicon solar cells (and for “micromorph” tandem solar cells).

In current *nip*-type solar cells, light scattering is, on the other hand, achieved mainly through the roughness of the underlying back contact layer, e.g. by using a rough silver layer as back contact; such a rough silver layer can be obtained by evaporation or sputtering at higher temperatures (around 400 °C, see [Banerjee-1991, Banerjee-1994] and [Terrazzoni-2006]). For *nip*-type solar cells on plastic substrates, the use of periodically textured polymer substrates in order to scatter the light is also being invesigated [Haug-2006, Haug-2009].

The goals of all these schemes is to make the light, especially that part of the absorbable light which has the longest wavelengths (and the smallest absorption coefficients), pass through the cell several times, before exiting from the cell again, or before being parasitically absorbed in the back reflector, or in the TCO-layer. Considering that the light may pass through the cell $m$ times, one can see in Figure 4.60 where – within the solar spectrum – the absorption edge of an amorphous silicon solar cell lies, for the 4 cases: $m = 1$ (no light trapping whatsoever); $m = 2$ (no light scattering, the light is just assumed to be ideally reflected by the back reflector), $m = 5$ (case generally obtained in laboratories at the moment by an efficient light trapping scheme) and $m = 16$ (considered by us to be the "theoretical" limit for light trapping, see below). The shaded areas in the Figure show the *additional* amount of light that can be absorbed, when the light passes $m$ times back and forth through the cell.

Figure 4.61 shows us, on the other hand, the difference in the irradiance absorbed by the solar cell between the two cases $m = 5$ and $m = 1$, both for a 300 nm

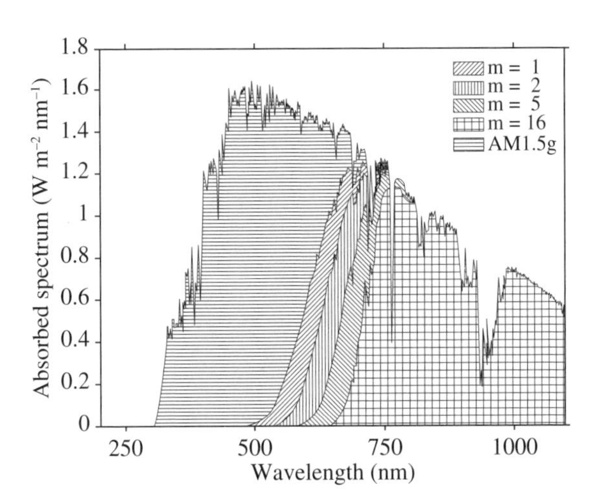


**Fig. 4.60** AM 1.5 solar spectrum as a function of wavelength and absorption edge of an amorphous silicon solar cell with an *i*-layer thickness $d_i = 300$ nm, for four cases: (1) $m = 1$ (no light trapping at all – light passes just once through the cell and is not reflected at the back contact); (2) $m = 2$ (light passes through the cell without being scattered and is ideally reflected at the back contact; (3) $m = 5$ (generally obtained in present solar cells, by scattering and reflection of the light); (4) $m = 16$ (considered here to be the "theoretical" limit for light trapping). Courtesy of Joelle Guillet and Michael Stückelberger, IMT PV Lab Neuchâtel.

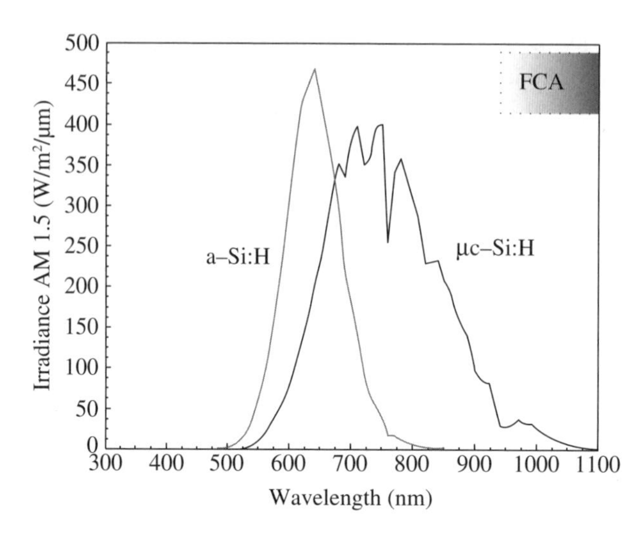


**Fig. 4.61** Gain in the absorption of the AM 1.5 solar spectrum between the cases $m = 1$ (light passes just once through the cell) and $m = 5$ (generally obtained with current solar cells, by scattering and reflection of the light), for (1) amorphous silicon and (2) microcrystalline silicon. Note that Figure 4.61 is not a SR/EQE curve but just shows the *additional* photo-generation thanks to light trapping; FCA denotes free carrier absorption; these are additional losses, caused by the TCO layers, in the near-infrared spectral region. Courtesy of Joelle Guillet, IMT PV Lab Neuchâtel.

thick amorphous silicon cell and for a 2 μm thick microcrystalline silicon cell. One notes that, in the case of amorphous silicon, light trapping is effective mainly in the narrow wavelength range between 580 and 700 nm, whereas for the microcrystalline silicon solar cell, light trapping has to be effective in a much broader spectral range, i.e. between 600 nm and 1000 nm.

As the light is reflected and scattered again and again, on either side of the thin-film silicon cell, it can be easily understood that the intensity of light steadily decreases, even if one assumes that the silicon layers themselves do not absorb any light. There exists therefore a theoretical maximum for $m$. Yablonovitch and Cody have derived the maximum value on the basis of statistical mechanics [Yablonovitch-1982]; they find for the silicon-air system $m_{max} = 4\,n^2 = 50$, where $n \approx 3.6$ is the refractive index of a-Si, assuming the back side to have a perfect reflector, and the front side to interface with air. In current thin-film silicon solar cells we usually cannot obtain such a high ratio between the indices of refraction (as between silicon and air); we typically have $SnO_2$ or ZnO in contact with the silicon on both sides; this gives us a value $m_{max} = 4(n_{Si}/n_{ZnO})^2 \approx 16$ and therefore the *maximum* path increase is smaller. Figure 4.62 shows schematically the basic situation for light trapping in thin-film silicon solar cells.

In fact, there is yet an additional limitation to be considered: TCO and back contact are absorbing; this again reduces the *actual, real* increase of the light path in silicon to values, that currently are typically between 5 and 10. In practice the values obtained for solar cells are, thus, much lower than 50, and even significantly lower

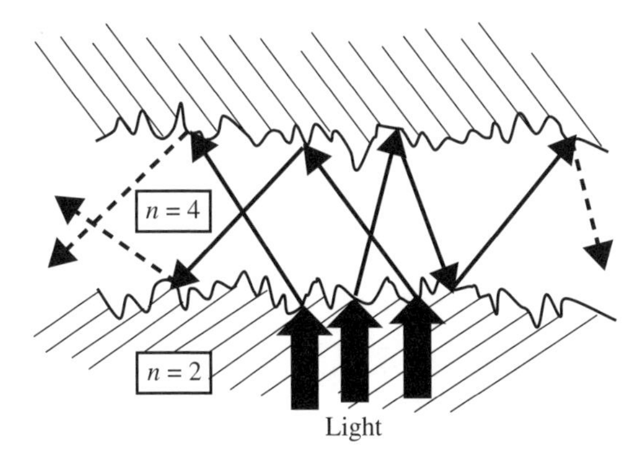


**Fig. 4.62** Basic situation for light trapping in thin-film silicon solar cells; this situation is somewhat different from the one studied by [Yablonovitch-1982].

than 16 (see also Chapt. 6, Sect. 6.4.6, in which "*m*" is introduced as "*N*"), so that one may consider that there is basically still good scope for improving light trapping in thin-film silicon solar cells.[11] Still, the problem of free carrier absorption (FCA) in the TCO layers remains a major problem in the case of light trapping for microcrystalline solar cells. (see also [Faÿ-2008, Vaneček-2002]). FCA can only be reduced if one succeeds in developing TCO materials with higher mobilities.

Light trapping has a direct bearing on the short circuit current density $J_{sc}$ of a thin-film silicon solar cell. A convenient and practical way of studying the effect of light trapping is therefore to plot the external quantum efficiency (EQE) curve as described in the preceding Section 4.11.1. Figure 4.63 shows EQE curves for *pin*-type microcrystalline silicon solar cells, with various degrees of light trapping. The cells have been deposited on glass substrates covered with ZnO layers deposited by low-pressure chemical vapour deposition (LP-CVD). Such ZnO layers have a very pronounced surface texturing; the surface gets rougher as the layer is made thicker [Fay-2007]. In Figure 4.63(a), the thickness of the LP-CVD ZnO layer is increased, in steps, from 0.1 to 4 μm, and, thus, the light scattering is also enhanced correspondingly. In Figure 4.63(b), the back reflector is modified. Note that the best performance is obtained with a "dielectric paste" as back reflector.

### 4.11.3 Limits for the efficiency η in *pin*-type thin-film silicon solar cells

To date, the best small-area (1cm$^2$) single-junction amorphous silicon solar cells produced in the laboratory have overall conversion efficiencies that are just over

[11][Berginski-2008] gives a rather comprehensive overview on the question of light trapping in thin-film silicon solar cells, both from the theoretical and experimental point of view. This paper indicates theoretical studies, such as [Tiedje-1984] and [Deckmann-1983], where the losses *are* included/studied; it also gives information on some of the latest practical results with amorphous and microcrystalline single-junction and "micromorph" tandem solar cells. A description of these rather complex phenomena goes, however, well beyond the scope of the present book.

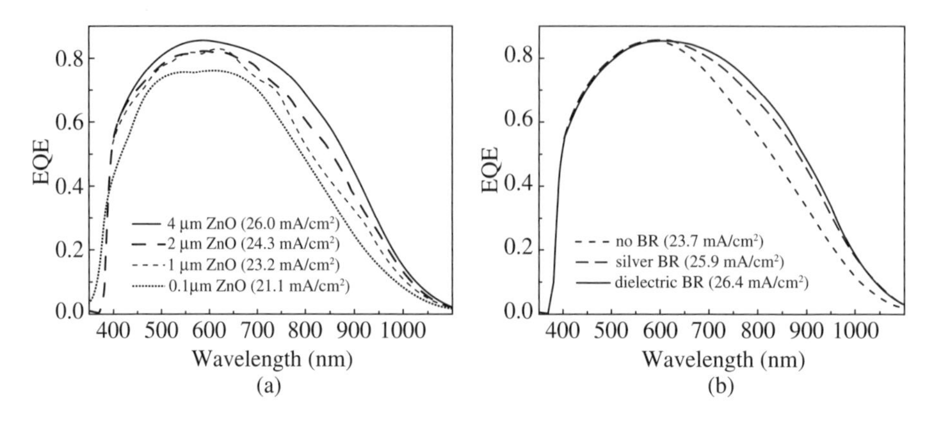


**Fig. 4.63** External quantum efficiency (EQE) plots for microcrystalline silicon *pin*-type solar cells, with varying degrees of light trapping: (a) variation of light scattering through variation of the thickness of the front TCO (LP-CVD ZnO) layer; (b) variation of the back reflector (BR). Courtesy of Peter Cuony, IMT PV Lab, Neuchâtel.

10 % in the stabilized state, after light-induced degradation [Benagli-2009]. The best small-area single-junction laboratory microcrystalline solar cells also have overall conversion efficiencies of around 10 % (see e.g. [van den Donken-2007]).

Can one expect these performances to be significantly improved in the coming years?

- if one looks at the (semi-empirical) limit values obtained from the bandgap considerations, according to Section 4.4, certainly yes; but
- if one takes into consideration additional limiting factors due to basic material properties, as well as those due to the necessity to use *pin*- or *nip*-type diodes, the expected improvements are certainly much lower. Table 4.7 recapitulates the situation.

There are, as illustrated in Table 4.7, four types of limitations, for thin-film silicon solar cells:

(a) The necessity to use a *pin-type solar cell design*, rather than a *pn*-type diode. From Figure 4.27(c) in Section 4.4, we can see that we will have, for the fill factor, a loss of about 6 % (in the case of amorphous silicon; gap 1.75 eV) and of about 10 % (in the case of microcrystalline silicon; gap 1.1 eV) when $n = 2$, as is the case for ideal *pin* cells. The *pin*-type solar cell also has more recombination than the *pn*-type solar cell; this leads to a loss in $V_{oc}$, which, for present material quality can be estimated to be about 200 mV for amorphous silicon and about 50 mV for microcrystalline silicon solar cells.

(b) Limitations due to the *amorphous nature of the intrinsic layer*. These limitations only affect amorphous silicon solar cells. Because of the bandtails (which are a direct consequence of the atomic "disorder" in the material), we are not able to push the Fermi level in doped a-Si:H near to the bands, loosing thereby with present a-Si:H layers about 200 mV for $V_{oc}$. As the fill factor is directly

**Table 4.7** Limiting factors for small-area single-junction thin-film silicon solar cells and their presumable effect on cell conversion efficiency.

| Type of cell | Consideration | Limitations | $J_{sc}$ | $V_{oc}$** | *FF* | η | Comments |
|---|---|---|---|---|---|---|---|
| (a) Amorphous Silicon (a-Si:H) | | | | | | | |
| a-Si:H; $E_g = 1.75$ eV | A/c to $E_g$ | none | 21 | 1.4 | 90 | 26 | rounded |
| | taking into account limitations | (a) *pin*-cell design | 21 | 1.3 | 84 | 23 | estimated |
| | | (b) amorphous material* | 21 | 1.0 | 80 | 17 | bandtails |
| | | (c) SWE effect* | 21 | 1.0 | 70 | 14 | improved material |
| | | (d) light trapping* | 18 | 1.0 | 65 | 12 | Yabl-Cody limit |
| | | present record cell | 17.2 | 0.88 | 66 | 10 | small area |
| (b) Microcrystalline Silicon (μc-Si:H) | | | | | | | |
| μc-Si:H $E_g = 1.12$ eV | A/c to $E_g$ | none | 44 | 0.7 | 85 | 26 | rounded |
| | taking into account limitations | (a) *pin*-cell design | 44 | 0.65 | 75 | 21.5 | estimated |
| | | (b) light trapping* | 35 | 0.65 | 75 | 17 | Yabl-Cody limit |
| | | (c) economical* (*i*-layer < 1.5 μm) | 28 | 0.65 | 75 | 13.5 | w/o addl. TCO losses (due to high $J_{sc}$) |
| | | present record cell | 22 | 0.6 | 74 | 10 | Small Area |

*cumulative
**semi-empirical, based on a diode reverse saturation current according to Equation 4.6

linked to the $V_{oc}$-value, this leads to a fill factor loss of around 5 %. In addition, we loose about 10 % in fill factor, because of light-induced degradation.

(c) Limitations in *light trapping*. We are, at present, not able to extract the full current density potential from the solar cell. We expect to make progress here through improved light trapping. However, the progress is limited: by both materials properties, such as absorption in the near-infrared spectral range, and by more fundamental limits, such as the Yablonovitch and Cody limit [Yablonovitch-1982]. It is very difficult to assess what those limits mean in terms of achievable current density. The values given here are pure intuitive guesswork.

(d) *Economical limitations*. The thickness $d_i$ of the intrinsic layer in microcrystalline silicon solar cells has a strong and direct impact on manufacturing costs: both on the costs of silane and hydrogen used for microcrystalline silicon deposition, and also on the depreciation costs of the very expensive PE-CVD deposition equipment generally used here. At present, one considers $d_i = 1.5$ μm to be an upper limit for the production of a commercially feasible microcrystalline silicon solar cell. This may, of course, change in future, if new high-rate deposition methods become available.

Table 4.7 does not have the ambition of providing a precise prediction on what may be obtained in future for the various single-junction solar cell parameters. Its use is rather to illustrate the different types of limitations that do play a role, and to show that each of these limiting factors contribute to reduce the achievable solar cell efficiency by about 3 to 4 %. Thus, the reader may be able to judge the impact and significance of results communicated in future by the various research and development laboratories:

- Are the given results a sign that one of the limiting factors can be reduced in its effect or even overcome?
- Or do they simply represent an exercise in parameter optimisation within the existing limitations?

If we look at corresponding *modules*, there exist at present only amorphous silicon single-junction modules. Their overall total area efficiencies are at best slightly over 6 % in the stabilized state, after light-induced degradation. Taking into account that by experience commercial module efficiencies are, in general, about 2/3 of the value of record cell efficiencies, we can estimate that single-junction amorphous modules may be able to reach 7 % in the near future (with present materials and present solar cell design methods); and that they be able to attain between 8 % and 9 % with improved materials and improved methods. The improvement in solar module efficiency that can be expected here, even in the best of circumstances, is, thus, rather modest.

Single-junction microcrystalline solar cell modules however do not exist. This is due to the high current densities prevailing in single-junction microcrystalline cells: This means that the TCO losses will become exorbitantly large, when we go over to the module level (see Chap. 6, Sect. 6.5.2)

At present, the strategy followed by most R & D groups is therefore not so much to improve single-junction cells and modules, but to develop tandem and multi-junction cells and modules, especially "micromorph" modules, which combine a microcrystalline bottom cell with an amorphous top cell. This topic will be treated in chapter 5.

## 4.12 Summary and conclusions

In order to understand properly the functioning of solar cells, we should keep in mind the following aspects:

(1) the physical nature of sunlight; it spectrum, its intensity and its angle of incidence – and how all this is modified by the geographical location, the time of the day and the season of the year, by weather and by other environmental conditions (pollution, dust, snowfall, fog, sandstorms, etc.);

(2) the fundamental physical principle underlying the photoelectric process (the latter was first demonstrated by Edmond Becquerel in 1839, but physically explained only about 70 years later by Albert Einstein, for which Einstein was granted the Noble Prize in 1921; the first technical devices based on the photoelectric effect were built by two German high school teachers, Julius Elster and Hans Geitel, around 1891);

(3) the rules prevailing for the absorption of light within solids (based on quantum physics, as established in the 1920s) and for the generation of electrons and holes (the latter concept having been introduced only much later);

(4) the functioning of classical semiconductor diodes, as established by William Shockley and co-workers in the 1950s (see S. M. Sze, *Modern Semiconductor Device Physics*, Wiley Interscience); the separation of electrons and holes through the internal electric field prevailing in the depletion region of the classical *pn*-diode.

If one has a solid understanding of the four aspects mentioned above, one will also easily be able to grasp the functioning of classical solar cells, in other words of the *pn*-type crystalline silicon solar cells, which constitute the overwhelming majority of the solar cells used today. Many other types of solar cells, such as GaAs-based solar cells (as used for space applications), function according to the very same principles. Even all thin-film solar cells, except for thin-film *silicon* solar cells, are also simply *pn*-diodes and function in the same way.

Sections 4.1 to 4.4 of the present Chapter give a concise summary of these basic principles, and describe all that one needs to know. Section 4.4 additionally discusses the limits for the solar-cell conversion efficiency $\eta$ and for the three key parameters (short-circuit current density $J_{sc}$, open-circuit voltage $V_{oc}$, and fill factor $FF$) of a solar cell; we have pointed out how far one has come today in approaching these limits. This has been discussed for standard test conditions, “STC” (standard AM 1.5 solar spectrum and 100 mW/cm$^2$ intensity, operating temperature of 25 °C). The situation is much more complicated when the prevailing conditions, such as spectrum, light intensity and cell operating temperature, are modified. We have only given here a few hints about what can happen; much experimental work remains to be done. The actual power output of a photovoltaic solar module is thus very dependent on geographical location, weather and time of the day. This will be briefly discussed again in Chapter 7. But a reliable answer can only be obtained through extensive field tests.

*Thin-film silicon* solar cells are, in general, not *pn*-diodes, but *pin*-diodes. The theory of *pin*-diodes is much more difficult to understand; in fact, it has not even been fully worked out so far. There are, however, some "pillars of comprehension" on which we can rely, so as to obtain at least an intuitive picture of the functioning of a *pin*-type solar cell. They are treated in Sections 4.5 to 4.10 of this Chapter and can be summarized as follows:

(a) the fact that for photoelectric conversion the essential part of the *pin*-type solar cell is the *i*-layer (intrinsic layer); it is here that photo-generation of electrons and holes takes place; it is here that these charge carriers are separated, thanks to the internal electric field prevailing within the whole *i*-layer;
(b) the principles governing the formation of this internal electric field;
(c) the fact that charge transport within the *pin*-type solar cell is governed mainly by drift (and not by diffusion, as is the case for classical *pn*-type solar cells);
(d) the resulting basic density profiles, within the *i*-layer, for free electrons and holes;
(e) the fact that in amorphous and microcrystalline silicon there are not only free electrons and holes, but also electrons and holes that are trapped, within localized states given by the bandtails of the material;
(f) the fact that there is a certain proportionality between free electrons and trapped electrons, on one side, and between free holes and trapped holes on the other side;
(g) the role of dangling bonds in determining recombination and in limiting the collection of electrons and holes;
(h) the fact that dangling bonds are "amphoteric", i.e. they can be either neutral or positively charged, or negatively charged;
(i) the fact that the charge state of the dangling bonds, within the *i*-layer of an illuminated *pin*-type solar cells is given by the ratio between free holes and free electrons;
(j) the deformation of the internal electric field within the *i*-layer, by trapped charge in the bandtails and by charged dangling bonds;
(k) the effect of light-induced degradation on all this.

No doubt that this is a rather complicated situation; nevertheless it is just a simplified model for what is actually happening within the *pin* solar cell. Of course, more detailed and accurate information can be obtained by computer simulation – such programs do exist. But although they allow us to calculate quite precisely what is happening in a given well-defined situation, they do not give us much insight into the basic mechanisms and dependencies. We have therefore in this book preferred an intuitive, analytical approach.

An equivalent circuit for *pin*-type solar cells has been introduced; it works surprisingly well over a wide range of operating conditions and for many different categories of *nip* or *pin* type, thin-film silicon solar cells, both amorphous and microcrystalline. There is probably an underlying theoretical reason for this, but so far no researcher has been able to shed light on these aspects.

A particular accent has been given here on methods for performing defect diagnosis. This is one of the first published attempts in that direction, and is therefore, by nature, very incomplete. There is a need for more information here and this can only be obtained through further experience in the fabrication of thin-film silicon solar cells and modules.

Section 4.11 is devoted to spectral response measurements and to light trapping issues: these two aspects are of cardinal importance for amorphous and microcrystalline silicon solar cells; the considerations given here, allow us, in fact, to make a detailed and probably quite realistic estimate of the future scope for efficiency improvement in thin-film silicon solar cells. One may expect, over the next few years, only a rather moderate efficiency increase for single-junction thin-film silicon solar cells and modules. The main thrust of improvement will, therefore, have to come from tandem and multi-junction cell concepts. This is the topic of Chapter 5.

## 4.13 REFERENCES

[Banerjee-1991] Banerjee, A. and Guha, S. 1991. "Study of back reflectors for amorphous silicon alloy solar cell application", (1991), *Journal of Applied Physics*, Vol. **69**, pp. 1030-1035

[Banerjee-1994] Banerjee, A.. Yang, J., Hoffman, K., Guha, S., "Characteristics of hydrogenated amorphous silicon alloy solar cells on a Lambertian back reflector", (1994), *Applied Physics Letters*, Vol. **65**, pp. 472-474

[Beck-1996] Beck, N., Wyrsch, N.. Hof, C., Shah. A., "Mo bility lifetime product - a Tool for Corrlating a-Si:H Film Properties and Solar Cell Performances", (1996), *Journal of Applied Physics*, Vol. **79**, pp. 9361-9368

[Benagli-2009] Benagli, S., D. Borrello, E. Vallat-Sauvain, J. Meier, U. Kroll, J. Hoetzel, J. Bailat, J. Steinhauser, M. Marmelo, G. Monteduro and L. Castens "High-efficiency amorphous silicon devices on LP-CVD ZnO prepared ion an Industrial KAI-M R & D Reactor" (2009), presented at the 24th European Photovoltaic Solar Energy Conference & Exhibition, Hamburg, 21-25 September 2009, Paper 3BO.9.3

[Berginski-2008] Bergisnki, M., Hüpkes, J., Gordijn, A., Reetz, W., Wätjen, T., Rech, B., Wuttig, M.,"Experimental studies and limitations of the light trapping and optical losses in microcrystalline silicon solar cells", (2008), *Solar Energy Materials and Solar Cells*, Vol. **92**, pp. 1037-1042

[Deckmann-1983] Deckman, H.W., Wronski, C.R. Optically enhanced amorphous silicon solar cells, Appl. Phys. Lett. 42 (1983) 968.

[Deng-2005] Deng, J., Wronski, C.R., "Carrier recombination and differential diode quality factors in the dark forward bias current-voltage characteristics of a-Si :H solar cells", (2005), *Journal of Applied Physics,* Vol. **98**, Paper No. 024509

[Durisch-2000] Durisch, W., Robert K., Meier. J., "Effect of Climatic Parameters on the Performance of Silicon Thin File Multijunction Solar Cells" GlobeEx 2000, Conference and Tradeshow, July 23 – 28, 2000, Las Vegas, Nevada (USA)

[Durisch-2001] Durisch, W., Leutwyler, B., "Performance and Spectral Efficiency of a Copper-Indium-Gallium-Diselenium Module", (2001), *Proceedings of the Sharjah Solar Energy Conference, Sharjah, U.A.E,* 174-PVOA04

[Durisch-2004] Durisch, W., Bittnar, B., Shah, A., Meier, J., "Impact of Air-Mass and Temperature on the Efficiencies of three Commercial Thin-Film Modules", (2004), *Proceedings of the 19th European Photovoltaic Solar Energy Conference, Paris, France,* pp. 2675-2677

[Durisch-2006] Durisch, W., Mayor, J.-Cl., Lam, K.-h., Close, J., Stettler, S, "Performance and Output of a Polycrystalline Photovoltaic Module under Actual Operating

Conditions", *Proceedings of the 21st European Photovoltaic Solar Energy Conference, Dresden, Germany,* pp. 2481–2484

[Durisch-2007] Durisch, W., Bittnar, B., Mayor, J.-Cl., Kiess, H., J.-Cl., Lam, K.-h., Close, "Efficiency Model for Photovoltaic Modules and Demonstration of the Application to Energy Yield Estimation" (2007), *Solar Energy Materials and Solar Cells,* Vol. **91**, pp. 79-84

[Durisch-2008] Durisch, W., Mayor, J.-Cl., Lam, K.-h., Close, J., Stettler, S, "Efficiency and Annual Output of a Monocrystalline Photovoltaic Module under Actual Operating Conditions", (2008), *Proceedings of the 23rd European Photovoltaic Solar Energy Conference, Valencia, Spain,* pp. 2992-2996

[Durisch-2010] Durisch, W., Mayor, J.-C., Dittmann, S., "Climate Impacts on the Efficiency of a CdTe PV Module and Annual Output under Real Working Conditions" submitted to the *25th European Photovoltaic Solar Energy Conference and Exhibition*, 06-10 September 2010, Valencia, Spain

[Droz-2000] Droz, C., Goerlitzer, M., Wyrsch, N., Shah, A., "Electronic Transport in Hydrogenated Microcrystalline Silicon: Similarities with Amorphous Silicon", (2000), *Journal of Non-Crystalline Solids,* Vol. **266-269**, pp.319-324J.

[Dylla-2006] Dylla, T. Reynolds, S. Carius, R., Finger F., "Electron and hole transport in microcrystalline silicon solar cells studied by time-of-flight photocurrent spectroscopy" (2006), *Journal of Non-Crystalline Solids,* Vol. **352**, pp.1093-1096

[Ellmer-2008] Ellmer, K., Klein, A., Rech, B., "*Transparent Conductive Zinc Oxide*", Springer Series in Materials Science, Springer Verlag, Berlin 2008

[Faÿ-2007] Faÿ, S., Steinhauser, J., Oliviera, N., Vallat-Sauvain, E., Ballif, C., "Opto-electronic properties of rough LP-CVD ZnO:B for use as TCO in thin-film silicon solar cells", (2007), *Thin Solid Films*, Vol. **515**, pp. 8558-8561

[Fischer-1993] Fischer, D., Wyrsch, N., Fortmann, C. M., Shah, A., "Amorphous Silicon Solar Cells with Low-Level Doped i-Layers Characterised by Bifacial Measurements", (1993), *Proceedings of the 23rd IEEE Photovoltaic Specialists Conference*, Louisville, pp. 878-884

[Gray-2003] Gray, J.L., "*The Physics of the Solar Cell"*, Chapter 3, in A. Luque and S. Hegedus, *"Handbook of Photovoltaic Engineering"*, J. Wiley and Sons, 2003

[Green-1982] Green, M. A., "Solar Cells, Operating Principles, Technology and System Applications", Prentice Hall, Englewood Cliffs, N.J., 07632, USA, 1982, *ISBN* 0 85823 580 3

[Green-1983] Green, M. A., "Accuracy of Analytical Expressions for Solar Cell Fill Factors", (1983), *Solar Cells*, Vol. **8**, pp 3-16

[Green-1987] Green, M. A., Editor, "High-Efficiency Silicon Solar Cells" Trans Tech, Aedermannsdorf (Switzerland), 1987, ISBN 0878495371

[Green-1995] Green, M. A, "Silicon Solar Cells, Advanced Principles and Practice", University of New South Wales Sydney, N.S.W. 2052, 1995, ISBN 0-7334-0994-6

[Green-2003] Green, M. A., "General Temperature Dependence of Solar Cell Performance and Implications for Device Modelling", (2003), *Progress in Photovoltaics: Research and Applications*, Vol. **11**, pp. 333-340

[Green-2010] Green, M. A., Emery, K., Hishikawa, Y., Warta, W., "Solar Efficiency Tables (Version 35)", (2010), *Progress in Photovoltaics: Research and Applications*, Vol. **18**, pp. 144-150

[Haug-2009] Haug, F. J., Söderström, T., Python, M., Terrazzoni-Daudrix, V., Niquille, X., and Ballif, C., *Sol. Energy Mater. Sol. Cells* Vol. **93**, 884 _2009_.

[Hubin-1992] Hubin, J., Shah, A., Sauvain, E., "Effects of Dangling Bonds on the Recombination Function in Amorphous Semiconductors", (1992)., *Philosophical Magazine Letters,* Vol. **66**. pp. 115-125;

[Kilper-2009] Kilper, T., Beyer, W., Bräuer, G., Bronger, T., Carius, R., van den Donker, M. N., Hrunski, D., Lambertz, A., Merdzhanova, T., Mück, A., Rech, B., Reetz, W., Schmitz, R., Zastrow, U., and Gordijn, A., "Oxygen and nitrogen impurities in microcrystalline silicon deposited under optimized conditions: Influence on material properties and solar cell performance" (2009*), Journal of Applied Physics*, Vol. **105**, Paper No. 074509

[Kiess-1995] Kiess, H., Rehwald, W., "On the ultimate efficiency of solar cells", (1995), *Solar Energy Materials and Solar Cells*, Vol. **38**, pp. 45-55

[King-2000] King, D.L., Kratochvil J. A., Boyson W. E., "Stabilization and performance characteristics of commercial amorphous-silicon PV modules", (2000), *Conference Record of the 28th IEEE Photovoltaic Specialists Conference, Anchorgae (AK), USA,* pp. 1446-1449

[Krc-2003a] Krc, J., Smole, F., Topic, M., "Analysis of light scattering in amorphous Si : H solar cells by a one-dimensional semi-coherent optical model", (2003), *Progress in Photovoltaics: Research and Applications*, Vol. **11**, pp. 15-26

[Krc-2003b] Krc, J., "Analysis and modeling of thin-film optoelectronic structures based on amorphous silicon", PhD Thesis, Univ. of Ljubljana, 2003

[Krc-2009] Krc, J., personal communication

[Kroll-1995] Kroll, U., Meier, J., Keppner, H., Littlewood, S.D., Kelly, I.E., Giannoulès, P., Shah. A., "Origins of Atmospheric Contaminations in Amorphous Silicon prepared by Very High Frequency (70 MHz) Glow Discharge", (1995), *Journal of Vacuum Science & Technology A*, Vol. **13**, pp. 2742-2746

[Kusian-1989] W. Kusian et al., "Buffer layer and light degradation of a-Si:H pin solar cells", (1989), Proceedings of the 9th European Photovoltaic Solar Energy Conference, pp. 52-55

[Lee-2006] Lee, C.-H., Sazonov, A., Nathan, A., Robertson, J., "Directly deposited nanocrystalline silicon thin-film transistors with ultra high mobilites", (2006), *Applied Physics Letters,* Vol., **89**, Paper No. 252101

[Maruyama-2006] Maruyama, E., Terakawa, A., Taguchi, M., Yoshimine, Y., Ide, D., Baba, T., Shima, M., Sakata, H., Tanaka, M., "Sanyo's challenges to the development of high-efficiency HIT solar cells and the expansion of HIT business", (2006), *Conference Record of the 2006 IEEE 4th World Conference on Photovoltaic Energy Conversion*, pp.1455-1460

[Meier-1997] Meier, J., Dubail, S., Platz, R., Torres, P., Kroll, U., Anna Selvan, J.A., Pellaton Vaucher, N., Hof, C., Fischer, D., Keppner, H., Flückiger, R., Shah, A., V. Shklover, V., Ufert, K.-D., "High Efficiency Thin-Film Solar Cells by the "Micromorph" Concept", (1997*), Solar Energy Materials and Solar Cells*, Vol. **49**, pp. 35-44

[Meier-1998] Meier, J., Keppner, H., Dubail, S., Ziegler Y., Feitknecht, L., Torres, P., Hof, C., Kroll, U., Fischer, D., Cuperus, J., Anna Selvan, J.A., Shah, A., "Microcrystalline and Micromorph Thin-Film Silicon Solar Cells" (1998), *Proceedings of the 2nd World Conference on Photovoltaic Energy Conversion, Vienna,* Vol. **1**, pp- 375-380

[Meier-2003] Meier, J., Spitznagel, J., Kroll, U., Bucher, C., Faÿ, S., Moriarty, T., Shah, A., "High-Efficiency Amorphous and "Micromorph" Silicon Solar Cells" (2003), *Proceedings of the 3rd World Conference on Photovoltaic Energy Conversion, Osaka, May 2003,* CD-ROM, S2O-B9-06, pp. 2801-2805

[Meillaud-2005] Meillaud, F., Vallat-Sauvain E., Niquille, X., Dubey, M., Bailat, J., Shah, A., Ballif, C., "Light-induced degradation of Thin-film Amorphous and Microcrystalline Silicon Solar Cells", (2005) *Proceedings of the 31st IEEE Photovoltaic Specialist Conference, Lake Buena Vista, FL, USA*, pp. 1412-1415

[Meillaud-2006] Meillaud, F., Shah, A., Bailat, J., Vallat-Sauvain E., Roschek, T., Rech, B., Dominé, D., Söderström, T., Python, M., Ballif, C., "Microcrystalline silicon solar cells: Theory and diagnostic tools", (2006), *Proceedings of the 3rd World Conference on Photovoltaic Energy Conversion, Kona Island, Hawaii, USA,* pp. 1572 -1575

[Meillaud-2008] Meillaud, F. Vallat-Sauvain, E., Shah, A., Ballif, C., "Kinetics of creation and of thermal annealing of light-induced defects in microcrystalline silicon solar cells", (2008), *Journal of Applied Physics*, Vol. **103**, Paper No. 054504

[Merten-1998] Merten, J., Asensi, J.M., Voz, C., Shah A. V., Platz, R., Andreu, J., "Improved equivalent circuit and analytical model for amorphous silicon solar cells and modules", (1998), *IEEE Transactions on Electron Devices,* Vol. **45**, pp. 423-429

[Miazza-2006] Miazza, C., "Photodétecteurs monolithiques par intégration verticale de couches de silicium amorphe hydrogéné", Ph. D. thesis, Faculté des Sciences, University of Neuchâtel, 2006

[Möller-1993] Möller, H. J., "Semiconductors for Solar Cells", Artech Houxe, Boston, 1993

[Moore-1984] Moore, A. R., "Diffusion length in undoped a-Si:H in "Hydrogenated Amorphous Silicon, Part C "Electronic and Transport Properties", ed. Pankove, J.I., Vol. **21B** of *Semiconductors and Semimetals*, Academic Press, Orlando, 1984

[Nann-1992] Nann, S, Emery, K., "Spectral Effects on PV-device rating, *Solar Energy Materials and Solar Cells*, (1992), Vol. **27**, pp. 189-216

[Pernet-2000] 326 P. Pernet, M. Hengsberger, C. Hof, M. Goetz, A. Shah *Growth of Thin μc-Si:H Layers for pin Solar Cells: Effect of the H2- or CO2- Plasma Treatments* Proceedings of the 16th EC Photovoltaic Solar Energy Conference, Glasgow, UK, pp. 498-501, May, 2000

[Powell-1992] Powell, M.J., van Berkel, C., Franklin, A.R., Deane, S.C., Milne, W.I., "Defect pool in amorphous-silicon thin-film transistors", (1992), *Physical Review B,* Vol. **45**, pp. 4160-4170

[Rose-1963] Rose, A., "Concepts in Photoconductivity and Allied Problems", Wiley-Interscience, John Wiley and Sons, New York, 1963; see also http://en.wikipedia.org/wiki/Albert_Rose

[Python-2008] Python, M., Vallat-Sauvain, E., Bailat, J., Dominé, D., Fesquet, L., Shah, A., Ballif, C., "Relation between substrate surface morphology and microcrystalline silicon solar cell performance" (2008), *Journal of Non-Crystalline Solids*, Vol. **354**, pp. 2258-2262

[Schiff-2004] Schiff, E.A., "Drift-mobility measurements and mobility edges in disordered silicons", (2004), *Journal of Physics: Condensed Matter,* Vol. **16**, pp- 55265-55275

[Schiff-2009a] Schiff, E. A., "Carrier drift-mobilities and solar cell models for amorphous and nanocrystalline silicon", (2009), *Proceedings of the Materials Research Society Symposia,* Vol. **1153**, Paper 1153-A15-01

[Schiff-2009b] Schiff, E. A., "Transit-time measurements of charge carriers in disordered silicons: Amorphous, nanocrystalline and porous", (2009), *Philosophical Magazine,* Vol. **89**, pp. 2505–2518

[Schockley-1961] Schockley, W., Queisser, H. J., "Detailed Balance Limit of Efficiency of p-n Junction Solar Cells", (1961), *Journal of Applied Physics*, Vol. **32**, pp. 510

[Sculati-Meillaud-2006] Sculati-Meillaud, F., "lMicrocrystalline silicon solar cells: theory, diagnosis and stability", Ph. D. thesis, Facuté des Sciences, University of Neuchâtel, 2006

[Shah-1995] Shah, A., Platz, R., Keppner, H., "Thin Film Silicon Solar Cells: A Review and Selected Trends", (1995), *Solar Energy Materials and Solar Cells*, Vol. **38**, pp. 501-520

[Shah-2004] Shah, A.V., Schade, H., Vaneček, M., Meier, J., Vallat-Sauvain, E., Wyrsch, N., Kroll, U., Droz, C., Bailat, J., "Thin-film Silicon Solar Cell Technology", (2004), *Progress in Photovoltaics: Research and Applications*, Vol. **12**, pp. 113-142

[Shah-2006] Shah, A., Meier, J. Buechel, A., Kroll, U., Steinhauser, J., Meillaud, F., Schade, H., Dominé, D., "Towards very low-cost Mass Production of Thin-film Silicon Photovoltaic (PV) Solar Modules on Glass", (2006), *Thin Solid Films,* Vol. **502**, pp. 292-299

[Shah-2010] Shah, A.V., Sculati-Meillaud, F.C., Berényi, Z.J., Kumar., R., "Diagnostics of thin-film silicon Solar Cells and Solar Panels/Modules with VIM (Variable

Intensity Measurements) ", submitted to *Solar Energy Materials and Solar Cells*, 2010.

[Spear-1988] Spear, W.E.., "Transport and tails state interactions in amorphous silicon", in "*Amorphous Silicon and related materials*", ed. Fritsche, H., World Scientific Publishing Company, Singapore, 1988

[Stückelberger-2010] Stückelberger M.-E., Shah, A.V., Krc, J., Sculati-Meillaud, F., Ballif, C., Depseisse, M., "Internal electric field and Fill Factor of amorphous silicon solar cells", submitted to the *35th IEEE Photovoltaic Specialists Conference (PVSC),* Waikiki, Hawaii, 20-25 June 2010

[Sze-2006] Sze, S.M., Kwok, K. Ng., "Physics of Semiconductor Devices", Wiley-Interscience, John Wiley and Sons, New York, 2006

[Tchakarov-2003] Tchakarov, S., Roca i Cabarrocas, P., Dutta, U., Chatterjee, P., Equer, B., "Experimental study and modeling of reverse-bias dark currents in *PIN* structures using amorphous and polymorphous silicon", (2003), *Journal of Applied Physics,* Vol. **94** pp 7317-7327

[Terrazzoni-2006] Terrazzoni Daudrix, V., Guillet, J., Freitas, F., Shah, A., Ballif, C., Winkler, P., Ferreloc, M., Benagli, S., Niquille, X., Fsicher, D., Morf, R., "Characterisation of Rough Reflecting Substrates incorporated into thin-film silicon solar cells", (2006), *Progress in Photovoltaics: Research and Applications*, Vol. **14**, pp. 485-498

[Tiedje-1984] Tiedje, T., Yablonovitch, E., Cody, G.D., Brooks, B.G., "Limiting efficiency of silicon solar cells", (1984), *IEEE Transactions on Electron Devices*, Vol. **ED-31**, pp. 711-716

[Torres-1996] Torres, P., Meier, J., Flückiger, R., Kroll, U., Anna Selvan, J., Keppner, H., Shah, A., Littlewood, S.D., Kelly, I.E., Giannoulès, P., "Device Grade Microcrystalline Silicon Owing to Reduced Oxygen Contamination", (1996), *Applied Physics Letters*, Vol. **69**, pp. 1373-1375

[van den Donker-2007] M. N. van den Donker, S. Klein, B. Rech, and F. Finger W. M. M. Kessels and M. C. M. van de Sanden "Microcrystalline silicon solar cells with an open-circuit voltage above 600 mV", (2007), *Applied Physics Letters*, Vol. **90**, Paper No. 183504

[Vaneček-2002] Vaneček. M., Poruba, A., "Fourier-transform photocurrent spectroscopy of microcrystalline silicon for solar cells", (2002) *Applied Physics Letters*, Vol. **80**, pp. 719-721

[Vallat-Sauvain-2006] Vallat-Sauvain, E., Shah, A., Bailat, J., "Advances in Microcrystalline Silicon Solar Cell Technologies", Chapter 4 (pp.133-165), in "*Thin Film Solar Cells*", eds. Poortmans, J., Arkhipov, V., *Wiley Series in Materials for Electronic & Optoelectronic Applications*, John Wiley and Sons, 2006

[Winer-1990] Winer, K., "Defect Pool Model of Defect Formation in a-Si,H", (1991), *Journal of Non-Crystalline Solids*, Vol. **137**, pp. 157-162.

[Wronski-1996] Wronski, C.R, "Amorphous silicon technology: Coming of age", (1996), Solar Energy. Materials and Solar Cells Vol. **41/42**, pp. 427-439]

[Wyrsch-1995] Wyrsch, N., Shah, A., "Drift Mobility and Staebler-Wronski Effect in Hydrogenated Amorphous Silicium", (1991), *Solid State Communications*, Vol. **80**, pp. 807-809

[Wyrsch-2010] Wyrsch, N., personal communication

[Yablonovitch-1982] Yablonovitch, E., Cody, G.D., "Intensity enhancement in textured optical sheets for solar cells", (1982) *IEEE Transactions on Electron Devices*, Vol. **ED 29** pp. 300-305

[Zhang-2010] Zhang, W., Bunte, E., Worbs J., Siekmann, H., Kirchhoff, J., Gordijn, A., Hüpkes, J., "Rough glass by 3d texture transfer for silicon thin film solar cells", (2010), *Physica Status Solidi* C, 1–4 / DOI 10.1002/pssc.200982773

CHAPTER 5

# TANDEM AND MULTI-JUNCTION SOLAR CELLS

*Arvind Shah and Martin Python*

## 5.1 INTRODUCTION, GENERAL CONCEPT

In Chapter 4 we have looked at the mode of functioning and at the conversion efficiencies of single-junction solar cells. Thereby we studied both *pn*-type solar cells, which are used for wafer-based crystalline silicon and for many other materials, as well as *pin*-type solar cells, which are used, in most cases, for thin-film silicon. However we looked exclusively at single-junction solar cells, i.e. at solar cells constituted by a single diode.

We implicitly assumed in Chapter 4 that all active layers, i.e. all layers which form the diode and/or contribute to photogeneration, are formed by the same material and have the same gap. Solar cells with such a structure are called "homojunction" solar cells.

A further degree of liberty arises when combining, within the same junction or diode, two different materials, preferably with two different bandgaps: we thus obtain so-called "heterojunction" solar cells. The theoretical study of this type of solar cell is more difficult; however, one basically does not increase the efficiency limits when compared to "homojunction" solar cells. Still, there may be at times technical reasons for combining materials of different bandgaps within the same diode. A very successful solar cell type was launched by Sanyo Corp. of Japan, composed of a crystalline silicon wafer and a thin amorphous silicon layer. It is called the "HIT" solar cell, where HIT stands for "Heterojunction with Intrinsic Thinlayer" [Taguchi-2005]. As it relies on the use of crystalline silicon wafers, we will not deal with it in the present book. Within solar cells made up entirely of thin-film silicon layers, one also extensively uses "heterojunctions", specifically when combining microcrystalline silicon and amorphous silicon within the same junction or when using different types of unalloyed silicon, silicon-carbon and silicon-germanium alloys within the same junction. The functioning of such thin-film heterojunctions is complex and is not fully understood, although in practice many advantages can be obtained.

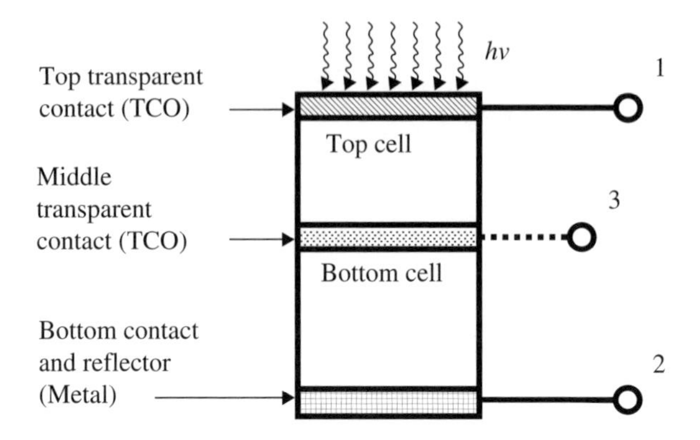


**Fig. 5.1** Principle of the dual-junction stacked cell or tandem cell.

In the present chapter, we shall look at a more fundamental way of combining two different materials: This is the so-called tandem concept, where two junctions (or diodes) are optically situated one on top of the other, in such a way that the light passes first through a top sub-cell (where it is partially absorbed) and thereafter through a bottom sub-cell (where the remaining part of the light has a possibility to be absorbed). The tandem concept can be extended to three junctions (triple-junction cells) or to even more than three junctions. The principle of the dual-junction or tandem cell is shown in Figure 5.1.

As shown in Figure 5.1, such a tandem cell can have three individual electrical contacts and is then called a "three-terminal tandem cell". However, because of the basic difficulty in creating a large-area, highly conductive and very transparent TCO layer for the middle contact, so far all tandem cells for thin-film silicon are two-terminal devices, i.e. the middle electrode (contact No. 3 in Figure 5.1) does not exist and the electrical current has to pass in series through both the bottom sub-cell and the top sub-cell. The middle transparent contact is replaced here by a "tunnel junction" or "recombination junction", as we shall see later. We are now dealing with "two-terminal tandem cells". The present Chapter 5 will be restricted to the treatment of these cells.

In Chapter 4 we have seen that amorphous silicon solar cells should be as thin as possible so as to ensure satisfactory collection and a high enough fill factor, even in the stabilized state, after light-induced degradation (see Section 4.5.16). Typically, *i*-layer thicknesses of 200 to 300 nm are used. The thinner the *i*-layer is, the lower will be the effect of light-induced degradation on cell efficiency. On the other hand, for very thin *i*-layers, we will also have very low values of absorbed light and of photogenerated current.

One way of overcoming this difficulty is to use tandem cells, where *top and bottom sub-cells have the same bandgap*, i.e. both are made of amorphous silicon layers (Figure 5.2). Thus, the total thickness of the two *i*-layers can be made quite large (say around 400 nm), so as to absorb (with suitable light-trapping schemes) a large part of the incoming sunlight; on the other hand, the *i*-layer of the top sub-cell can be kept relatively thin (say around 100 nm). Now it is precisely the top sub-cell that is subject

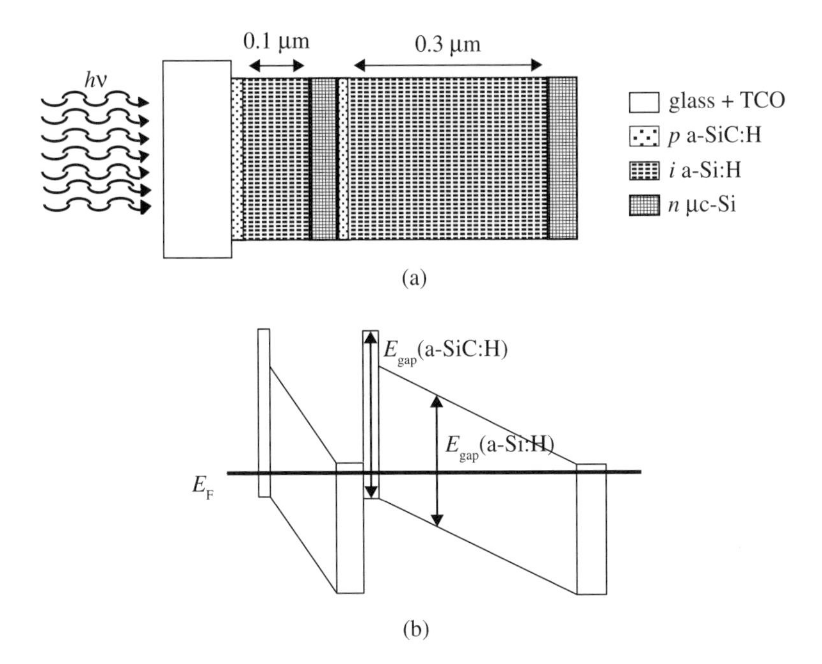


**Fig. 5.2** Two-terminal tandem cell with 2 sub-cells having approximately the same bandgap in the i-layer: (a) structure; (b) band diagram.

to the highest light intensity, and thus has the highest carrier generation density; therefore this sub-cell is basically more in danger of undergoing light-induced degradation. Because the top sub-cell is now very thin, it will hardly be affected; in fact, even after the light-induced degradation phase, the collection length (or drift length $l_{drift}$) within its *i*-layer will still remain much larger than its thickness $d_{i..top}$. Note that the drift length $l_{drift}$ itself is proportional to $1/d_i$, (see Sect. 4.5.11).

One of the difficulties which arises, then, is that the *i*-layer thicknesses of the two sub-cells would actually have to be very different in order to ensure current matching (see Section 5.2.1 below): the top sub-cell has to be very much thinner than the bottom sub-cell. This is technologically difficult to realize. Either the top sub-cell is extremely thin, and then it will be prone to shunts, or the bottom sub-cell is very thick and it will take a long time to deposit and also again have a higher degradation effect. In spite of these shortcomings, tandem cells using amorphous silicon alone have achieved remarkable results and are at present quite popular. There are various commercial solar modules in which such a tandem concept is currently used, with surprisingly high stabilized module efficiencies [Lechner-2008].

Theoretically, the preferred method for tandems (and for triple-junction cells) in thin-film silicon technology is *to vary the bandgap of the sub-cells*. The top sub-cell has a higher bandgap (and absorbs mainly light with short wavelengths); the bottom sub-cell has a lower bandgap (and absorbs light with longer wavelengths). (In the case of triple-junction cells, the middle cell will have an intermediate bandgap.)

*Bandgaps can be varied (but only slightly) by varying the deposition conditions* of the amorphous silicon layers (As an example, the "protocrystalline" and "polymorphous" layers, which were briefly mentioned in Section 2.1.2, have slightly higher bandgaps than standard amorphous silicon – maybe at the most about 100 to 200 meV higher.) Although some research work has indeed been carried out in this field, there is still plenty of scope for developing new types of amorphous silicon layers with modified bandgaps by varying the deposition conditions.

On the other hand, the *bandgap of amorphous silicon can be varied much more substantially by alloying silicon with germanium* and creating amorphous silicon-germanium alloys. These materials have much lower bandgaps than amorphous silicon, provided the germanium content is high enough. However, they unfortunately also have a much higher light-induced degradation effect and much higher recombination losses in the degraded state. It requires careful optimization of all layers and all cell thicknesses to produce efficient tandem and triple-junction solar cells based on silicon-germanium alloys.

Finally, the use of microcrystalline silicon (see Chapter 3) – as a complement to amorphous silicon – is one of the most promising ways of creating a tandem cell, because this material has a low bandgap (1.1 eV), is hardly affected by light-induced degradation and can be deposited with the same input gas as amorphous silicon. This leads to the "***micro***crystalline/a***morph***ous" or "***micromorph***" tandems, introduced first under the name of "mixed stacked a-Si:H/μc-Si:H cells" by IMT Neuchâtel in 1994 [Meier-1994]; the term "micromorph" cells was only coined later [Meier-1995].

## 5.2 PRINCIPLE OF THE TWO-TERMINAL TANDEM CELL

### 5.2.1 Construction of basic $J$–$V$ diagram: Rules for finding tandem $J_{sc}$, $V_{oc}$, $FF$

In a two-terminal tandem cell, both the top sub-cell and the bottom sub-cell are electrically in series. If we use the following symbols:

| | |
|---|---|
| $J_{top}$, $V_{top}$ | top sub-cell current density and voltage |
| $J_{bottom}$, $V_{bottom}$ | bottom sub-cell current density and voltage |
| $J_{tandem}$, $V_{tandem}$ | tandem cell current density and voltage |

we will then have:

$$J_{tandem} = J_{top} = J_{bottom} \tag{5.1}$$

$$V_{tandem} = V_{top} + V_{bottom} \tag{5.2}$$

The construction of the corresponding diagram for $J_{tandem}(V_{tandem})$ is shown in Figure 5.3. It is easy to convince oneself from this figure that :

$$J_{sc.tandem} \approx \text{Min}\ \{ J_{sc.top}, J_{sc.bottom} \} \tag{5.3}$$

$$V_{oc.tandem} \approx V_{oc.top} + V_{oc.bottom} \tag{5.4}$$

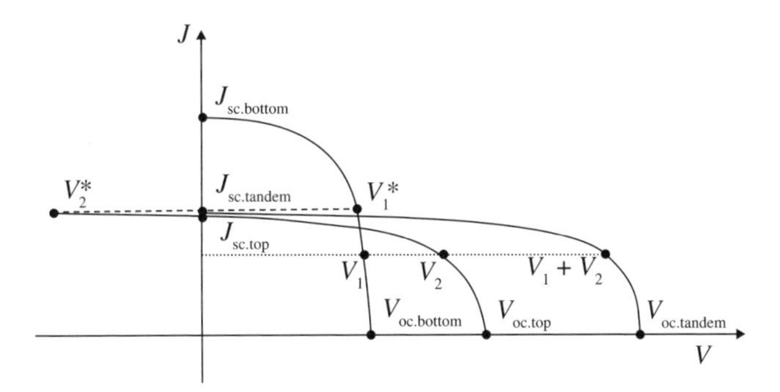


**Fig. 5.3** Construction of the *J-V* diagram for a two-terminal tandem solar cell. The short-circuit current density of the tandem is slightly higher than the short-current density of the top sub-cell. Actually, it corresponds to the current density of the top sub-cell in a reverse voltage condition ($V_1^* = -V_2^*$).

Equation 5.1 leads to the ***current-matching condition***; for optimal performance one should have:

$$J_{sc.top} \approx J_{sc.bottom} \tag{5.5}$$

If, on the other hand, there is a large difference between $J_{sc.top}$ and $J_{sc.bottom}$, the short-circuit current density of the tandem will be given by the smaller of the 2 short-circuit current densities of the sub-cells. The sub-cell with the higher short-circuit current density will then not be using its full current potential.

As far as open-circuit voltages are concerned, Equation 5.4 is only approximately fulfilled. In fact the open-circuit voltage of the tandem is always slightly lower than the sum of the open-circuit voltages of the two sub-cells. There are two reasons for this:

(a) Each sub-cell will be absorbing less light than an individual single-junction cell of the same gap, and this will lead, according to Equation 4.16, to a corresponding decrease in its $V_{oc}$. To obtain a rough estimate of this effect, look first, at a-Si/a-Si tandem cells. Here, the reduction in $V_{oc}$ will be $\ln(2) \times 1.8 \times$ 26 mV ($V_{oc}$ at half-illumination for a cell with ideality factor $n = 1.8$; $kT/q$ is approximately 26 mV at room temperature); we will, through this effect, have a decrease of about 32 mV for the tandem cell. For a typical "micromorph" tandem cell, the a-Si:H sub-cell will be having a $J_{sc}$ of 12 mA/cm$^2$ (instead of the ideal value of 18 mA/cm$^2$, in a single-junction a-Si:H cell); it, thus looses roughly 10 mV; the μ-Si:H cell can be estimated to loose about 15 mV. Thus, in a typical "micromorph" tandem cell,the total loss is roughly 25 mV.

(b) The recombination junction will also lead to a further reduction of the total tandem voltage; this reduction depends on the exact nature of the layers present in the recombination junction (see Section 5.2.2); it may be estimated to be in the range of 20 to 40 mV.

The fill factor *FF* of the tandem cell merits a special comment. If we do have current-matching, the *FF* may be considered to lie within the fill factors of the top and of the bottom sub-cells. However, if the currents are not matched, the *FF* of the tandem is very often quite a bit higher. On the other hand, we will, by such a current mismatch, certainly be losing more in current than we will be gaining in *FF*, so that it is not worthwhile to try to obtain mismatch conditions when attempting to increase tandem cell efficiency. It is in general a rather complicated task to predict the *FF* of tandem and multi-junction cells (see e.g. [Yan-2008]).

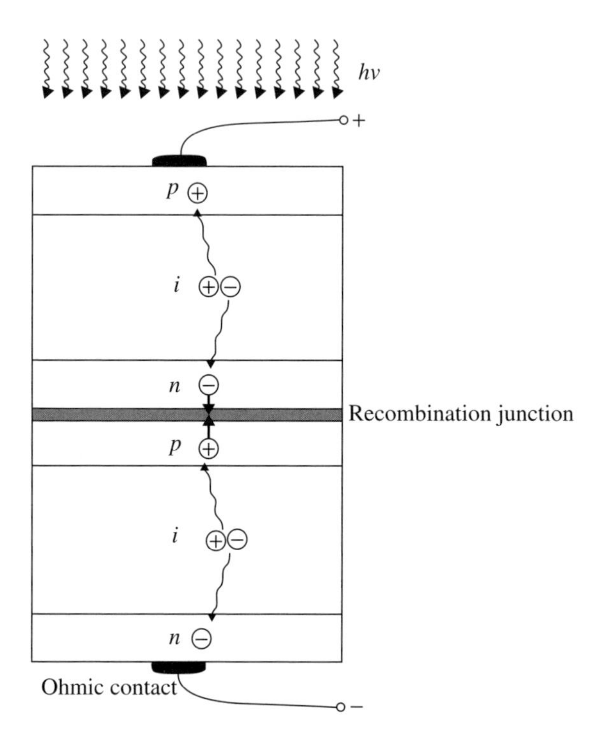


**Fig. 5.4** Principle of current continuity between top and bottom sub-cells for a two-terminal tandem solar cell, shown here for the example of two *pin*-type sub-cells. The current continuity condition requires that the total flux of electrons or holes exiting from one of the sub-cells has to recombine with the total flux of carriers (of opposite polarity) exiting from the other sub-cell. This requirement is valid for all types of two-terminal tandem cells, i.e. for *pin*-type and *pn*-type cells, and for both the superstrate (*pin*) and the substrate (*nip*) configurations.

## 5.2.2 Recombination (tunnel) junction

In Figure 5.4, the detailed structure of a thin-film silicon tandem solar cell is shown for the superstrate configuration. We can see that current continuity between the two sub-cells can only be maintained when electrons which exit from the *n*-layer on the bottom part of the top sub-cell ***recombine*** with holes exiting from the *p*-layer, or topmost layer of the bottom sub-cell. This means practically that it is advantageous to include, between the two sub-cells, a highly doped and very thin layer (*n*- or *p*-layer) with enhanced recombination, i.e. with a large density of "midgap defects" or "dangling bonds". If this layer is too thick, it will lead to optical losses by absorption.

In order for the electrons from the $n$-layer of the top sub-cell to recombine with the holes from the $p$-layer of the bottom sub-cell, (at least) one of the two charge carriers should be transported to the zone where the other type of carriers are in majority. With the help of tunnelling, this transport can be enhanced. Tunnelling becomes possible when the corresponding layer is heavily doped, i.e. when the Fermi level $E_F$ is shifted very near to the band edge. Such a shift is easier to obtain for microcrystalline silicon layers than for amorphous silicon layers. For this reason, at least one of the doped layers which form the tunnel/recombination junction within a thin-film silicon tandem solar cell should be microcrystalline (and not amorphous).

The general principle of the recombination junction is shown in Figure 5.4, and a detailed example for transport and recombination is shown in Figure 5.5.

If the recombination junction does not function properly, e.g. because neither of the adjoining doped layers is microcrystalline (and thus both adjoining doped layers are amorphous), we expect the tandem cell to have a somewhat lower open-circuit voltage $V_{oc.tandem}$ and, above all, a higher series resistance $R_{series.tandem}$.

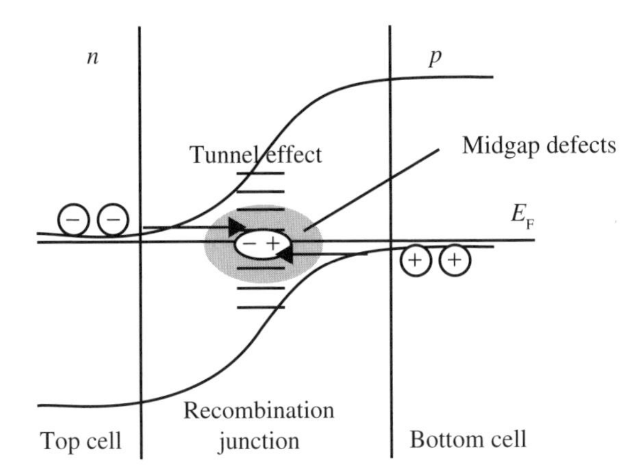


**Fig. 5.5** Example (band diagram) showing the flow of electrons and holes within a recombination (tunnel) junction and their recombination via midgap defects (dangling bonds) (see also Sect. 6.3).

## 5.2.3 Efficiency limits for tandems

We have derived in Section 4.4 the "semi-empirical" efficiency limit for a single-junction $pn$-type solar cell (Fig. 4.30). We thereby assumed that all photons with energy $h\nu$ higher than the bandgap energy $E_g$ are usefully absorbed. We also assumed that collection is perfect and no recombination occurs. We also assumed thereby that the open-circuit voltage $V_{oc}$ is given, as in Equation 4.17, according to the "semi-empirical" limit for the solar cell's reverse saturation current density $J_0$ introduced by Martin Green (Equation 4.6).

We can now extend our reasoning to the case of two-terminal tandem cells. Thereby we will first assume that the top sub-cell usefully absorbs all photons with energy $h\nu$ higher than its bandgap energy $E_{g.top}$ and that the bottom sub-cell usefully absorbs all *remaining* photons with energy $h\nu$ higher than its bandgap energy

$E_{g.bottom}$. This assumption corresponds to having very thick intrinsic layers in top and bottom sub-cells and to neglecting all recombination losses prevailing during short-circuit conditions. It also corresponds to taking the optimal values for the open-circuit voltages $V_{oc.top}$, $V_{oc.bottom}$ of the sub-cells according to Equation 4.17 and to assuming values for the *FF*, which are given by the limit values of Section 4.3.4. These are not very realistic assumptions for thin-film silicon, and especially not for amorphous silicon, but can be made, in order to have a first idea of how a thin-film silicon tandem functions.

Now, as a *first step*, let us *assume that the two bandgaps are identical*, i.e. that $E_{g.top} = E_{g.bottom}$, and let us neglect the loss in $V_{oc}$ according to Section 5.2.1 above [effects (a) and (b)].Then, one can easily be convinced that the tandem has a $V_{oc}$ value that is twice as high as the corresponding single-junction cell, but a $J_{sc}$ value that is only half that of the corresponding single-junction cell: One has not gained anything at all; the efficiency of the tandem is just the same as that of the corresponding single-junction cell.

In practice, however, the intrinsic layers will have only a very limited thickness; and therefore there are, notably in the case of amorphous silicon, two reasons for using such a tandem structure:

(a) the sub-cells can be thinner and they then have less light-induced degradation (as mentioned already above);
(b) the current density $J_{sc}$ will be only half the value for a single-junction cell and this will mean less Ohmic losses in the TCO layer, an advantage that is essential at the module level (see Sects. 6.4 and 6.5).

As *second step*, let us look at what happens to $V_{oc}$ and $J_{sc}$, when we modify the bandgaps, i.e. when $E_{g.top} \neq E_{g,bottom}$: The bandgaps of the two sub-cells now have to be properly adjusted. To see how to do this, look at Figure 5.6 (a). Set the "original", "central" bandgap $E_g^*$ somewhere in the range for an optimal single-junction bandgap according to Chapter 4, e.g. at 1.4 eV. This would give rise to an "original" single-junction cell with certain $V_{oc}^*$, $FF^*$ and $J_{sc}^*$ values. Now, select a top cell with a limited thickness and with $E_{g.top} > E_g^*$, and a bottom cell with a limited thickness and with $E_{g.bottom} < E_g^*$, placing $E_{g.top}$ and $E_{g.bottom}$ more or less symmetrically with respect to $E_g^*$. As you move $E_{g.top}$ to the right (towards higher $E_g$-values) the current balance between the two sub-cells is altered, thus increasingly favoring, the bottom sub-cell.

Returning again to the assumption that all photons with $I > E_g$ can be absorbed in the corresponding sub-cell, current matching with $J_{sc.top} = I_{sc.bottom}$ is reached when $E_{g.top} \approx 1.75$ eV and $E_{g.bottom} \approx 1.1$ eV, yielding, as shown in Figure 5.6 (b), $J_{sc.top} \approx 22$ mA/cm$^2$ and $J_{sc.bottom} \approx 22$ mA/cm$^2$. This is exactly the bandgap combination obtained with a-Si:H as top sub-cell and μc-Si:H as bottom sub-cell!

The resulting tandem will have the following limit performance:

$$J_{sc} = J_{sc.top} = J_{sc.bottom} = 22 \text{ mA/cm}^2$$

$$V_{oc} = V_{oc.top} + V_{oc.bottom} \approx 2 \times V_{oc}^* \approx 2.0 \text{ V}$$

$$FF \approx FF^* \approx 0.88 \text{ (for } n = 1);$$

$$FF \approx FF^* \approx 0.80 \text{ (for } n = 2)$$

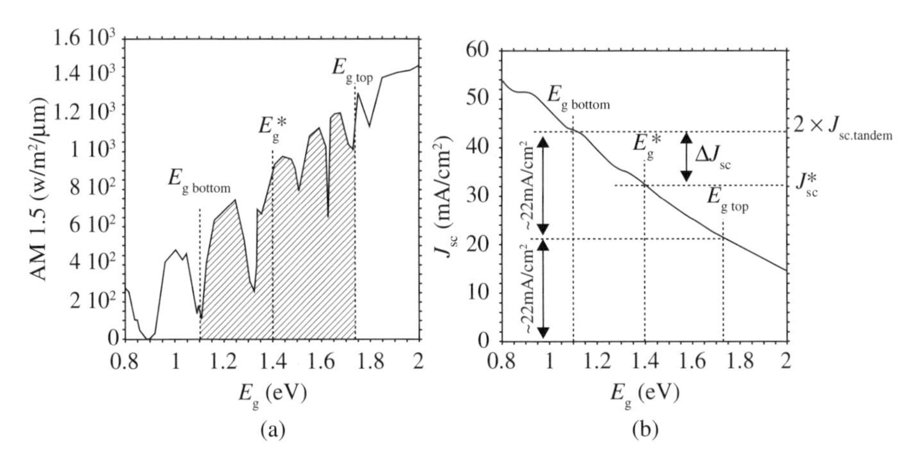


**Fig. 5.6** Basic principle illustrating tandem cell concept with bandgap shifting and its fundamental, theoretical limit: (a) AM 1.5 spectrum and bandgaps $E^*_g$, $E_{g.bottom}$ and $E_{g.top}$ of the orginal, "optimal" single-junction cell, of the bottom sub-cell and of the top sub-cell, respectively. The shaded area corresponds to the photogenerated current of the bottom sub-cell, under the assumption that all photons with $h\nu > E_g$ are absorbed by both sub-cells; (b) corresponding short-circuit current densities $J_{sc}$; the curve for $J_{sc}$ as a function of $E_g$ is the same as in Figure 4.27; the double-arrow denoted as "$\Delta J_{sc}$" shows the net current density gain which can be obtained with the tandem, compared to the "original" single-junction cell.

Here, the asterisk is used to denote the parameters of the corresponding optimal single-junction cell, and the limits for $V_{oc}$ and $FF$ have been taken according to the considerations of Chapter 4, Sections 4.3 and 4.4.

One can also quantify the gain in conversion efficiency obtained by this tandem structure, with $E_{g.top} \neq E_{g,bottom}$. Assuming, for the tandem $FF \approx FF^*$ and $V_{oc} \approx 2 \times V^*_{oc}$, we will have for the tandem [Figure 5.6 (b)]:

$$J_{sc.tandem} = 1/2 \times (\, J^*_{sc} + \Delta J_{sc})$$

The gain in conversion efficiency is, therefore, given by $(J^*_{sc} + \Delta J_{sc})/\, J^*_{sc}$.

It is evident that with tandem cells one reaches a complexity which can no longer be handled with simple calculations. It is therefore useful to resort to computer-assisted calculations. In [Shah-2004] the maximum efficiency of a two-terminal tandem cell was calculated, based on the assumption that the top sub-cell usefully absorbs all photons with energy $h\nu$ higher than its bandgap energy $E_{g.top}$ and that the bottom sub-cell usefully absorbs all *remaining* photons with energy $h\nu$higher than its bandgap energy $E_{g.bottom}$. For $V_{oc}$, $J_{sc}$ and *FF,* limits analogous to those given in Chapter 4, Section 4.4, were taken. The result is reproduced herein Figure 5.7 as a maximum efficiency plot. This plot gives us a rough idea of where to "place" the bandgaps of top and bottom cells, in order to obtain high efficiencies.

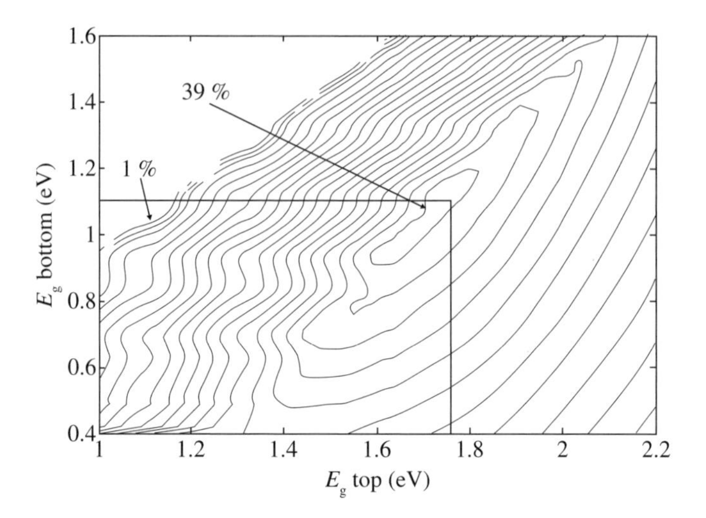


**Fig. 5.7** Maximum efficiency diagram for a two-terminal tandem cell under the following assumptions: $J_{sc}$ values for the sub-cells determined by assuming that all photons above the corresponding bandgap are usefully absorbed without loss; $V_{oc}$ values and $J(V)$ characteristics according to Equation (4.12) with the semi-empirical formula (4.16) for the reverse saturation current density $J_0$ (taken from [Shah-2004]).

The main message of Figures 5.6 and 5.7 is that, in principle, the bandgap combination 1.1 eV/1.75 eV, as given by the combination of microcrystalline silicon and amorphous silicon,would be an "ideal" bandgap combination

(a) *if light trapping* was perfect;
(b) *if carrier collection* was ideal;
(c) *and if the amorphous silicon*[1] *sub-cell would have the* $V_{oc}$ *value corresponding to its bandgap.*

One is, at present, striving to approach condition (a), as far as cell technology and as far as the theoretical limit allows (the Yablonovitch-Cody limit [Yablonovitch-1982], see also Sect. 4.5.15). On the other hand, condition (b) cannot be fulfilled, because the open-circuit voltage $V_{oc}$ in amorphous silicon cells is always significantly lower than the limit value as given by the bandgap (see Sect. 4.5.13).

## 5.3 PRACTICAL PROBLEMS OF TWO-TERMINAL TANDEM CELLS

### 5.3.1 Light trapping

In a single-junction thin-film silicon solar cell, the light enters through a textured transparent contact layer, is scattered there and transits through the *pin*-cell (or *nip*-cell),

[1] In microcrystalline cells, on the other hand, one obtains $V_{OC}$ values, which (almost) correspond to the "theoretical" limit as given by their bandgap.

it is partly absorbed; thereafter it reaches the back reflector and is sent back again ...gh the same *pin*-cell (or *nip*-cell). This is schematically shown in Figure 5.8(a).

In a tandem cell, there is no back reflector immediately after the top sub-cell; all light that is not absorbed within the top sub-cell passes through to the bottom sub-cell. It is only after passing through the bottom sub-cell that the remaining light will be reflected by the back reflector. If the top sub-cell consists of amorphous silicon (a-Si:H) and the bottom sub-cell of microcrystalline silicon (μc-Si:H), hardly any absorbable light will return to the top sub-cell.

Thus, in the first versions of the "micromorph tandem" the beneficial effect of light trapping on the top sub-cell was not very strong, and current matching could only be obtained with top sub-cells that were far too thick for tolerable light-induced degradation. For this reason, "intermediate reflectors" are presently being developed which are deposited between the top sub-cell and the bottom sub-cell in a "micromorph" tandem. The principle of such an intermediate reflector is shown in Figure 5.8(b).

The functioning of such an intermediate reflector is based on using a material with a refractive index ($n \approx 2$) lower than amorphous or microcrystalline silicon. First trials with intermediate reflectors involved the use of ZnO [Fischer-1996; Pellaton-1998]. Subsequently, it was found more useful, for fabrication reasons, to introduce an intermediate reflector that could be deposited in the same plasma-CVD system as the amorphous and microcrystalline layers. For this reason, other types of intermediate reflectors [Yamamoto-2006], especially doped layers based on silicon oxide [Buehlmann-2007], [Dominé-2008] were introduced.

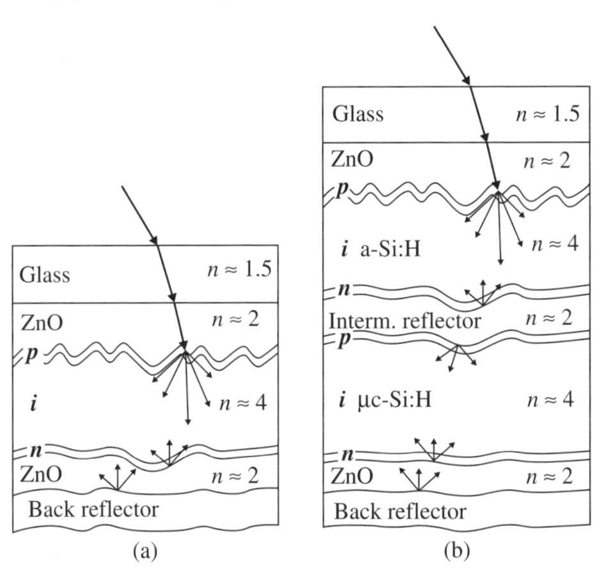


**Fig. 5.8** Schematic representation of light trapping: (a) in single-junction *pin*-type solar cells. The back reflector which follows immediately after the single-junction cell reflects the light back into the latter; (b) in a microcrystalline/amorphous ("micromorph") tandem cell. The light that is not absorbed within the amorphous top sub-cell will not return to the latter unless an "intermediate reflector" is placed (as shown) after the amorphous top sub-cell. (The drawing is not to scale.)

### 5.3.2 Efficiency variation due to changes in the solar spectrum

All solar cells are sensitive to variations in the spectrum of the incoming light. Such variations can occur both during the course of the day (the solar spectrum tends to shift towards longer wavelengths, i.e. towards the red, in the morning and in the evening. Variations also occur due to meteorological and seasonal changes (overcast skies or snow tend to shift the spectrum towards shorter wavelengths, i.e. towards the blue. A similar blue shift of the spectrum is seen when the solar module is located in the vicinity of a lake or the sea). What is important to note is that tandem and multi-junction solar cells are, in general, even more sensitive to variations in the solar spectrum than single-junction solar cells (see e.g. [Gottschalg-2005]). This is because of the condition of current continuity between the sub-cells. As the spectrum of the incoming light varies, the balance between the various sub-cells is modified. If the whole tandem (or the whole multi-junction solar cell) was designed to have current matching for a given incident spectrum, say for the AM 1.5 spectrum, the tandem (or multi-junction cell) will be in a state of current mismatch for a spectrum that is substantially different. The situation of current mismatch will lead to a further loss in efficiency.

It is difficult to predict what exactly will happen in a given tandem or multi-junction cell when the spectrum of the incoming light is modified. At present, the only reliable way to assess this is by outdoor testing.

### 5.3.3 Temperature coefficients

Photovoltaic modules operating outdoors, especially in hot climates, are usually subject to temperatures which are quite a bit higher than the "standard test conditions (STC)" or "laboratory conditions", under which they are generally evaluated. It is not uncommon to have operating temperatures in the range 60°C to 80°C (or even higher, in tropical countries). All solar cells have a conversion efficiency that drops as the temperature increases. Basically, the lower the bandgap of the solar cell, the larger is its relative temperature coefficient. Thus, as already noted in Section 4.4.3., wafer-based crystalline silicon (c-Si) solar cells (with a bandgap of 1.1 eV) have a temperature coefficient that is about twice the temperature coefficient of amorphous silicon (a-Si) solar cells (with a bandgap of 1.7 to 1.85 eV). The relative drop in the efficiency of c-Si solar cells is about –0.5 %/°C and that of a-Si solar cells is about –0.2 %/°C. This means that the output power of an amorphous silicon module will show a less pronounced decrease as the operation temperature is increased. Microcrystalline silicon solar cells have a temperature coefficient that lies somewhere between that of wafer-based crystalline silicon and that of amorphous silicon [Meier-1998].

The temperature coefficient of tandem and multi-junction cells depends therefore on the bandgap of the sub-cells involved. In general, a tandem will have a temperature coefficient that lies somewhere between the temperature coefficients of the two sub-cells.

There is yet another effect to be considered. If there is current matching at 25°C, the currents *may* become mismatched at higher operating temperatures. Often, the sub-cell with the lowest bandgap (usually the bottom sub-cell) will have the highest temperature coefficient and become the current-limiting cell. However, as higher temperatures lead mainly to a reduction in $V_{oc}$ but not to a significant reduction in either $FF$ or in $J_{sc}$, this effect is usually not very pronounced.

Note that at higher operating temperatures the light-induced degradation effect present in amorphous silicon will be reduced, because the balance between generation of new dangling bonds and annealing of dangling bonds becomes more favourable.

### 5.3.4 Pinholes and Shunts

If the bandgaps of the two sub-cells are exactly the same, then (as mentioned above) the condition of current matching will only be obtained if the thickness $d_1$ of the top sub-cell is smaller than the thickness $d_2$ of the bottom sub-cell, i.e. if $d_2 >> d_1$ (see Fig. 5.11, below). Such a high $d_2$: $d_1$ ratio is difficult to obtain; it means that either the top sub-cell will have to be very thin ($d_1 < 100$ nm), leading to frequent shunts (in connection with "pinholes" and other "fissures"(see Fig. 5.9a), or the bottom cell will have to be very thick ($d_2 \approx 500$ nm), leading to a pronounced light-induced degradation effect in the bottom cell

Therefore, in a-Si:H/a-Si:H tandems (see Sect. 5.4.1), the bandgap of the amorphous silicon top sub-cell is, in general, designed to be slightly higher than the bandgap of the amorphous silicon bottom cell. The desired variation in bandgaps can be obtained by depositing the top sub-cell at a lower temperature than the bottom sub-cell. It can also be obtained by using a higher value of hydrogen dilution for the deposition of the top sub-cell than for the deposition of the bottom cell (Note: To obtain more information about the influence of hydrogen dilution on the bandgap of a-Si:H layers, the reader may refer to Sections 2.1.2, 2.6.8 and 6.1.5).

A reasonable compromise for a-Si:H/a-Si:H tandems is obtained with the parameters $d_1 \approx 100$ nm and $d_2 \approx 300$ nm and with a slight amount of variation in bandgaps.

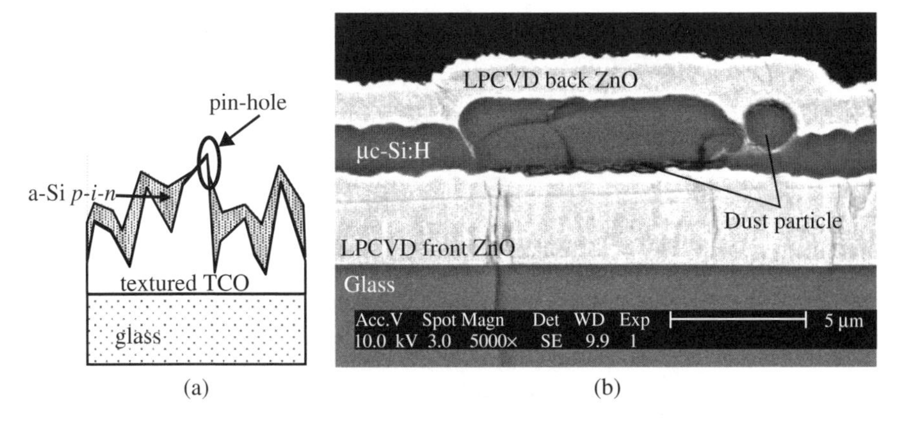


**Fig. 5.9** "Pinholes" and "shunts" in the solar cells in the cases of rough TCO, a very thin intrinsic layer and presence of dust particles: (a) principle of shunt formation due to rough TCO and very thin *i*-layer; (b) electron micrograph, showing shunt formation, due to dust particle, from [Python-2009].

### 5.3.5 Cracks

In tandems and triple-junction cells, microcrystalline silicon is, as mentioned above, a good candidate for the bottom sub-cell, i.e. for the sub-cell with the lowest bandgap. Indeed, its gap is close to 1.1 eV, degradation is hardly present [Torres-1996], and the same deposition process as for amorphous silicon can be used. Nevertheless, the deposition of microcrystalline silicon on rough substrates leads to large-size defects in the formation of the layer. Zones of non-dense material, called "cracks", are formed, perpendicular to the growth direction, between the main structures of the substrate (see Section 4.5.11). These cracks decrease the $V_{oc}$ and the $FF$ of the solar cells and act as "bad diodes" in the equivalent electrical circuit [Python-2008].

Cracks in single-junction solar cells can be avoided by an appropriate modification of the substrate morphology. In a "superstrate"-type (*pin*-type) tandem or triple-junction cell, the microcrystalline sub-cell is, in general, deposited *on top* of an amorphous sub-cell as in Figure 5.10.

If the μc-Si:H sub-cell is deposited *directly* on top of the a-Si:H sub-cell (without an intermediate reflector), then it is basically no more possible to modify the surface without deterioration of the sub-cell deposited before. It is then necessary to modify the deposition conditions in order decrease the crack density; this can be done by increasing the substrate temperature during growth [Python-2009].

Another option, in the case of tandem cells with an intermediate reflector [see Fig. 5.8(b)], is to change the morphology of the intermediate reflector: one then uses a particular type of intermediate reflector, which has been called an "asymmetric" intermediate reflector. Such "asymmetric" reflectors have been studied by [Söderström-2009].

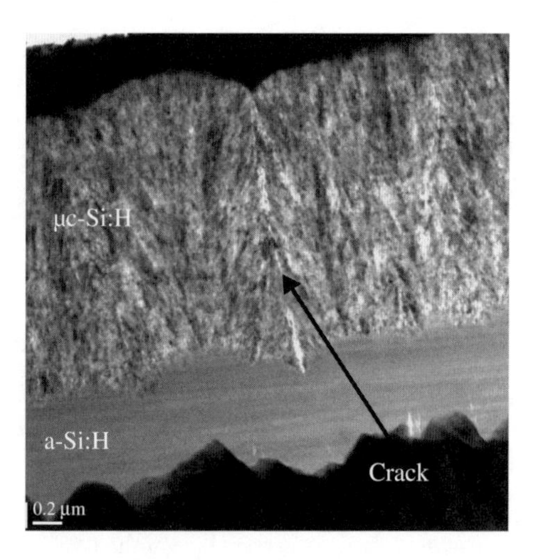


**Fig. 5.10** "Cracks" in the microcrystalline sub-cell within a "micromorph" tandem cell (electron micrograph, from [Python-2009]).

# 5.4 TYPICAL TANDEM AND MULTI-JUNCTION CELLS

## 5.4.1 Amorphous tandem cells a-Si:H/a-Si:H

The simplest tandem structure within silicon thin-film technology is a tandem with an amorphous silicon top sub-cell sitting on an amorphous silicon bottom sub-cell (Fig. 5.11).

It is evident that a-Si:H/a-Si:H tandems will suffer from the Staebler-Wronski effect (SWE). To reduce the impact of the SWE, as mentioned above, the a-Si:H top sub-cell must be very thin (*i*-layer thickness $d_i \approx 100$ nm) and the a-Si:H bottom sub-cell relatively thin (*i*-layer thickness $d_i \approx 300$ nm). The top sub-cell receives a higher light intensity and would degrade severely, if it were not kept very, very thin. The bottom sub-cell receives only about half the light intensityand thus the amplitude of the SWE is reduced in this sub-cell. However, the bottom sub-cell may take longer to approach a "stabilized" state because, generally, if the light intensity is lower, the "time to reach the onset of saturation (of the SWE) is found to be approximately reciprocal to the square of the carrier generation rate" (i.e. to the square of the light intensity) [Park-1989]. *If the bottom sub-cell is the limiting sub-cell in the tandem*, then the whole tandem cell can also take longer to approach "stabilisation". Such a long drawn-out degradation process would certainly not be desirable from a commercial point of view. Therefore generally the following guidelines should be adopted when designing a-Si:H/a-Si:H tandems:

(1) Keep the *i*-layer thickness $d_i$ for the top sub-cell as low as possible (below a critical value however, there will, be too many shunts in the cell).

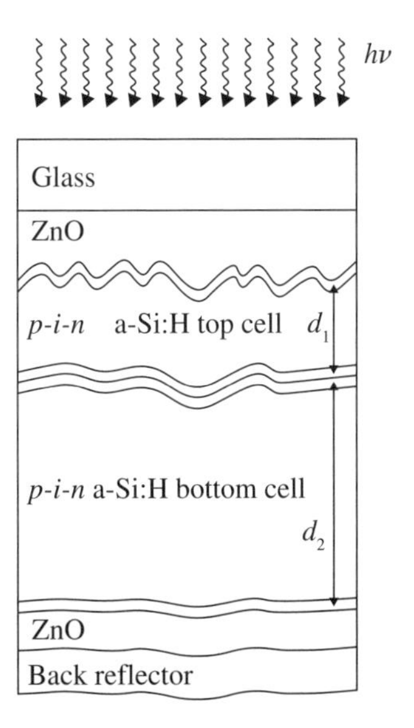


**Fig. 5.11** Sketch of an a-Si:H/a-Si:H tandem cell.

(2) Adjust the bottom sub-cell thickness in such a way that the tandem is limited by the top sub-cell.

In [Von Der Linden-1995] the Utrecht group reports on simulation studies and experimental results with a-Si:H/a-Si:H tandems of different sub-cell thicknesses deposited on a glass + TCO substrate (actually a "superstrate"). The results showed that in order to obtain current matching, a specific thickness ratio $r_d = d_{bottom}/d_{top}$ must be selected. For the particular case of the cells investigated by [Von Der Linden-1995], where the deposition conditions, and hence the bandgaps of the two *i*-layers, were the same, and where the TCO used corresponded to "generally available" TCO of the early 90's, current matching was obtained for $r_d =$ 1:6. If the bandgaps are varied between depositions of the two *i*-layers, then one will obtain a different thickness ratio $r_d$. At present, strong hydrogen dilution is often used to render the bandgap of the top sub-cell slightly higher than the bandgap of the bottom sub-cell (see Sects. 5.3.4 and 2.2.3). This leads to a lower value of $r_d$ if current matching is desired. At present, it is preferable to adjust the sub-cell thicknesses in such a way as to obtain a slightly top-limited state, rather than current matching. Under all these conditions and with present state-of-the art TCO's, we may estimate the optimal value of $r_d$ to be between 1:3 and 1:4.

The firm SCHOTT Solar Thin Film GmbH (formerly Phototronics) fabricates a-Si:H/a-Si:H tandems based on these principles. They have recently reported stabilized total-area module efficiencies of approximately 7 % and a relatively rapid degradation process, as shown in Figure 5.12 [Lechner-2008].

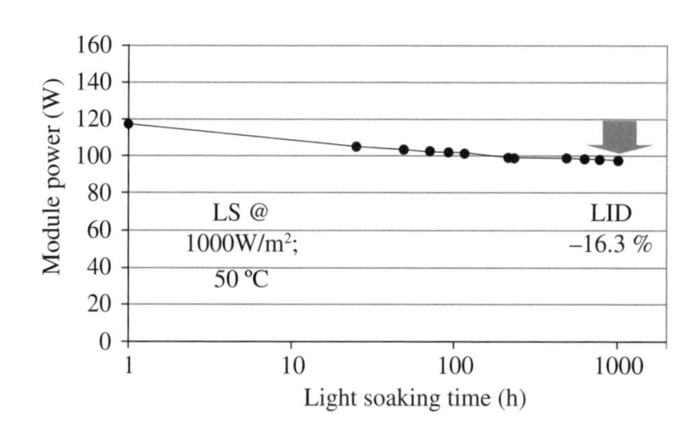


**Fig. 5.12** Typical curve for LID (Light-Induced Degradation) of an a-Si:H/aSi:H tandem as fabricated by SCHOTT Solar Thin Film GmbH, with limitation by top sub-cell and a relatively rapid degradation process. (Courtesy of Peter Lechner).

### 5.4.2 Triple-junction amorphous cells with germanium

[Yang-1997] have successfully made triple-junction amorphous cells. These are laboratory cells with an active-area initial efficiency of 14.6 % and a stabilized efficiency of 13.0 %, which were confirmed by the National Renewable Energy Laboratory

(NREL). The firm United Solar Ovonic also sells commercial modules based on this triple-cell concept (Figure 5.13), although the commercial module efficiencies are at present not much higher than for modules with tandem cells using only amorphous silicon (without silicon-germanium alloys).

The use of amorphous silicon-carbon alloys enables one to obtain higher bandgaps. Tandem solar cells with a top sub-cell having an *i*-layer based on amorphous silicon-carbon have as yet not lead to very high efficiencies, because as soon as the carbon content becomes sufficiently high and thus, begins to have a significant effect on the bandgap, the quality of the layers suffers drastically. Nevertheless, quite interesting results were obtained in the early stages of thin-film silicon development [Delahoy-1989].

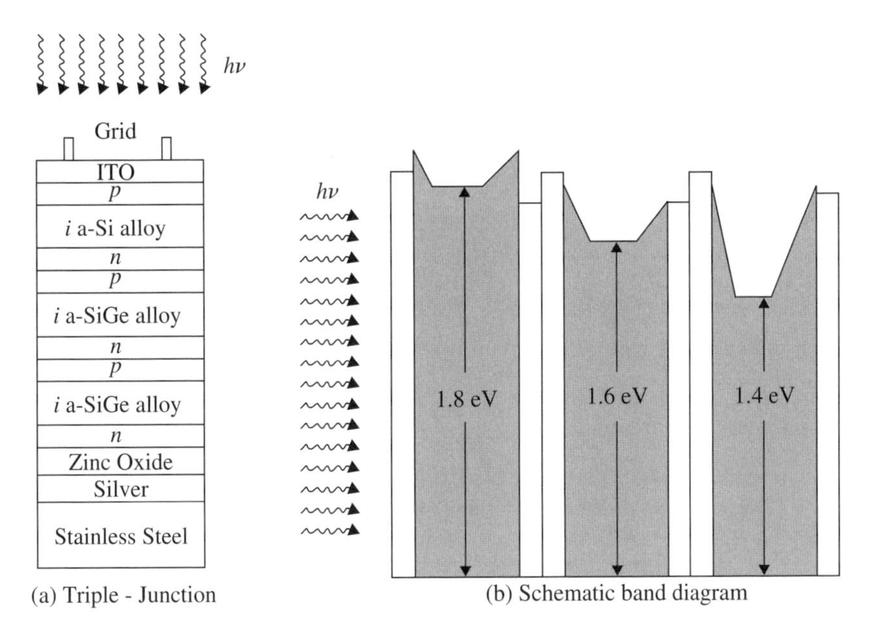


**Fig. 5.13** (a) Triple-junction a-Si:H/a-SiGe:H/a-SiGe:H solar cell structure; (b) corresponding schematic band diagram. The triple-junction structure leads to astable 13.0 % active-area efficiency for laboratory cells; it is also used in the current production lines of United Solar Ovonic, albeit with considerably lower commercial module efficiencies [Yang-1997].

### 5.4.3 Micromorph (a-Si:H/μc-Si:H) tandem cells

One of the most promising ways to obtain a tandem cell with materials having two different and very distinct bandgaps is to combine an amorphous silicon (bandgap 1.7 to 1.85 eV) top sub-cell with a microcrystalline silicon (bandgap 1.1 eV) bottom sub-cell. As mentioned in the Section 5.1, this type of tandem was introduced by the University of Neuchâtel in 1994 [Meier-1994] and is called the "micromorph" tandem (Figure 5.14). The best cells have achieved so far a confirmed stabilized efficiency of 11.7 % [Yoshimi-2003].

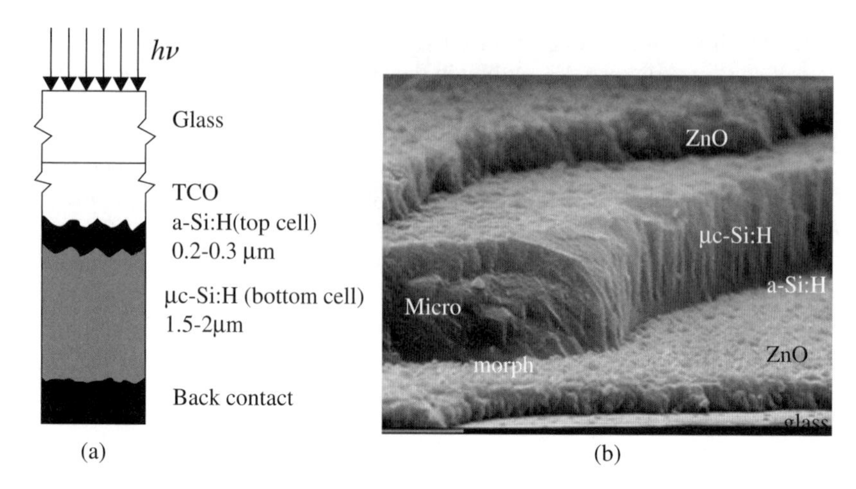


**Fig. 5.14** a-Si:H/μc-Si:H or "micromorph" tandem solar cell: (a) basic structure; (b) electron micrograph.

One would, of course, wish to increase the stabilized efficiencies of "micromorph" tandem cells. Figure 5.7 seems to indicate that an efficiency of over 30 % should be possible. Why then, are "micromorph" tandem cells at present limited, even in the laboratory, to the range of 10 to 12 %? Table 5.1 indicates the reasons.

From Table 5.1 we can immediately see what should be done in order to increase the stabilized efficiencies of "micromorph" tandem cells:

(1) improve light trapping;

(2) substitute, whenever possible, the amorphous top sub-cell with a microcrystalline top sub-cell of similar bandgap (1.6 to 1.8 eV);

(3) if that is not possible, reduce the light-induced degradation in the amorphous top sub-cell by;
   - reducing the amorphous top sub-cell thickness (possible only with improved light trapping),
   - optimizing the deposition conditions of the amorphous top sub-cell, so as to obtain amorphous intrinsic layers with less degradation.

Light trapping is thus a central issue if we wish to increase the stabilized efficiencies of "micromorph" tandem cells. It is also a complex and very difficult issue for R and D work, as it involves all layers within the cell, as well as their optical and electronic properties.

### 5.4.4 Triple-junctions with microcrystalline silicon

Triple-junction cells having a microcrystalline bottom sub-cell, an amorphous silicon-germanium middle sub-cell and an amorphous silicon top sub-cell have also been tested in the laboratory, with noteworthy results [Yan-2010]. Another interesting

**Table 5.1** Characteristic parameters of "micromorph" tandem cells (and their sub-cells): "theoretical" limit value (as well as the equation used to determine the limit value; experimental record results obtained so far in the laboratory; reasons for present experimental limitations.

| Parameter | Equation | Limit value | Value reached for laboratory cells | Limited by |
|---|---|---|---|---|
| $J_{sc}$ | Min $\{J_{sc.top}, J_{sc.bottom}\}$ | 22 mA/cm$^2$ | 11.5 mA/cm$^2$ | insufficient light trapping, in μc-Si:H sub-cell |
| $V_{oc}$ | $V_{oc.top} + V_{oc.bottom}$ | $\approx$ 1.4 V + 0.6 V $\approx$ 2.0 V | 0.9 V + 0.5 V = 1.4 V | $V_{oc}$ of amorphous sub-cell |
| $FF$ | $\approx$ Aver$\{FF_{top}, FF_{bottom}\}$ | $\approx$ Aver{80 %, 75 %} $\approx$ 76 % | $\approx$ 70 % | light-induced degradation, in amorphous sub-cell |
| $\eta$ | | $\approx$ 39 % | 11.7 % | |

combination is a triple-junction cell with a microcrystalline bottom sub-cell, an amorphous silicon middle sub-cell and a further amorphous silicon top sub-cell [Yan-2006, Yan-2007].

All these multi-junction cells are at present very much in the R and D stage and one can expect interesting new results to emerge in the next few years, especially if new materials such as microcrystalline silicon-germanium and microcrystalline silicon-carbon become available.

## 5.5 SPECTRAL RESPONSE (SR) AND EXTERNAL QUANTUM EFFICIENCY (EQE) MEASUREMENTS

### 5.5.1 General principles

Spectral response (SR) and External Quantum Efficiency (EQE) measurements are a powerful tool used to investigate the "interior" of a solar cell. In single-junction cells (see Section 4.5.17), it is used to identify *p*-layers having too high absorption, and especially to probe different regions of the *i*-layer. In tandem and triple-junction cells, it is, on the other hand, mainly used to evaluate the current density contributions of each sub-cell. It will thus allow us to identify which sub-cell is the limiting one. In this manner it is an excellent tool to experimentally attain current-matching conditions.

The spectral response (SR) is the ratio of the photogenerated current at the output of the solar cell to the energy of incoming monochromatic light, as a function of wavelength; it is also designated by the term "photosensitivity", and is typically given in mA/mW. The external quantum efficiency (EQE) is a ratio without physical dimensions; it is the ratio of the flux of photogenerated and collected electron/hole pairs (at the output of the solar cell) to the flux of incoming photons at a specific wavelength. SR and EQE curves are related to each other: one goes from the EQE-curve to the SR-curve by dividing by the photon energy (which in turn depends on wavelength) and multiplying by the unit charge (charge of an electron). SR- and EQE-curves are obtained with practically the same measurement set-up; compared to the EQE-curve, the SR-curve is "blown-up" at lower wavelengths, and also has a different ordinate scale .

A typical measurement set-up is composed of a halogen or xenon lamp followed by a monochromator, which separates light into the different wavelengths, and permits only a narrow range ($\lambda_0 \pm 10$ nm) of the spectrum to pass through and reach the solar cell. The intensity of the incoming light on the measured cell is calibrated using a reference cell.

One generally uses a chopped "probe" light beam of variable wavelength and relatively low intensity, produced by the halogen lamp and the monochromator, and one often superimposes on to this probe light a suitable steady-state *bias light* of high intensity. SR- and EQE-curves are determined by a lock-in amplifier (synchronized with the chopper) and an electronic ampere meter. In single-junction cells, a white light beam is often used as bias light, allowing the solar cell to operate under similar conditions (with respect to gap states' occupation) as under AM 1.5 light. In tandem and multi-junction cells, on the other hand, "coloured" bias light beams are used in order to separate the contributions of the different sub-cells.

The total short-circuit current density is established by multiplying the spectral response with the AM 1.5 spectrum and integrating it over the whole measurement spectrum. In the case of single-junction cells, this provides the short-circuit current density $J_{sc}$ of the whole solar cell. In the case of tandem and triple-junction cells, integrating separately over each of the two or three SR-curves provides the short-circuit current density $J_{sc}$ of the corresponding sub-cells.

An *optional reverse bias voltage* is applied to the whole solar cell:

In ***single-junction cells***, the reverse bias creates an additional electrical field superposed on the electrical field present within the *pin-* (or *nip-*) device. This it favours, when it is switched on, the collection of electron and holes for lower-quality cells.

In the case of ***tandem and triple-junction cells***, a fixed forward bias should be applied, together with the bias light, so as to ensure that the sub-cell being measured has zero voltage lying on it. In fact, the other sub-cells (which are saturated by the bias light beams) will be forced into forward bias and therefore they tend to create a reverse bias voltage for the specific sub-cell being evaluated. This reverse bias voltage must be compensated by applying a forward bias voltage to the whole tandem (see Sect. 5.5.7 below).

### 5.5.2 Use of "colored" bias light beams for SR/EQE-measurements on tandems and triple-junction cells

In case of multi-junctions solar cells, we want to evaluate each of the sub-cells separately. In order to measure each sub-cell separately, it is necessary to "saturate" all the other sub-cells. ("Saturation" means that the sub-cell receives enough absorbable light to photo-generate a current which is clearly higher than the current of the sub-cell one wants to measure; the current through the entire device is then limited by the sub-cell under investigation.)

The light absorbed in every sub-cell will depend on the bandgap of the material and on the position of the sub-cells with respect to the incoming light (top – middle – bottom). This is shown for the example of a triple-junction cell in Figure 5.15. In this example, the high energy photons (blue light) are absorbed in the top sub-cell and can "saturate" it, the green light is absorbed in the middle sub-cell and can "saturate" it, whereas the red light is absorbed in the bottom cell and can "saturate" it.

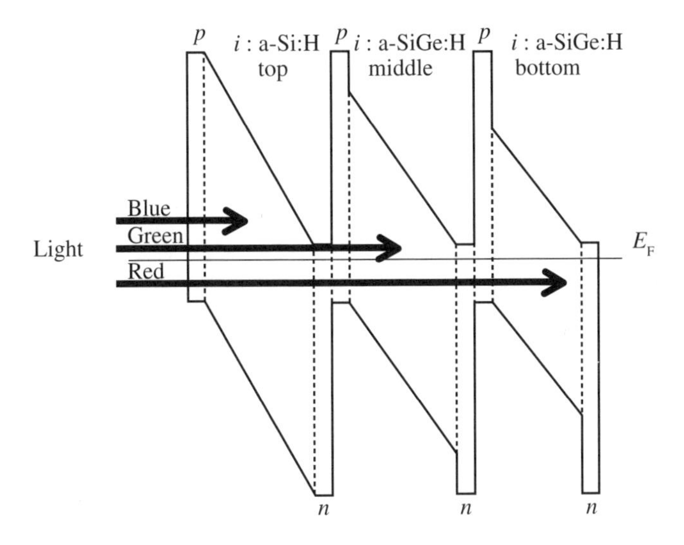


**Fig. 5.15** Principle of EQE measurement and schematic simplified band diagram, for an a-Si:H/a-SiGe:H/a-SiGe:H triple-junction solar cell; here, the a-Si:H top sub-cell absorbs the blue light, the a-SiGe:H middle sub-cell absorbs the green light, and the a-SiGe:H bottom sub-cell absorbs the red light; two of the three bias light beams (blue, green and red beams, shown as arrows in the Figure) can be switched on, as described in the text, so as to "saturate" the corresponding two sub-cells and obtain the EQE curve for the the third cell.

In a practical measurement situation, only two of the three "saturating" light beams are switched on at any moment, i.e. during any of the three measurement phases. The SR/EQE-curve of the specific sub-cell whose corresponding "saturating" light beam is *not* switched on can be determined during that specific measurement phase.

The current-matching condition formulated by Equation 5.5 postulates that the current density must be the same in each sub-cell in order to avoid losses and to maximize the efficiency of a multi-junction solar cell. Spectral response measurements enable us to check whether the current-matching conditions are satisfied or not. They also allow us to install a deliberate amount of current mismatch, as may be specified by our design requirements (see e.g. Sections 5.3.3 and 5.4.1).

We shall now look at individual examples of multi-junction cells.

### 5.5.3 SR/EQE measurements for a-Si:H/a-Si:H tandem cells

For example, in the case of a-Si:H/a-Si:H tandem solar cells, a bias light at long wavelengths is applied during the measurement of the top sub-cell. In this manner, the bottom sub-cell is "saturated" and the sub-cell limiting the current during this measurement phase is the top cell. Conversely, a bias light at short wavelengths is applied during the measurement of the bottom sub-cell. Now the top sub-cell is "saturated" and the sub-cell limiting the current during this second measurement phase is the bottom sub-cell. It is important to choose a bias light that is not absorbed in the sub-cell one wants to measure. Typical filters used for generating the bias light in this example are: a long wavelength pass filter with light transmission from 630 nm upwards (typically the RG630 filter from Schott) for the measurement of the top sub-cell; and a short-wavelength pass filter with light transmission from 500 nm downwards (typically a combination of the BG23 and KG1 filters from Schott) for measurement of the bottom sub-cell. The filters are fitted on-to a halogen lamp, which constitutes a white light source. The transmittance curves of these two filters are shown in Figure 5.16.

Typical results for External Quantum Efficiency (EQE) and *J*(*V*) curves of a a-Si:H/a-Si:H tandem solar cell are given in Figure 5.17. Here, the thickness of the amorphous top cell is ≈ 100 nm and that of the amorphous bottom sub-cell is ≈ 300 nm. The current-matching condition is almost achieved.

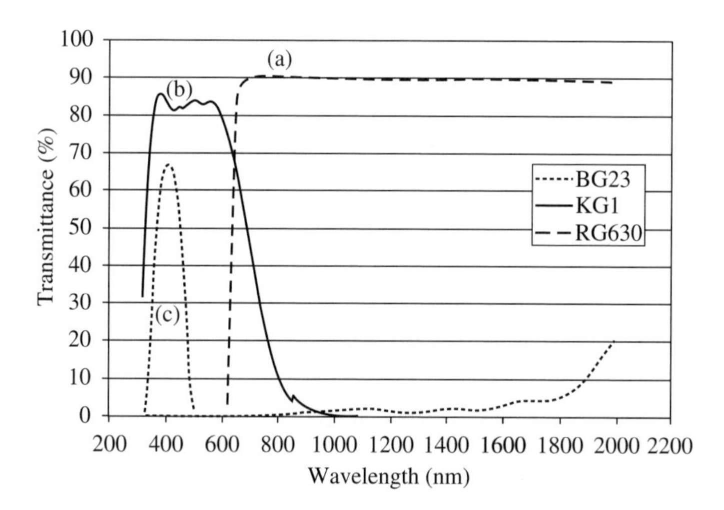


**Fig. 5.16** (a) Long-wavelength pass filter for saturation of the bottom sub-cell (RG630 from Schott); (b) and (c) short-wavelength pass filters for saturation of the top sub-cell.

## 5.5.4 Shunt detection in sub-cells by SR/EQE measurements

An interesting *additional measurement* is done *without any bias light,* and allows one to *detect shunts in sub-cells*. For high-quality tandem solar cells without any shunts in the sub-cells, this third measurement curve (the *dotted* curve in Figure 5.18) should now follow, at every wavelength, the lower of the two other curves obtained for each of the two sub-cells during the preceding two measurement phases. If an additional signal appears, as in Figure 5.18, one can deduce that the corresponding sub-cell contains shunts.

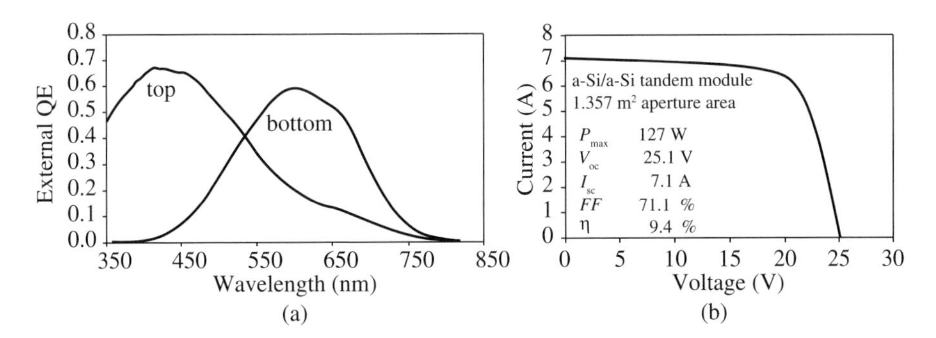


**Fig. 5.17** (a) External quantum efficiency of a typical a-Si:H/a-Si:H tandem solar cell; (b) *J-V* characteristics of the corresponding module (courtesy of P. Lechner, SCHOTT Solar Thin Film GmbH Putzbrunn; see also [Lechner-2008]).

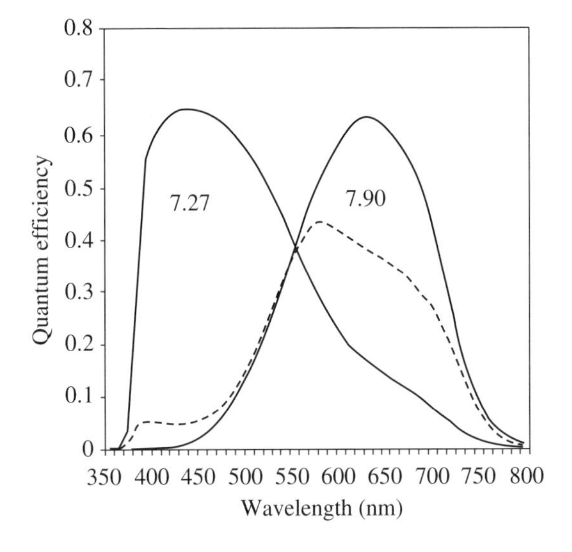


**Fig. 5.18** External quantum efficiency of a tandem amorphous cell. The numbers under the two continuous EQE-curves denote the current densities in mA/cm$^2$ of the top and bottom sub-cells, respectively. The dotted EQE-curve permits us to detect shunts in the sub-cells. Here, the top sub-cell is shunted (see text). (Data from measurements of IMT Neuchâtel).

As an example, from Figure 5.18 we can conclude that the top sub-cell is shunted. The reasoning is as follows [Löffler-2005, Rubinelli-2006]:

(1) The dotted curve corresponds to a measurement without bias light. If there were no shunts in the sub-cells, we would have a dotted curve which follows the lower of the other two (solid line) curves. Furthermore, the top sub-cell in this tandem is only capable of absorbing the short wavelength light. So basically, in the long wavelength region only a very low EQE-signal, corresponding to the individual EQE-curve of the top sub-cell, should appear.

(2) However, the dotted curve in Figure 5.18 in the long wavelength region is clearly higher than the individual EQE-curve of the top sub-cell. This increase in current density, and consequently in quantum efficiency, must be due to shunts in the top sub-cell.

### 5.5.5 SR/EQE measurements for triple-junction cells

For triple-junction solar cells (see also Figure 5.15), another combination of filters is used, and the halogen lamp is generally complemented with a lamp consisting of blue LEDs. Typically, the measurement of the top sub-cell is done with a bias light obtained from white light (halogen lamp) with an RG630 filter (this bias light saturates the middle and bottom sub-cells); the middle sub-cell is measured with bias light obtained from a blue LED light source and a RG830 filter (see Figure 5.19) (this bias light saturates the top and the bottom sub-cells); the bottom sub-cell is measured under bias light obtained from white light with KG1 and BG23 filters. The use of blue LEDs is generally necessary, because the intensity in the blue region is often not high enough to saturate the top sub-cell if one uses a halogen lamp and a short-wavelength pass filter.

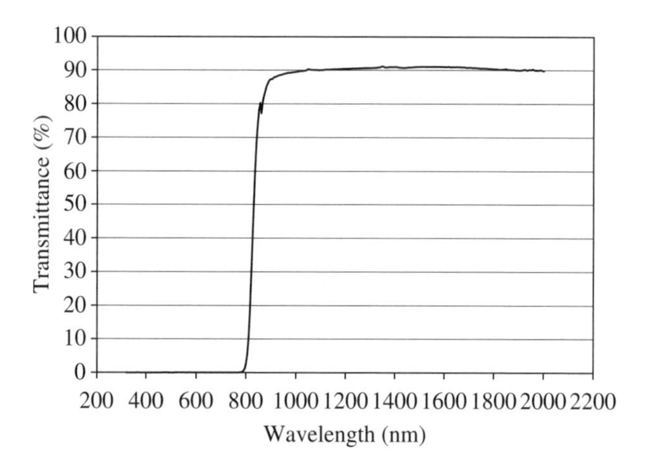


**Fig. 5.19** Long-wavelength pass filter for saturation of the bottom sub-cell (RG830 from Schott).

The amorphous silicon solar cells which have so far given the highest (active-area) stabilized efficiencies are the triple-junction cells from United-Solar. These are laboratory cells with a stabilized efficiency of 13.0 %[Yang-1997]. Their $J(V)$- and EQE-curves are given in Figure 5.20. The total current density (26.88 mA/cm$^2$) is separated into the individual contributions of 8.57, 9.01 and 9.30 mA/cm$^2$, for top, middle and bottom cells. The current-matching condition is almost achieved.

### 5.5.6 SR/EQE measurements for "micromorph" tandem cells

The measurement procedure used here is similar to that one used for tandem a-Si:H/a-Si:H cells. The bias light applied for the measurement of the bottom sub-cell is obtained from white light by using a short-wavelength pass filter with light transmission from 500 nm downwards, whereas for the top sub-cell a long-wavelength pass filter with light transmission from 830 nm upwards (RG830 filter from Schott) is employed. The example represented in Figure 5.21 shows that in such a "micromorph" tandem cell, photogeneration extends up to 1100 nm instead of only up to 800 nm, as is the case for amorphous/amorphous tandem cells. This extension of the useful spectral range is based on the fact that the bandgap of the microcrystalline silicon solar sub-cell is 1.1 eV instead of approximately 1.75 eV, as is obtained for an amorphous sub-cell. Figure 5.21 shows the $J(V)$ and EQE curves for both the initial and the stabilized states of the solar cell. The tandem-cell parameters, as well as the current densities which can be attributed to the individual sub-cells, are given in Table 5.2.

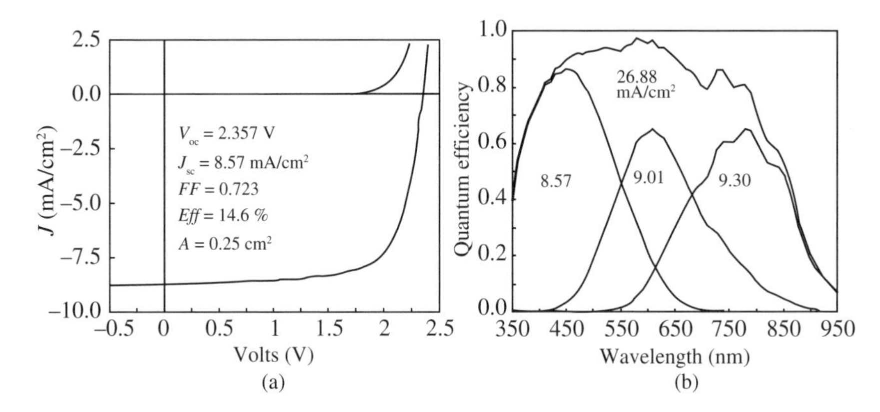


**Fig. 5.20** (a) Initial characteristics of an a-SiGe:H/a-SiGe:H/a-SiGe:H triple junction solar cells, (b) typical $J$-$V$ characteristics. The stabilized active area efficiency is 13 % (as confirmed by NREL) (courtesy of Subhendu Guha, United Solar Ononic).

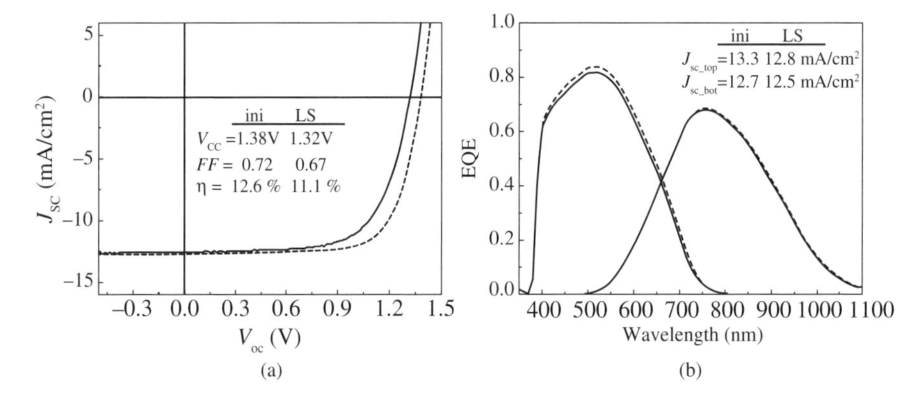


**Fig. 5.21** Characteristics of a typical "micromorph" (a-Si:H/μc-Si:H) tandem cell: (a) $J$-$V$ characteristics; (b) external quantum efficiency (EQE) curves of top and bottom sub-cells. The dotted line denotes the cell in the initial state (before light soaking); the solid line denotes the cell in the stabilized state (after 1000 hours of light-soaking under white AM1.5-near light at 50 °C cell temperature and under open-circuit conditions) (Courtesy of IMT Neuchâtel).

**Table 5.2** Example of a typical "micromorph" (a-Si:H/μc-Si:H) tandem cell. Parameters of the tandem cell, determined from the *J-V* curve in Figure 5.21 (a), and short-circuit current densities of individual sub-cells, as evaluated from the EQE data in Figure 5.21 (b).

| Parameters of tandem cell from *J(V)* curve | Before light-soaking (initial) | After light-soaking ("stabilized") |
|---|---|---|
| Open-circuit voltage $V_{oc}$ | 1.38 V | 1.32 V |
| Fill Factor *FF* | 72 % | 67 % |
| Efficiency $\eta$ | 12.6 % | 11.1 % |
| Short-circuit current density $J_{sc}$ | 12.68 mA/cm$^2$ | 12.54 mA/cm$^2$ |
| Short circuit current densities $J_{sc}$ of sub-cells, from EQE curves | | |
| Top sub-cell (a-Si:H) | 13.3 mA/cm$^2$ | 12.84 mA/cm$^2$ |
| Bottom sub-cell (μc-Si:H) | 12.68 mA/cm$^2$ | 12.54 mA/cm$^2$ |

In "micromorph" tandem cells one generally aims at having a slight current mismatch (with current limitation by the bottom sub-cell) in the initial state, so that current matching is obtained *after* light soaking. One can see from Table 5.2 that this condition is over-fulfilled for the specific cell reported here. The tandem is still bottom-limited, even after light soaking. As light soaking takes up a lot of time (1000 hours to really perform light soaking according to standard test conditions), it is indeed quite an arduous task to design and fabricate cells which have exact current matching in the stabilized state.

One may also wish to set the current mismatch of "micromorph" tandem cells in such a way that the temperature coefficient of the tandem is kept low. This would probably call for limitation by the top sub-cell. Another criterion for setting the current mismatch would be optimal performance under varying spectral conditions, as given by the time of the day, by seasonal changes and by the specific location of the modules. It is, however, extremely difficult to predict which type of current mismatch would be best here. There is thus ample scope for further research work in this sector.

### 5.5.7 Necessity for voltage correction (with bias voltage)

When the spectral response of tandem and triple-junction cells is measured using the "bias light method", a suitable overall voltage bias must be applied on the external electrodes of the whole cell, in order to ensure that the sub-cell being measured really attains short-circuit conditions. Indeed, a "saturated" sub-cell does not have 0 V over it, but develops a forward bias. Thus, if the overall voltage over the whole tandem is kept at 0 V, the measured sub-cell will be under a reverse bias voltage, whose value corresponds approximately to the sum of the $V_{oc}$ values of the "saturated" sub-cells (see Figure 5.22). With such a reverse bias voltage, the current density measured will be overestimated. In order to correct this effect, a forward bias is applied on the external contacts of the whole tandem or triple-junction cell [Burdick-1986, Meusel-2003].

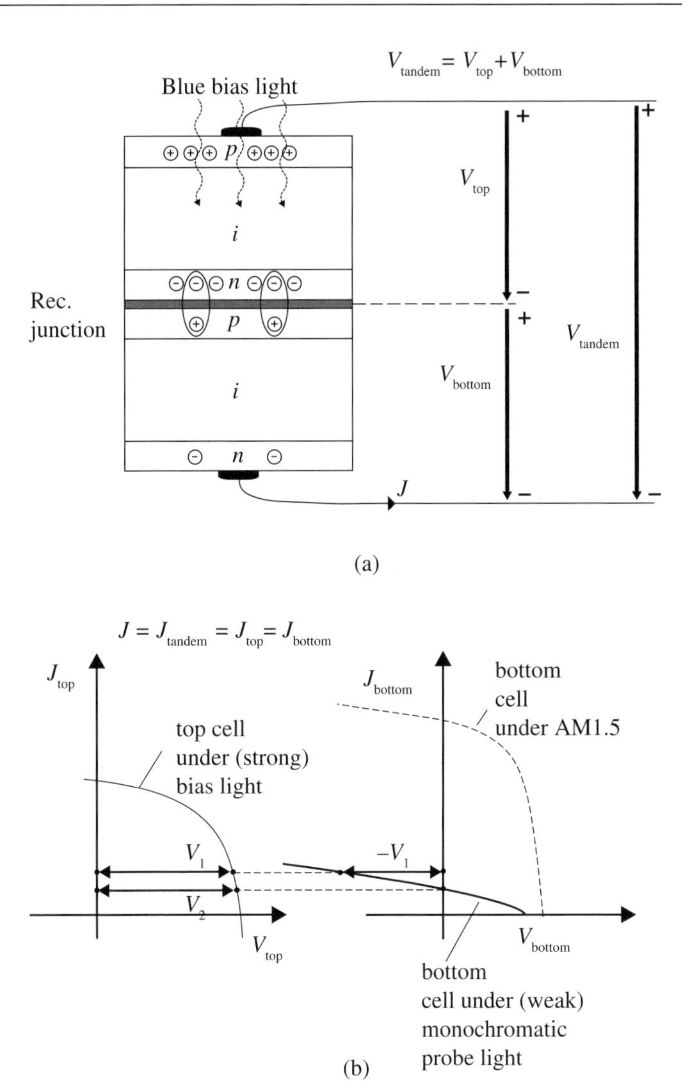


**Fig. 5.22** (a) Example of a tandem solar cell under illumination with a blue bias light ("saturating" the top sub-cell) and a monochromatic probe light of variable wavelength (measuring the EQE of the bottom sub-cell): tandem cell structure and voltage definitions (b) *J-V* diagrams of sub-cells. $V_1$ is here the undesired reverse bias voltage which would lie on the bottom sub-cell if no bias voltage is applied to the tandem; $V_2$ is the forward bias voltage which has to be applied to the tandem.

According to [Meusel-2003], the forward bias $V_{bias}$ to be applied on the external contacts of the whole tandem or triple-junction cell can be calculated with help of Equation 5.6:

$$V_{bias} = \frac{1}{2}\sum_i \left(V_{oc}^i + V_{MPP}^i\right) \tag{5.6}$$

**Table 5.3** Example of the bias voltages necessary to stay at short-circuit conditions during SR or EQE measurements, for the case of a "micromorph" tandem cell.

| Sub-cell | $V_{oc}$ (V) | $V_{MPP}$ (V) | $V_{bias}$ (V) |
|---|---|---|---|
| Bottom | 0.550 | 0.400 | ½ (1 + 0.82) ≈ 0.91 |
| Top | 1.000 | 0.820 | ½ (0.55 + 0.4) ≈ 0.475 |

**Table 5.4** Example of the bias voltages necessary to stay at short-circuit conditions during SR or EQE measurements, for the case of an a-Si:H/µc-Si:H/µc-Si-H triple-junction cell.

| Sub-cell | $V_{oc}$ (V) | $V_{MPP}$ (V) | $V_{bias}$ (V) |
|---|---|---|---|
| Bottom | 0.550 | 0.400 | ½ (0.9 + 0.735 + 1 + 0.82) ≈ 1.73 |
| Middle | 0.900 | 0.735 | ½ (0.55 + 0.4 + 1 + 0.82) ≈ 1.38 |
| Top | 1.000 | 0.820 | ½ (0.55 + 0.4 + 0.9 + 0.735) ≈ 1.3 |

where $V_{oc}^{i}$ and $V_{MPP}^{i}$ are the open-circuit voltage and the maximum power point voltage, respectively, summed up over all the sub-cells "*i*" not being measured.

As first example, the voltage correction for an a-Si:H/µc-Si:H tandem cell ("micromorph" cell) is presented in Table 5.3. As second example, the voltage correction for an a-Si:H/µc-Si:H/µc-Si:H triple-junction cell is presented in Table 5.4.

***Note:*** In the case of tandem cells, one cannot easily apply a ***reverse bias voltage*** to "strengthen the electric field within the *i*-layer, so as to basically avoid almost all recombination", as one does with single-junction cells (see Section 4.5.17). In fact, if one (or both) of the sub-cells is partially shunted, then a reverse bias voltage applied to the whole tandem will be distributed on the two sub-cells in a way that cannot be predicted, but essentially depends on the shunting conditions. Evidently, this remark is also valid for triple-junction cells.

## 5.6 CONCLUSIONS

Tandems and multi-junction cells are a natural extension of the simple single-junction solar cell concept.

They are extensively used for ***space cells***; in fact, for satellites and other space vehicles it has now become common to use triple- and even quadruple-junction solar cells based on the GaAs materials system. Thereby, the bandgap of each of the sub-cells which make up the multi-junction solar cell must be very carefully chosen, by varying the materials composition of ternary (and even quaternary) alloys, such $Al_xGa_yAs_z$ These are solar cells that have very high efficiencies (well over 30 %), but also exceedingly high fabrication costs.

A further extension of the multi-junction solar cell concept involves the use of spectrum-splitting optics, which furnishes the light to completely separate individual solar cells. With such laboratory set-ups, efficiencies over 40 % have been so far obtained. This type of set-up has, at present, no relevance for practical use, but only serves as a demonstration of the ultimate limit of photovoltaic (PV) technology.

Here, we are primarily concerned with low-cost large-area PV cells and modules, and in particular with those PV cells and modules that are fabricated with thin-film silicon.

In thin-film silicon solar cells it has, indeed, also become quite common to use tandem and triple-junction solar cells. Just as is the case for GaAs-based space cells, the solar cell conversion efficiency can thereby be increased, provided the bandgaps of the sub-cells are adequately chosen. There is, however, much less liberty for choosing the bandgaps in the thin-film silicon material system than in the GaAs materials system.

The first attempts were based on amorphous alloys of silicon and germanium and of silicon and carbon. As we have seen in Chapter 2 (Sect. 2.7), the electronic quality of the alloy becomes very poor, especially in the light-soaked state (after degradation), as soon as the bandgap of the material significantly differs from the bandgap of unalloyed amorphous silicon. Therefore, the efficiencies of these tandem and triple-junction cells are often just marginally higher than those of corresponding single-junction cells

There are, however, two other reasons why it is of special interest to use tandem and triple-junction cells, in the case of large-area thin-film silicon modules intended for outdoor applications:

- By going from a single-junction cell to a tandem cell, the voltage is basically doubled and the current density basically halved. This is specially advantageous in connection with the transparent contact layers (TCO layers, see also Chapter 6, Section 6.4): there will be less resistive losses due to the resistance of the TCO contact layers, if current densities are lowered. In the case of triple-junction cells the effect is, of course, even more pronounced;
- Tandems and triple-junction cells lead to thinner sub-cells; such cells suffer less from light-induced degradation effect (see also Chapter 4, Sect. 4.5.16).

On the other hand, the design and fabrication of multi-junction thin-film silicon solar cells is quite delicate, as the thicknesses of the sub-cells must be carefully adjusted to ensure current matching between the sub-cells. As one is also quite limited with respect to the bandgaps which can be obtained (without an inacceptable reduction in material quality), it is not at all straightforward to fabricate multi-junction solar cells based on thin-film silicon, with high efficiencies.

In general, sophisticated "*light management methods*" (textured transparent contact layers for light scattering, back reflectors and intermediate reflectors) have to be employed. World-wide, a substantial R and D effort is being consecrated to these optical methods, for their use in tandem and multi-junction thin-film silicon solar cells. One may, therefore, expect substantial further progress in this sector in future, but first the theoretical understanding of these rather complex structures must be improved.

The experimental testing of tandem and multi-junction cells is another important but difficult issue. A key method here is the measurement of EQE (external quantum efficiency). This has been described in this Chapter in detail and with many examples.

One of the most promising avenues for thin-film silicon tandems is the combination of a microcrystalline silicon bottom cell with an amorphous silicon top cell, i.e. the so-called "micromorph" tandem cell. From the theoretical point of view, and based on the bandgaps of microcrystalline silicon (1.1 eV) and amorphous silicon (1.75 V), such a tandem should be able to attain efficiencies of over 30 %. However, as of today, stabilized efficiencies of barely 12 % have been obtained in the laboratory. The two main factors that, in practice, limit efficiency for a "micromorph" tandem cell are:

- The fact that amorphous silicon solar cells have an open-circuit voltage which is always considerably lower (about 30 %) than the theoretical limit value which should be reached according to solar cell theory, for a bandgap of 1.75 eV; this limitation could, in principle, be overcome, if microcrystalline alloys were developed with corresponding bandgaps of around 1.7 to 1.8 eV; it is, however, not at all clear, whether such alloys will be found in the foreseeable future.
- The practical limitations in current densities, especially in the microcrystalline bottom cell where losses are about a factor 2; one is back to the light management problem described above: The task of improving the light management within thin-film silicon solar cells, is, on the other hand, a task that can certainly be accomplished during the coming years – we may, indeed, expect substantial progress here.

## 5.7 REFERENCES

[Buehlmann-2007] Buehlmann, P., Bailat, J., Dominé, D., Billet, A., Meillaud, F., Feltrin, A., and Ballif, C., "*In situ* silicon oxide based intermediate reflector for thin-film silicon micromorph solar cells", (2007), *Applied Physics Letters*, Vol. **91**, pp. 143505-1-143505-2.

[Burdick-1986] Burdick, J., Glatfelter, T., "Spectral response and I-V measurements of tandem amorphous silicon alloy solar cells", (1986) *Solar Cells*, Vol **18**, pp. 301-314.

[Dominé-2008] Dominé, D., Buehlmann, P., Bailat, J., Billet, A., Feltrin, A., and Ballif, C., "Optical management in high-efficiency thin-film silicon micromorph solar cells with a silicon oxide based intermediate reflector", (2008), *Physica status solidi (RRL) 2*, Vol. **4**, pp. 163–165.

[Delahoy-1989] Delahoy, A.E., "Recent developments in amorphous silicon photovoltaic research and manufacturing at Chronar corporation", (1989), *Solar Cells*, Vol. **27**, pp.39-57.

[Fischer-1996] Fischer, D., Dubail, S., Anna Selvan, J. A., Pellaton Vaucher, N., Platz, R., Hof, Ch., Kroll, U., Meier, J., Torres, P., Keppner, H., Wyrsch, N., Goetz, M., Shah, A., Ufert, K.D., "The "micromorph" solar cell: extending a-Si:H technology towards thin film crystalline silicon", (1996), *Proceedings of the 25th IEEE Photovoltaic Specialists Conference*, pp. 1053-1056.

[Gottschalg-2005] Gottschalg, R., Betts, T. R., Infield, D. G., Del Cueto, J. A., "Seasonal Performance of A-SI Single- and Multifunction Modules in Two Locations", (2005), *Proceedings of 31st IEEE-PVSC*, pp. 1484–1487.

[Lechner-2008] Lechner, P., Frammelsberger, W., Psyk, W., Geyer, R., Maurus, H., Lundszien, D., Wagner, H., Eichhorn, B., "Status of performance of thin film silicon solar cells and modules", (2008), *Proceedings of the 23rd European Photovoltaic Solar Energy Conference*, pp. 2023-2026.

[Löffler-2005] Löffler, J., Gordijn, A., Stolk, R., Li, H., Rath, J.K., Schropp, R.E.I., "Amorphous and'micromorph' silicon tandem cells with high open-circuit voltage", (2005) *Solar Energy Materials & Solar Cells*, Vol. **87**, pp. 251–259.

[Meier-1994] Meier, J., Dubail, S., Flückiger, R., Fischer, D., Keppner, H., Shah, A., "Intrinsic Microcrystalline Silicon (μc-Si:H) - a Promising New Thin Film Solar Cell Material", (1994) *Proceedings of the 1st World Conference on Photovoltaic Energy Conversion*, pp. 409-412.

[Meier-1995] Meier, J., Dubail, S., Fischer, D., Anna Selvan, J. A., Pellaton Vaucher, N., Platz, R., Hof, Ch., Flückiger, R., Kroll, U., Wyrsch, N., Torres, P., Keppner, H., Shah, A., Ufert, K. D., "The ,Micromorph' Solar Cells: a New Way to High Efficiency Thin Film Silicon Solar Cells", (1995), *Proceedings of the 13th EC Photovoltaic Solar Energy Conference*, pp. 1445-1450.

[Meier-1996] Meier, J., Torres, P., Platz, R., Dubail, S., Kroll, U., Anna Selvan, J. A., Pellaton Vaucher, N., Hof, Ch., Fischer, D., Keppner, H., Shah, A., Ufert, K.-D., "High Efficiency Thin-Film Solar Cells by the "Micromorph" Concept", (1996), *Technical Digest of the 9th PVSEC*, pp. 653-654.

[Meier-1998] Meier, J., Keppner, H., Dubail, S., Ziegler, Y., Feitknecht, L., Torres, P., Hof, Ch., Kroll, U., Fischer, D., Cuperus, J., Anna Selvan, J. A., Shah, A., "Microcrystalline and Micromorph Thin-Film Silicon Solar Cells", (1998), *Proceedings of the 2nd World Conference on Photovoltaic Energy Conversion*, Vol **I**, pp. 375-380.

[Meusel-2003] Meusel, M., Baur, C., Létay, G., Bett, A. W, Warta, W., Fernandez, E., "Spectral response measurements of monolithic GaInP/Ga(In)As/Ge triple-junction solar cells: measurement artifacts and their explanation", (2003), *Progress in Photovoltaics: Research and Applications*, Vol. **11**, pp. 499-514.

[Park-1989] Park, H. R., Liu, J. Z., Wagner, S., "Saturation of the light induced defect density in hydrogenated amorphous silicon", (1989), *Applied Physics Letters*, Vol. **55** (25), pp. 2658-2660.

[Pellaton-1998] Pellaton Vaucher, N., Nagel, J.-L., Platz, R., Fischer, D., Shah, A., Light management in tandem cells by an intermediate reflector layer, (1998), *Proceedings of the 2nd World Conference on Photovoltaic Energy Conversion*, Vol **I**, pp. 729-731.

[Python-2008] Python, M., Vallat-Sauvain, E., Bailat, J., Dominé, D., Fesquet, L., Shah, A., Ballif, C., "Relation between substrate surface morphology and microcrystalline silicon solar cell performance", (2008), *Journal of Non-Crystalline Solids,* Vol. **354**, pp. 2258–2262.

[Python-2009] Python, M., Madani, O., Dominé, D., Meillaud, F., Vallat-Sauvain, E., Ballif, C., "Influence of the substrate geometrical parameters on microcrystalline silicon growth for thin-film solar cells", (2009), *Solar Energy Materials and Solar Cells*, Vol. **93**, pp. 1714-1720.

[Rubinelli-2006] Rubinelli, F.A., Stolk, R.L., Sturiale, A., Rath, J. K., Schropp, R. E. I, "Sensitivity of the dark spectral response of thin film silicon based tandem solar cells on the defective regions in the intrinsic layers", (2006), *Journal of Non-Crystalline Solids,* Vol. **352**, pp. 1876–1879.

[Shah-2004] Shah, A., Vanecek, M., Meier, J., Meillaud, F., Guillet, J., Fischer, D., Droz, C., Niquille, X., faÿ, S., Vallat-Sauvain, E., Terrazzoni-Daudrix, V., "Basic efficiency limits, recent experimental results and novel light-trapping schemes in a-Si:H, μc-Si:H and "micromorph tandem" solar cells", (2004), *Journal of Non-Crystalline Solids*, Vol. **338-340**, pp. 639-645.

[Söderström-2009] Söderström, T., Haug, F.-J., Niquille, X., Terrazzoni, V., and Ballif, C., "Asymmetric intermediate reflector for tandem micromorph thin film silicon solar cells", (2009), *Applied Physics Letters*, Vol. **94**, pp. 063501-1-063501-3.

[Taguchi-2005] Taguchi, M., Terakawa, A., Maruyama, E., and Tanaka, M., "Obtaining a Higher Voc in HIT Cells", (2005), *Progress in Photovoltaics: Research and Applications*, Vol. **13**, pp. 481–488.

[Torres-1996] Torres, P., Meier, J., Flückiger, R., Kroll, U., Anna Selvan, J., Keppner, H., Shah, A., Littlewood, S. D., Kelly, I. E., Giannoulès, P., "Device Grade Microcrystalline Silicon Owing to Reduced Oxygen Contamination", (1996), *Applied Physics Letters*, Vol. **69**, pp. 1373-1375.

[Von Der Linden-1995] Von Der Linden, M.B., Hyviirinen, J., Loyer, W., Schropp, R.E.I, "Thickness optimization of aSi:H/aSi:H tandem modules", (1995), *Proceedings of the 13th European Photovoltaic Solar Energy Conference,* pp. 284-287

[Yablonovitch-1982] Yablonovitch, E., Cody, G.D., "Intensity Enhancement in Textured Optical Sheets for Solar Cells", (1982), *IEEE Transactions on Electron Devices,* Vol. **ED-29**, pp. 300-305.

[Yamamoto-2006] Yamamoto, K., Nakajima, A., Yoshimi, M., Sawada, T., Fukuda, S., Suezaki, T., Ichikawa, M., Koi, Y., Goto, M., Meguro, T., Matsuda, T., Sasaki, T., and Tawada, Y., "High Efficiency Thin Film Silicon Hybrid Cell and Module with Newly Developed Innovative Interlayer," (2006), *Conference Record of the 4th World Conference on Photovoltaic Energy Conversion*, Vol. **2**, pp. 1489-1492.

[Yan-2006] Yan, B., Yue, G., Owens, J. M., Yang, J., Guha, S., "Over 15 % Efficient Hydrogenated Amorphous Silicon Based Triple-Junction Solar Cells Incorporating Nanocrystalline Silicon", (2006), *Conference Record of the 4th World Conference on Photovoltaic Energy Conversion*, Vol. **2**, pp.1477-1480.

[Yan-2007] Yan, B., Yue, G., Guha, S., "Status of nc-Si:H Solar Cells at United Solar and Roadmap for Manufacturing a-Si:H and nc-Si:H Based Solar Panels", (2007), *Proceedings of the Materials Research Society Symposium* Vol. **989**, pp. 359-346.

[Yan-2008] Yan, B., Yue, G., Yang J., Guha, S., "Correlation of current mismatch and fill factor in amorphous and nanocrystalline silicon based high efficiency multi-junction solar cells", (2008), *Proceedings of the 33rd IEEE Photvoltaics Specialists Conference,* Paper no. 275.

[Yan-2010] Yan, B., Yue. G., Xu, X., Yang, J., and Guha, S., "High efficiency amorphous and nanocrystalline silicon solar cells", (2010), *Physica. Status Solidi A*, Vol. **207,** pp. 671-677.

[Yang-1997] Yang, J., Banerjee, A., Guha, S., Triple-junction amorphous silicon alloy solar cell with 14.6 % initial and 13.0 % stable conversion efficiencies, (1997), *Applied Physics Letters*, Vol. **70**, pp. 2975-2977.

[Yoshimi-2003] Yoshimi M., Sasaki T., Sawada T., Suezaki T., Meguro T., Matsuda T., Santo K., Wadano K., Ichikawa M., Nakajima A., Yamamoto K., "High efficiency thin film silicon hybrid solar cell module on 1 $m^2$-class large area substrate", (2003), *Conference Record, 3rd World Conference on Photovoltaic Energy Conversion*, pp. 1566–1569.

CHAPTER 6

# MODULE FABRICATION AND PERFORMANCE

*Horst Schade, with contributions from Jean-Eric Bourée*

## 6.1 PLASMA-ENHANCED CHEMICAL VAPOR DEPOSITION (PECVD)

The deposition of thin films has several attractive features, namely

- large active areas;
- low consumption of material;
- low deposition temperatures, and hence the possibility of using low-cost substrate materials.

In general, thin films may be fabricated either by physical vapor deposition (e.g., evaporation or sputtering) of the bulk material, or by chemical vapor deposition from the decomposition of suitable gases that contain the desired film material.

In Chapters 2-4 the photovoltaically relevant properties of Si-based thin films have been described. In the context of depositing such films, the role of hydrogen is essential.

Historically both evaporation and sputtering of elemental silicon were used to obtain thin films of amorphous silicon. However, these films contained high defect concentrations, mostly caused by poorly coordinated Si atoms, i.e. by so-called "dangling bonds" (see Sect. 2.1). Due to these high defect concentrations, the gap state density between the valence and conduction bands far exceeds the dopant densities that can be typically achieved. Therefore, this material is not suitable for use in an electronic semiconductor device, which inherently requires the presence of *n*- and/or *p*-type regions.

The high density of defect states can be somewhat reduced through post-hydrogenation by heating the films in a hydrogen atmosphere, or, in the case of sputtering, by adding hydrogen to the sputtering gas. However, these methods of film preparation have not proved successful in the further development of suitable PV-grade materials. Since it was recognized that silicon films deposited at low temperatures require the presence of hydrogen to "saturate" or "passivate" the dangling bonds, Chemical Vapor Deposition (CVD) methods have been chosen with a source gas containing both silicon and hydrogen, which is then decomposed by supplying thermal or electrical energy to generally form hydrogenated amorphous silicon (a-Si:H) films. Thus, in 1969, the decomposition of silane ($SiH_4$) in a glow discharge yielded hydrogenated amorphous silicon films with good photoconductivity [Chittick-1969], indicating a defect concentration well below the generated photocarrier concentration. Adding a dopant gas to the silane, glow discharge deposition has opened the way for the effective doping of a-Si:H thin films [Spear-1975], and hence for the possibility of creating electronic thin-film devices.

Since then the development of Si-based thin films for PV applications has remained based on the decomposition of silane as the central source material. Both forms of energy supply have been developed further, namely

- electrical energy, supplied in a glow discharge (plasma) for plasma-enhanced chemical vapor deposition (PECVD);
- thermal energy by heating a filament, for thermo-catalytic chemical vapor deposition (Cat-CVD), more commonly called Hot-Wire CVD (HWCVD).

To date, only PECVD has reached industrial relevance, whereas HWCVD has so far not gone beyond industrial feasibility studies. Here, we will mainly describe the basic features of PECVD, and only summarize the principles of HWCVD (see Sect. 6.2).

PECVD is conducted in a plasma reactor that usually consists of

- a vacuum chamber, equipped with an inlet for the reaction gases, and pumps to remove unreacted and reacted gases;
- a pair of parallel electrodes, one grounded and acting as the support for the substrate to be coated, and the other connected to an electrical power supply, and possibly also acting as a shower head to uniformly distribute the reaction gases;
- a gas handling system;
- substrate heating.

This arrangement, schematically shown in Figure 6.1, facilitates the sequential deposition of the entire semiconductor structure, either in a single-junction or a multi-junction (tandem or triple cell) configuration, as *pin* or *pin/pin* etc. To this end, not only silane is used (as is basically sufficient for the intrinsic *i*-layer), but also doping gases containing boron and phosphorus, such as diborane ($B_2H_6$) or trimethylboron [$B(CH_3)_3$], and phosphine ($PH_3$), for depositing *p*- and *n*-layers, respectively. Additionally, the energy gaps of the semiconductor film may be varied by adding gases containing other group-IV elements (see Sect. 2.7), specifically

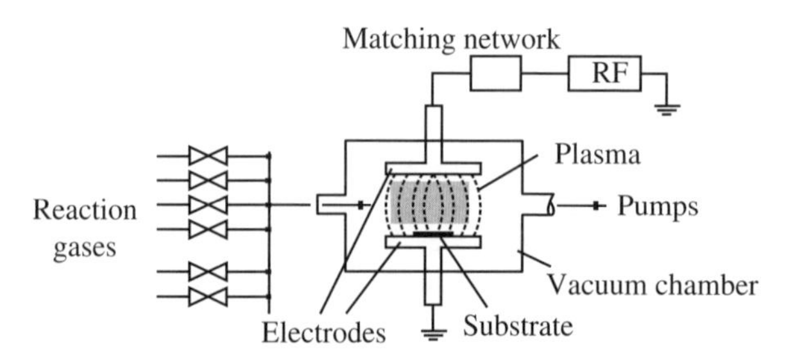


**Fig. 6.1** Schematic of a PECVD reactor.

- methane ($CH_4$) in order to alloy with carbon and, thus, to increase the energy gap of the *p*-layer to create a window layer (see Sect. 6.3);
- germane ($GeH_4$) in order to alloy with germanium and, thus, to decrease the energy gap of the *i*-layer for enhanced red-response, particularly for multi-junction cell structures (see Chapter 5).

PECVD is an extremely complex process involving a multitude of interactions between the plasma, chemical reactions of the gases and their reaction products, as well as interactions with the surrounding surfaces, ultimately including the substrate to be coated.

The application of a sufficiently high RF voltage (typically at the industrial frequency of 13.56 MHz, see also Sect. 6.1.2) to a capacitive configuration of parallel electrodes in a container filled with a gas at low pressure leads to the generation of a low-pressure plasma, also designated as "cold plasma" or "glow discharge (GD)". The plasma reactions occur while the gas and the parts exposed to the plasma remain at relatively low temperatures. The plasma is ignited by the generation of electrons and ions, due to the ionization of gas molecules (initiated by ever-present cosmic rays), followed by subsequent secondary electron emission from the electrodes, and further ionizations with charge carrier multiplication. The ignition of the plasma may sometimes require high voltage pulses supplied from a Tesla coil via a high-voltage feedthrough.

The plasma contains electrons, positive and negative ions, as well as neutral atoms, molecules, free radicals, and metastable species. Basically, the described formation of electrons and ions leads to various interactions with the gases, and gives rise to chemical reactions and the formation of radicals and other reactive molecular fragments. Both concentration gradients and electrical fields support a mass transport by diffusion and drift, respectively. Thereby interactions of various species with the substrate surface are initiated, and specific temperature-dependent sticking coefficients determine the composition and structure of the film that is deposited. A compilation of such processes is visualized in Figure 6.2.

Obviously, a quantitative description of these interrelated parameters is only possible for defined specifications of the plasma reactor system. The fundamental literature is abundant and dates back more than 20 years. As a representative illustration of the complexity, [Mataras-1997] may be consulted. Rather than to base one's work on theoretical calculations, one generally prefers to optimize the thin-film deposition process through experimental trials. Therefore, we shall limit ourselves here to some very general considerations.

Based on the use of the source gases silane ($SiH_4$, also called, more specifically, "monosilane") and hydrogen ($H_2$), the main chemical reactions may be grouped into three categories [Strahm-2007a]:

- electron impact dissociation, with the reactions

$$SiH_4 + e \rightarrow SiH_2 + 2\,H$$

$$SiH_4 + e \rightarrow SiH_3 + H$$

$$H_2 + e \rightarrow 2\,H$$

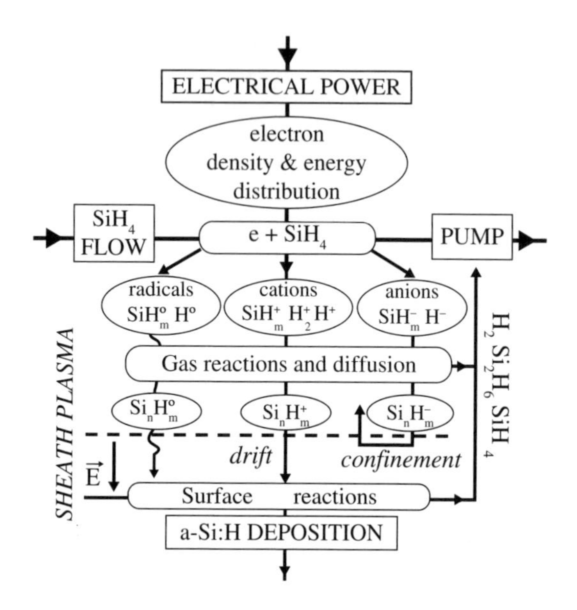


**Fig. 6.2** Interactions and reactions during PECVD (with permission [Perin-1991]).

- reactions with radicals

$$SiH_4 + H \rightarrow SiH_3 + H_2$$

$$SiH_3 + H \rightarrow SiH_3 + H_2$$

$$SiH_4 + SiH_2 \rightarrow Si_2H_6$$

- surface reactions that occur on the substrate, as well as on plasma-exposed surfaces in the reactor; these reactions refer to the deposition of silicon radicals, to etching by atomic hydrogen, and to hydrogen recombination at the surface:

$$SiH_2 \rightarrow Si_{surf.} + H_2$$

$$SiH_3 \rightarrow 2\ Si_{surf.} + 3\ H_2$$

$$2\ H \rightarrow H_2$$

$$4\ H + Si_{surf.} \rightarrow SiH_4$$

$Si_{surf.}$ refers to an a-Si:H or a μc-Si:H layer surface. One should note that the amount of hydrogen needed to saturate dangling bonds is typically only in the range of a few percent. The amount of hydrogen being incorporated into the deposited silicon film may be as high as 10 %. However, the actual hydrogen content in the deposited film ($Si_{surf.}$) is not apparent in the chemical reactions given above. For the surface reactions, the availability of $SiH_2$ and atomic hydrogen mainly determines the balance between film growth and etching. This balance governs the degree of crystallinity from amorphous to microcrystalline silicon (see Sect. 6.1.6).

These reactions, in addition to those between higher silanes $Si_nH_m$ not shown here, take place with different rate constants that are partly pressure- and temperature-dependent. In this context, the formation of ions and their energetic interaction with the growing film are not included (see Sect. 6.1.2).

### 6.1.1 Electrical plasma properties

The voltage required for plasma ignition depends on the gas pressure $p$ and on the electrode spacing $d$; it is determined by the Paschen curve; a curve which features a minimum for the ignition voltage at a specific value of $p \times d$.

A potential distribution between the electrodes is established; it depends on the applied RF voltage $U_{RF}$, and on the areas of the electrodes (including grounded shields or reactor parts exposed to the plasma). The potential distribution is given by the energy distributions of the electrons and ions that are exposed to the electric fields between the electrodes. These electric fields are set up in the vicinity of the electrodes. The basic mechanism is as follows: electrons have much higher thermal velocities than ions, and can, thus, reach the electrodes faster, leaving the ions behind. To preserve overall charge neutrality and to render net currents zero, electric fields develop near the electrodes, retarding electrons and accelerating ions. These fields in front of the electrodes extend over relatively small distances (typically in the order of a fraction of a millimeter), which are determined by the resulting space charge densities. These regions in front of the electrodes are positively charged, are composed mostly of ions, and represent the so-called "sheaths", as illustrated in Figure 6.3. The sheaths are responsible for ion bombardment. At higher plasma excitation frequencies the sheaths become thinner; this also means lower energies for the ion bombardment (see Sect. 6.1.2).

As a result of the generated fields, the following time-averaged potentials are assumed (see Fig. 6.4):

- the so-called self-bias $V_{sb}$ for the RF-powered electrode;
- zero potential for the grounded electrode; it is the grounded electrode on which one normally fixes the substrate to be deposited. The substrate surface is separated from the grounded electrode surface by the substrate thickness (i.e. usually by the glass thickness), and takes on (in case of an electrically isolating substrate) a certain relatively small potential, which is not specifically shown in Figure 6.4;

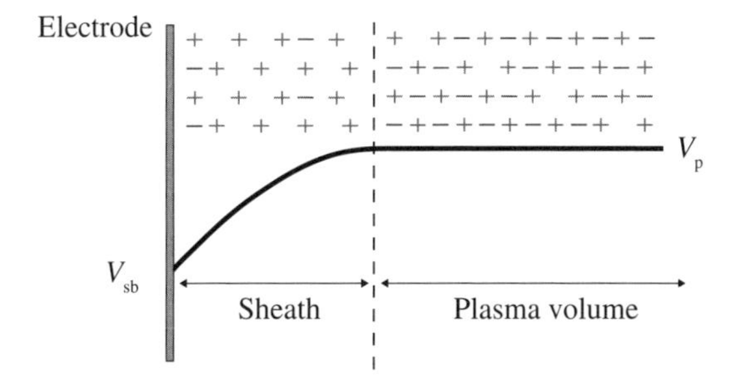


**Fig. 6.3** Positive space charge region (so-called sheath), due to faster electrons entering the electrode and leaving the slower positive ions behind, and field-free plasma volume (also called plasma bulk). The potentials $V_{sb}$ and $V_p$ are defined below and in Figure 6.4.

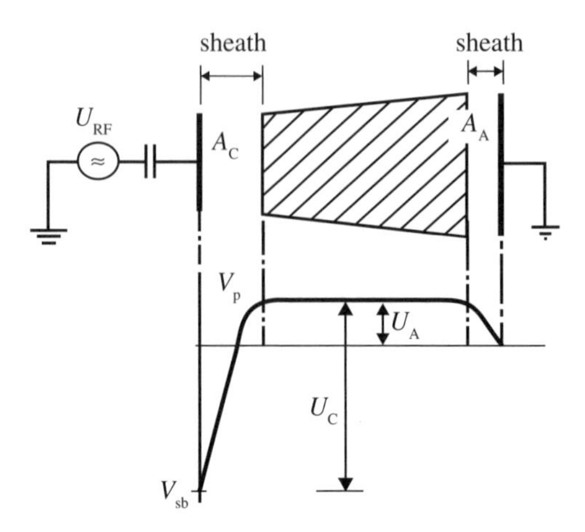


**Fig. 6.4** Potentials and sheath voltages in a plasma reactor (with permission [Kasper-1992]).

- the plasma potential $V_p$; since, due to their thermal velocities, more electrons than positive ions leave the plasma, the bulk of the plasma between the sheaths assumes a positive plasma potential $V_p$;
- the potential differences across the sheaths (sheath voltages); these are $U_C$ in front of the RF-powered electrode, and $U_A$ in front of the grounded electrode, respectively.

Note that the electron and ion concentrations within the plasma volume are sufficiently high to render it conductive. Thus the plasma volume essentially does not cause a voltage drop. Therefore, for all practical purposes, the applied RF voltage $U_{RF}$ is distributed only across the two space charge regions of the sheaths, which electrically represent two capacitances, i.e.

$$U_{RF} = U_C + U_A. \tag{6.1}$$

As already mentioned above, the potentials depend on the areas $A_A$, $A_C$ of the electrodes, and more specifically on the ratio of the electrode areas. One generally assumes the relationship:

$$U_C/U_A = (A_A/A_C)^n. \tag{6.2}$$

The value of the exponent $n$ has been the subject of theoretical and experimental investigations [Kasper-1992]; reasonable agreement has led to $n \approx 2$.

In the special case of equal areas (called the symmetrical case), which is approached in most large-area reactors, the self-bias tends towards zero. In this case both electrodes approach zero potential, and both sheath voltages are equal at $U_C = U_A = U_{RF}/2$. In this context it is important to note that the potential differences between the plasma potential and the electrode potentials, i.e. $U_C$ and $U_A$, are proportional to the maximal ion energies with which the electrodes are bombarded. If one does not have symmetry, as in small research reactors or in large-area reactors with antenna-type electrodes (see Sect. 6.10), then the grounded electrode accommodating the substrate to be deposited tends to be larger than the RF-powered electrode ($A_A > A_C$, asymmetrical case), and the sheath voltage in front of the substrate becomes smaller ($U_A < U_{RF}/2$) compared to the symmetrical case and, thus, also the ion energies affecting the deposited film quality tend to be smaller than in the symmetrical case.

The reactor may be electrically characterized by electrical circuit diagrams as shown in Figure 6.5. These circuit diagrams refer to the reactor, with or without a plasma running. The circuit components represent the following: $Z_{pl}$ is the impedance of the space between the RF electrode and substrate (with no plasma switched on, $Z_{pl}$ is just the parallel-plate capacitance $C_{pp}$ formed by the RF electrode and the substrate surface; with the plasma switched on, $Z_{pl}$ is composed of the dark-space capacitances $C_{sh}$ and the sheath and plasma resistances $R_{sh}$ and $R_{pl}$ of the plasma discharge). $C_{sg}$ represents the capacitance of the glass plate in series with the vacuum gap between the glass and the substrate electrode. $C_e$ typically may represent a stray capacitance formed between the edge of the RF electrode and the plasma box. $R$ represents a series resistance formed by the electrodes. Note that at RF frequencies, the skin depth of common electrode materials is only of the order of 10 $\mu$m, and hence the current flow is confined to the surface regions of the electrodes so that $R$ can be relatively high. $L$ is the inductivity of the RF connector(s). It becomes qualitatively obvious that the source power which is externally fed into the reactor differs from the actual plasma power, i.e. from the power dissipated in the plasma and consumed to dissociate the feed gases.

The components of the diagram shown in Figure 6.5 can be partly calculated from the geometrical dimensions of the reactor and, wherever that is not possible, they can be derived from impedance measurements under vacuum (without plasma discharge). From additional measurements of current, voltage and phase angle, with

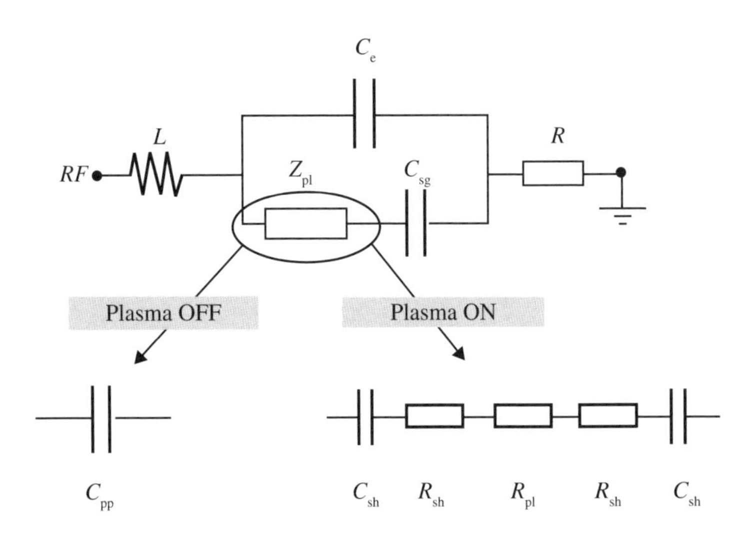


**Fig. 6.5** Equivalent circuit diagrams of a parallel-plate reactor, with plasma switched off or on; the abbreviations for the circuit components are defined in the text (with permission [Ossadnik-1995]).

the plasma switched on, the actual voltage distribution between the plasma sheaths and the plasma bulk, as well as the effective and apparent power, can be determined. Figure 6.6 is a result of such procedures, and serves as an illustration for a typically sharp transition from a capacitive to a resistive plasma regime. The different symbols of the experimental points identify, for different values of externally supplied powers, the flow rates (in sccm) of silane, with or without an additional dilution with non-reactive helium. In this particular experimental series, helium dilution was used in order to address the role of partial silane pressures, without changing the reactive gas species. Note that there is no correlation between the externally applied powers and the plasma regime transition. Instead, it is the power distribution within the plasma and the silane partial pressure that determine this relatively sharp transition. Resistive plasma regimes are attributed to the formation of charged particles trapped in the plasma and, thus, indicate powder formation due to gas-phase polymerization (see also Sect. 6.1.2).

The capacitive and resistive plasma regimes are consistent with earlier experiments [Perrin-1998] in which a sharp increase of the deposition rate as a function of the silane pressure had been observed, provided the externally supplied RF power was sufficiently high. This phenomenological transition (also called $\alpha/\gamma$ transition) from low to high deposition rates has been found to be linked to the onset of powder formation in the plasma, and it also implies a distinct reduction in the quality of the deposited film. It had already been recognized [Perrin-1998] that this more or less drastic

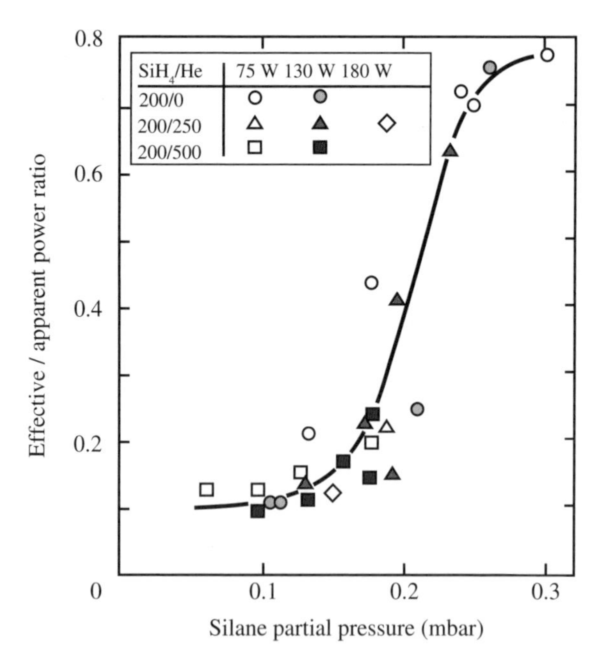


**Fig. 6.6** Ratio between effective and apparent power consumption as a function of the silane partial pressure (with permission [Ossadnik-1995]).

transition had to be related to the RF power dissipation in the plasma discharge, and hence to the electric field distribution in the space between the two electrodes.

### 6.1.2 VHF plasma excitation

In the preceding discussion, the electrical power supplied to the plasma reactor was implicitly assumed to be at a frequency of 13.56 MHz (generally called "RF", radio frequency). This frequency corresponds to the standard frequency reserved by the International Electrotechnical Commission (IEC) for Industrial High-Frequency Applications. It is therefore also the frequency that has traditionally been most widely used. However, benefits in applying higher frequencies extending into the range of 30-300 MHz (generally called "VHF", very high frequencies) have been recognized for quite some time [Curtins-1987a]. As a result, the use of higher excitation frequencies has led to higher deposition rates, without sacrificing film qualities. The underlying reasons, as well as implications, will be outlined in this section.

A parallel plate reactor is typically represented by an electrical circuit diagram as shown in Figure 6.5. The circuit impedances, and also the RF matching networks used externally to adjust for zero reflected power, are frequency-dependent. Experimental and theoretical studies [Beneking-1990, Kroll-1995] on capacitively coupled RF discharges have shown that both the resistive and capacitive contributions to the plasma impedance, i.e. the real and imaginary values Re [$Z_{pl}$] and Im [$Z_{pl}$], decrease with the excitation frequency (see Fig. 6.7).

These values reflect properties of the plasma bulk and the sheaths, respectively. The power spent in the plasma bulk is used for the decomposition of the feed gases. The power spent in the sheaths is essentially considered as a loss; the thickness and voltage drops across the sheaths affect the transport of the ionic species to the electrodes, and in particular to the growing film. These voltage drops depend on the voltage amplitude $V_{pp}$ at the powered electrode, which has also been shown to substantially decrease with the frequency (see Fig. 6.8a). As already noted in Section 6.1.1, the source power differs from the plasma power. It is the latter that actually determines the deposition rate. Therefore, in order to assess the effect of the excitation frequency on the deposition rate, the plasma power at various frequencies must be kept constant, while the source power is adjusted depending on the frequency-dependent circuit components. This condition has been met for the data shown in Figure 6.8(a).

From the frequency dependences of the plasma impedance and electrode voltage (in Figs. 6.7 and 6.8a), the following conclusions with regard to the deposition rate and quality of the deposited films may be drawn:

#### *Plasma bulk*

The decrease in Re [$Z_{pl}$] indicates an increase in the concentration of charge carriers, i.e. of electrons and ionized species. In fact, this can be attributed to an increase in the electron density and a change in the electron energy distribution towards higher electron energies, which is caused by changes in the power coupling mechanism from ohmic to plasma surface heating with increasing frequency [Abdel-Fatah-2003]. As a

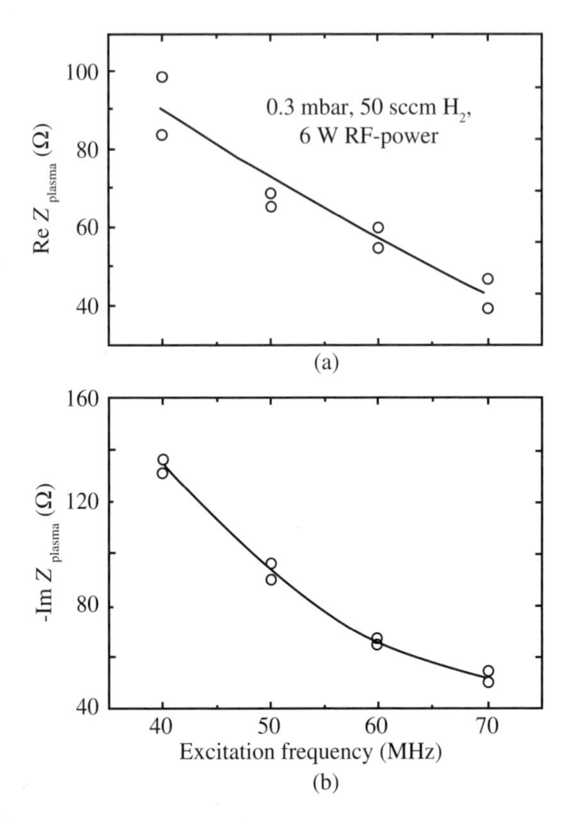


**Fig. 6.7** (a) Real and (b) imaginary part of the plasma impedance as a function of the excitation frequency (with permission [Meier-2004]).

result, a significant enhancement of the dissociation of the feed gases (see Fig. 6.8b), and hence of the deposition rate, for both amorphous and microcrystalline silicon, occurs.

### *Sheaths*

The decrease in Im $[Z_{pl}]$ indicates an increase in the capacitances that represent the sheaths. This translates into thinner sheaths, and thus enhances the mass transport from the plasma to the electrodes, including the growing film on the substrate. The voltage amplitude $V_{pp}$ is related to the time-averaged plasma potential i.e. $V_p = V_{pp} /4$ [Howling-1992], and hence to the voltage drops across the sheaths, as shown in Figure 6.4 (for large-area reactors the self bias may be neglected). The ions reaching the growing film are generated at the plasma potential, and are thus subject to the electric field across the sheath. They attain a maximal ion energy of about $V_{pp}/4$. In accordance with Figure 6.8(a), the ion energies become lower with higher plasma frequencies (see Fig. 6.9), while at the same time higher deposition rates are achieved.

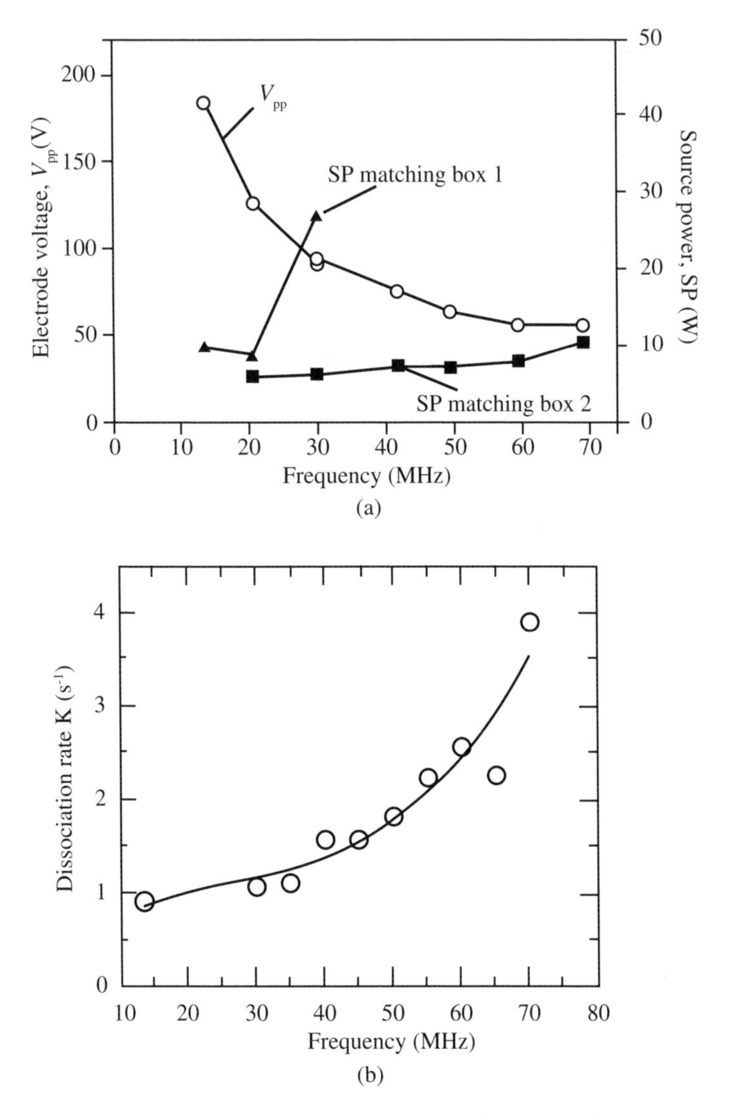


**Fig. 6.8** (a) Source powers and electrode voltages as a function of frequency for constant plasma power (5 W in this particular arrangement). The matching boxes 1 and 2 cover different frequency ranges (with permission [Howling-1992]). (b) Dissociation rate of silane as a function of frequency (with permission [Sansonnens-1998]).

Depending on the ion energy, ion bombardment affects the quality of the growing film in two ways:

- For ion energies larger than the threshold energy for defect generation (approximately 15 eV), which typically applies to reactor operation at the standard RF, ion bombardment creates damage by breaking bonds, and by forming weak bonds and microvoids.

- At higher frequencies, ion energies remain lower, i.e. below the threshold energy for defect generation. These lower ion energies give rise to an increased surface mobility of the deposited species on the growing film, and thus provide for rearrangements of deposited species and more orderly film growth.

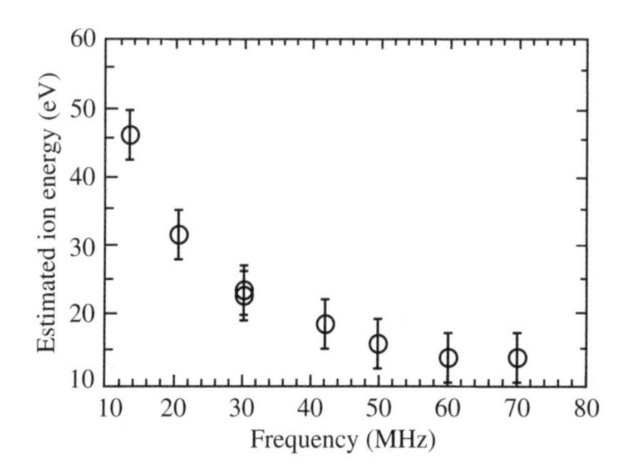


**Fig. 6.9** Ion energy vs. plasma-excitation frequency (with permission [Dutta-1992]).

Obviously, these effects are further dependent on the ion flux, as well as on the composition of the entire mass flow of both ions and neutral radicals (the latter due to concentration gradients). These mass fluxes are generally higher at higher frequencies; this may also be concluded from the increased dissociation rate (see Fig. 6.8b).

The combined effects based on the frequency dependencies for plasma bulk and sheaths result in a substantial increase of the deposition rate, as shown in Figure 6.10, without adversely affecting the quality of the deposited film.

In addition to the effects on the deposition rate, VHF differs from the standard RF excitation at 13.56 MHz in the following two aspects:

*Powder formation*

All attempts to raise the deposition rate by increasing the plasma power are generally limited by the onset of powder formation within the gas phase. It has already been noted in Section 6.1.1 that the $\alpha/\gamma$ transition between plasma regimes corresponds to a transition from a capacitive to a resistive plasma impedance, and is linked to the onset of powder formation. This onset is generally detectable by an abrupt change of the electrical plasma properties, as seen also in the dependence of the electrode voltage as a function of the plasma power for example [Dorier-1992]. Thereby a power threshold is defined, below which the deposition rate can be raised within a powder-free plasma. Higher temperatures, as well as higher plasma excitation frequencies, lead to a rise in this maximal power threshold (see Fig. 6.11). At higher frequencies the electric fields of the plasma discharge are smaller, and also the composition of the radicals may be less conducive to gas-phase polymerization. In effect higher deposition rates at VHF are clearly compatible with powder-free plasmas.

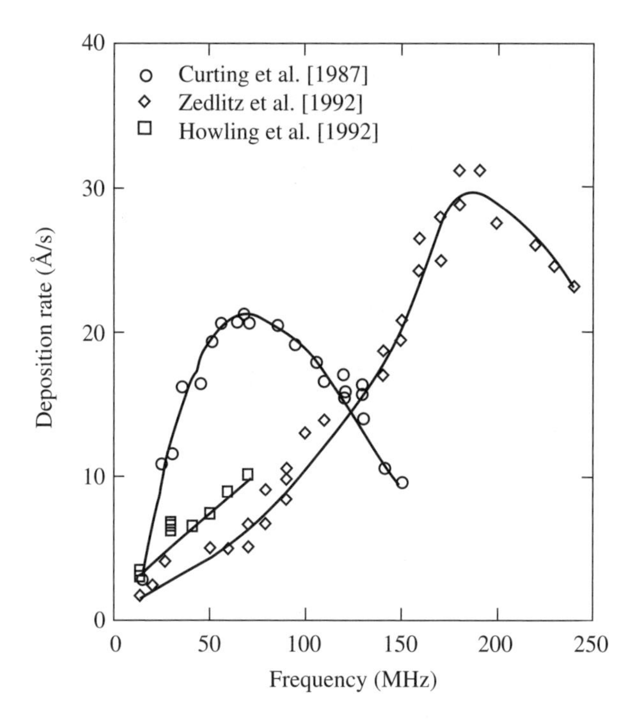


**Fig. 6.10** Deposition rate vs. plasma excitation frequency, as obtained in different reactors (with permission [Meier-2004]); data from independent research groups: [Curtins-1987b, Zedlitz-1992, Howling-1992]. The observed maxima of the deposition rate are correlated with the specific reactor design, and not further explained in this context.

### *Deposition uniformity*

To obtain uniform deposition across large areas, the excitation frequencies should not be exceeded, once the electrode dimensions reach about a quarter of the frequency-equivalent vacuum wavelength (i.e. $\lambda/4 = 5.53$ m at 13.56 MHz, $\lambda/4 = 0.75$ m at 100 MHz). The vacuum wavelength determines the voltage distribution. With the deposition rate varying proportionally to the square of the voltage distribution [Sansonnens-1997], the deposited film thickness varies across the substrate according to the voltage distribution across the substrate.

In a simplified model of a one-dimensional linear electrode (with length coordinate $x$), the voltage distribution (with the phase $\theta$) is

$$U = U_0 \cos(2\pi x/\lambda + \theta), \tag{6.3}$$

and the thickness distribution is

$$d = d_0 \cos^2(2\pi x/\lambda + \theta). \tag{6.4}$$

In the worst case, the voltage maximum occurs at the edge of the electrode ($x = 0$, $\theta = 0$), and the maximum thickness variation is

$$\Delta d/d_0 = 1 - \cos^2(2\pi \ell/\lambda), \tag{6.5}$$

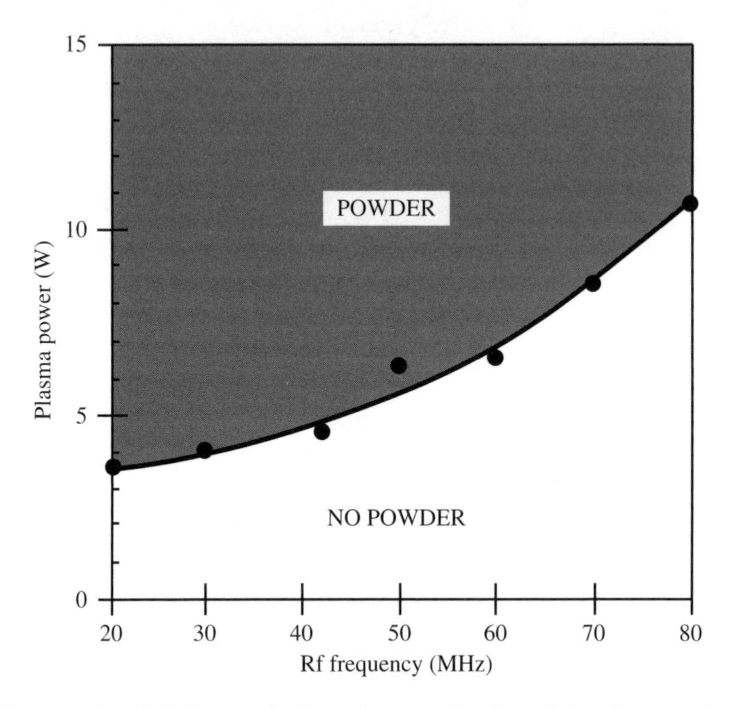


**Fig. 6.11** Power threshold for powder formation as a function of the plasma excitation frequency (with permission [Dorier-1992]).

where $\ell$ is the length of the electrode. For an electrode length of 1 m, the thickness variation obtained with the two excitation frequencies given above (13.56 MHz and 100 MHz) amounts to 8 % and 75 %, respectively, i.e. it is clearly unacceptable in the latter case. A considerable improvement of the thickness distribution can be obtained by changing the position of the voltage connection from the end to the center of the linear electrode. In this case ($x = \ell/2$, $\theta = -\pi\ell/\lambda$), the corresponding thickness variation is

$$\Delta d/d_0 = 1 - \cos^2 (\pi\ell/\lambda), \tag{6.6}$$

i.e. it amounts to only 2 % and 25 %, compared to 8 % and 75 % with edge feeding, respectively.

A two-dimensional case more realistically illustrates the effects of excitation frequency and its feeding location to the electrode system (see Fig. 6.12).

Several approaches aim at combining the higher deposition rates, due to higher plasma excitation frequencies, with acceptable deposition uniformities over large areas:

- For areas up to about 1.5 $m^2$, excitation frequencies up to 40.68 MHz still satisfy uniformity requirements, provided the feed-in point for the voltage is placed at the centre of the electrode.
- So-called "dielectric lenses" [Sansonnens-2006], where the shape of the electrode is modified in such a way as to counteract the non-uniform voltage distribution otherwise to be expected.

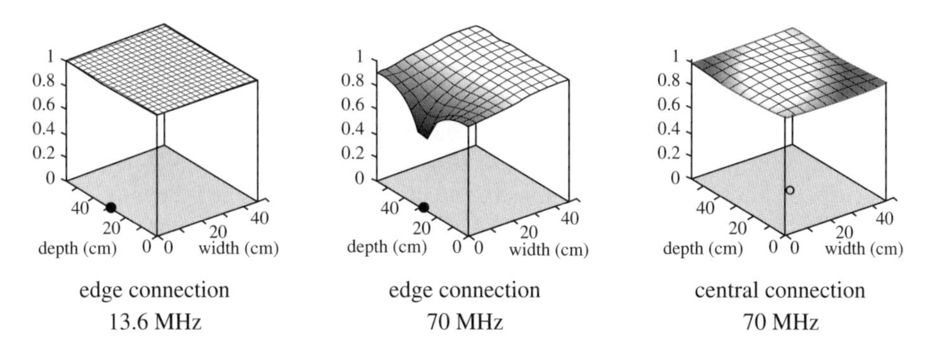


**Fig. 6.12** Interelectrode voltage distribution for the edge and the central RF connection cases. The points indicate the positions of the RF and ground connections (adapted from [Sansonnens-1997a]).

- VHF line sources [Strobel-2008], with deposition onto a moving substrate; here uniformity is required only in the direction of the line-shaped electrode, whereas it is inherently obtained in the direction of the substrate movement. However, in order to achieve the same throughput and film quality, compared to static depositions with areal electrodes, the "static" deposition rate of the line electrode must be larger by the ratio of the substrate length (in the direction of movement) and the width of the line electrode.
- Antenna-type electrodes [Takatsuka-2006, Takagi-2008], in combination with phase-modulated inputs to the elements of the electrode (see Sect. 6.1.10).

The required film properties, described in Section 6.1.3 below, depend on a multitude of interrelated plasma parameters that are further discussed in Section 6.1.4.

## 6.1.3 Device-grade material

PECVD applies to both the doped and intrinsic layers of the solar cell structure. Doped layers are deposited by additionally feeding a doping gas into the plasma discharge: *p*-type layers are obtained by the incorporation of boron, either from diborane ($B_2H_6$) or now, more commonly, from the more stable gas trimethylboron [$B(CH_3)_3$]; *n*-type layers are obtained by the incorporation of phosphorus from phosphine ($PH_3$). Both types of doped layers are relatively thin (around 10 nm and 40 nm, respectively) compared to the intrinsic layers (around 300 nm for a-Si:H, and 1500 nm for μc-Si:H). Particularly for the *p*-layer, which serves also as a "window" layer, certain additional deposition conditions must be taken into account (see Sect. 6.3). Here, we shall focus on the *i*-layers. For present state-of-the-art PV module technology, one employs four different types of *i*-layer material (see also Chap. 5):

- a-Si:H, contained in single-junction and tandem cell structures;
- a-(Si,C):H, contained as top cell material in some tandem cell structures;
- a-(Si,Ge):H, contained in triple-junction cell structures, in combination with a-Si:H as top cell material;
- μc-Si:H, contained as bottom cell material in "micromorph" tandem cell structures.

The majority of thin-film silicon-based module manufacturers rely on a-Si:H *i*-layers, and increasingly on using a combination of a-Si:H and μc-Si:H *i*-layers for micromorph modules. In establishing guidelines for the required film properties, it is imperative to include the aspect of "photostability", i.e. the stability of layer properties during exposure to light (see Sect. 2.2.3). The requirements for PECVD-dependent properties relating directly to the film structure are expressed by the following:

- Urbach energy, $U^\circ < 50$ meV, which is a characteristic energy value defined by the slope of the exponential energy dependence of the optical absorption edge (see Sect. 2.3.3), and is a measure of the atomic order within the amorphous network. Lower values of $U^\circ$ indicate better structural order, as well as fewer voids (see below).
- Microstructure parameter $r < 0.20$; this parameter (see Sect. 2.2.3) is given by the ratio

$$r = I_{2090} / (I_{2000} + I_{2090}) \qquad (6.7)$$

of the integrated intensities $I_{2000}$ and $I_{2090}$ of the infrared (IR) absorption bands at 2000 and 2090 cm$^{-1}$ wavenumbers, representing different bonding forms of hydrogen: as monohydride SiH, and as dihydride $SiH_2$, and as higher silicon hydrides, respectively. The IR absorption peak at 2000 cm$^{-1}$ corresponds to stretching modes of isolated silicon-hydrogen bonds, and the IR absorption peak at 2090 cm$^{-1}$ to stretching modes of clustered $SiH_2$ and also to $SiH_n$-clusters located at the inner surfaces of voids.

The above two properties are directly dependent on the deposition rate (see Sect. 6.1.5), which, in turn, is a function of a combination of deposition parameters (see Sect. 6.1.4). These parameters influence the time frame for the arrangement and accommodation of the deposited species. Additionally, hydrogen, either inherently present according to the chemical reactions specified above or added to the silane fed into the reactor, may also etch the deposited film. Thereby especially weak bonds of $SiH_n$ contained mainly in voids and clusters are being removed.

As an illustration, Figure 6.13 shows, as a general trend, how the deposition rate is limited if device-grade film quality is to be maintained. Here, film quality is expressed in terms of the microstructure parameter, which should be kept at $r < 0.20$. This parameter is also widely used for meeting the required values of the other device-grade properties. Figure 6.13 applies to the deposition of a-Si:H *i*-layers at an excitation frequency of 40.68 MHz in a research reactor with a substrate area of 0.3 m$^2$; the two sets of data refer to electrode spacings of 10 mm and 25 mm, respectively. More or less independently of the various combinations of power, pressure and hydrogen dilution, not further specified here, a smaller electrode spacing favors higher deposition rates for device-grade material.

The following additional film properties also count as criteria for obtaining device-grade material:

- Low defect density, $<1 \times 10^{15}$ cm$^{-3}$; this parameter refers mainly to the dangling bond concentration and is measured as sub-band gap absorption (see Sect. 2.3.5). Its magnitude should not exceed a certain value, namely

$$\alpha(1.2 \text{ eV}) < 3 \text{ cm}^{-1} \text{ for a-Si:H, and } \alpha(0.8 \text{ eV}) < 0.3 \text{ cm}^{-1} \text{ for } \mu\text{c-Si:H.}$$

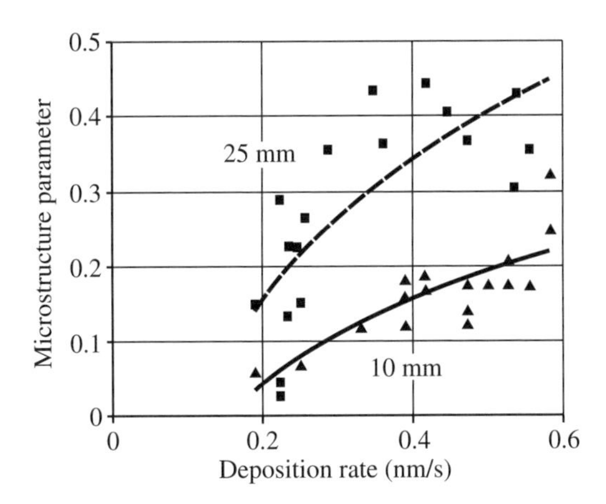


**Fig. 6.13** Microstructure parameter as a function of the deposition rate, for electrode spacings of 10 mm and 25 mm (data from [Frammelsberger-2004]).

- Low oxygen contamination, leakage from the vacuum chamber, desorption from reactor surfaces, and/or leakage from the gas supply may cause the incorporation of oxygen into the deposited film. However, the device-grade properties are only affected above a critical oxygen concentration. For *i*-layers of a-Si:H it should not exceed $2 \times 10^{19}$ cm$^{-3,}$ for *i*-layers of μc-Si:H it should be lower and not exceed about $5 \times 10^{18}$ cm$^{-3}$.
- Low dopant contamination; this is particularly critical for single-chamber processing. In this case, the entire *pin* cell structure is sequentially deposited in a single chamber, and inadequate gas-handling procedures following the deposition of the doped layer may cause traces of the preceding doping gas to be incorporated into the *i*-layer. By so-called boron tailing, the interface formation, such as the *p/i* interface, is affected, which in turn reduces the required electric field in the *i*-layer and, thus, causes the carrier collection to deteriorate (see also Sect. 4.5.17).
- Fill factor *FF* of the resulting solar cell after photostabilization; one should be able to obtain $FF > 0.65$ for a-Si:H, and $FF > 0.7$ for μc-Si:H. Note that it is often more practical to assess the quality of the *i*-layer by the *FF* of a corresponding cell structure, rather than by the above-mentioned criteria for individual *i*-layers.

*Plasma box*

The contamination with impurities is frequently mitigated by a specific reactor design that involves a so-called "plasma box" [Bubenzer-1990]. With this concept, schematically shown in Figure 6.14, the plasma volume is confined in a gas-tight enclosure installed within the vacuum chamber of the reactor. The processing gases are admitted only to this enclosure, i.e. to the plasma box, and are separately pumped by the process pump, whereas outside the plasma box a lower pressure is maintained by a high-vacuum pump. Due to this pressure differential, the plasma volume is not

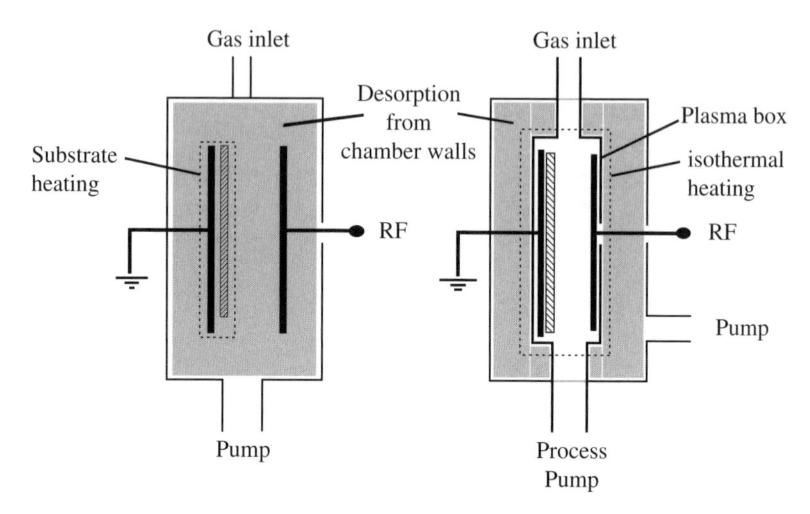


**Fig. 6.14** Schematics of PECVD reactors, without and with plasma box.

contaminated by impurities desorbed from the walls of the vacuum chamber. In addition to the plasma confinement and isolation from desorbed impurities, the plasma box also provides for isothermal conditions and, thus, fosters deposition uniformity.

Since gas residence times (see Sect. 6.1.5) are better defined for the conditions of a plasma box, they can in this case be chosen sufficiently short in order to achieve rapid changes of the processing gases, and to obtain thereby single-chamber processing of *pin* cell structures with sharp interfaces.

In this case only the substrate and the inside of the plasma box are exposed to the plasma. Nevertheless, these limited surface areas still need to be regularly cleaned to remove deposits. This can easily be achieved by dry in-situ plasma etching, while a suitable pressure differential is maintained to protect the outer vacuum chamber from the corrosive reaction with the etching gas. Traditionally these etching gases are fluorocarbons that are considered as greenhouse gases with large global warming potential and long atmospheric life-time. The most hazardous of these, sulfur hexafluoride ($SF_6$), already has been widely replaced with the environmentally preferable substitute nitrogen trifluoride ($NF_3$). More recently, the use of on-site generated fluorine gas ($F_2$) is being investigated, considering the scope of handling issues involved [Schottler-2008]. Also, a mixture of fluorine and inert gases has been tested for plasma etching of silicon nitride and silicon oxide, as well as for amorphous silicon [Oshinowo-2009].

### 6.1.4 Deposition parameters

Besides the film properties discussed above, PECVD must provide

- sufficiently high deposition rates, in order to reach economically viable throughputs and costs (see Sect. 6.1.5).
- film uniformity over large areas (>1 $m^2$) in terms of thickness, composition and structure; the implications for uniform deposition over large areas are discussed in Section 6.1.2.

These properties are interrelated and depend on the PECVD parameters, many of which are also interdependent, and thus cannot be freely chosen. They are governed by the:

- reactor geometry, including
  - gas flow arrangements (determined by the gas inlets and pumping ports);
  - substrate size;
  - electrode spacing;
  - plasma volume (basically determined by the previous two);
- temperatures, in particular the substrate temperature;
- gas flow and partial pressures;
- electrical power input; here, two parameters are mainly involved:
  - the excitation frequency, one typically uses the "standard" industrial radio frequency (RF) of 13.56 MHz, or, more recently, also higher frequencies, e.g., 27.12 MHz or 40.68 MHz. So-called very high frequencies (VHF, see Sect. 6.1.2), ranging from 30 MHz to 300 MHz, were introduced in 1987 [Curtins-1987a]. Thereby higher deposition rates can be achieved, while maintaining similar electronic film properties as with conventional RF;
  - the power consumed within the plasma volume.

This enumeration of plasma parameters indicates the complexities involved.

### 6.1.5 Deposition rate

Among the deposition parameters, the deposition rate (deposited film thickness per unit of time) is of central importance, since it directly affects the film quality, but it also affects the throughput. Higher deposition rates increase the throughput, and this, in turn, facilitates lower manufacturing costs [Shah-2006]. In a module production line, the PECVD process is typically the throughput-limiting process, i.e. the process requiring the longest duration per processed substrate area. Besides procedural time requirements (for loading, unloading, pump- down, reactor venting), the actual duration of the silicon deposition, which is given by the required film thickness divided by the deposition rate, predominantly determines the PECVD throughput. Comparing a-Si:H and μc-Si:H, the quest for high deposition rates applies even more pressingly to μc-Si:H because of its lower absorption coefficient and hence thicker *i*-layer requirements, and, depending on the applicable deposition rate, longer deposition durations especially for high values of the hydrogen dilution ratio $R$ (Eq. 6.9).

Generally, the deposition rate is determined by the rate of $SiH_4$ decomposition in the plasma, and thus also by the volumetric $SiH_4$ flow rate supplied to the discharge chamber, which is measured in standard cubic-centimeters per minute (sccm, at 0 °C and 1 bar = $1.0 \times 10^5$ Pa = 750 Torr). The conversion from volumetric to mass flow rate is: for $SiH_4$, 1 sccm = $8.99 \times 10^{-5}$ g/min, and for $H_2$, 1 sccm = $1.43 \times 10^{-3}$ g/min. The masses of the gases that are consumed to deposit the cell layer structure follow from the mass flow rates and the corresponding durations of the deposition. These quantities then determine the cost of the feed gases per module area. As a rule-of-thumb, the feed gas prices contribute about 10 % to the module manufacturing cost per square meter. One may note that the various schemes with hydrogen dilution discussed below may significantly impact the overall cost of the gases.

The ratio between the deposition rate and the gas flow rate physically determines the silane utilization for the a-Si:H deposition process. However, for practical considerations, the silane utilization is preferably expressed as the ratio between the amount of silicon deposited onto the substrate and the amount of silicon originally contained in the quantity of silane that was consumed during the deposition. This latter more practical definition of gas utilization takes into account that not only the substrate but also the electrode and other surfaces of the reactor exposed to the plasma are unavoidably coated with silicon. Silane utilizations tend to be slightly higher for VHF depositions, as compared to RF, but they typically do not exceed 30 %. More accurate values can only be determined for specific deposition conditions.

In order to obtain high deposition rates for device-grade material, and ultimately also uniformity over large areas, the following parameters need to be taken into account. In doing so, their mutual dependencies must also be considered.

### *Electrical power*

The rate of silane decomposition and usually also the deposition rate are increased by increasing the power coupled into the discharge. Depending on other plasma parameters, such as partial pressures, gas flow pattern, hydrogen dilution (see below) and electrode spacing, this power is used to generate reactive radicals that may interact with the substrate and plasma-exposed surfaces to form deposited silicon, and/or to form powder (gas-phase polymerization). Note that this power differs from the power externally fed into the reactor, due to the voltage and phase relations deriving from the electrical circuit diagram that applies to the given reactor (see Sect. 6.1.1). Besides raising the deposition rate, increasing the power may affect the film quality by

- higher ion bombardment in terms of both higher ion energy and ion concentration; these effects may be mitigated by
  - an increase in the plasma excitation frequency (see Sect. 6.1.2), and
  - an increase in the gas pressure, leading to a loss of ions due to more collisons;
- gas-phase polymerization, which results in powder formation and clusters being incorporated into the film, as well as in powdery deposits on reactor surfaces (which, in turn, can lead to pinholes and shunts within the completed device).

### *Gas residence time*

The gas residence time $t_{res}$ is the time the gas molecules spend in the plasma volume before undergoing chemical reactions or being pumped away. It is defined as the ratio

$$t_{res} = V\,p/\Phi, \tag{6.8}$$

where $V$ is the plasma volume (basically given by the substrate area and the electrode spacing), $p$ the partial gas pressure, and $\Phi$ the gas flow rate. The gas residence time needs to be considered in relation to the chemical decomposition times $t_{decomp}$. Whereas the gas residence time depends on the chosen deposition conditions and can, within limits, be varied, the decomposition time is given by the specific chemical reactions.

For favorable chemical reactions leading to film growth, the gas residence time should match or even exceed the chemical decomposition time, i.e. $t_{res} \geq t_{decomp}$. In this case, $t_{res}$ may be reduced to the limit of $t_{decomp}$ by increasing the gas flow rate $\Phi$ (although such an increase may entail a lower silane utilization and hence higher cost of materials). Thereby more molecules, still with sufficient gas residence time, are supplied for decomposition; hence one will also have a higher deposition rate. In order to avoid the reduction of $t_{res}$ to values below $t_{decomp}$, the gas pressure $p$ may additionally have to be raised, in conjunction with the gas flow rate rate $\Phi$, according to Equation 6.8.

In the case of deposition conditions with gas residence times shorter than the decomposition time, i.e. $t_{res} \leq t_{decomp}$, an increase of $t_{res}$ to match $t_{decomp}$ may be desirable. This can be easily achieved by increasing the pressure. As a consequence, the probability of more molecules being decomposed rises, and basically leads to a higher deposition rate. However, higher pressures imply higher collision rates, which may also result in gas phase reactions, and hence in powder formation. Thereby fewer reacted species reach the growing film, and this, in turn, tends to lower the deposition rate. In the case of relatively small electrode spacings, the probability of chemical decomposition at the growing film surface, rather than in the plasma volume, may be higher and, thus, the deposition rate will be less adversely affected by powder formation.

The variations described above apply to a given reactor with a defined plasma volume. Additionally, they are influenced by the gas flow geometry, which depends on the construction of the gas-feeding system and the relative location of the pumping arrangement. In this context issues of gas depletion and deposition uniformity must be considered.

Obviously, the plasma volume $V$, approximately determined by the substrate area and the electrode spacing, decisively enters into the gas residence time (Eq. 6.8). This aspect needs to be taken into account when upscaling from small-area research reactors to large-area production reactors (see Sect. 6.1.7).

*Hydrogen dilution*

The deposition of silicon, as indicated in the introduction of Section 6.1, is based on the decomposition of silane. The decomposition of pure silane leads to the generation of a certain amount of atomic hydrogen. If hydrogen is additionally mixed to the silane fed into the reactor, a larger amount of atomic hydrogen will be generated within the plasma. As a consequence, the gas phase and plasma/surface interactions will lead to different reactive radical distributions and fluxes to the growing film surface, and the resulting quality of the deposited film may be substantially influenced. The amount of dilution is defined by the ratio $R$ of the hydrogen and silane flow rates, $[H_2]$ and $[SiH_4]$, i.e. by

$$R = [H_2]/[SiH_4]. \qquad (6.9)$$

High dilution ratios will give rise to a high flux of atomic hydrogen to the surface. This leads to an etching of weak bonds, and also to an increased passivation of dangling bonds. Thereby, the structure of the deposited film becomes more ordered, and tends to incorporate extremely small crystallites. If the dilution is increased further, a substantial volume fraction of microcrystallites will be incorporated. At even higher dilutions, full microcrystallinity develops (see further details in Sect. 6.1.6). However, the more ordered network formation, involving both etching and reconstruction of the growing film, implies a slower film growth and hence a lower deposition rate (see Fig. 6.15b) if the plasma power is kept constant.

Besides lowering the deposition rate, hydrogen dilution, within the range for amorphous film growth, affects the film quality in two ways: it increases the bandgap of a-Si:H, and it leads to a reduction in the light-induced degradation. Both effects are related to enhanced atomic order and more stable hydrogen bonds within the atomic network. These structural improvements may be widely attributed to the increase of atomic hydrogen at the growing film surface with increasing hydrogen dilution. The effects of hydrogen dilution are extensively described in Section 2.6.

As an example, the variation of the optical bandgap with the hydrogen dilution ratio is shown in Figure 6.15(a). The increase of the bandgap up to a maximal value, followed by a decrease, indicates a transition from more and more order within the amorphous network to the incorporation of an increasing fraction of crystallites. For these, the smaller bandgap of crystalline silicon causes the observed decrease in the bandgap when the dilution ratio is increased further. The results

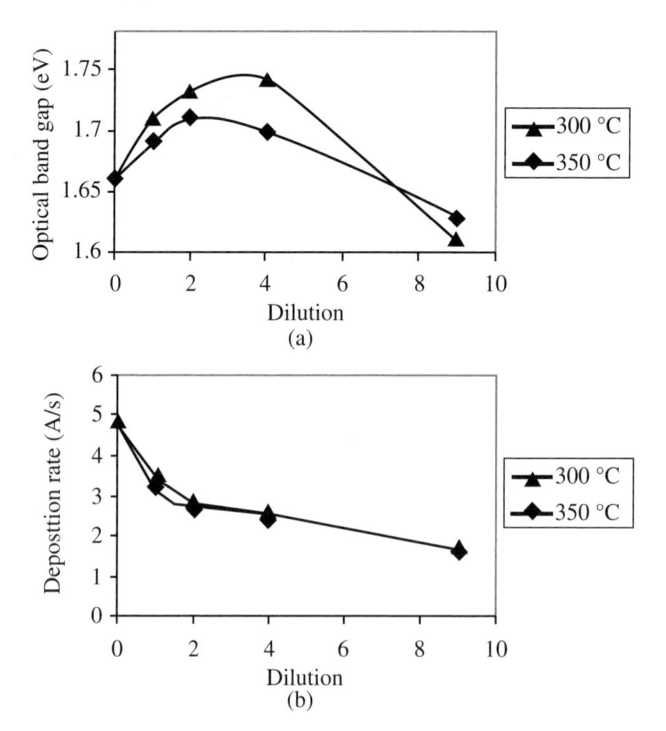


**Fig. 6.15** (a) Optical band gap, and (b) deposition rate as a function of the hydrogen dilution (with permission [Daudrix-2000]).

shown in Figure 6.15 were obtained with VHF depositions at 70 MHz (for further experimental details see [Daudrix-2000]). For both RF at 13.56 MHz and VHF at 70 MHz, the hydrogen content ($C$H[at.%]) increases with the dilution ratio $R$. This increase is much more pronounced for VHF than for RF [Platz-1998]. Therefore, the transition to smaller bandgaps, shown in Figure 6.15 to occur at $R \approx 4$, will occur at higher dilution ratios for RF. This is consistent with the general correlation of the bandgap rising with the hydrogen content, as shown in Figure 2.30 (for more information about the influence of hydrogen dilution on the bandgap of a-Si:H layers, see Sects. 2.2.3; 2.6.3; 2.6.8).

The essential reason for the increase of the bandgap with hydrogen dilution is enhanced order in the atomic network. The Urbach energy $E^0$ (see Sects. 2.2.1; 2.3.3; 6.1.3) is generally determined by the degree of disorder. It has been postulated [Cody-1981] that there also exists a link between bandgap and disorder, such that the bandgap increases when the atomic network becomes more ordered. The incorporation of hydrogen into the atomic network results in less disorder; this leads to a lower value of the Urbach energy $E^0$, and also to a larger bandgap. The link between disorder and bandgap, as established by [Cody-1981], is based on the effects of both thermal and structural disorder on the electronic structure of a-Si:H. Accordingly, it takes into account structural effects that are caused not only by the various types of atomic binding and coordination, but also by thermal disorder due to "lattice vibrations", giving rise to mean-square atomic displacements. The correlation thus derived applies to the bandgap dependence on both temperature and disorder; it covers bandgaps in the range from 1.6 eV to 1.8 eV, with Urbach energies in the range from 80 meV to 40 meV, respectively. Note that the described link between the bandgap and temperature refers to the ambient temperature, not to be confused with the temperature of the layer deposition.

The increase in the bandgap with enhanced film quality implies a trade-off for practical applications: a higher bandgap reduces optical absorption and hence also the carrier generation. This may be compensated by either increasing the film thickness or by increasing the deposition temperature to reduce again the bandgap. If the higher bandgap is due to less disorder, an increase in thickness, usually entailing higher light-induced degradation, may in this case not be affected (see below).

As mentioned above, hydrogen dilution also mitigates the light-induced degradation (Staebler-Wronski effect), due to more stable bonding configurations (see Sects. 2.2.3 and 2.6.10). Therefore, increasing the layer thickness to enhance the carrier generation is, in this case, acceptable. In fact, moderate hydrogen dilution is generally applied by most module manufacturers. Specifically, in conventional a-Si:H/a-Si:H tandem cell structures, it is beneficial to use hydrogen dilution for the top sub-cell, so as to allow for an increase in its thickness. Otherwise, the condition for current-matching might require the top sub-cell to be too thin, which could easily lead to shunting (see Sect. 5.4).

Hydrogen dilution also prevents gas-phase polymerization (powder formation). Aiming at a high silane utilization implies choosing a long residence time for silane, and hence a low silane flow rate. In this case the plasma power may suffice to widely dissociate the silane into reactive radicals (case of so-called silane depletion).

### 6.1.6 Deposition regimes for a-Si:H and μc-Si:H

So far, qualitative correlations of PECVD parameters and film properties have been described. In this section, guidelines for the deposition conditions leading to microcrystalline silicon μc-Si:H will be presented. The "parameter space" for PECVD basically offers the possibility of depositing a wide range of Si-based film material, ranging from amorphous to microcrystalline structures of different quality. This is schematically illustrated in Figure 6.16.

The invention of the "micromorph" tandem cell by the Neuchâtel group [Meier-1994] has opened up a novel way of achieving significantly higher stabilized efficiencies, as compared to conventional structures, based solely on amorphous materials. This work [Meier-1994] was done with a VHF PECVD system; in fact, it turned out that VHF deposition was specially suitable for the fabrication of device-grade μc-Si:H layers and high-quality cells. However, the VHF technology originally employed is not applicable in a direct straightforward way to economically viable large-area production (see Sect. 6.1.10). Therefore, lower plasma frequencies ranging from 13.56 MHz to 40.68 MHz have been explored for deposition of the microcrystalline material to be used in the bottom cells of micromorph tandem cells. As indicated in Figure 6.16, the deposition of microcrystalline material is favored by a high hydrogen dilution $R = [H_2] / [SiH_4]$, i.e. by low silane concentrations $SC = [SiH_4] / ([H_2] + [SiH_4])$. This can be explained by simultaneous etching of poorly coordinated bonds during the deposition, a process that is favored by high hydrogen dilution ratios. In order to compensate for the low deposition rates due to simultaneous etching, higher plasma powers then need to be used.

The plasma deposition conditions for microcrystalline growth have been discussed in more detail in a relatively simple plasma chemistry model [Howling-2000]. Accordingly, a ratio $Q$ between deposition and etch rates is determined, namely: $Q < 0.5$ for a-Si:H deposition, $0.5 < Q < 1$ for μc-Si:H deposition, and $Q > 1$ for etching. This ratio $Q$ depends on three dimensionless parameters:

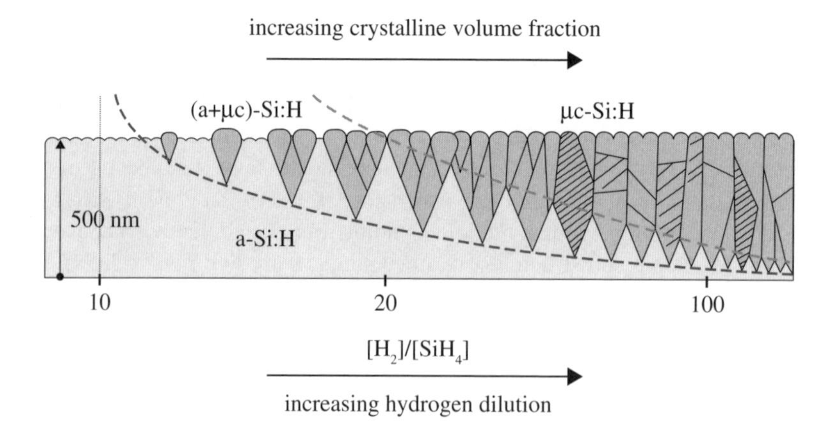


**Fig. 6.16** Range of film structures (schematic), obtained with different PECVD parameters (with permission [Collins-2002]); the dashed lines indicate the a → (a + μc) and (a + μc) → μc transitions, respectively.

- the hydrogen dilution ratio $R = [H_2] / [SiH_4]$;
- the ratio between the silane dissociation rate and the pumping speed, which equals the ratio $t_{res} / t_{decomp}$ of gas residence time over decomposition time (see Sect. 6.1.5). Since for RF frequencies at 13.56 MHz the silane dissociation rates are much smaller than for VHF (i.e. $t_{decomp}$ is larger), the pumping speed must also be chosen correspondingly smaller. This leads to longer residence times which are achievable in a given reactor with higher pressures at a given gas flow (see Eq. 6.8);
- the ratio between the dissociation rate of $SiH_4$ (for Si deposition) and the dissociation rate of $H_2$ into atomic hydrogen (for Si etching).

The ratio $Q$ is not explicitly derived here; it depends on the three ratios given above. Specifically for $Q = 1$, when deposition and etching are balanced, the hydrogen dilution ratio assumes the value $R = 50$. This value, being the result of a simplified model [Howling-2000], cannot be applied generally. Nevertheless, it represents the order of magnitude of the hydrogen dilutions that are required for microcrystalline film growth in the low-pressure plasma regimes, as opposed to those required for μc-Si:H film growth in the high-pressure depletion regime, mentioned further below. The plasma chemistry model outlined here has subsequently been further expanded by including additional plasma and surface reactions, such as secondary reactions between radicals, and surface diffusion and hydrogen surface recombination, respectively [Strahm-2007a].

Microcrystalline growth is not entirely determined by the above three parameters, but also by the actual gas species present in the plasma, by the plasma power density, and by the reactor geometry. Furthermore, high power and high pressure may pose limits due to powder formation. These considerations demonstrate the complexities of PECVD that arise from the various interdependencies of the plasma parameters involved.

In order to find out whether the chosen deposition conditions have yielded an amorphous or microcrystalline layer, one of the three following methods may be used:

- Raman spectroscopy; this is by far the most reliable and informative method from which the degree of crystallinity may be determined (see Sect. 3.2.1, as well as Figure 6.17 and related text below).
- Evaluation of the absorption coefficient of an individual layer (see Sect. 2.3.4); specifically, the parameters $E^{03}$ or $E^{04}$ are a measure of the optical gap, since low values of $E^{03}$ or $E^{04}$ indicate microcrystallinity.
- Evaluation of the Spectral Response (SR)/External Quantum Efficiency (EQE) for completed *pin* (or *nip*) solar cells (see Sect. 4.5.17); if the SR/EQE plot extends well over 800 nm, then the *i*-layer of the completed solar cell is microcrystalline.

Figure 6.17 shows typical Raman spectra for crystalline, microcrystalline, mixed amorphous/microcrystalline, and amorphous films. In particular, the spectra representing mixed phases may be fitted by three Gaussian peaks centered at 480 cm$^{-1}$ for the amorphous phase, around 500 cm$^{-1}$ for a defective crystalline phase associated with grain boundaries, and 520 cm$^{-1}$ for the crystalline phase. The "degree of crystallinity" can be defined by the intensity ratio of these peaks, i.e. by the ratio

$$X_C^{RS} = I_c / (I_c + I_a), \tag{6.10}$$

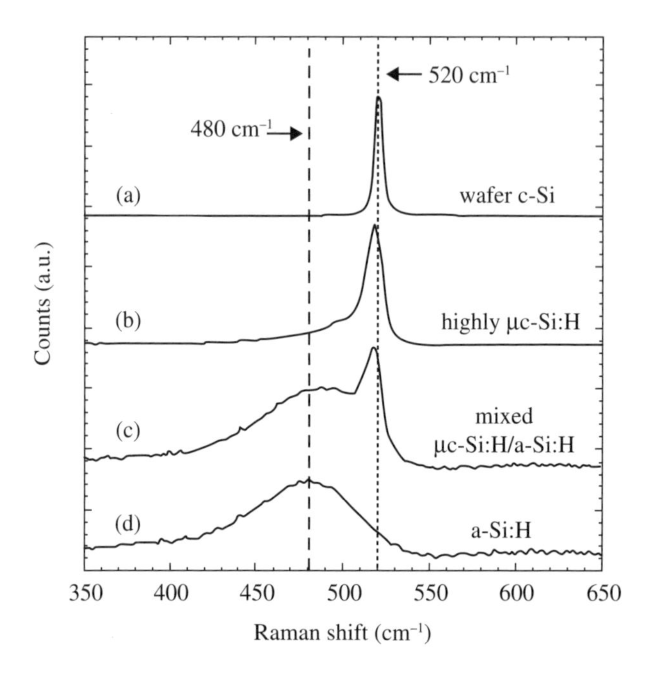


**Fig. 6.17** Typical Raman spectra for crystalline, microcrystalline, mixed amorphous/microcrystalline, and amorphous silicon (with permission [Droz-2003]).

where the $I_c$ is the total area under the Gaussians associated with the crystalline phase (at 520 $cm^{-1}$ and 500 $cm^{-1}$), and $I_a$ is the area under the Gaussian for the amorphous phase (at 480 $cm^{-1}$). This ratio $X_C^{RS}$ is also called the "Raman crystallinity"; it is slightly different from the actual crystalline volume fraction [Droz-2004].

It is of practical relevance that *pin* solar cells with *i*-layers of a mixed amorphous/microcrystalline phase exhibit maximal efficiencies (arising from higher short-circuit currents and fill factors) only within a rather narrow range.

The shift of the transition to a lower silane concentration as a consequence of a higher pressure [compare parts (a) and (b) of Fig. 6.18] may be explained by considering the parameter ratios described above: Assuming the same *Q*-value at the amorphous/microcrystalline transition, a higher pressure (increased here from 5 Torr to 10 Torr, i.e. from 6.7 mb to 13.3 mb) implies a longer residence time for $SiH_4$ and, thus, a lower gas flow rate, i.e. a lower pumping speed. In order to keep the ratio of the $SiH_4$ dissociation and the pumping speed constant, the silane concentration $SC = SiH_4/(SiH_4 + H_2)$ needs to be reduced (from 2.3 % to 1.5 %). In this way sufficient etching is maintained for microcrystalline film growth.

From Figure 6.18 it may be concluded that high efficiencies for microcrystalline cells are obtained only within a narrow range of deposition conditions (here the silane concentration) close to the microcrystalline/amorphous transition. This is further illustrated in Figure 6.19, where the characteristic solar cell parameters (taken from $I(V)$ data) are shown as a function of both the silane concentration and the deposition pressure.

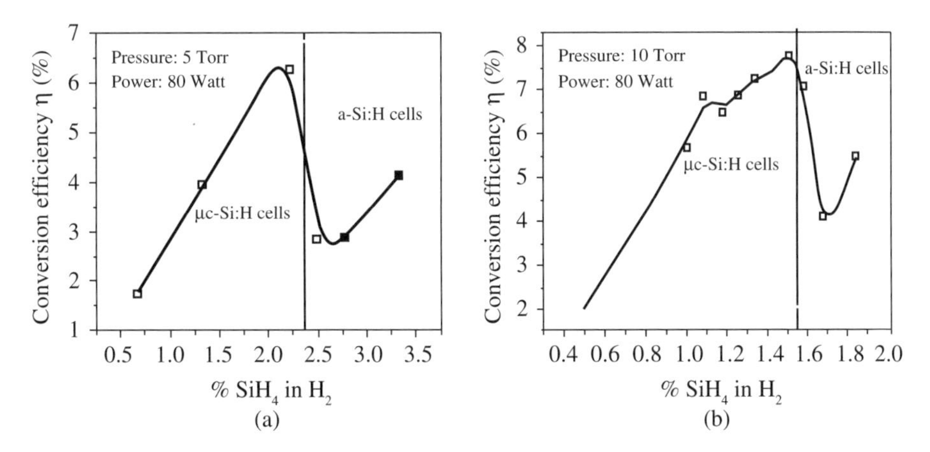


**Fig. 6.18** Cell efficiencies of *pin* cells, deposited at constant plasma power, as a function of the silane conentration of the gas mixture at a total pressure of (a) 5 Torr and (b) 10 Torr (with permission ([Amanatides-2005]).

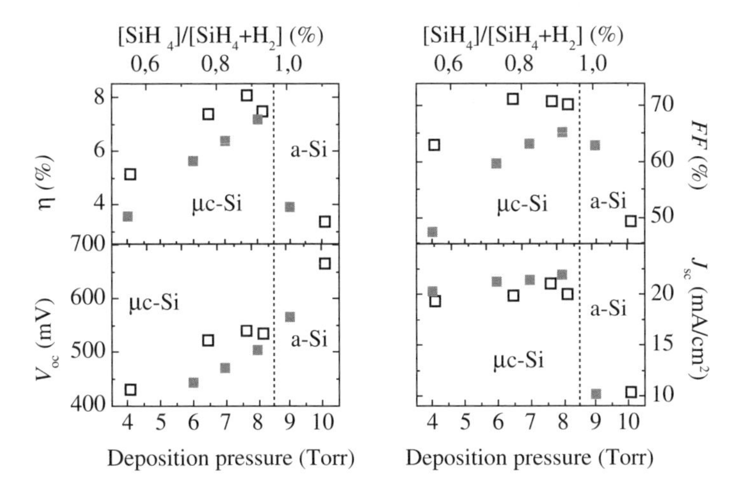


**Fig. 6.19** Efficiencies, fill factors, open-circuit voltages, and short-circuit currents of *pin* cells for various deposition conditions of their *i*-layers (with permission [Roschek-2002]): as a function of the silane concentrations, at a constant deposition pressure of 8 Torr (10.6 mb) and plasma power of 0.3 W/cm$^2$ (open squares), and as a function of the deposition pressures (1 Torr = 1.3 mb), at a constant silane concentration of 1 % and plasma power of 0.4 W/cm$^2$ (filled squares).

Furthermore, as already indicated in Figures 6.18 and 6.19, the best efficiencies for microcrystalline cells are obtained in a relatively narrow range of deposition conditions close to the transition to amorphous growth. In this range the films are composed of crystallites embedded in an amorphous matrix, exhibiting a Raman crystallinity of the order of about 50 %. For such films, the defect concentration, measured by subbandgap absorption, reaches a minimal value as a function of the crystallinity across the amorphous/microcrystalline transition. This is shown in Figure 6.20 for values before and after light-soaking. One may recall that the first μc-Si:H cells were reported to be entirely free of light-induced degradation [Meier-1994]. However, these were probably cells having a relatively high value of Raman crystallinity, whereas cells deposited with intermediate crystallinity close to the transition to amorphous structure are subject to moderate light degradation. These cells with intermediate crystallinity still have higher stabilized efficiencies after light-soaking than the non-degrading efficiencies of cells with higher crystallinity. The efficiency degradation observed for such "optimal" cells with intermediate crystallinity typically amounts to about 10 % for cells with an *i*-layer thickness of 2 $\mu$m. It is, thus, distinctly lower than for typical cells with amorphous *i*-layers. Since in practice microcrystalline cells are (almost always) used as bottom cells in tandem configurations, and are thus exposed only to the long-wavelength portion of the illumination, their light-induced degradation is even further reduced.

Besides the approach for depositing μc-Si:H based on high hydrogen dilution at relatively low pressures, an alternative method has emerged that relies on relatively high pressures of almost undiluted silane (so-called high-pressure depletion (HPD) method [Matsuda-2004]). The underlying idea for HPD is to maintain a high concentration of atomic hydrogen relative to the concentration of silane radicals. Such a relatively high concentration is needed at the growing film surface to etch dangling bonds. The atomic hydrogen may otherwise be scavenged by reactions with excessive silane molecules prior to reaching the film surface. This loss of atomic hydrogen, according to the reaction $SiH_4 + H \rightarrow SiH_3 + H_2$, is prevented by supplying just sufficient $SiH_4$, for high Si growth rates but without leaving excessive $SiH_4$ to support the scavenging reaction. This growth condition is called "silane depletion"; it is defined by the fraction

$$D = (c - c_p)/c, \tag{6.11}$$

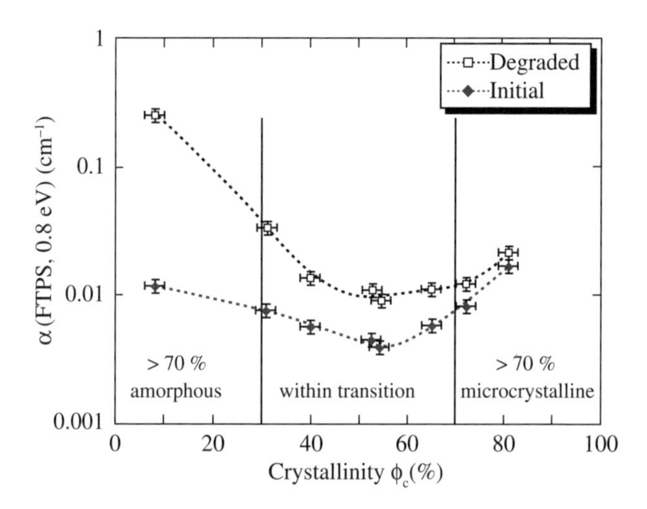


**Fig. 6.20** Defect-related sub-band gap absorption as a function of crystallinity before and after light-induced degradation (Copyright © [2005] IEEE, with permission[Meillaud-2005]).

where $c$ is the silane concentration supplied to the reactor, and $c_p$ is the remaining silane concentration that is not consumed by the plasma reactions. The concentrations are proportional to the partial pressures, i.e. $c = p_0/p$ and $c_p = p_p/p$, where $p_p$ and $p_0$ are the partial pressures of silane with and without the plasma, respectively, and $p$ is the total pressure maintained during the deposition [Strahm-2007b]. Accordingly, the silane depletion may also be expressed by

$$D = (p_0 - p_p)/p_0. \quad (6.12)$$

A highly depleted plasma consists mainly of hydrogen, and thus behaves similarly to a low depletion plasma with high hydrogen dilution. To reach high silane depletion, the dissociation rate and the gas residence time have to be high [Strahm-2007b]. Accordingly, the plasma power density, as well as the pressure, must be chosen high, but not too high (so as to avoid powder formation). The required flow rate of silane is determined by the deposition rate. The specific qualities of the deposited film are dependent on the specific interactions of the chosen plasma parameters that apply to any given reactor arrangement. Pressure and electrode spacing influence the mass transport to the growing film surface.

Generally, a transition between amorphous and microcrystalline growth can be achieved by a combination of different variations of plasma power, silane concentration, pressure, excitation frequency, and electrode spacing. Their interplay and consequences for deposition rates and uniformities, and film properties are still the subject of ongoing research and development (see, e.g., [Parascandolo-2009]).

From the relatively narrow window of deposition parameters for optimum microcrystalline cell performance, as exemplified in Figures 6.18-6.20, one must conclude that the deposition process is rather sensitive to any variation or drift of the processing parameters. This situation has to be taken into account and represents an added challenge to the task of upscaling (see Sect. 6.1.7), for which good performance must be reached with sufficient area uniformity, in combination with high reproducibility (yield) and high throughput.

## 6.1.7 Upscaling

The experience in PECVD originates in R&D, and is obtained with small-area laboratory equipment. At this stage, the main efforts focus on establishing links between the plasma conditions and the resulting film properties; they also aim at an understanding of the underlying mechanisms. Additional considerations must exploit the experience obtained with small-area equipment, but include more practical aspects in view of commercial applications. These applications basically rely on an upscaling of the deposition equipment for large areas. Addressing this task is not just an engineering problem to be solved simply by scaling and by providing the same area-related deposition conditions. In practice, this may inherently be impossible (e.g., in the case of the residence time, see below). Fundamental differences between small- and large-area deposition refer to two main factors, stemming from the reactor-specific design, namely:

- the electrical circuit characteristics; these are affected by the operating frequency, and also by reactor dimensions, particularly by the electrode area, by

the electrode spacing, by the geometry and location of the electrical power feeding, and by the grounding;
- the gas flow dynamics; these are determined by the substrate size, the arrangements of the gas inlet and pump exit, and the electrode spacing.

The operating frequency is a unique property that, independently of other parameters, directly affects upscaling in terms of large-area deposition uniformity. Compared to conventional RF-PECVD at 13.56 MHz, VHF-PECVD (at 30 MHz-120 MHz) has indeed been demonstrated to yield high-quality a-Si:H and μc-Si:H films at higher deposition rates [Meillaud-2009]. This is due to higher gas decomposition rates and "softer" ion bombardment, as discussed in Section 6.1.2. In order to increase the deposition rate with conventional RF, the power input to the reactor would have to be increased, and thereby the ion energies and hence the damage by ion bombardment are also increased. However, as an alternative to raising the plasma excitation frequency to the VHF range, one may increase the pressure in the reactor in order to reduce the effect of ion bombardment [High-Pressure Depletion method (HPD), see Sect. 6.1.6]. Some large-area deposition systems, like that of Mitsubishi Heavy Industries, Ltd. (MHI), Japan, are operated by utilizing a combination of both methods, VHF and HPD (see also Sect. 6.1.10).

The deposition parameters cannot independently be scaled to large-area depositions and still guarantee the same film properties as obtained in smaller research-type reactors. Although the film properties must ultimately be optimized empirically for any given reactor, some parameter changes as a function of reactor size can be anticipated and, accordingly, be adapted. In order to preserve the film properties with upscaling from small-area to large-area depositions, one should, as a general rule, strive to keep the following quantities constant:

- The partial pressures of the gases.
- The gas residence times (see also Sect. 6.1.5, Eq. 6.8); the gas residence times, most importantly for silane and hydrogen, determine the balance of the reaction equilibria. Thus an individual residence time, pumping speed and partial pressure apply also to hydrogen [even without being separately fed into the reactor, hydrogen is generated from the decomposition of silane (see Sect. 6.1)]. There may be an inherent practical limitation to upscaling given by the pumps: the requirement of keeping the gas residence time and the pressure constant with increasing plasma volume can only be satisfied, up to the point when the pumps can no longer handle the required correspondingly higher gas flow.
- The composition of feed gases (degree of silane depletion or hydrogen dilution), aiming for a high gas utilization; the design of the gas flow geometry, such as, for example, shower head or laminar flow, is of prime importance in this context.
- The ratio of plasma power and gas flow; this ratio determines the silane usage for the various reaction products and, thus, also the deposition rate.
- The areal power density ($W/cm^2$), provided this power is uniformly distributed over the deposition area. Note that this plasma power differs from the power fed into the reactor at the input to the matching network outside of the reactor, as explained in Section 6.1.1. Part of the power is lost within the matching

network, and another part is lost within the sheaths, so that only a fraction of the power originally fed-in at the input of the matching network actually reaches the bulk of the plasma (plasma volume) itself. There exist, however, special techniques [Howling-2005] to evaluate that part of the electrical power which is actually supplied to the plasma volume.
- The ratio of effective and apparent power consumption within the plasma; this ratio basically refers to the power distribution between the plasma sheaths and the plasma volume. The "effective" power is the power dissipated within the "bulk" (volume) of the plasma; the "apparent" power is the power dissipated in the whole volume, i.e. in bulk plus sheaths. This ratio also serves to distinguish between plasma regimes; it must be kept below a certain threshold to avoid powder formation (see Sect. 6.1.1).

As noted, most of the parameters listed above cannot be kept constant with upscaling to large deposition areas. In the end, gas flow, pressure and electrical power must be chosen so as to reach the required uniformity with an acceptable deposition rate.

### 6.1.8 Deposition systems

The incorporation of the PECVD process into a thin-film silicon-based module production line is determined by the types of modules that are planned to be produced. The module type is, in its turn, linked to the applications that are envisaged, and also to cost factors and market considerations. Depending on the kind of substrate material used, there are basically two categories of modules and corresponding deposition systems, namely for

- glass substrates (rigid substrates); these are precoated with TCO and deposited with one or more layer sequences of *pin* layers; here, different arrangements of the plasma reactors are possible (see below);
- flexible substrates, based on metal [Izu-2003, Hamers-2008] or plastic foils [Ballif-2007] these concepts involve roll-to-roll deposition; they have individually characteristic features that are unique and will be briefly discussed in Section 6.1.9.

The vast majority of PECVD-based production lines, both those already in operation and those now being announced, use large-area glass/TCO substrates. Various layouts and design features of the PECVD equipment have to be considered in order to achieve high throughput and high production yield and, thus, to ultimately attain competitive manufacturing costs.

Two basic approaches are being pursued. Common to most approaches are load-lock chambers to handle the entry and exit of the substrates and, at the same time, to provide preheating and cool-down:

- Single-chamber deposition: here the entire layer sequence of the cell structure is deposited in the same chamber. Cross contamination of sequential layers (so-called "dopant tailing") can be avoided by special procedures, which are often kept as proprietary know-how by the manufacturing companies involved.

Throughput requirements are satisfied by parallel concurrent processing (batch processing) of several substrates in a single chamber or in a number of identical chambers.

- Multi-chamber deposition: this method is applied to conduct the deposition of doped and intrinsic layers in separate chambers, and thus to avoid cross-contamination. Generally, the multi-chamber approach involves additional transport mechanisms for the substrates and, hence, leads to more complex equipment, which takes up more floor space in the production facility and is typically also more expensive. For optimal throughputs, the different processing durations for the doped and intrinsic layers (on the order of minutes vs. tens of minutes, respectively) are taken into account by providing correspondingly more chambers for the parallel processing of the *i*-layers. Multi-chamber systems fall into two sub-categories:
  - Cluster systems, which consist of a central transfer chamber surrounded by various deposition chambers. These can be loaded from the central transfer chamber in practically any sequence and frequency. Thereby a high degree of flexibility is obtained, and one can easily adapt to the processing durations of the various layers in different cell structures.
  - In-line systems are, in contrast, governed by the same line speed for all chambers, thus allowing different processing durations to be determined only by the length of the corresponding deposition zones, or by the number of corresponding chambers that are needed to obtain the required thickness of the various cell layers at the given deposition rates and line speed. This means that the different processing durations for the doped and intrinsic layers are achieved by providing correspondingly more chambers in-line for parallel processing of the *i*-layers. The substrate motion may be accomplished either continuously for dynamic deposition, or stepwise from chamber to chamber for static depositions [Klein-2007]. Dynamic deposition implies an advantage in terms of the deposition uniformity in the moving direction. However, failure of any chamber along the line will usually cause the shutdown of the entire line, compared to only a reduced throughput in a cluster system, where individual chambers may easily be bypassed. One may note here that in-line systems also form the basis for roll-to-roll deposition, as used for flexible substrates (see Sect. 6.1.9).

In order to choose the most suitable approach for production, one has in all cases to carry out a complete evaluation of the entire production line and of the products one intends to produce.

### 6.1.9 Roll-to-roll depositions

Roll-to-roll deposition systems inherently belong to the category of in-line systems. Their general features have been outlined above in Section 6.1.8. The additional feature of unwinding a web from a roll to feed into processing, and then of rewinding it onto a second roll after processing, implies the use of flexible substrate materials. The final product may thus result in flexible, long, lightweight thin-film modules, which cannot be realized with their glass-based counterparts. How far these special module properties are exploited and/or how much the manufacturing costs, involving equipment, materials, throughput and yield, offer advantages compared to more conventional approaches cannot generally be stated; it must be evaluated individually.

The flexible web materials suited for roll-to-roll processing are either metal foils or plastic foils. Being either opaque or optically inadequate, these materials are generally not suitable for superstrate cell structures, where for optimal collection of the low-mobility carriers (holes) the light enters through the substrate and TCO layer from the *p*-side of the cell structure TCO/*pin*/back-contact. Therefore, these web materials serve as substrates for the inverse cell layer sequence, back-contact/*nip*/TCO. In this context, both cell layer sequences may also stand as dual- and triple-junction cell structures.

Except for a unique special approach briefly described below, the use of metal foils excludes the monolithic series connection of cells (see Sect. 6.5). Plastic foils do allow for monolithic series connections; moreover, recent developments may offer the following features:

- Texturing: to provide light scattering and light trapping, processes are being developed to replicate optimal textures by hot-embossing or nano-imprinting lithography [Terrazzoni-2006].
- Optical suitability as superstrate web material: novel plastic substrate materials based on acrylic polymers have been demonstrated that can be used as superstrates for the deposition of TCO/*pin*/back-contact cell structures [Katsuma-2007]. These plastic web materials exhibit high heat resistance (>200 °C), high transmission (> 90 % at 410 nm), and a light-scattering texture that replicates glass/TCO combinations of proven optical requirements.

At present there are about five different roll-to-roll approaches being pursued by different thin-film Si module producers. For three of them their specific features are summarized below.

*Energy Conversion Devices/Uni-Solar*

Since the early 1980s, Energy Conversion Devices, Inc. (ECD) has developed and commercially utilized a continuous roll-to-roll manufacturing technology for the production of a-Si/a-Si,Ge alloy solar cells. Advancing this technology from a small-scale pilot machine to large-scale production machines, ECD commissioned a 30 MW per year machine for United Solar Systems Corp. (United Solar Ovonic) in Auburn Hills, Michigan, in 2002 [Izu-2003]. Since then, a second production line has gone into operation at the same location and, furthermore, two more installations with capacities of 60 MW per year each have become operational at Greenville, Michigan.

Unlike transparent glass substrates using the deposition sequence TCO/*pin*/back-contact, ECD employs an opaque stainless-steel web as substrate material, which implies the opposite deposition sequence: back-contact/*nip*/TCO. Specifically, ECD deposits triple-junction cells consisting of nine various Si alloy layers, where the *i*-layers for the bottom, middle, and top cell consist of a-SiGe:H with low bandgap (1.4-1.5 eV), a-SiGe:H with intermediate bandgap (1.60-1.65 eV), and a-Si:H with high bandgap (1.80 - 1.85 eV), respectively, to optimally adapt to the spectrum of the solar irradiance.

The production line basically consists of four roll-to-roll machines used for substrate washing, sputter deposition of the back contact, PECVD of the *nip/nip/nip* triple-junction cell structure, and deposition of ITO (indium-tin oxide) as TCO. With each of these machines, the roll is unwound to be fed to the corresponding process, and wound up again in a special take-up chamber after processing. The stainless steel web typically is 130 $\mu$m thick, 0.36 m wide, and up to 2.6 km long. The PECVD machines are designed to deposit simultaneously on the webs of six coils at a web speed of 1 cm/s [Izu-2003].

Following the last roll-to-roll process for the TCO deposition, further processing of the long 0.36 m wide triple-junction cells into solar modules comprises cutting to cell lengths, providing a top grid for current collection and reduction of contact resistance, series connection of cells to form modules, and encapsulation by foil materials, such as ethylene vinyl acetate (EVA) and ethylene-tetrafluorethylene (Tefzel®).

Note that, due to the conductive stainless steel substrate material, unlike conventional module fabrication on non-conductive glass substrates, monolithic series connection of cells (see Sect. 6.5) is not possible. Therefore, the various deposition steps are not interrupted by any arrangements for the delineating of cell stripes. Instead, the cells are formed by 0.36 m wide sections, cut from the long web, and are series-connected externally, similarly to the stringing of crystalline Si cells. As a consequence, Uni-Solar modules typically exhibit high currents (typically 2 A, therefore the above-mentioned grid) and low voltages (typically 2.3 V), compared to conventional glass-based a-Si modules of similar area.

In 2009, Xunlight Corporation, Toledo, Ohio, has reported on a further development for roll-to-roll production lines. This development is also based on triple-junction technology on a stainless steel web with a length of 1600 m, featuring, however, a wider deposition width of 0.92 m.

*VHF Technologies/Flexcell*

Founded in 2000, VHF Technologies, Yverdon, Switzerland, is now ramping up a production line for an annual output of 25 MW. The applied roll-to-roll technology is based on VHF at 80 MHz deposition of *nip* a-Si:H single-junction cell structures on a plastic foil web. The plastic foil (typically polyethylene naphthalate, PEN, or polyethylene terephthalate, PET) is about 50 $\mu$m thick and 53 cm wide. Following a metal coating on the web, the *nip* layer sequence is deposited in a multi-pass process performed in a single-chamber VHF deposition system. Further processing involves deposition of TCO, patterning for monolithic series connection of cells, and encapsulation by lamination on different back-sheets and flexible polymer encapsulants on the front side. Products manufactured with this technology carry the brand name Flexcell®.

Originally the products from a smaller pilot line were intended for portable light-weight flexible consumer applications, such as battery chargers and recreational uses (see also Sect. 7.2). More recently, the larger production capacity is increasingly utilized for BIPV installations (see Sect. 7.1), particularly for solar roofing. Here light-weight flexible modules may be quite straightforwardly incorporated with conventional building elements, for example by laminating with metal, thermoplastic olefin (TPO), or bitumen roof sheeting. No special support structures are required like those that are usually needed for heavier glass-based modules.

For roofing applications, the present Flexcell® basic module size is $0.47 \times 3.35$ m²; three of these modules are laminated side by side and connected in series to form a PV roofing element. The stabilized power output of such elements with a module aperture area of $1.40 \times 3.35$ m² is specified as 200 W (i.e. the stabilized aperture area efficiency is 4.3 %). Research programs are underway to extend the Flexcell® technology to a-Si/a-Si and micromorph tandem cell structures [Ballif-2007].

*Helianthos/Nuon*

Since the mid-nineties, Helianthos b.v., a subsidiary of Nuon n.v., The Netherlands, has been developing a unique roll-to-roll fabrication of thin-film Si-based flexible modules [Middelman-1998]. Unlike other roll-to-roll approaches, the technology of Helianthos is based on superstrate deposition, using a *pin* deposition sequence for the a-Si cell structure. However, the initial superstrate is an aluminum (Al) foil, and only serves as a temporary substrate. After the various roll-to-roll deposition steps in separate dedicated machines, from the TCO as front contact to the *pin* structure, and further on to the back-contact, the coated Al web is laminated with the back-contact side to a permanent flexible substrate material. Then, the original Al superstrate is removed by chemical etching, and the front TCO contact is exposed. Monolithic series connection of cell stripes, accomplished by a company-private laser process, allows one to obtain $I(V)$ characteristics within the range of conventional glass-based a-Si modules. Module finishing is concluded by providing contact stripes, cable connections and front encapsulation by foil.

In this context, it is worth recalling the importance of the TCO front contact. As described in Section 6.5, the TCO material must satisfy both electrical and optical properties. Particularly for TCO based on $SnO_2$:F, many module producers rely on glass/TCO combinations that are produced on floatlines from various glass manufacturers. While the electrical and optical specifications of these on-line manufactured products are specifically adapted to PV applications, they still may not be optimal. By depositing $SnO_2$:F off-line on Al foil, Helianthos is in the position to optimally adjust the TCO properties to the specific requirements of its cell structure [Bartl-2006].

To date, this approach by Helianthos has been applied to 0.35 m wide Al foil, and finished modules as long as 6 m have been demonstrated. Initial aperture efficiencies with the presently used *pin* single-junction a-Si technology have reached stabilized aperture efficiencies of 6.0 % [Hamers-2008]. A pilot line yielding this product is in operation, and it is intended to particularly address the market for building integration in roofs and facades. Developments towards tandem-junction technology (both a-Si/a-Si and a-Si/μc-Si) have been demonstrated. Furthermore, large-scale equipment for the deposition on a web width of 1.2 m has been installed. Efforts to upscale production to an annual output of 1 million square meters (around 60 MW/year) are in progress.

### 6.1.10 Novel deposition systems

In Section 6.1.2 several approaches were mentioned that aim to apply the advantages of VHF plasma excitation frequencies to uniform deposition across large areas (order of magnitude of 1 m²). Compared to conventional RF at 13.56 MHz, deposition with VHF (30 to 300 MHz) enables deposition at higher deposition rates with lower ion bombardment damage and, thereby, still results in obtaining device-grade material. This is particularly interesting for the deposition of microcrystalline silicon. Here one needs significantly higher deposition rates in order to maintain a high production throughput in spite of the layer thickness, which is typically about 5 times higher than for amorphous silicon. However, the advantages of using VHF may normally only be exploited when the wavelength corresponding to the used frequency is sufficiently large (for quantitative estimates, see Sect. 6.1.2).

Below we briefly describe an approach for which the compatibility of VHF with "static" uniform large-area deposition has been accomplished, as opposed to approaches with moving substrates (see Sect. 6.1.2). The key to this achievement is to modify the electrical supply to the electrode, which in addition to the applied voltage and frequency employs time-variable phase relations. To this end the electrode is subdivided into an array of linear rods. These are individually supplied with VHF power in such a way as to form standing electromagnetic waves along their length. By changing the phase relations, the nodes and antinodes of the voltage are shifted in position. Since the square of the voltage determines the rate of silane decomposition (see Sect. 6.1.2), the locations of maximal deposition move according to the time-dependent phase relations. As a consequence, the interplay of the time-averaged phase modulations along the various electrode rods that form the areal electrode can, if properly chosen, provide for uniform large-area deposition at the selected VHF.

A specific realization of this principle is shown in Figure 6.21. The rods of the ladder-shaped electrode are individually connected at each end with a VHF power supply (60 MHz), i.e. on one side directly, and on the other side via a phase-modulator (typically ± 120° with a modulation frequency of 20 kHz). Depending on the phase

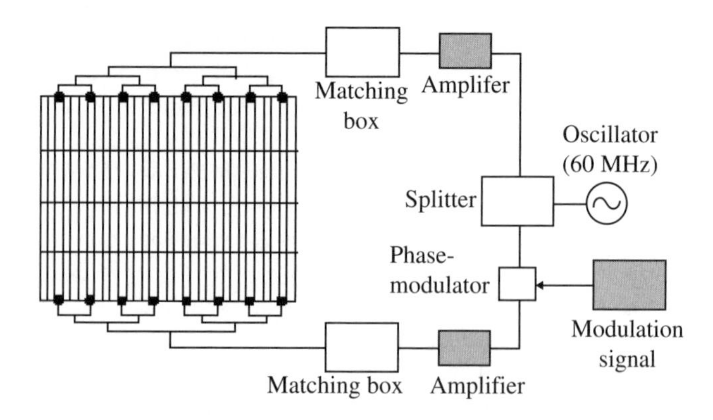


**Fig. 6.21** Conceptual principle of multi-rod (ladder-shaped) electrode and phase modulation method (with permission [Yamauchi-2005]).

difference at the two ends, the antinodes of the standing wave move to different length coordinates.

Based on this principle, Mitsubishi Heavy Industries, Ltd. (MHI), Japan, has obtained a uniform film thickness within ±15 % across substrates of $1.1 \times 1.4\ m^2$ for μc-Si:H, deposited at 2.5 nm/s [Kawamura-2006, Takatsuka-2006]. For micromorph tandem modules, of size $42 \times 47\ cm^2$, taken from $1.1 \times 1.4\ m^2$ substrates, initial efficiencies of 12.8 %, estimated to be stabilized at 11.5 %, have been reached. These results were obtained with a plasma regime at high pressure (several mb) and high power density. Since the plasma tends to become more localized at these pressures, one has to use smaller electrode spacings between the powered electrode and the grounded electrode on which the substrate is fixed. This, together with the required power density, would entail excessive heating of the substrate. Such excessive heating is avoided by circulating a coolant through the electrode rods.

This phase modulation method has been extended to be applied micromorph tandem modules by developing more sophisticated power feeding systems. By supplying two VHF frequencies in superposition, e.g., 58 MHz and 60 MHz, at the two ends of the antenna rods (not shown in Fig. 6.21), a much lower beat frequency (2 MHz for the example) is generated. With this beating effect, in conjunction with a duty cycle weighting of the standing waves, a time-averaged power distribution for uniform deposition can be accomplished [Sansonnens-2005].

Unlike symmetric large-area parallel-plate reactors, the arrangement of an array antenna electrode opposite a large-area substrate must be considered to be a highly asymmetric configuration (see Sect. 6.1.1). This results in a relatively low plasma potential and also in a low voltage drop $U_A$ across the sheath in front of the grounded substrate, and in a high self-bias (see Fig. 6.4). As a consequence, the ion bombardment energies affecting the growing film remain very low. This explains the high state-of-the-art efficiencies reached for deposition rates that are considerably higher than those obtained with conventional systems.

With regard to overall throughput considerations with commercial reactors, it has been stated [Kawamura-2006], that in-situ plasma-etch cleaning in a $NF_3$ plasma has been accomplished with uniform etch rates of about 5 nm/s by using the same phase modulation principle described above. As already mentioned in Section 6.1.3, serious efforts are independently underway [Schottler-2008, Oshinowo-2009] to eliminate the use of such fluorocarbon gases which are greenhouse gases and have high global warming potential.

MHI is on record to start a production line for micromorph tandem modules during 2007, with a nominal annual output of 40 MW. The stabilized module power is specified as 130 W, which corresponds to a stabilized total-area efficiency of 8.3 %.

A different type of an array antenna electrode has been developed by the equipment manufacturer Ishikawajima-Harima Heavy Industries, Co., Ltd. (IHI Corporation), Japan, [Takagi-2008]. In this approach, the array consists of U-shaped rods that are fed on one side with a VHF of 85 MHz, whereas the other end of each rod is grounded. Phase relations are applied by feeding adjacent rods with a phase difference of 180°. Although potential merits in terms of large-area uniformity and throughput have been demonstrated, this approach has not yet been applied to module manufacture.

## 6.2 HOT-WIRE CHEMICAL VAPOR DEPOSITION (HWCVD) (by Jean-Eric Bourée)

### 6.2.1 Introduction

As an alternative to PECVD, Hot-Wire Chemical Vapor Deposition (HWCVD), also known as catalytic CVD (Cat-CVD) has emerged since the eighties as a new method to deposit thin-film silicon and related materials. For a complete historicel review, the reader may refer to [Mahan-2003]; for a detailed report on HWCVD of amorphous and microcrystalline silicon the reader may consult [Schroeder-2003]. Concerning the industrial implementation of HWCVD technology (essentially in Japan), see [Matsumura-2008].

### 6.2.2 Description of the HWCVD technique

In the HWCVD technique, the source gases ($SiH_4$, $C_2H_2$, $H_2$,...) are introduced into a vacuum chamber, and a metallic filament (so far generally tungsten) is heated up to a high temperature (1700 °C to 1900 °C) providing thereby the surface for heterogeneous thermal decomposition of the gases into radical species. The filament acts as catalyst, and therefore the process is also called "catalytic chemical vapor deposition (Cat-CVD)" [Matsumura-1986]. The radicals thus generated induce film growth on a temperature-controlled substrate, which is facing the filament. A view of a hot-wire deposition chamber is shown in Figure 6.22. The wire temperature is determined by using a single-wavelength, disappearing-filament optical pyrometer (by levelling the brightness of the incandescent object with that of the inner lamp), with corrections made for effective emissivity. As will be explained in more detail below, different spectroscopic techniques, such as threshold ionization mass spectrometry, single-photon ionization or laser-induced fluorescence, can be attached to the reactor for probing the radicals produced by hot-wire decomposition.

Let us begin by summarizing the most important advantages of the HWCVD process, as compared to the PECVD process:

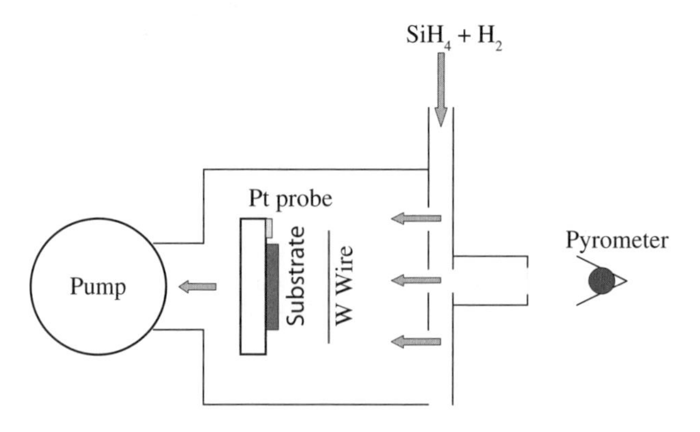


**Fig. 6.22** Schematic arrangement of a HWCVD reactor.

- Since the technique is thermal and catalytic in nature, and relies on a heated metal filament to decompose the gas species, there are no ions and electric fields present. Therefore, the substrate and the growing film are not damaged by energetic ion bombardment. This is important for depositing passivation or gas barrier films on organic devices (see Sect. 6.2.8). This can also be important when depositing the first part of the *i*-layer in *pin*-type solar cells, so as to avoid damage of the critical *p*/*i*-interface.
- Due to its high decomposition efficiency, the technique has the potential for high deposition rates (up to 10 nm/s).
- Due to the low pressures that can be used for gas phase decomposition (some mTorr), a high flux of atomic hydrogen can be obtained; this can be useful for material etching, for removing dangling bonds, and for enhancing the abstraction of H atoms from the growing surface (leading, thereby, to a low H-content in the films). At the same time, the formation of $SiH_3$, as growth precursor, is favored due to H abstraction from $SiH_4$: $H + SiH_4 \rightarrow SiH_3 + H_2$. This leads to a-Si:H films with reduced light-induced degradation [Kondo-2006].
- Since no plasma is needed, the substrate is decoupled from the deposition process, enabling substrates to be easily introduced and removed from the deposition chamber without disturbing the deposition. Moreover, step coverage (conformal film deposition) is excellent [Wang-2004], and uniformity can easily be optimized.
- It is a method that is easily scalable by expanding the spanned area of the catalyst; this fact is particularly important for industrial implementation, as will be discussed below.

### 6.2.3 Filament materials

Over the last years, tungsten (W), which was previously the filament material most frequently used, has gradually been replaced by tantalum (Ta), essentially when silane flow is used. The replacement was motivated by the following considerations: to avoid filament aging, it is important to avoid the formation of silicides on the filament surface. Now, the formation of Ta-silicide takes place at lower temperatures (only up to 1750 °C under silane flow) when compared to the formation of W-silicide (up to 1850 °C) [Honda-2008]. Apart from these two materials, ruthenium (Ru), rhenium (Re), iridium (Ir), molybdenum (Mo), graphite (C) and a nickel-chromium alloy (NiCr) have also been used occasionally as filament materials.

### 6.2.4 Types of materials deposited by HWCVD

The great majority of research activities on Hot-Wire deposition have so far been devoted to gas phase and deposition chemistry of silicon-related materials: hydrogenated amorphous silicon, microcrystalline silicon, polysilicon, epitaxial silicon, silicon alloys with carbon, nitrogen and germanium. This research has also been done in view of applying these materials, for instance, in solar cells and thin film transistors. Since 2004, an increasing variety of thin-film materials have been obtained with the HWCVD method, namely: silicon dioxide, aluminum oxide, aluminum nitride, Si-O C, Si-N-C alloys, diamond, carbon nanotubes, nanowalls or nanoparticles. Moreover, transition metal oxide nanoparticles have been synthesized for applications like gas sensors or electrochromic windows. A novel HWCVD variant, called "initiated CVD (iCVD)", appeared in 2006, and has spread rapidly [Lau-2008]. It was demonstrated to be a convenient single-step fabrication method to produce high-quality polymer thin films.

### 6.2.5 Mechanisms of the deposition process

Looking specifically at the deposition of a-Si:H layers, several researchers have attempted to undertake a systematic study of the deposition process mechanisms for HWCVD [Doyle-1988, Molenbroek-1996, Duan-2001, Holt-2002]. Optimum conditions for the deposition of amorphous silicon films have been obtained [Molenbroek-2006] for a particular value of the gas phase parameter $p \times L$, i.e. for $p \times L \approx$ 20-75 mTorr$\cdot$cm, where $p$ is the silane chamber pressure in mTorr, and $L$ is the distance in cm between filament and substrate. They postulated that this optimum results from the need to increase the pressure to a level where most Si atoms react with $SiH_4$ before reaching the substrate, while avoiding excessive gas phase reactions that can lead, at still higher pressures, to the formation of large radicals, such as $Si_2H_6$, $Si_3H_8$ ... We point out here that $SiH_3$ is considered to be a "good" film growth precursor (see [Kondo-2006]) because it has high surface mobility and has the longest lifetime among the various $SiH_n$-type species; in contrast to $SiH_2$, it does not lead to the formation of large radicals.

The formation of radicals (especially of those radicals that are growth precursors) is an essential point when studying the deposition of thin-film silicon layers. Radical formation has been studied by various methods, as follows:

The nature and the flux of radicals desorbed from a hot filament depend on the filament temperature and on the gas pressure in the reactor. In the case of a tungsten filament heated at 1900 °C in a low pressure of silane, most of the authors agreed that, using different detection methods, one does detect Si, H and $SiH_3$. Using threshold ionization mass spectrometry, H and Si were found as primary radicals with a small contribution of $SiH_3$ [Doyle-1988, Zheng-2006] for a large range of temperatures (1450 to 2700 °C) and for $H_2$ pressures up to 0.1 Torr, whereas using single-photon ionization with a vacuum ultraviolet laser, gas-phase species identified were Si, $SiH_3$ and $Si_2H_6$ [Duan-2001]. In the latter case H was not detected because the ionization potential for H exceeded the photon energy of the laser. Moreover, considering newly replaced ("virgin") filaments, the small activation energy (8 kcal/mol) then observed for $SiH_3$ formation suggested that the process was "catalyzed" [Holt-2002].

The absolute density of H-atoms in the gas phase has been determined by combining a two-photon laser-induced fluorescence technique and a vacuum ultraviolet (Lyman $\alpha$) absorption technique [Umemoto-2002]. At high temperatures of the tungsten wire ($\approx$1930 °C), the absolute density of H-atoms reached values that were as high as $1.0 \times 10^{14}$ cm$^{-3}$. These values are two orders of magnitude higher than those obtained by PECVD under comparable conditions. The effective enthalpy for the H-atom formation from $H_2$ on the catalyst surface was determined to be 57.1 kcal/mol [Umemoto-2002] and should be compared with the effective bond dissociation energy of gas-phase $H_2$ molecules, which is 109.5 kcal/mol. This result established clearly the *catalytic nature of the hot-wire CVD process.*

As concerns the mechanism of H-atom formation by HWCVD, it was demonstrated that this mechanism proceeded via dissociative adsorption at bare sites followed by desorption at hydrogenated sites on the hot wire surface [Comerford-2009].

### 6.2.6 Filament aging

It has been suggested [Mahan-2000] that the electronic properties of a-Si:H deposited by HW-CVD are related to radical chemistries, and that the differences observed in thin-film properties are linked to the differences between a "virgin" wire (newly replaced wire) and an aged wire. For an aged wire, the $SiH_3$ signal exhibited an activation energy of 106 kcal/mol: a value that is much higher than the 8 kcal/mol observed for a "virgin" wire. This suggests that the aging of the wire leads to a drastic reduction of its catalytic activity [Holt-2002]. Also SiH and $SiH_2$ are now more abundant than $SiH_3$, in contrast with the results obtained when a "virgin" filament is used [Holt-2002].

To gain insight into the nature of the changes occurring at the wire surface, scanning electron micrographs and Auger electron spectroscopy were used to characterize the surface morphology of heat-treated and aged wires, and to measure the Si concentrations at the surface and in the interior of the wire, respectively. A Si concentration of 15 at.% was observed at the surface of an aged wire, corresponding to a two-phase equilibrium between tungsten silicide ($W_5Si_3$) and tungsten, whereas approximately 2 at.% Si concentration was measured in the interior of the wire, a value that is comparable to the equilibrium solid solubility of Si in W [Holt-2002].

Using a tantalum filament exposed to a silane pressure of 0.25 Torr [van der Werf-2009], the formation of a $Ta_5Si_3$ shell (as determined by X ray diffraction) of 20 μm thickness was observed. After 4 h of annealing of the filament in vacuum at high temperature (2100 °C), the tantalum silicide was completely removed. This regeneration procedure was shown to greatly enhance the lifetime of the Ta filaments.

For the future, a more fundamental study of the physico-chemical processes occurring at the filament surface as a function of temperature (metal evaporation in competition with metal silicide formation) seems necessary.

### 6.2.7 Amorphous and microcrystalline silicon films, and microcrystalline silicon carbide alloys

Amorphous silicon films have been deposited at ultra-high deposition rates (> 10 nm/s) under a wide range of silane depletion conditions (the silane depletion condition can be evaluated here by dividing the deposition rate of the films measured at the substrate holder by the total silane flow [Mahan-2002]). By increasing the deposition rate (when increasing the flow rate for a fixed substrate temperature), changes occurred in the atomic structure of the films (an increase of the Urbach energy, indicating an increase in atomic network disorder), correlated with an increase in the small angle X-ray scattering signal, which, in its turn, corresponds to a large increase of microvoid density. Fortunately, this increased microvoid density did not lead to an enhanced Staebler-Wronski effect [Mahan-2002]. This is due to the low H-content in HWCVD films (≈ 3 at.%), in comparison with the high H-content (10-12 at.%) observed in PECVD films. The low H-content is attributable to the high concentration of H atoms generated in the gas phase during the HWCVD process [Umemoto-2002]: the high concentration of H atoms enhances the abstraction of H atoms from the growing surface ($H + H \rightarrow H_2$ + heat) and, thus, induces local heating. Based on these considerations, one understands how it is possible to deposit a-Si:H films on a substrate of $0.4 \times 0.96$ $m^2$ size at an average deposition rate as high as 32 nm/min (see [Ishibashi-2003]).

Concerning μc-Si:H films, good progress has been reported: using a low substrate temperature (250 °C) and a low filament temperature (<1800 °C), μc-Si:H solar cells were obtained, close to the transition from crystalline to amorphous growth. With a deposition rate of 0.4 nm/s, an AM1.5 efficiency of 9.4 % and open-circuit voltages up to 600 mV were achieved ($J_{sc}$ = 22 mA/cm$^2$, $FF$ = 71 %) [Klein-2003]. Two comments should be added at this point:

- Compared to the μc-Si :H solar cells synthesized by PECVD, the Staebler-Wronski effect has been minimized, even though the optimum conditions for this type of cell necessitates a relatively low crystalline volume fraction (50 %). However, the H-content is low (≈ 3 at.%), as in the case of the a-Si:H layers described above.
- The μc-Si:H material deposited by HWCVD is disordered in such a highly inhomogeneous way that a concentration of W as high as $2 \times 10^{18}$ at./cm$^3$ can be accommodated: Such a high W-concentration, associated with a filament temperature of 1800 °C, and measured in the film by secondary ion mass spectrometry, has, in fact, no real effect on the electronic properties of the μc-Si:H film [Bourée-2003].

Microcrystalline silicon-carbide (μc-SiC:H) thin films (20 nm) have been prepared, in stoichiometric crystalline form, from monomethylsilane/hydrogen mixtures at temperatures below 300 °C. This material has high optical transparency, has an *n*-type character, and is highly conductive due to unintentional doping (by nitrogen or oxygen impurities) [Huang-2007]. Therefore, it has been used as a window layer in *n*-side illuminated μc-Si solar cells in the *n-i-p* configuration [Finger-2009]. With an *i*-layer thickness of 2 μm, an AM1.5 efficiency of 9.2 % associated with a short-circuit current density of 28 mA/cm$^2$ was obtained ($V_{oc}$ = 512 mV, $FF$ = 64 %).

For multijunction cell structures prepared by HWCVD, such as *pin/pin,* or *nip/nip* tandems, or triple-junction *nip* cell structures, the reader is referred to the literature [Schroeder-2003, Schroeder-2008, Stolk-2008].

Recently, fabrication of a-Si:H thin film solar cells by HW-CVD on flexible substrates, i.e. at low temperatures, has been attempted [Alpuim-2008, Villar-2009]. An initial efficiency of 4.6 % was achieved.

The first industrial implementation of HWCVD technology concerning deposition of a-Si:H has been realized in Japan [Matsumura-2008]. Here, by using a large size, vertical-type HWCVD apparatus, a 13 % film thickness uniformity has been demonstrated within an area of $1 \times 0.6$ m$^2$ [Asari-2008].

### 6.2.8 Silicon nitride and silicon oxynitride films

Besides applications for thin-film solar cells, HW-CVD is also employed, with considerable success, to deposit anti-reflection, passivation and barrier layers for other PV technologies, and for various other thin-film devices, such as thin-film transistors. $SiN_x$ films, particularly pursued by the Japanese industry [Matsumura-2008, Asari-2008, Oku-2008], are prepared with a gas mixture of $SiH_4$, $NH_3$ and $H_2$. They are highly transparent, dense (2.93 g/cm$^3$), associated with a low H-content, and exhibit good dielectric properties [Masuda-2006, Alpuim-2009]. A very high deposition rate (7 nm/s) was achieved for device-quality films [Verlaan-2009]. The $SiN_x$ films can, therefore, be used as passivation layers and as encapsulation films against moisture for organic electronics [organic solar cells, organic light-emitting diodes (OLEDs)], or as gate dielectric films in thin-film transistors [Heya-2009, Ogawa-2008]. Such properties have attracted several groups who have focused their efforts on developing specific HWCVD apparatus [Matsumura-2009].

Silicon oxynitride films have also been deposited by the HWCVD method [Ogawa-2008]. In the future, $SiN_x/SiO_xN_y$ stacked films covering organic solar cells could efficiently protect the latter against oxygen and moisture.

## 6.3. DOPED LAYERS

As mentioned in Section 6.1, doped *p*- and *n*-layers are obtained during the course of PECVD of the cell structure, usually by adding trimethylboron $B(CH_3)_3$, and generally phosphine $PH_3$ to the plasma discharge, respectively. Depending on single- or multi-chamber processing, the procedures of gas handling differ. For single-chamber processing, the transition to the *i*-layer must occur without boron contamination (so-called boron tailing) of the *i*-layer, which may be achieved by gas flushing, pumping procedures, or special proprietary treatments. In multichamber systems, there are individual chambers that are solely used for the deposition of doped layers. In this case, the difference in deposition durations between the relatively thin doped and the much thicker *i*-layers is taken into account by providing more chambers for *i*-layer deposition than for doped layers. Thereby high throughput and high equipment utilization can be satisfied.

The main function of the doped layers adjacent to the *i*-layer is the establishment of an internal electric field within the *i*-layer. This is a consequence of establishing thermal equilibrium upon joining the *p*-, *i*- and *n*-layers whereby the Fermi levels of the doped layers assume the same potential. Since the doped layers are not photoactive, they cause optical absorption and electrical series resistance losses. Both may be minimized by keeping the layers thin.

The optical absorption is measured in terms of the absorbance $1 - \exp(-\alpha\, d)$, where $\alpha(\lambda)$ is the spectral dependence of the absorption coefficient, and $d$ is the layer thickness. The series resistance losses arise from carrier transport across the layers. With typical resistivities of the order of $10^{-6}$ S/cm and $10^{-2}$ S/cm, and typical thicknesses of 10 nm and 25 nm for B-doped a-SiC (see below) and for P-doped a-Si, respectively, the series resistance losses of the *n*-layer are negligible, but for the *p*-layer they may reach the order of 1 % of the generated power, and thus may become critical with respect to the *p*-layer thickness. The importance of the *p*-layer is further discussed in the following section.

### 6.3.1 *p*-layers

For the best performance of a-Si-based solar cells, light is required to enter the cell through the *p*-layer. This requirement stems from the difference in electron and hole mobilities in intrinsic a-Si (see Sect. 2.4). Accordingly, the drift lengths of holes are shorter than those for electrons. Since more electron/hole pairs are generated closer to the entrance of light, more holes can be collected the closer they are to the *p*-layer, i.e. if the light enters through the *p*-layer. The longer drift lengths of the electrons extend all the way across the *i*-layer to the further distant *n*-layer, and thus facilitate electron collection as well.

The *p*-layer thus must act as a "window" layer, i.e. it should transmit the incoming light without any further absorption in addition to that of the glass/TCO (for superstrate structures, see Sect. 6.4) or of the encapsulation/TCO (for substrate structures, see Sect. 6.1.9). In order to pass the blue portion of the irradiation in particular, the energy gap of the *p*-layer is increased by alloying with carbon. The addition of methane ($CH_4$) to the basic processing gas silane ($SiH_4$) results in the deposition of carbon-containing amorphous silicon, i.e. a-$Si_{1-x}C_x$:H (in short a-SiC). However, the incorporation of carbon leads to distinct changes in the optical and electrical material properties, which in the context of boron-doped *p*-layer material pertain to the following:

- *The energy gap* this rises with the fraction $x$ of the incorporated carbon. The energy gap may be deduced from optical absorption measurements by evaluating so-called Tauc plots [Tauc-1966], namely the dependence of the absorption coefficient $\alpha$ on the photon energy $h\nu$,

$$\alpha = C^2(h\nu - Eg)^2/h\nu, \text{ for } h\nu > E_g, \tag{6.13}$$

  where $E_g$ is the energy gap, and $C = 7.7 \times 10^4$ (cm eV)$^{-1/2}$ is an empirical constant [Schade-1985a] for $h\nu > 1.95$ eV (see also Sect. 2.3). It is evident from the above relationship that for a given photon energy $h\nu > E_g$ the absorption coefficient, and hence the absorbance for a given layer thickness, decreases with increasing energy gap. This is demonstrated in Figure 6.23 for two values of the energy gap, 1.7 eV for a-Si, and 2.1 eV for a-SiC, the latter being approximately the maximal value still compatible with a sufficiently high electrical conductivity (see below).

- *The electrical conductivity*: generally the electrically active doping level, and hence the electrical conductivity, decreases with increasing energy gap. For boron concentrations above $5 \times 10^{19}$ cm$^{-3}$ the energy gap even decreases, i.e. sufficiently high conductivities are obtainable only up to energy gaps around 2.1 eV. The opposite trends, higher energy gaps with lower conductivities, are shown in Figure 6.24, in which the envelope curve represents the combinations of energy gap and electrical conductivity which could experimentally be reached

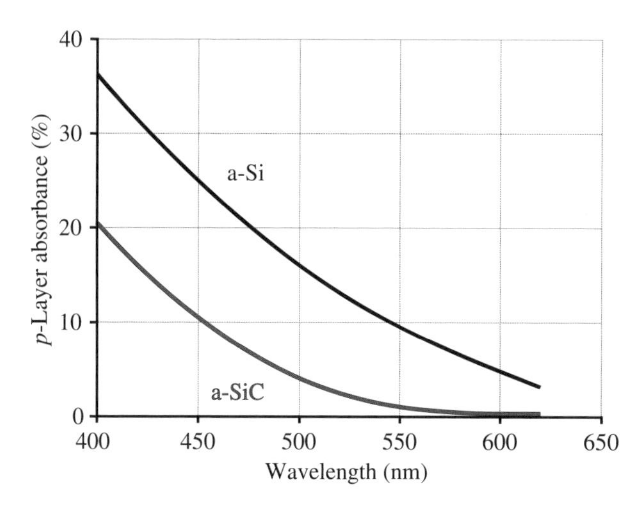


**Fig. 6.23** Optical losses (*p*-layer absorbance) of 12 nm thick *p*-layers, made from a-Si with an energy gap of 1.7 eV, or from a-SiC with an energy gap of 2.1 eV, respectively. The absorbance values are calculated for the nominal layer thickness, i.e. for vertical light incidence without scattering.

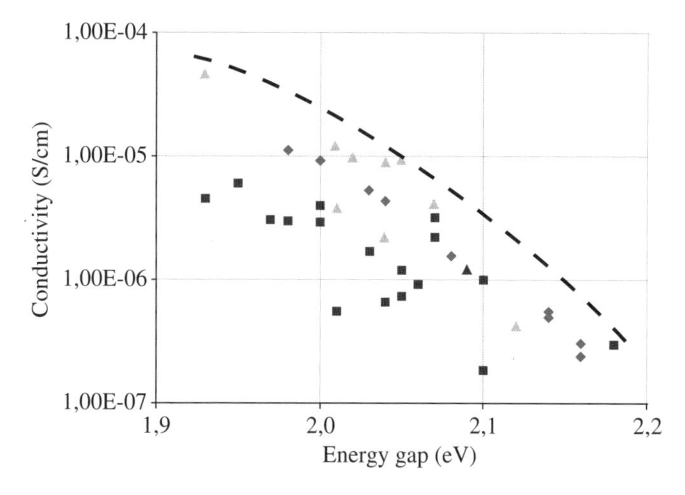


**Fig. 6.24** Energy gap dependence of the conductivity for boron-doped a-SiC *p*-layers of various carbon contents and doping concentrations, deposited in different PECVD reactors [Frammelsberger-2004].

in the best cases. Whereas higher energy gaps result in lower conductivities and hence, in view of low losses, even more strongly call for thinner *p*-layers, the thickness, on the other hand, must remain sufficiently high to facilitate the full development of the built-in voltage within the cell structure. The minimal thickness must exceed the space charge width that develops within the *p*-layer upon joining with the *i*-layer.

In summary, estimates of the optical and electrical losses indicate the predominant contribution resulting from the increasing optical absorbance towards the blue spectral region. Based on typical data for an a-SiC *p*-layer with an energy gap of 2.1 eV and a thickness of 12 nm, the optical absorption loss, and hence the relative loss in the quantum efficiency, at 400 nm amounts to 20 % (see Fig. 6.23). This loss agrees quite well with the *p*-layer loss shown in Figure 6.39. The electrical loss, based on $1 \times 10^{-6}$ S/cm, is estimated to be around 1.2 % of the generated power.

### 6.3.2 Doped microcrystalline layers

In order to avoid or reduce the Staebler-Wronski effect (see Sect. 4.5.16), single-junction microcrystalline cells, and more frequently tandem (micromorph) cells, have been developed lately. Doped microcrystalline layers have been incorporated into these cell structures, but advantages can also be realized, particularly in tandem cell structures, with amorphous *i*-layers (e.g., a-Si/a-Si). Both *p*- and *n*-type μc-Si offer lower optical absorption (for wavelengths $\lambda < 700$ nm [Shah-2004]) and higher conductivities (of the order of $10^{-3}$ S/cm and 1 S/cm, respectively), compared to conventional a-Si. These properties are exploited as follows:

- *Lower optical and electrical losses*, as described above. However, for single-junction cells based on superstrates, the use of microcrystalline *p*-layers requires further qualification. In this case, the *p*-layer should act as a "window" layer. For the blue spectral region, the absorption coefficients for a-SiC derived from the relationship given in Section 6.3.1 are smaller than those for μc-Si [Shah-2004]. Therefore, in regard to lower optical losses in the blue spectral region, the absorption coefficient of a microcrystalline *p*-layer must be at least as low as that of a-SiC. Using μc-SiC, instead of μc-Si, may prove beneficial, since both higher energy gaps and higher conductivities have been observed [Schade-1989]. However, depositing μc-SiC on TCO typically requires an incubation layer to fully develop microcrystallinity [Vallat-Sauvin-2001]. Therefore, the advantage of a lower absorption coefficient may be reduced by the thickness requirement.
  In tandem junction cells, the microcrystalline *p*-layer of the bottom cell is exposed only to the relatively weakly absorbed longer wavelengths of the remaining irradiation that is not absorbed in the top cell. In order to assess the optical losses for the bottom cells, the advantage of using microcrystalline Si *p*- and *n*-layers, as opposed to a-Si, is clearly demonstrated in Figure 6.25. Unlike Figure 6.23, here the shown absorbances are based on absorption coefficients obtained more recently [Shah-2004]. Furthermore, due to the higher conductivities of the doped microcrstalline layers, the corresponding series resistance losses of the bottom cell are negligible.

- *n/p tunnel-recombination junctions in tandem cells*. The series connection of the top and bottom cell in tandem cell structures is established by the *n/p* junction between the *n*-layer of the top cell and the *p*-layer of the bottom cell. For

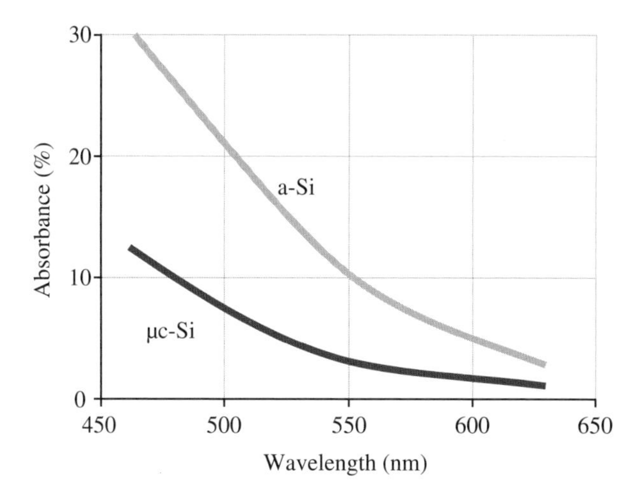


**Fig. 6.25** Optical losses (*p*-layer absorbance) of 12-nm-thick *p*-layers, made from a-Si or from μc-Si, respectively. The absorbance values are based on experimental absorption coeffcients, and are calculated for the nominal layer thickness, i.e. for vertical light incidence without scattering.

solar cell operation this *n/p* junction is reversely biased and, in principle, represents a series resistance and hence an electrical loss. The electron current, photogenerated in the top cell and collected by the *n*-layer of the top cell, must be taken over by the hole current, photogenerated in the bottom cell and collected by the *p*-layer of the bottom cell. This is accomplished by recombination of the electrons and holes arriving at the *n/p* junction. In a conventional *n/p* junction under reverse bias, the recombination would occur on either side of the junction after thermally activated transport of electrons and holes to the *p*-and *n*-side, respectively. Obviously, this mechanism would imply a huge electrical loss due to the series resistance of a conventional reverse-biased junction. Instead, the recombination may be obtained via defect states located at or very close to the junction. Doped a-Si, and even more so doped μc-Si, contain high concentrations of, and thus closely spaced, defect states. Depending on their charge state, electrons and holes are trapped, and then recombine by tunneling between these closely spaced states. In that sense the *n/p* junction is not a classical tunnel junction, which requires degenerate doping for direct tunneling of electrons from the conduction into the valence band across the *n/p* junction. To account for recombination via tunneling, the *n/p* junctions here are called tunnel-recombination junctions. A sufficiently high concentration of defect states at the *n/p* junction is obtained by incorporating a highly boron-doped *p*-layer between the *n*- and *p*-layers of the top and bottom cell, where at least one of these doped cell layers is μc-Si [Hegedus-1995]. This is generally easily accomplished with *n*-type μc-Si layers, whereas the nucleation of a *p*-type μc-Si layer may not be sufficiently close to the *n/p* junction to result in the required concentration of interface states [Pellaton Vaucher-1997].

# 6.4. TRANSPARENT CONDUCTIVE OXIDES (TCO) AS CONTACT MATERIALS

Generally thin-film solar cells incorporate a transparent conductive oxide (TCO) layer as the front contact material. The choice of such materials is rather limited, since they must satisfy a number of requirements concurrently. These comprise both physical and chemical properties, but also the availability on large areas (order of 1 $m^2$), which counts as one of the attractive features of thin-film photovoltaics, at an economically viable cost.

The physical properties distinctly contribute to the solar module efficiency due to the

- optical properties that affect the overall light intensity being absorbed within the cell structure; and the
- electrical properties that affect the generated cell current to be conducted to the external load. This aspect must be considered for cells in series connection to form a module (see Sect. 6.5).

Besides the TCO properties per se, possible interactions between the TCO and the adjoining semiconductor structure must be considered. In the case of substrate-based cells, deposition of the TCO layer must be compatible with the semiconductor layers underneath. In particular the deposition temperature should not exceed temperatures (typically > 150 °C) harmful to the semiconductor device structure. In the case of superstrate-based cells the semiconductor layers are deposited onto the TCO. This deposition process, usually by PECVD, involves the presence of atomic hydrogen that under certain conditions may lead to an onset of chemical reduction of the TCO (specifically $SnO_2$, which is partly chemically reduced to metallic tin, and hence suffers a loss in optical transmission [Schade-1984]).

## 6.4.1 Glass substrates and specific TCO materials

The following discussion applies to specific TCO materials that are deposited onto glass, and thus are used in superstrate cell configurations, which are representative of a major portion of thin-film silicon modules being manufactured. Even independent of the TCO material, the type and thickness of the glass substrate imply different optical transmissions that tend to decrease with increasing wavelengths in the red portion of the incident spectrum. There are two types of soda lime glasses that are produced in floatline processes [Gerhardinger-1996] and that, depending on their $Fe_2O_3$ content, significantly differ in their optical transmission (see Fig. 6.26).

The difference between the two types of glass is noticeable from the color of the fracture surfaces, i.e.

- green glass, containing > 0,5 % $Fe_2O_3$, and
- white glass, containing < 0,05 % $Fe_2O_3$.

The present developments of "next-generation" cell structures, like the a-Si/µc-Si tandem concept (micromorph concept, see Sect. 5.4.3) are only meaningful if the extended red response of these structures is not significantly counteracted by a

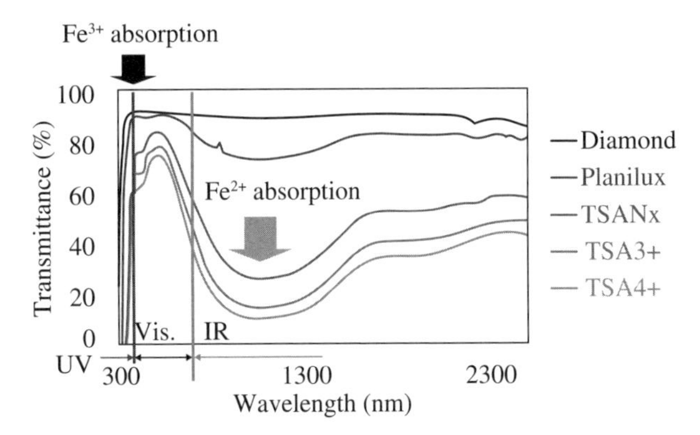


**Fig. 6.26** Transmittance of various floatline glasses with different Fe content; $Fe^{2+}$ causes a wide absorption band in the near infrared, and $Fe^{3+}$ an absorption shoulder near 450 nm; the specified types of glass refer to products of Saint-Gobain Glass Company (with permission [Lehmann-2006]).

decreasing glass transmission in the extended red spectral distribution. In view of these developments, the choice of glass/TCO combinations must involve white glass.

The following materials are being employed, or are potentially applicable, as TCO, namely

- $SnO_2$:F, deposited on-line or off-line by atmospheric-pressure chemical vapor deposition (APCVD). The on-line process is performed at a certain position along the commercial float-line for glass production. Therefore, it is a very cost-effective process, which mostly, however, applies only to green-glass float-lines, with limited glass thickness possibilities.
- ZnO:B, deposited by low-pressure chemical vapour deposition (LPCVD).

These two materials feature an as-grown surface texture that is exploited for light scattering and light trapping (see below). Furthermore, a third material has proven to be suitable, namely

- ZnO:Al, deposited by sputtering, typically from ceramic Al-doped ZnO targets, followed by subsequent texture-etching to provide the desired light scattering conditions.

From these materials, only $SnO_2$:F can be produced both on-line and off-line, whereas the ZnO materials are always produced off-line. Off-line production offers a choice of glass type and glass thickness, but is generally less cost-effective. In practice, the added cost for optimal glass/TCO properties must be weighed against the improved performance in terms of cost/W.

## 6.4.2 Qualification of TCO materials

The suitability of TCO materials to be used in large-area modules may be evaluated in terms of optical and electrical losses, where in a first step the effects of surface texture are not considered. The optical losses are quantified by the absorbance, i.e. by the light intensity lost in the TCO material due to its optical absorption. The absorbance $A$ is obtained from measuring the transmission $T$ and reflection $R$ of the experimental set-up air/glass/TCO/air, i.e.

$$A = 1 - T - R. \tag{6.14}$$

However, due to the surface texture of the TCO, the obtained value of $A$ is incorrect, since light scattering from the surface texture introduces angular light distributions within the glass/TCO, which cause total internal reflections, whereby the optical path lengths and hence the absorption are increased. Therefore, the experimental set-up must be changed to air/glass/TCO/immersion liquid/glass/air, where the refractive index of the immersion liquid approaches that of the TCO material and thus practically eliminates light scattering from the TCO texture. The refractive index of TCO is about 1.9; diiodomethane ($CH_2I_2$) features the highest refractive index (1.75) to be found among liquids.

In order to obtain the true absorbance of the TCO separately, the absorbances of the glass substrate (after removing the TCO) and the glass slide adjoining the immersion liquid must be determined separately and subtracted. Figure 6.27 compares the absorbances of two glass/TCO combinations, as well as those of the individual TCO layers separately. The increasing influence of the glass substrate material towards the red spectral region is demonstrated. In this example, the effect of the glass derives mainly from a difference in thickness, rather than in glass quality: both glasses are green glasses; the on-line deposited TCO is on 3 mm glass, while the off-line produced TCO from Asahi (type U) is deposited on 1 mm glass. There still remains a distinct difference in the absorbance of the two TCO materials shown, particularly in the blue and green region of the spectrum.

The electrical losses are calculated as described in Section 6.5. Besides the cell data at the maximum power point and the cell width, the electrical losses directly depend on the sheet resistance of the TCO layer, which is simply measured by a

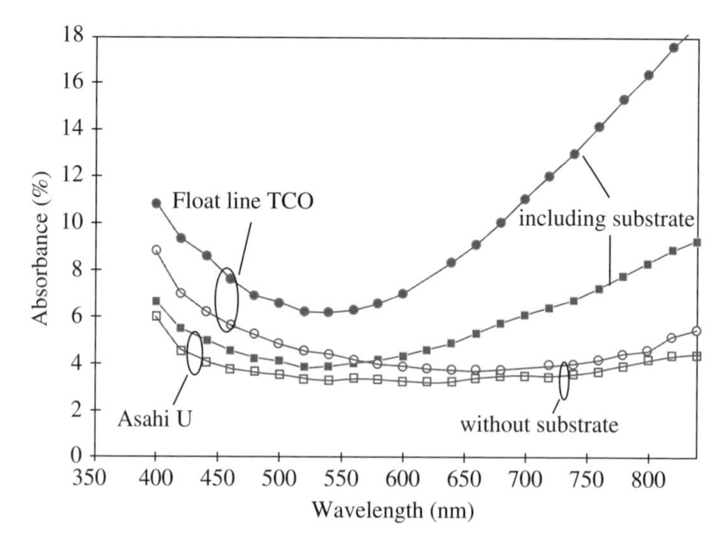


**Fig. 6.27** Absorbances of two different TCO materials for the combination glass/TCO and TCO separately (with permission [Beneking-1999]).

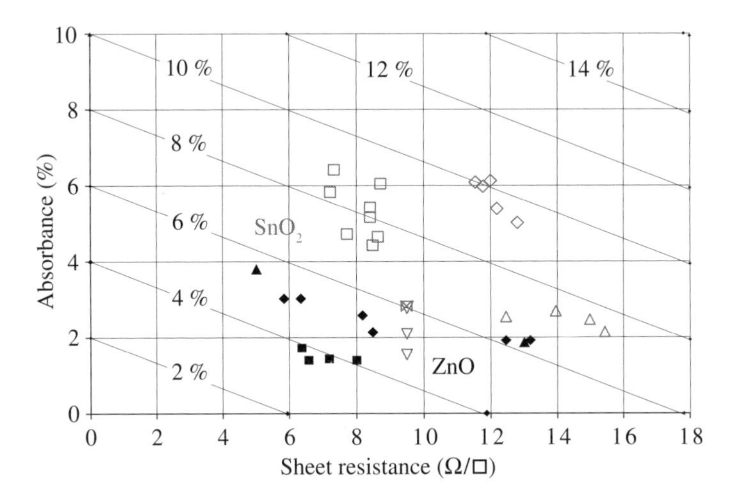


**Fig. 6.28** Loss diagram to rank TCO materials according to their absorbance and sheet resistance; shown are various sample series for both $SnO_2$ and ZnO; electrical losses are based on the cell impedance of an a-Si/a-Si tandem-junction module with cell width of 1.6 cm (with permission [Müller-2001]).

four-point probe. The following evaluation of TCO losses (see Fig. 6.28) assumes an a-Si/a-Si tandem module, with a cell width of 1.6 cm.

For different TCO materials, the absorbances vs. the sheet resistances are plotted in a diagram, together with a series of "equi-loss" lines. Figure 6.28 shows that the reduction of the absorbance tends to increase the sheet resistance. This is a direct consequence of the opposite thickness dependencies, i.e. the absorbance is

$$A = 1 - \exp(-\alpha\, d), \tag{6.15}$$

where $\alpha$ is the absorption coefficient, while the sheet resistance is

$$R_{sh} = \rho/d, \tag{6.16}$$

where $\rho$ is the specific resistivity of the TCO and $d$ its thickness.

With regard to the red spectral region, it is desirable that the TCO materials feature a low carrier concentration combined with a high mobility. Thereby the free-carrier absorption may be reduced, while the specific resistivity is not increased. Thus the sheet resistance may be maintained at a lower absorbance or, conversely, the absorbance may be maintained at a lower sheet resistance. These possibilities need to be evaluated in view of other, often practical, aspects as, for example, the cell width and the resulting output voltage of the module.

### 6.4.3 Surface texture of TCO

Besides low optical absorption and low sheet resistance, the TCO should provide additional optical functions that comprise:

- reduced reflection of the incident light, due to refractive index grading; this antireflection (AR) effect applies to the entire wavelength range of the spectral response;

- light scattering and subsequent light trapping in the silicon absorber; this second effect applies to weakly absorbed light that penetrates to the reflective back contact of the cell. This effect is absolutely essential for thin-film silicon solar cells, and particularly for μc-Si cells that are preferentially used in a-Si/μc-Si micromorph cell structures.

Both functions are achievable by a suitable surface texture of the TCO. As already mentioned above, surface texture is generated either during the deposition of the TCO layer, i.e. as-grown, or it is generated by a chemical etch process in diluted acid [Kluth-2003] following the deposition by sputtering. Depending on these two origins of surface texture, there are two general kinds of texture details that may be typified as follows:

- pyramids, arising on as-grown films, such as $SnO_2$:F deposited by APCVD, or ZnO:B grown by LPCVD;
- craters, formed by wet-etching of sputter-deposited ZnO:Al.

Examples of these two types are shown in Figures 6.29 and 6.30, and Figure 6.31, respectively.

With all these TCO textures, their features are randomly distributed, and their size is comparable to the longer weakly absorbed wavelengths. Therefore, the light is subject to scattering rather than to diffraction or reflection at well-defined interface planes. Without going into further details on the underlying scattering mechanisms,

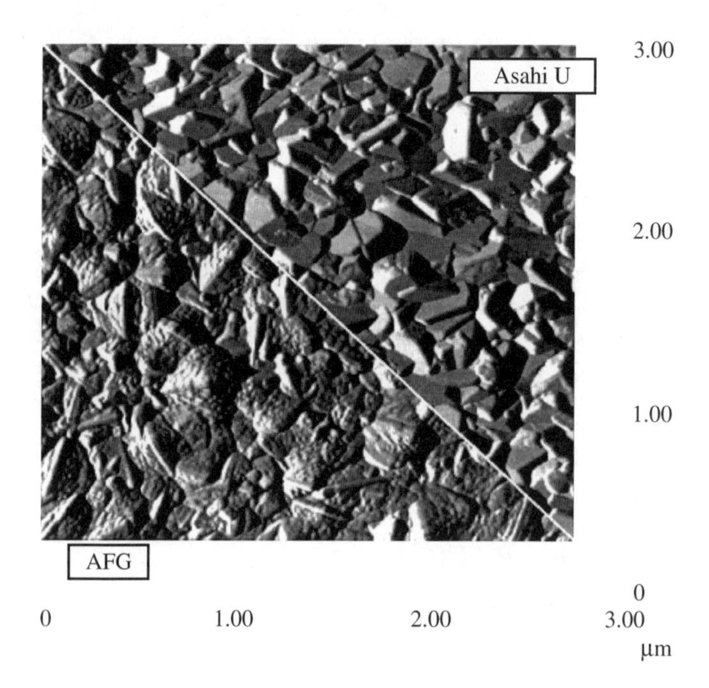


**Fig. 6.29** Atomic force microscopy of surface textures for two products of $SnO_2$:F (courtesy of [Nebel-2000]).

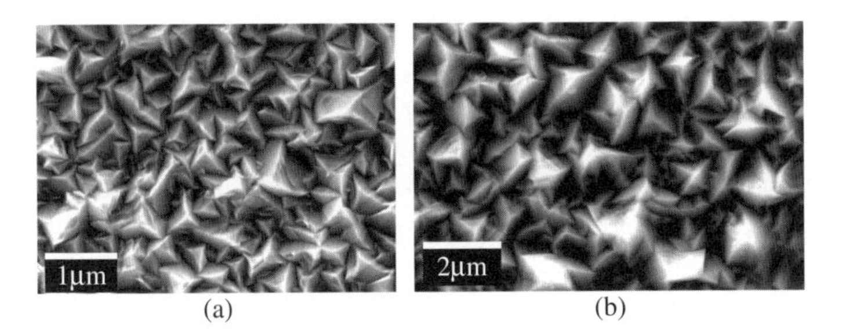


**Fig. 6.30** Scanning electron micrographs of ZnO:B, deposited by LPCVD, and optimized for (a) a-Si cells, (b) μc-Si cells (with permission [Ballif-2006]).

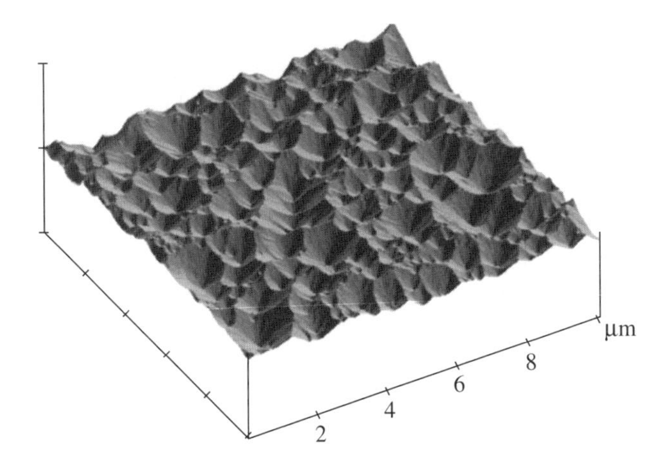


**Fig. 6.31** Atomic force micrograph of texture-etched ZnO:Al (with permission [Kluth-2004]).

the following quantities are used to characterize the TCO texture in efforts to optimize the enhancement of cell currents:

- Root mean square roughness $\delta_{rms}$, which is defined from a linear surface profile by

$$\delta_{rms} = \left\{ \frac{1}{L} \int_0^L [y(x)]^2 \, dx \right\}^{\frac{1}{2}} \tag{6.17}$$

  where $y(x)$ is the deviation from an average profile height at the position $x$ along the linear surface profile of sufficient length $L$. Note that $\delta_{rms}$ is a measure of only the "vertical" variations on a surface texture; it does not contain any information on the distribution of the feature heights and their lateral spacings along the profile, i.e. "horizontal" information.
- Haze is a macroscopic optical measure of surface roughness. It is defined as the ratio between scattered (diffuse) and total transmission; it is usually obtained

from these two transmission values as a function of the wavelength, by using an integrating sphere (so-called Ulbricht sphere). The glass industry more simply characterizes haze by using a haze-meter that renders an avergage value for the visual spectrum weighted with the human eye response. This value closely corresponds to the haze measured at 550 nm.

The haze correlates with the root-mean-square roughness. Within the experimental uncertainties, this correlation holds for both pyramid and crater types of TCO, as shown in Figure 6.32. Theoretically, the wavelength dependence of the haze $H$ may be described by a scalar scattering theory [Bennet-1989], namely by

$$H = 1 - \exp([-(4 \pi C \delta_{rms} (n_{TCO} - n_{air})/\lambda)^2]), \tag{6.18}$$

where $C = 0.5$ is an empirical constant, $n$ refers to the applicable refractive indices (here scattering from TCO into air). This wavelength dependence is in satisfactory agreement with experimental data for scattering from crater-type TCO. For pyramid-type TCO, the above $\lambda^2$ dependence needs to be replaced by a phenomenological $\lambda^3$ dependence, in order to agree with experimental data. The comparison of these different wavelength dependencies (see Fig. 6.33) is based on identical roughness values $\delta_{rms} = 55$ nm.

This difference in the wavelength dependence of the haze is an indication that the surface texture is not adequately described solely by one statistical parameter ($\delta_{rms}$) for the surface morphology, as already noted above.

Further evidence for the inadequacy of only one morphology parameter to fully determine the scattering behavior comes from angle-resolved scattering. Although measurements of the angular distribution of light scattered from textured TCO can be performed only in air, as opposed to the real situation of scattering into the semiconductor structure of a cell, they are nevertheless representative for a comparison of different TCO materials. Even though the $\delta_{rms}$ values of pyramid-and crater-type TCO

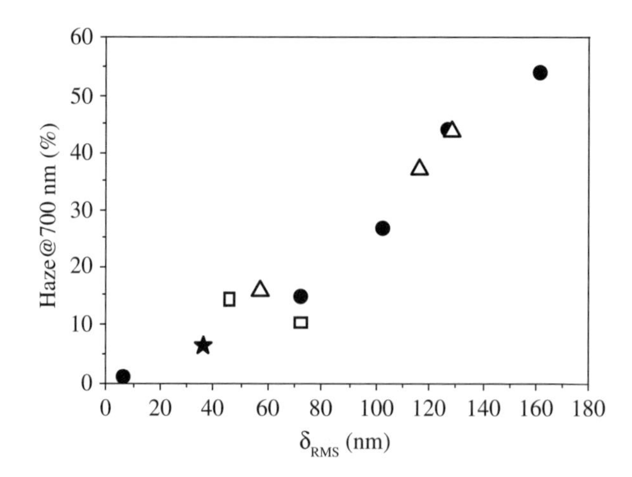


**Fig. 6.32** Haze (700 nm) and root mean square roughness for a variety of differently textured TCO films (with permission [Lechner-2004]).

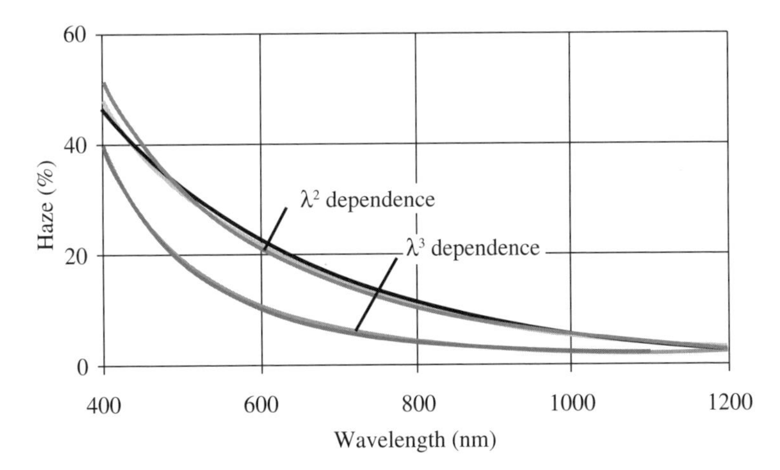


**Fig. 6.33** Wavelength dependencies of the haze for crater- and pyramid-type TCO textures; $\lambda^2$ and $\lambda^3$ dependence for upper and lower curve, respectively, measured under transmission in air (Copyright © [2005] IEEE, with permission [Schade-2005]).

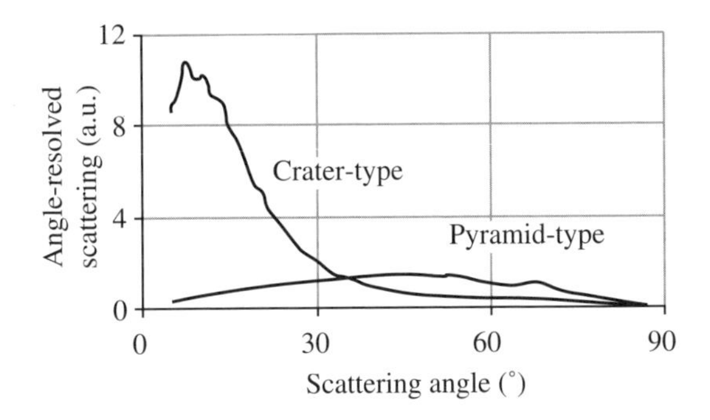


**Fig. 6.34** Angle-resolved scattering distributions for crater- and pyramid-type TCO textures, measured under transmission in air ($\lambda$ = 594 nm) (Copyright © [2005] IEEE, with permission [Schade-2005]).

may be equal, the corresponding angular scattering distributions are substantially different (see Fig. 6.34):

- Lambertian (cosine)-like distributions for pyramid-type textures; and
- predominantly small-angle scattering for crater-type textures.

The distributions shown refer to the angularly resolved power scattered into a certain scattering angle, integrated over all azimuthal scattering angles; isotropic scattering by the substrates was verified.

Here it can only be stated that generally angular scattering distributions depend on both vertical and horizontal dimensions of the surface morphology, and thus more adequately derive from decisive features of the actual surface texture. This aspect will gain importance in the context of correlating specific surface textures with light trapping and hence current enhancements.

### 6.4.4 Cell optics

The effects of rough (textured) TCO in finished cells, compared to smooth TCO, are shown schematically in Figure 6.35 by the reflectances from cell structures for different spectral ranges. Throughout the entire wavelength range the reflectance is reduced due to TCO with haze. For shorter wavelengths, the lower reflectance is due to index grading that arises from the roughness of the front TCO/*p*-layer interface. For the longer and weakly absorbed wavelengths reaching the back contact, a large fraction is reflected from the back contact. Compared to cells with smooth TCO, the reflected portion is reduced in cells with rough TCO, due to a combination of both index grading also at the back contact and light trapping. Since the roughness of the TCO is at least partly replicated at the *n*-layer/back contact interface, index grading and hence reduced reflectance must be assumed to occur there also. Light trapping arises from the light scattered at the rough front and back interfaces, and from the occurrence of total internal reflections at the various interfaces that adjoin media of lower refractive indices (a-Si/TCO with typical refractive indices 3.2/1.9, TCO/glass with 1.9/1.5, and glass/air with 1.5/1.0). From reflectance measurements alone, as shown in Figure 6.35, the individual contributions of back contact reflectance and light trapping cannot be distinguished. The cell reflectance for weakly absorbed red light may be lowered by both reduced reflectance at the index-graded back contact

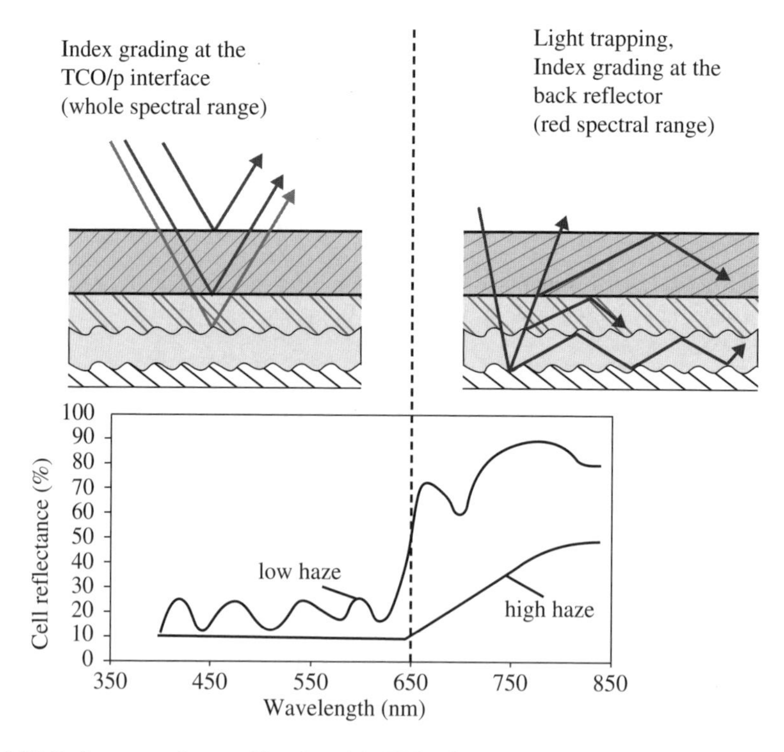


**Fig. 6.35** Reflectances from a-Si cells with TCO of low and high haze, respectively (with permission [Shah-2004]).

interface and/or by light trapping. The degree of both effects is interrelated, i.e. light trapping is enhanced by high back-contact reflectivity.

However, a qualitative separation of these effects, i.e. reduced reflectance or light trapping, can be obtained from additonal measurements of the quantum efficiency. An enhancement of the quantum efficiency in that spectral range (see Fig. 6.36) clearly indicates the presence of light trapping, and cannot be explained solely by a lower reflectance. In this context note that light scattering due to the TCO surface texture leads to a loss of optical phase relations, and hence to the disappearance of optical interferences.

The optical characterization described so far comprises both the individual glass/TCO substrates (wavelength dependence of the true absorbance) and the finished solar cells (wavelength dependence of the reflectance). The latter yields information on the light reflected from the cell or, conversely, on the light input into the cell and, for longer wavelengths, on light trapping. Based on these optical measurements, the so-called effective transmission [Sato-1990] into the cell structure can be defined by

$$T_{\text{eff}} = 1 - A_{\text{glass/TCO}} - R_{\text{cell}}, \tag{6.19}$$

where $A_{\text{glass/TCO}}$ is the true absorbance of the glass/TCO, and $R_{\text{cell}}$ is the cell reflectivity. $T_{\text{eff}}$ represents the fraction of irradiation that is transmitted through the glass/TCO substrate into the cell structure, and at best is available for carrier generation. In comparison, the quantum efficiency is the fraction of light that is actually converted into electrical current. As shown in Figure 6.36 for two TCO materials with different haze (roughness), for high haze the quantum efficiency is substantially increased over the whole spectral range.

In order to further evaluate the enhancement of the quantum efficiency, particularly in the long-wavelength range, a comparison is made between the spectral

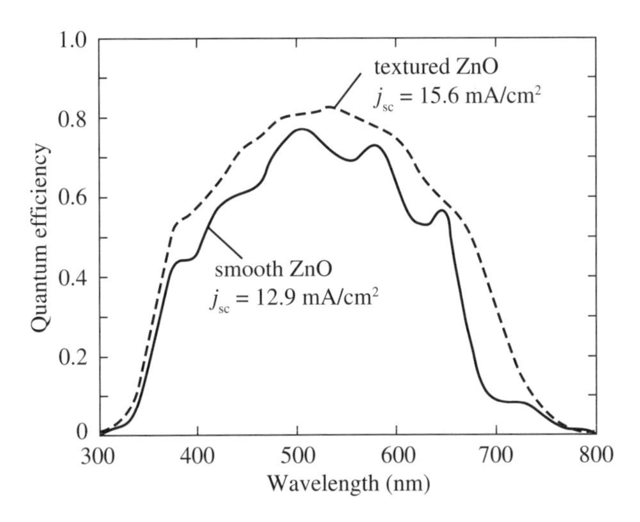


**Fig. 6.36** Quantum efficiencies of two identically deposited *pin* cells on smooth and rough TCO, respectively (with permission [Beneking-1999]).

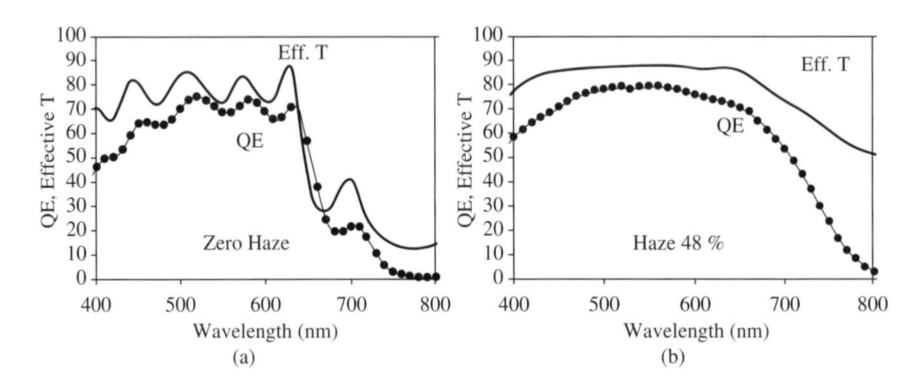


**Fig. 6.37** Effective transmission and quantum efficiency for cells with (a) smooth and (b) textured TCO, i.e. of zero and high haze, respectively (with permission [Lechner-2000]).

dependencies of the effective transmission of the cell structure and the corresponding quantum efficiencies. These dependencies are shown for two extreme haze levels in Figure 6.37. For practically zero haze, the two quantities feature quite similar spectral dependencies, besides distinct optical interferences. The relatively small shift with respect to each other is ascribed to optical losses in the non-photoactive cell layers. These losses can be accounted for by the absorption in the $p$- and $n$-layers under normal incidence (see Sect. 6.3). For high haze, qualitatively different behavior emerges:

- The optical interferences are strongly reduced due to light scattering.
- Compared to the case of low haze, the quantum efficiency is distinctly increased over the whole spectral range, particularly towards the red spectral region. This increase arises from the increased light input into the cell as a consequence of the AR-effect at the textured TCO/$p$ interface, due to index-grading. For longer wavelengths, the quantum efficiency is additionally raised, and the reflectance from the cell significantly reduced, due to light trapping. However, the haze-induced changes in quantum efficiency and reflectance do not fully correspond to each other. The underlying reason must be sought in the next point, namely:
- For higher haze, a substantial difference between the effective transmission and quantum efficiency opens up in the red spectral region. This large difference indicates higher optical losses, which paradoxically result from light trapping. The origin and identification of these losses are discussed further below.

### 6.4.5 Light management in cells

As we have seen in Figures 6.36 and 6.37, the quantum efficiencies and hence the cell currents can be substantially enhanced by increased absorption in the photo-active layers. Such increases follow from textured TCO that gives rise to light scattering and light trapping. Two aspects of light management in cells need to be considered:

- a correlation between texture properties of the TCO and current enhancement, to “tailor” the TCO for optimal properties in different cell types;

- identification and reduction of optical losses within the cell (for modules electrical losses due to the TCO are additional (see Sect. 6.5).

It is generally difficult to unambiguously evaluate texture-related effects in terms of a higher quantum efficiency and hence short-circuit current, since the degree of texture may inherently depend on the TCO film thickness and hence on its absorbance. This generally applies to the as-grown textures obtained for $SnO_2$:F deposited by APCVD and ZnO:B deposited by LPCVD. Alternatively, the use of sputtered ZnO offers the possibility of producing surface texture independently of the ZnO absorbance, and thus allows one to separately assess the efficacy of light scattering, provided identical cells are deposited on these TCO films of different texture but equal absorbance [Frammelsberger-2000]. As an illustration, three TCO substrates with equal absorbance but different surface textures, characterized by their roughness $\delta_{rms}$ and haze *H,* are shown in Figure 6.38 [Kluth-2004]; also shown are the corresponding short-circuit currents of identical μc-Si cells deposited on these substrates. While *H* and $\delta_{rms}$ correlate as shown in Figure 6.32, no correlation appears between the short-circuit currents $J_{sc}$, and $\delta_{rms}$ or *H*, i.e. the same currents occur for quite different surface parameters, and the same surface parameters lead to different currents. These examples are representative of the general observation that there is no unique correlation between the simple surface parameters, $\delta_{rms}$ or *H,* and short-circuit currents $J_{sc}$. This is a consequence of fundamental reasons requiring one to take into account both vertical and horizontal statistical dimensions of the texture details, as already described in Section 6.4.3.

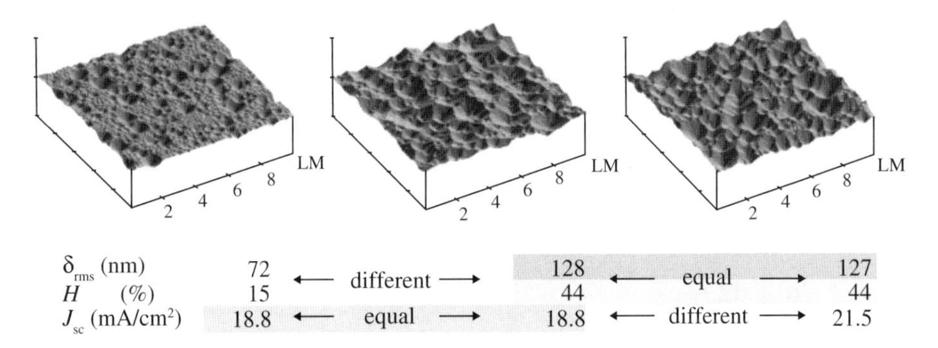


**Fig. 6.38** Short-circuit currents of identical μc-Si cells deposited on TCO substrates with different surface texture parameters (adapted from [Kluth-2004]).

Investigations addressing these findings include evaluations of angular scattering distribution and power spectral density [Schade-2005, Dekker-2005, Stiebig-2006]. Thus geometrical features and scattering-specific properties are combined and tend to yield a unique relationship between the parameters of surface texture and current enhancement. Without going into further detail, the following reasoning is given:

- Angular scattering distribution: light trapping relies on scattering into angles larger than the critical angle for total internal reflections. Thus only the light intensity integrated over angles beyond the so-called escape cone contributes to an enhancement of the currents. Since the angular scattering distributions (measured in air) are almost independent of the wavelength [Stiebig-2006], they apply to all incident wavelengths, and not specifically to those that are predominantly trapped, and hence contribute to the light trapping. These long wavelengths $\lambda_{trap}$ relevant for trapping span a relatively narrow range: shorter wavelengths cannot be trapped because they are absorbed within the cell thickness prior to incurring multiple passes; longer wavelengths require multiple passes back and forth across the cell thickness, but are subject to cumulated losses increasing with the number of passes (see below).

- Power spectral density (PSD): Optical scattering depends on the size of the texture details, which may be derived from atomic force micrographs (AFM). The PSD is part of the evaluation package of AFM, and it describes the decomposition of a texture profile into sine profiles and measures the strengths of various spatial wavelengths. Based on the assumption of Mie scattering [Schade-1985b], one would expect scattering, and hence light trapping, to be most effective for profile components described by a spatial sine wave of a wavelength (within the silicon) comparable to $\lambda_{Si} = \lambda_{trap}/n$, where $n$ is the refractive index of silicon. An evaluation of surface textures in terms of PSD at the spatial wavelength $\lambda_{Si}$ is thus taken as a measure of the trapping-relevant scattering efficiencies of the various TCO samples. Using these PSD values as weighting factors for the large-angle light scattering may lead to a satisfactory general correlation between surface texture and current enhancement.

It is worth noting that the feature sizes (heights and base widths of "pyramids", or depths and diameters of "craters"), as seen in Figures 6.29 - 6.31, approach, or are within the order of magnitude of, the wavelength $\lambda_{Si} = \lambda_{trap}/n$. From the efficiency of scattering stated above, it may be concluded that the surface texture ought to be adapted to the wavelength range to be trapped, around $\lambda_{trap}$, which is shifted more to the red spectral range for μc-Si compared to a-Si. This is also noticeable in Figure 6.30(b), where the average feature size, chosen for μc-Si, is distinctly larger.

### 6.4.6 Optical losses

Although TCO surface texture leads to higher absorbance in the *i*-layer, and hence to more current generation, it also causes higher absorption losses in the non-photoactive layers and higher reflection losses at the back contact.

Light scattering causes longer optical path lengths due to an angular distribution of the light input. Compared to the optical losses based on the path lengths of the nominal layer thicknesses, the path lengths are now longer and given by $d_{n,p}/\cos\theta$, where $\theta$ is an average scattering angle. For the longer wavelengths ($\lambda > 600$ nm), penetrating to the back reflector and being reflected, the light scattering leads to light trapping, which involves:

- multiple light passes in the semiconductor silicon; only the path length within the *i*-layer contributes to the current generation, whereas the portions through the *n*- and *p*-layer cause absorption losses;

- multiple light passes in the TCO and glass, as a result of total internal reflections at the Si/TCO, TCO/glass and glass/air interfaces, which represent transitions to a lower refractive index, i.e. 3.2/1.9 for Si/TCO, 1.9/1.5 for TCO/glass, and 1.5/1.0 for glass/air. Here the absorbance values of glass/TCO, as exemplified in Figure 6.27, multiply with the number of light passes.
- multiple reflections at the back contact; reflection losses occur with the reflectance to the power of the number of reflections.

The sum of the losses is represented by the difference between the effective transmission and the quantum efficiency, as shown in Figure 6.37. The individual contributions to these losses can be determined separately; their wavelength dependences are shown in Figure 6.39, illustrating the case of a-Si *pin* cells on TCO with high haze:

- In the blue-green spectral region ($\lambda < 600$ nm), mainly the *p*-layer absorbance accounts for the losses (see Sect. 6.3.1).
- Towards the red spectral region ($\lambda > 600$ nm), losses from the non-photoactive layers increase with wavelength due to light trapping. The effective optical path lengths extend beyond the nominal layer thicknesses, and may amount for a number of passes across the layers. The number of light passes are estimated from a comparison of the quantum efficiency $QE(1)$, calculated for 1 light pass with the nominal path length equal to the *i*-layer thickness $d_i$, and the measured quantum efficiency $QE(N)$, arising from $N$ passes with a total path length of $N\,d_i$. For $\lambda > 600$ nm, these quantum efficiencies are:

$$QE(1) = QE_{\max}(1 - \exp(-\alpha\, d_i)), \tag{6.20}$$

and

$$QE(N) = QE_{\max}(1 - \exp(-\alpha\, N\, d_i)), \tag{6.21}$$

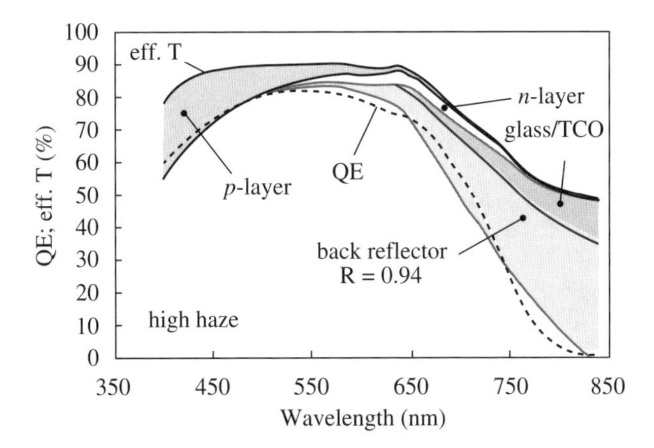


**Fig. 6.39** Spectral dependence of the optical loss contributions for a-Si cells deposited on TCO with high haze (Copyright © [2000] IEEE, with permission [Lechner-2000]).

where $QE_{max}$ is the maximum of the spectral dependence of the quantum efficiency, measured typically at a shorter wavelength that is fully absorbed within the *i*-layer thickness.
For very low haze (no data shown here), $N$ saturates with the wavelength to the value 2, which corresponds to two light passes (incident plus reflected once), whereas for high haze $N$ considerably increases with the wavelength, as shown in Figure 6.40.

- The same number $N$ of light passes as deduced above also applies to the non-photoactive *n*-layer, and also to the glass/TCO. At least for single-junction cells discussed here, *p*-layer losses are not enhanced by multiple light passes, since the absorbance of longer wavelengths is practically negligible due to the higher band gap of a-SiC. Although there is no enhancement of *p*-layer losses due to light trapping, their relative contribution at shorter wavelengths from 400 to 650 nm is quite appreciable (see also Sect. 6.3.1).

In order to account for the angular light distribution, the nominal layer thicknesses $d$ must be replaced by effective thicknesses $d/\cos\Phi$, where now $\Phi$ is an average scattering angle larger than the corresponding critical angle for internal total reflection. Furthermore, these optical losses must be weighted with the corresponding fractions of light intensity that are subject to total internal reflections at the respective interfaces.

- The reflectance $R$ of the back contact may be reduced by some index grading analogous to the AR effect at the front surface, since the TCO texture may be partly replicated at the back contact interface. With multiple reflections, the reflectance is amplified with the power of half the number of passes (one pass back and forth to the next reflection), i.e. the effective reflectance is $R^{N/2}$, and the resulting reflection losses are $1 - R^{N/2}$.

This short description of a detailed reasoning for the quantum efficiency and optical losses is meant only to convey the complexities encountered in cell optics,

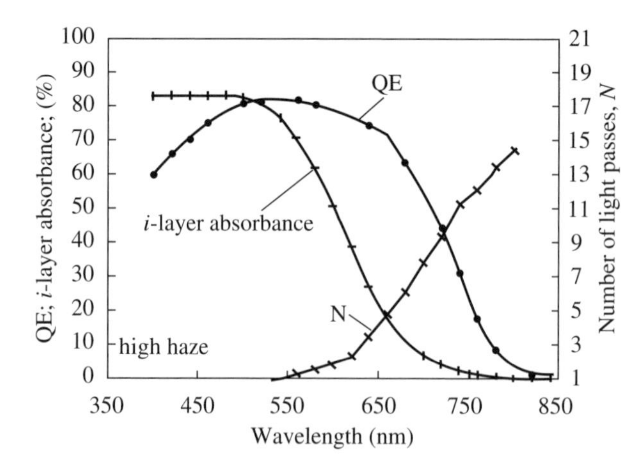


**Fig. 6.40** Quantum efficiency, *i*-layer absorbance and number $N$ of light passes in a-Si cells deposited on TCO with high haze (Copyright © [2000] IEEE, with permission [Lechner-2000]).

even under various simplifying approximations. Nevertheless, it may serve as a guide to identify key properties to optimize the solar cell efficiencies.

## 6.5 LASER SCRIBING AND SERIES CONNECTION OF CELLS

In addition to materials-specific properties, thin-film PV technologies possess the following advantages:

- cells can be manufactured on large-area substrates, with substrate sizes up to 5.7 $m^2$;
- two-terminal stacked cell structures that allow for higher voltages and lower currents;
- incorporation of different bandgaps for extended spectral response;
- monolithic series connection of cells to modules, which is highly reliable and does away with the need for external cell-to-cell solder joints, as is commonly required for wafer-based crystalline-silicon technologies, and also for stainless-steel-based thin-film roll-to-roll technology (see Sect. 6.1.9).

This section deals with the last point, i.e. how cells on large-area substrates are formed and electrically connected. This series connection is to be accomplished with minimal series resistance losses, and minimal losses of photoactive area.

### 6.5.1 Cell interconnection scheme

The functional layers of a thin-film silicon solar cell basically comprise

- a transparent front contact, typically consisting of a transparent conductive oxide (TCO);
- the semiconductor diode structure;
- the back contact, which may typically involve silver or aluminum and should also act as an optical reflector.

The thin-film solar modules are formed by cells in the form of parallel stripes of width $w$, separated by a narrow photo-inactive regions of width $\Delta w$ between the cells. These regions comprise three adjoining channels, in each of which one of the functional layers is removed in order to electrically connect the front and back contact layers of adjacent cells (see Fig. 6.41). In a typical embodiment, which applies to glass-based superstrate structures, this cell interconnection scheme is realized as follows:

The TCO layer is subdivided into parallel stripes by laser scribing. During this step the cell width is defined. Subsequently, deposition of the semiconductor structure covers the stripe-patterned TCO layer, leaving the semiconductor without a conducting front contact along the scribe widths of the TCO layer. Now the semiconductor layer is also subdivided by laser scribing in such a way that these new scribes are positioned with a slight offset from the TCO scribes. As a result, the TCO is exposed along the scribe widths of the semiconductor layer, and subsequently brought into contact with

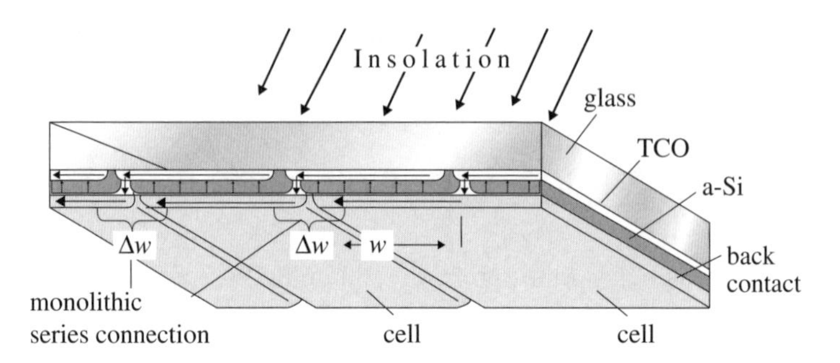


**Fig. 6.41** Monolithic series connection of thin-film cells (photo-active cell width $w$, photo-inactive interconnection width $\Delta w$) [Schade-2001].

the back contact layer upon its final deposition. In order to prevent this conducting back contact layer from forming a common contact to all cell stripes, it needs to be subdivided along the individual cell stripes as well, by removing the back contact layer along narrow channels. The latter are again slightly offset from the semiconductor scribe.

By subdividing the functional cell layers to form narrow channels slightly offset from each other, electrical contacts between the front TCO layer and the back contact layer of adjacent cell stripes are established in the course of depositing the respective cell layers. Generally, such an assembly of series-connected cells represents a solar module. While for technologies based on crystalline silicon this series connection can only be obtained by solder-joining individual cells, thin-film technologies allow for a monolithic series connection of cells as described above. The solar module thus formed requires only two separate lead connections that are attached to the first and last cell stripes on the module, and are led to an external junction box. Within tolerable power losses (see below), the module voltage can be chosen: it is determined by the number of series-connected cells, and thus depends on the chosen cell width and the number of cells that can be accommodated lengthwise or crosswise on a rectangular substrate.

As an example for laser scribing, Figure 6.42 shows a portion of the interconnect area between two adjacent cells in plan view.

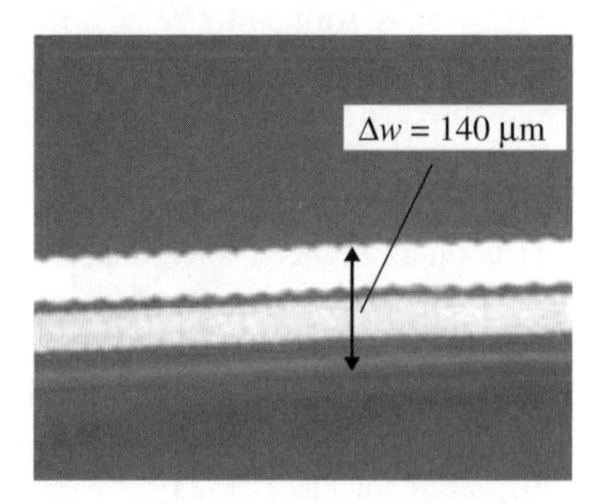


**Fig. 6.42** Laser scribes to establish series connection between thin-film solar cells, shown in plan view; from top to bottom: separation of back contact, of semiconductor layers, and of front TCO layer, respectively (Copyright © [2005] IEEE, with permission [Meier-2005]).

In this particular case, the interconnect width is $\Delta w = 140\,\mu m$. The electrical contact between the two cells occurs along the central scribe; the neighboring scribes represent the separation of the TCO front layer and the back contact layer, respectively. Technical details on laser scribing are rarely published [Golay-2000]. Lately, specialized equipment manufacturers offer dedicated laser patterning systems, also for large-area modules.

### 6.5.2 Power losses due to the series connection of cells

The series connection of cells to form a module inevitably results in power losses, so-called inherent losses, that are due to

- area losses: obviously, the interconnect areas cannot be photoactive, and hence represent area losses given by the ratio $\Delta w/(w + \Delta w)$, where $w$ is the photoactive cell width, and $\Delta w$ is the interconnect width, comprising the widths of the three laser scribes and the two interjacent spacings;
- Joule losses: mainly resulting from the sheet resistance of the TCO layer, leading to series resistance losses.

Both types of losses are a function of the cell width $w$, and their sum can be minimized, as shown below.

In series-connected cells the photo-generated current of each cell must be conducted across the contact layers through the contact at the central scribe to the next cell, i.e. across the TCO layer through the line contact between the TCO and the back contact to and across the back contact layer of the adjacent cell. Joule losses arise from the sheet resistances of the contact layers. For frequently used metallic back contact layers, the electrical resistivity ρ and also the sheet resistance, defined by $R_{sh} = \rho/d$, where $d$ is the layer thickness, are sufficiently small, and thus lead to negligible Joule losses. However, this does not apply to the front TCO layer, for which Joule losses must be taken into account. These are calculated as follows.

At the maximum power point, each cell generates a uniform current density $j_{mp}$ across the cell stripe. This current density is collected by the TCO layer and conducted within the TCO layer as an increasing current

$$J(x) = j_{mp}\,\ell\,x \tag{6.22}$$

towards the line contact to the adjacent cell, where $\ell$ is the length of the cell stripe, and $x$ the coordinate of the cell width (see Fig. 6.43). The total power generated by this current across the cell stripe follows from an integration in the $x$-direction, namely

$$\Delta P = \int J(x)^2\,\mathrm{d}R \tag{6.23}$$

$$\Delta P = \int (j_{mp}\ell x)^2 (R_{sh}/\ell)\,\mathrm{d}x = j_{mp}^{\ 2} w^3 \ell R_{sh}/3 \tag{6.24}$$

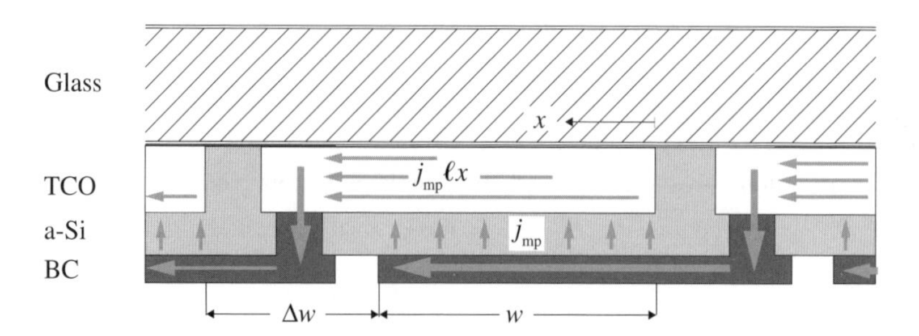


**Fig. 6.43** Schematic cross section of current flow through series-connected cell stripes (photo-active cell width $w$, cell length $\ell$; photo-inactive interconnection width $\Delta w$, current density at the maximum power point $j_{mp}$, back contact BC).

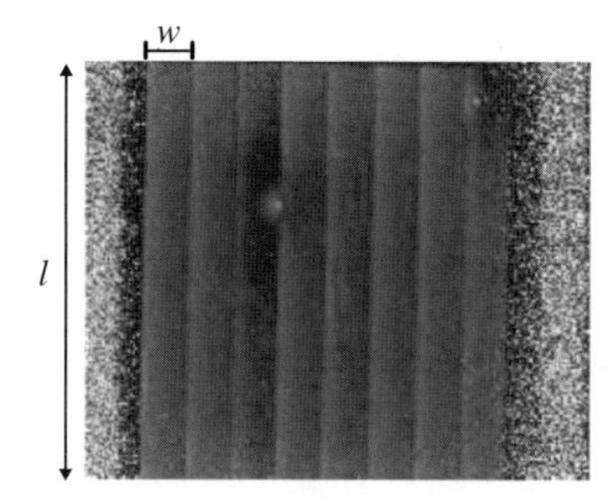


**Fig. 6.44** Thermographic image of a module with series-connected cells (courtesy of [Rech-2005]).

This width-dependent power loss across the TCO layers of cells is qualitatively illustrated by a thermographic image from a portion of an illuminated module (see Fig. 6.44): the increasing brightness towards the cell interconnects is proportional to the generated heat $\Delta P$, which according to Equation 6.24 increases with the cube of the cell width.

The *relative* power loss, due to the TCO sheet resistance, at the maximum power point of the module is

$$(\Delta P/P)_{mp} = j_{mp}\, w^2\, R_{sh} / 3\, V_{mp}, \tag{6.25}$$

where $V_{mp}$ is the cell voltage at the maximum power point. This relation indicates significant differences in electrical losses, depending on the type of cells chosen for the module. For double-junction cells, such as or example a-Si/a-Si, the current densities are about one half, and the cell voltages about twice, compared to single-junction cells. Thus, the relative electrical losses would only be one quarter, if the cell width were kept constant; however, for various practical reasons (such as module voltage, production throughput) one may choose wider cell widths and still keep losses lower than those in single-junction cells.

In addition to the electrical losses just described, there is also a relative area loss, caused by the photo-inactive interconnect area; it amounts to

$$(\Delta P/P)_{area} = \Delta w/(w + \Delta w) \quad (6.26)$$

While the relative electrical losses increase with $w^2$, the area losses are proportional to $1/w$ (see Fig. 6.45); thus, the combination of these two loss contributions leads to an optimal cell width, for which the resulting losses are minimal (see Fig. 6.46). The curves shown in these two figures apply to typical $I(V)$ data of single-junction a-Si cells, namely $j_{mp} = 12$ mA/cm$^2$, $V_{mp} = 0{,}76$ V; the interconnect width is assumed to be a typical value of $\Delta w = 250$ $\mu$m. The important influence of the TCO front contact layer is illustrated with three different sheet resistance values

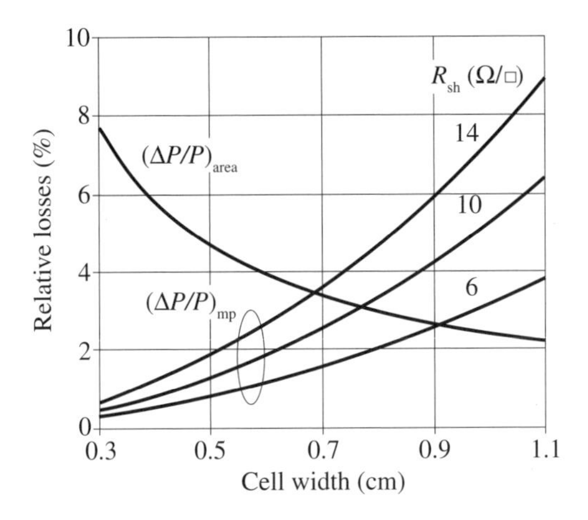


**Fig. 6.45** Relative losses of generated module power resulting from area losses due to photo-inactive interconnection areas, and Joule losses due to the TCO sheet resistance.

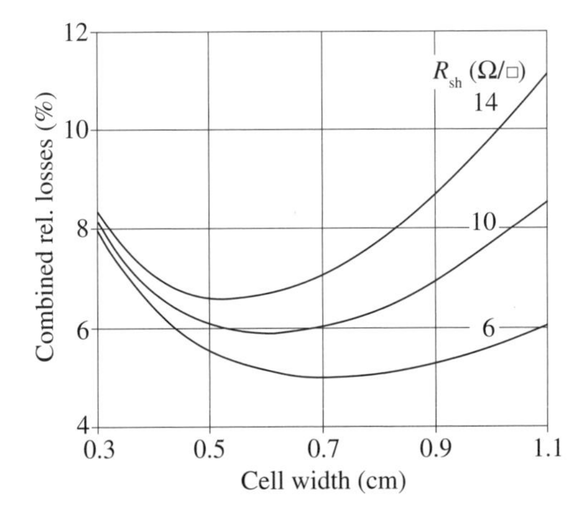


**Fig. 6.46** Combined relative losses of generated module power due to series connection of cells.

**Table 6.1** Optimal cell widths for solar modules based on single- and double-junction cell structures (a-Si/a-Si and a-Si/μc-Si) for various sheet resistance values of the TCO front layer.

| Cell type | $V_{mp}$ [V] | $j_{mp}$ [mA/cm$^2$] | $R_{sh}$ (Ω/□) | | |
|---|---|---|---|---|---|
| | | | 6 | 10 | 14 |
| | | | Optimal cell width (cm) | | |
| a-Si | 0.76 | 12 | 0.72 | 0.61 | 0.54 |
| a-Si/μc-Si | 1.10 | 10 | 0.86 | 0.73 | 0.65 |
| a-Si/a-Si | 1,52 | 6 | 1.15 | 0.97 | 0.86 |

$R_{sh}$ = 6, 10, 14 Ω/□. These sheet resistance values cover the range for typical TCO materials.

The optimal cell widths derived from the above data are shown in Table 6.1; also shown in this table are the optimal cell widths for tandem cell structures, both a-Si/a-Si and a-Si/μc-Si (micromorph structures). Note that such tandem cell structures are very common in thin-film silicon photovoltaics (see Sect. 5.4).

Unlike the cases discussed so far where the sheet resistance of the bottom contact layer could be assumed to be negligible, other cell structures are used where the back contact layer also consists of TCO material. In these cases, additional losses need to be considered, i.e. the optimal cell widths must now be determined by the sum of the sheet resistances of the front and back contact layers.

## 6.6 MODULE PERFORMANCE

### 6.6.1 Efficiencies

All applications in photovoltaics, based on either crystalline silicon technologies or thin-film technologies, basically involve a number of solar cells electrically connected in series to form the essential constituents of a solar module (so-called "raw module"). Additional process steps, such as the attachment of leads and encapsulation for protection against external influences will be discussed in Section 6.7.

For the user of a solar module it is important to compare the properties of solar modules derived from different technologies, and from various manufacturers. The standard method used so far to express such comparisons is conversion efficiency, measured under standard test conditions (STC), i.e. at 25 °C with 1 kW/m$^2$ irradiation under normal incidence with an AM 1.5 g spectrum [Air Mass 1.5 g stands for global (direct plus diffuse) solar irradiation transmitted through the 1.5-fold thickness of the atmosphere, i.e. at a zenith angle of 48.2 °]. For valid efficiency comparisons, it is absolutely imperative to refer to equivalent conditions (e.g., not to compare efficiencies of laboratory cells and commercial modules, see below). Based on *record cell* efficiencies, *average module* efficiencies may be projected. This is needed for any estimate of manufacturing costs, balance-of-system costs, and, hence, market acceptance, and ultimately the financial development and outcome of an investment.

The efficiency value that is relevant for a given product is the statistical average of the module efficiencies, referred either to the aperture area or to the total module area. For all PV technologies, there is, generally, a substantial discrepancy of 30 to 40 % between record efficiencies achieved with laboratory cells and efficiencies available in commercial modules. The reasons for this discrepancy are as follows:

- Inherent "losses": The transition from individual cells to large-area modules of interconnected cells implies a monolithic series connection of cells (see Sect. 6.5), which leads to Joule losses due to the sheet resistance of the TCO and contact resistances, and to area losses due to the photo-inactive cell interconnect and contact areas. Furthermore, the large-area deposition leads to areal non-uniformities due to non-uniform potential distributions and feedgas flow distributions in the PECVD reactor.
  Specific to the superstrate configuration are the constraints in the commercially available large-area TCO substrate materials (see Sect. 6.4), which may exhibit non-optimal optical properties, compared to, for example, research-grade type U or HU material [Taneda-2008] supplied by AGC Asahi, or texture-etched sputtered ZnO [Kluth-2003]. Provided the module manufacturer has not decided to produce his own TCO, he must rely on purchased TCO material, which then possibly must be counted as an inherent cause for reduced efficiency.
- Non-compatible objectives: the achievement of record cell efficiencies rests on the use of special effects and/or processes to raise the efficiency, irrespective of the complexity and expense, while module efficiencies must rely on production-relevant procedures and economic considerations that must satisfy throughput, yield and direct-cost criteria, and hence economic viability.

Table 6.2 summarizes the inherent and process-related effects, and corresponding efficiency reduction factors. Except for the TCO influence, these factors

**Table 6.2** Typical efficiency reduction factors: from record research cell to production module.

| Effect | Efficiency Reduction factor | | Device | Area |
|---|---|---|---|---|
| | Per effect | Cumulated | | |
| **Record research cell** | 1.00 | **1.00** | cell | cell |
| Non-optimal TCO | 0.90 | 0.90 | cell | cell |
| Non-uniform deposition | 0.96 | 0.86 | module | photo-active |
| Production-relevant processes | 0.90 | 0.78 | module | photo-active |
| Series resistance | 0.96 | 0.75 | module | photo-active |
| Cell interconnect area loss | 0.96 | 0.72 | module | aperture |
| Peripheral film removal | 0.95 | 0.68 | module | total |
| **Production average** | 0.95 | **0.65** | **production module** | total |

qualitatively also apply to other thin-film technologies. It is assumed that the reference record cell efficiency corresponds to the stabilized condition. The assumed reduction factors are typically representative. It must be noted that only the cell interconnect area loss will differ, depending on single-junction or tandem-junction cell structures which require different optimal cell widths for minimal losses (see Sect. 6.5.2). Given a certain interconnect width, the relative cell interconnect area losses will differ; this difference is neglected in the present context.

Table 6.2 thus shows that the cumulative effect of the reduction factors amounts to typically 0.65, i.e. the average total-area efficiency of production modules is about 35 % below the coresponding record research cell. The inherent losses mainly stem from the monolithic series connection of cells (see Sect. 6.5.2), i.e. from both

- Joule losses due to the sheet resistance of the TCO and from contact resistances; and
- area losses due to the cell interconnect and contact areas.

Furthermore, large-area deposition leads to area non-uniformities due to non-uniform potential distributions and feed gas flow distributions in the plasma reactor used for PECVD of the silicon-based cell layers. Moreover, all procedures of the entire processing sequence are subject to the effícts of statistics. Hence, an additional inherent efficiency reduction factor arises from the averaging. While these effects on efficiency reductions are more or less inherent, the influences of the TCO quality and/or production-relevant processes may be reduced by choosing alternative approaches. However, this usually entails higher costs, which must be weighed against the achievable efficiency increase in terms of the resulting cost/W.

Representative stabilized module efficiencies, corresponding to data sheet specifications of several manufacturers, are compiled in Table 6.3. From the module dimensions and the stabilized power outputs, the efficiencies referring to the total module area (total-area efficiencies) are derived. In cases where the specifications distinguish between several classes of power output, only the top-of-the-line product is included in Table 6.3.

### 6.6.2 Energy yield

Besides comparisons in terms of efficiencies referred to STC, the user may judge, in a more pragmatic way, the performance of modules in terms of the energy yield under a given irradiation. The energy yield (kWh/m$^2$year) is the energy generated per square meter over a given period of time at a certain location. Since the cost of a PV installation usually refers to the nominal power, the user may be particularly interested in the so-called relative performance (kWh/kWp year), which is defined as the energy yield per installed nominal power, rather than per square meter. The relative performance indicates the number of hours per year the modules would generate energy at the nominal power. Still another practical quantity, derived from these energy data, is the so-called effective efficiency. Unlike efficiency values normally referred to STC, the effective efficiencies are the ratios between the energy yields and the corresponding irradiations averaged over one year; they include all variations of temperature, solar irradiation intensity and spectral distribution, which affect the power generation

**Table 6.3** Stabilized power outputs (at STC) and total-area efficiencies, according to data sheet specifications of representative module manufacturers.

| Manufacturer | Cell structure | Power (W) | Module size | | Efficiency (%) |
|---|---|---|---|---|---|
| | | | Dimensions (m × m) | Total area ($m^2$) | |
| Bosch Solar Thin Film GmbH | a-Si | 90 | 1.300 × 1.100 | 1.43 | 6.3 |
| Inventux | a-Si | 94 | 1.300 × 1.100 | 1.43 | 6.6 |
| Mitsubishi Heavy Industries | a-Si | 100 | 1.414 × 1.114 | 1.58 | 6.3 |
| Signet Solar | a-Si | 400 | 2.200 × 2.600 | 5.72 | 7.0 |
| EPV Solar | a-Si/a-Si | 58 | 1.321 × 0.711 | 0.94 | 6.2 |
| SCHOTT Solar Thin Film GmbH | a-Si/a-Si | 103 | 1.308 × 1.108 | 1.45 | 7.1 |
| Bosch Solar Thin Film GmbH | a-Si/a-Si | 115 | 1.300 × 1.100 | 1.43 | 8.0 |
| Kaneka | a-Si/μc-Si | 110 | 1.008 × 1.210 | 1.22 | 9.0 |
| Inventux | a-Si/μc-Si | 130 | 1.100 × 1.300 | 1.43 | 9.1 |
| Mitsubishi Heavy Industries | a-Si/μc-Si | 130 | 1.414 × 1.114 | 1.58 | 8.3 |
| Sharp | a-Si/μc-Si | 135 | 1.409 × 1.009 | 1.42 | 9.5 |
| Sunfilm (Sontor) | a-Si/μc-Si | 150 | 1.684 × 1.056 | 1.78 | 8.4 |
| Sunfilm | a-Si/μc-Si | 490 | 2.200 × 2.600 | 5.72 | 8.6 |
| Uni-Solar | a-Si/a-SiGe/a-SiGe | 144 | 5.486 × 0.394 | 2.16 | 6.7 |

[Schade-1998] differently for each product and technology at the specific location of the installation. Thus the following dependencies are involved:

- *Temperature*: the temperature coefficients of the efficiency, e.g., that for a-Si is only one half of that for c-Si.
- *Irradiation intensity*: for a-Si the efficiency remains higher at lower light levels compared to c-Si.
- *Spectral composition of the irradiation*: due to its higher energy gap, a-Si is more sensitive to the blue portion of the spectrum compared to c-Si. The outdoor irradiation (insolation) contains higher contributions in the blue spectral region, both daily and seasonally (i.e. around noon and summer time, respectively), which coincides with periods of higher irradiation intensity and, therefore, favors the relative energy yield of a-Si. In this context, it should be noted that the spectral benefit might be weakened for tandem- and even more for triple-junction cell structures, since for optimal performance they are required to be current-matched. However, this requirement can be reached with a time-dependent spectral distribution only over relatively shorter periods compared to single-junction cell structures.

Table 6.4 [Jardine-2002] shows energy yields of different cell technologies, represented by modules placed in two locations in Europe with substantially different insolation

**Table 6.4** Energy yields and relative performances of different cell technologies, installed in locations of different insolation.

| **Cell type** | **Mallorca (1700 kWh/m$^2$ year)** | | | | | **Oxford (1022 kWh/m$^2$ year)** | | | | |
|---|---|---|---|---|---|---|---|---|---|---|
| | Efficiency STC [%] | Energy yield [kWh/m$^2$ yr] | Relative performance [kWh/kWp yr] | Effective efficiency [%] | Converter [%] | Efficiency STC [%] | Energy yield [kWh/m$^2$ yr] | Relative performance [kWh/kWp yr] | Effective efficiency [%] | Converter [%] |
| Tandem: a-Si/a-Si | 5.3 | 84 | 1585 | 5.3 | 92.7 | 5.3 | 51 | 959 | 5.4 | 91.7 |
| Triple: a-Si/ a-SiGe/a-SiGe | 6.3 | 87 | 1380 | 5.7 | 90.6 | 6.3 | 54 | 859 | 5.9 | 90.5 |
| CIGS | 9.7 | 150 | 1553 | 9.5 | 93.4 | 9.7 | 99 | 1025 | 10.5 | 92.2 |
| Mono-cryst.-Si | 13.5 | 188 | 1389 | 11.6 | 95.3 | 13.4 | 117 | 872 | 12.4 | 92.4 |
| Multi-cryst.-Si | 11.0 | 150 | 1354 | 9.1 | 97.0 | 11.5 | 99 | 859 | 10.4 | 92.8 |

(Mallorca and Oxford). Obviously, the energy yields per square meter are approximately proportional to the average insolation, modified by the individual effects of temperature and spectral distribution, as well as by the conversion factor of the applied converter.

More importantly, the energy yields per module power, i.e. the relative performances, are higher for the thin-film-based, compared to the crystalline-Si, technologies. This difference applies to both locations, and it is particularly significant for the examples of the a-Si-based tandem and CIGS modules. While the examples shown are not yet sufficient to claim the general validity of the observed differences, additional evidence for higher relative performances of a-Si based modules, compared to crystalline-Si, is shown in Figures 6.47 and 6.48.

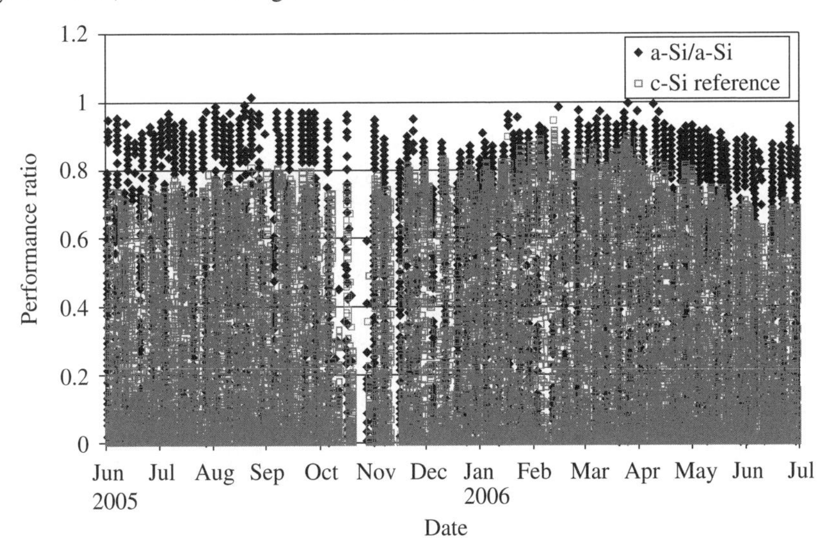


**Fig. 6.47** Performance ratios (effective efficiency/STC efficiency), measured in Florida, for a-Si/a-Si compared to c-Si (Copyright © [2006] IEEE, with permission [Jansen-2006]).

From the energy yield considerations above, the following conclusion emerges. In spite of lower STC efficiencies, and correspondingly larger installation areas, a-Si-based modules distinctly outperform c-Si modules in terms of the generated energy per nominal power, and hence generally also in terms of module cost.

## 6.6.3 Partial shading

External conditions, such as partial shading by adjacent buildings, snow cover or deposited leaves, may affect the unimpeded irradiation of modules, and thus result in losses of the module power output. Generally, these losses cannot be accounted for solely by the size of the shaded area, but must be evaluated in terms of the changed current voltage characteristic, $I(V)$, of the partly shaded module. The following discussion is restricted to an individual module that typically consists of a series connection of a number of cells. This applies to both c-Si based modules and monolithically interconnected thin-film modules (see Sect. 6.5). In both cases, the $I(V)$ characteristic under illumination is drastically changed by reduced irradiation. This reduction

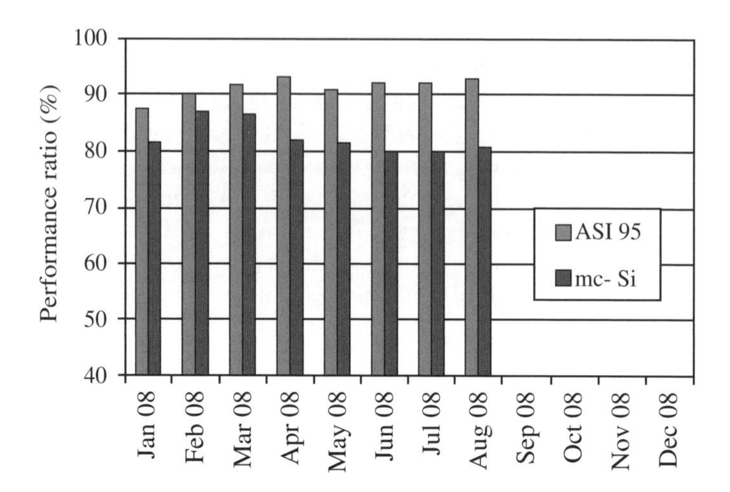


**Fig. 6.48** Performance ratios (effective efficiency/STC efficiency), measured in Munich, for a-Si/a-Si compared to c-Si (with permission [Lechner-2008]).

mainly arises from the series connection of irradiated and only partly irradiated cells, which are all required to support the same current. In series connection this implies that the partly irradiated cells will assume reverse bias to support the required current, depending on the specific "dark" reverse characteristics of the affected cells. In many instances, the arising reverse biases become sufficiently high to cause dielectric breakdown, and thus lead to the formation of irreversible local shunts (so-called hot spots). Hot-spot formation usually originates at local non-uniformities or defects within the cell structure. As a result, the cell is irreversibly shunted and represents a load resistance in series with the remaining unaffected cells.

In order to further quantify the effect of shading on the module power output, $I(V)$ characteristics are shown below for an assumed module with 15 series-connected cells under full irradiation, and for the same module with one of its cells shaded by 50 %. The $I(V)$ characteristics used here are calculated on the basis of a diode model, combined with a drift length model for the photo-current, i.e.

$$I = I_0\,(\exp(qV/nkT) - 1) - I_{\mathrm{ph}}(L_{\mathrm{c}}/d_i), \tag{6.27}$$

where $I_0$ is the diode saturation current, $q$ the elementary charge, $n$ the diode quality factor (typically 1.5), $I_{\mathrm{ph}}$ the photocurrent, $L_{\mathrm{c}}$ the voltage-dependent carrier drift length, and $d_i$ the *i*-layer thickness.

The specific assumptions used to calculate the $I(V)$ curves shown below are typical for a-Si *pin* cells, but are not further detailed in this context.

The series connection of 15 fully irradiated cells is obtained by adding the voltages of 15 individual cells as a function of current. Shading of one cell by 50 % is taken into account by reducing the current of one cell by 50 %. The $I(V)$ characteristic of the partially shaded module is obtained by adding the voltages of 14 individual cells to the voltage of one 50 % shaded cell, as a function of the same current to be supported by all cells. For the shaded cell, this requires an assumption about the behavior of the dark characteristic extending to reverse voltages of approximately the open-circuit voltage of the number of unshaded cells (i.e. to about –14 V under the above assumption of 14 unshaded cells).

Although the reverse characteristics will decisively determine the module behavior under shading, they are usually not measured. This may be partly justified, since at reverse voltages towards the order of 10 V breakdown phenomena may occur, which are quite sensitive to local imperfections and their areal distribution within the cell structure. Thus the breakdown may be localized, reversible, or may lead to irreversible changes, depending on the specific cell structure. There appears to be experimental evidence that for a-Si based cells these breakdown phenomena already occur at lower voltages with a relatively gentler rise in current, compared to c-Si cells. While for the illustration of the power losses due to shading these differences between a-Si and c-Si cells are qualitatively not essential, they are, nevertheless, important in regard to the incorporation of by-pass diodes (see below).

For the illustration here, the reverse characteristic of the partly shaded cell is assumed to be governed by one of two different scenarios:

- a "gentle" breakdown, which phenomenologically may be characterized by a power dependence as $I(V) = C V^4$, where $C$ is a matching constant;
- a shunt resistance that has formed as a result of irreversible breakdown of the shaded cell. The value of the shunt resistance is assumed to be 100 = $\Omega cm^2$ (the shown $I(V)$ characteristics refer to current densities, and hence power densities, $mA/cm^2$ and $W/cm^2$, respectively).

Figures 6.49 and 6.50 show the $I(V)$ characteristics for:

- a module of 15 series-connected cells under full irradiation;
- 14 unshaded cells in series;
- 1 cell shaded by 50 %;
- the partly shaded module, as a result of summing the voltages of the 14 + 1 cells, as a function of the current;
- the shaded cell is assumed either to function under 50 % shading and "gentle" breakdown (Fig. 6.49), or as a shunt after irreversible breakdown (Fig. 6.50).

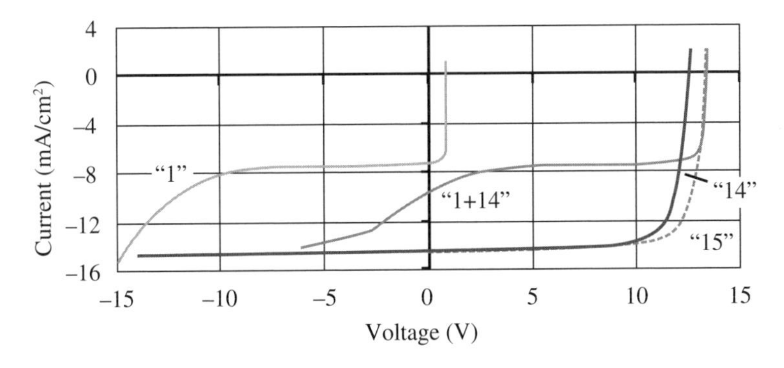


**Fig. 6.49** $I(V)$ characteristics for: a module of 15 series-connected unshaded cells ("15"), one cell 50 % shaded ("1"), 14 series-connected unshaded cells ("14"), and a module of 15 series-connected cells, where one cell is 50 % shaded ("1 + 14").

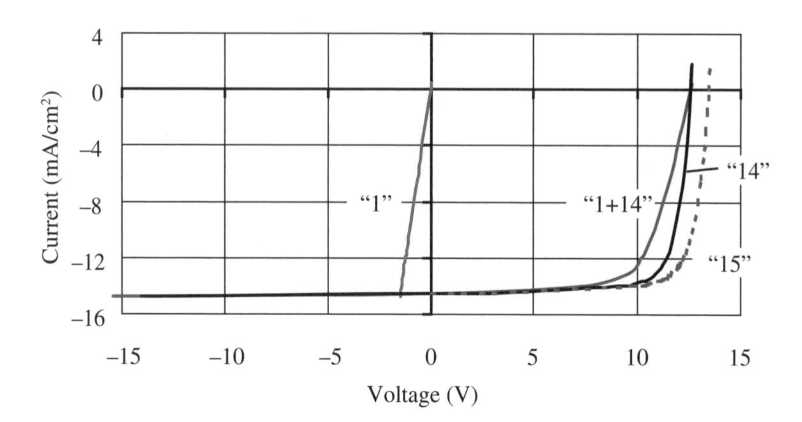


**Fig. 6.50** $I(V)$ characteristics for: a module of 15 series-connected unshaded cells ("15"), one cell irreversibly shunted to result in 100 Ωcm² ("1"), 14 series-connected unshaded cells ("14"), and a module of 15 series-connected cells, where one cell is shunted ("1 + 14").

**Table 6.5** Calculated power outputs of an unshaded and partly shaded a-Si module (since for the calculations the currents are expressed as current densities (mA/cm²), the calculated powers and resistances are also referred to cm²; however, in order to apply to 1 cell, both quantities must be divided by the number of series-connected cells).

| | | Module unshaded | Module partly shaded | |
|---|---|---|---|---|
| | | 15 cells 50 % shaded | 14 + 1 cell shunted | 14 + 1 cell |
| Power (MPP) | [mW/cm²] | 152 | 89 | 125 |
| Load resistance (MPP) | [Ωcm²] | 876 | 1857 | 738 |
| Power ($R_{15}$) | [mW/cm²] | | 49 | 121 |
| Power loss formshading | [%] | | –68 | –21 |

In Table 6.5, the power outputs of the unshaded and partly shaded module are calculated from the corresponding $I(V)$ characteristics. The power output of the unshaded module refers to the maximum power point (MPP), whereas for the partly shaded module the power output is calculated for the same load resistance, $R_{15}$, that applies to the maximum power of the 15 series-connected unshaded cells.

Table 6.5 illustrates that even for relatively low shading (in the above example 0.5/15 = 3.3 %) reverse biasing of the partly shaded cell causes a substantial power loss of 68 %. It is further shown that irreversible breakdown of the shaded cell, provided it is not accompanied by additional damage to the module (as for example by rendering the encapsulation ineffective due to excessive temperatures), may lead to a smaller power loss. However, when full irradiation is restored, this power loss remains, whereas in the first case full power will be restored.

While the primary effect of shading generally leads to reverse-biasing of the shaded cell(s), secondary effects leading to irreversible breakdown and possibly further damage are quite specific to the particular cell structure, and hence cannot be universally predicted. Since these secondary effects usually do not occur uniformly over the cell area, but rather locally, they are accompanied by significant temperature increases. As a result, so-called hot spots may be formed. In order to prevent hot-spot formation, by-pass diodes may be incorporated into the module. Their function is explained below.

*By-pass diodes*

The destructive effects of hot-spot heating may be circumvented through the use of by-pass diodes. By-pass diodes are conventional *p-n* or Schottky diodes, which are connected in parallel but with opposite polarity to one or more solar cells, or even across an entire module. Under normal operation, each solar cell will be forward-biased and, therefore, the by-pass diode will be reverse-biased and will effectively not conduct current. However, if solar cells are driven into reverse bias due to shading, the by-pass diode connected in parallel assumes forward bias to conduct the photocurrent generated in the unshaded cells of the module. Thus the maximal reverse bias of the affected cells is reduced to the forward bias of the by-pass diode, and hot-spot heating is prevented.

A quantitative assessment on the incorporation of by-pass diodes goes beyond the scope of this discussion, particularly since it depends on the specific reverse characteristics of the solar cells, the properties of the encapsulation and related heat transfer mechanisms. Generally, the application of by-pass diodes is handled differently for c-Si and a-Si-based modules. This can be explained by differences in the reverse $I(V)$ characteristics of c-Si and a-Si cells, as measured in representative modules and shown in Figure 6.51.

For c-Si cells, further independent experiments have shown a wide spread in reverse breakdown behavior, characterized by the breakdown voltage (–10 to –21 V) and hot-spot temperature (60 to 125 °C) [Herrmann-1998]. Depending on the output voltage and the tolerance to reverse voltage of the cells, a by-pass diode may be placed across every few cells, typically 15 cells per by-pass diode.

For a-Si-based cells, the reverse dark $I(V)$ characteristic is essentially current-limited, i.e. the dark current of the shaded cell reaches the value of the module current at the maximum power point, $I_{MP}$, at a reverse voltage that is considerably smaller than the module voltage at the maximum power point, $V_{MP}$. The corresponding $I(V)$ characteristic would correspond to intermediate behavior between the two cases illustrated by the graphs "1" in Figures 6.49 and 6.50.

Due to the reasons described above, the reverse $I(V)$ characteristics of a-Si-based cells may be considered to be more tolerant towards shading. This is reflected in the use of by-pass diodes, which differs for the two major types of a-Si-based modules, namely for superstrate-based modules with monolithically series-connected cells, and for cells deposited on stainless steel and individually series-connected to modules (see Sect. 6.1.9).

Based on the typical reverse characteristics of a-Si cells, superstrate-based modules frequently do not incorporate by-pass diodes. Furthermore, the effect of shading on an individual module may be mitigated by parallel-connecting two or more portions of series-connected cells [Lechner-2008], whereby the value of $V_{mp}$ is reduced. More

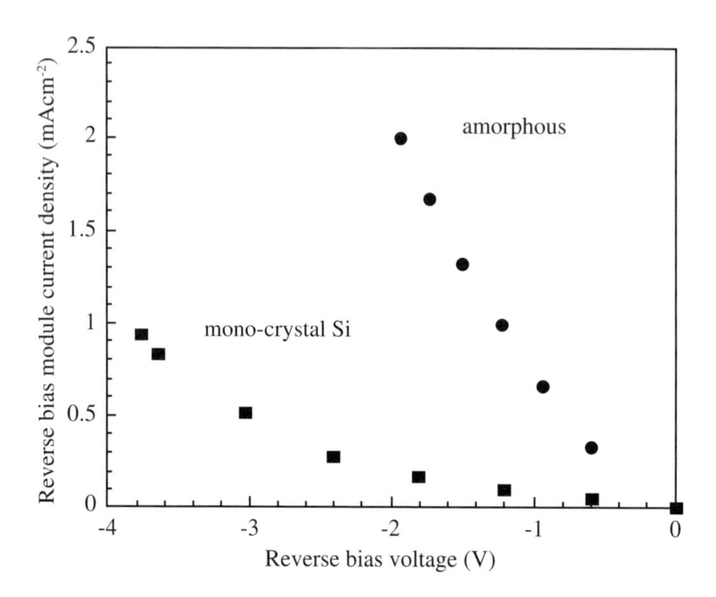


**Fig. 6.51** Reverse $I(V)$ characteristics of a c-Si and an a-Si module (Copyright © [1991] IEEE, with permission[Molenbroek-1991]).

recently, with module sizes exceeding 1 $m^2$, a by-pass diode has been shown to protect the entire module that is contained in a string of many modules connected in series.

Modules manufactured by roll-to-roll deposition on stainless steel feature large-area cells (about 960 $cm^2$), which are individually series-connected to form a module (see Sect. 6.1.9). These cells, unlike monolithically interconnected cells, are directly accessible to the placement of by-pass diodes. This, in fact, is practiced for this type of product, where each cell is provided with an individual by-pass diode.

### 6.6.4 Shunting

Shunts act as current paths in parallel to the current of the ideal solar cell. The amount of shunting is derived from the $I(V)$ characteristic (Equation 6.27), namely as shunt resistance, defined as the slope of the $I(V)$ curve at $I_{sc}$:

$$R_{sc} = (dV/dI)I_{sc} \tag{6.28}$$

This value is determined by two effects described below, which are caused by so-called photo-shunts and dark shunts, respectively.

*Photo-shunts*[1]

Photo-shunts (also called inherent shunts or recombination shunts) depend on the effectiveness of the carrier collection under illumination, which is a function of

[1] "Photo-shunts" are defined as "recombination current sinks" in Sect. 4.9 (with an equivalent circuit in Fig. 4.51). "Dark shunts" introduced below are called "shunts" in Chap. 4.

their transport and recombination properties, and may be expressed by the ratio of a drift length and the *i*-layer thickness [Faughnan-1984, Meillaud-2006]. Unlike dark shunts, discussed below, photo-shunts are not necessarily directly related to specific "external" imperfections of the cell structure, but rather to the quality of the *i*-layer. Based on $I(V)$ characteristics (see Equation 6.27), composed of a diode current density according to the drift/diffusion diode model, and a collection current density given by a voltage-dependent photocurrent [Crandall-1982], Figure 6.52 shows $I(V)$ characteristics for various *pin* single-junction cells with different drift length/*i*-layer thickness ratios $L_c/d_i$. Here it is assumed that no additional shunt is active (see below). It can be shown that the shunt resistance determined according to Equation 6.28 is inversely proportional to the illumination intensity and hence to the short-circuit current density, which explains the term photo-shunt. For the illumination with 1 sun, shunt resistances, as well as fill factors, calculated from the above $I(V)$ characteristics, are summarized in Table 6.6.

Typical photo-shunt values for state-of-the-art tandem or micromorph cells track with the corresponding ratios of $V_{oc}/I_{sc}$. Compared to about 2 kΩ cm$^2$ for single-junction a-Si cells, as shown in Table 6.6, tandem-junction a-Si/a-Si cells, with about twice the open-circuit voltage and half the short-circuit current, exhibit about 8 kΩ cm$^2$, and micromorph a-Si/μc-Si with typical values for $V_{oc}$ and $I_{sc}$, exhibit about 4 kΩ cm$^2$.

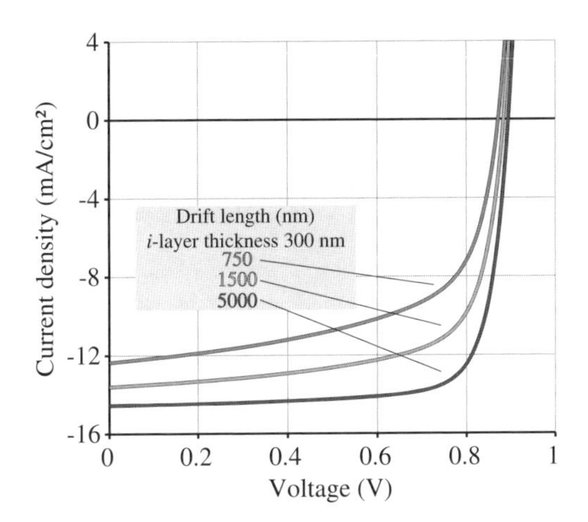


**Fig. 6.52** Calculated $I(V)$ characteristics of *pin* cells with *i*-layer thickness of 300 nm and different drift lengths.

**Table 6.6** Calculated photo-shunt resistances and fill factors of single-juntion *pin* cells with *i*-layer thickness of 300 nm and different drift lengths.

| $L_c(V = 0)$ nm | $L_c/d_i$ = 300 nm | $R_{sc}$ Ω cm$^2$ | $FF$ |
|---|---|---|---|
| 750 | 2.5 | 461 | 0.60 |
| 1500 | 5.0 | 804 | 0.69 |
| 5000 | 16.7 | 2433 | 0.78 |

*Dark shunts*

Dark shunts arise from current paths in parallel to the photo-diode current. These shunt currents are independent of the illumination, and are given by $V/R_{sh}$, where $V$ is the cell voltage and $R_{sh}$ the shunt resistance. Based on the same model as used above, the corresponding $I(V)$ characteristics are given by

$$I = I_0\,(\exp(qV/nkT) - 1) - I_{ph}(L_c/d_i) + V/R_{sh}, \tag{6.29}$$

The effect of low dark shunt resistances is shown in Figure 6.53 (for simplicity, the photo-current has been assumed to be voltage-independent, i.e. the drift length is assumed to be sufficiently long to rule out any carrier recombination).

Dark shunts are essentially local shorts between the front and back contact layer of the cell. Their resistances may be ohmic, as shown in Figure 6.53, or non-linear (voltage-dependent), depending on the properties of the materials in local contact. There are mainly two origins for dark shunts:

- Local imperfections in the areal uniformity of the cell layers, whereby so-called pinholes may be formed. These may arise, for example, from dust particles on the TCO layer (possibly also from powder formation in the plasma, see Sect. 6.1), which are covered by the a-Si deposition, but during subsequent processing flake off to leave holes through which the back contact layer makes contact to the TCO.
- Laser patterning in conjunction with the monolithic series connection of cells (see Sect. 6.5); here imperfect separation of the front and/or back contact layer (pattern 1 and/or pattern 3) results in so-called bridging, whereby a cell may be partly shorted. The separation of the a-Si layer (pattern 2) may lead to local recrystallizations of the a-Si, and thus also provide local low-resistance paths.

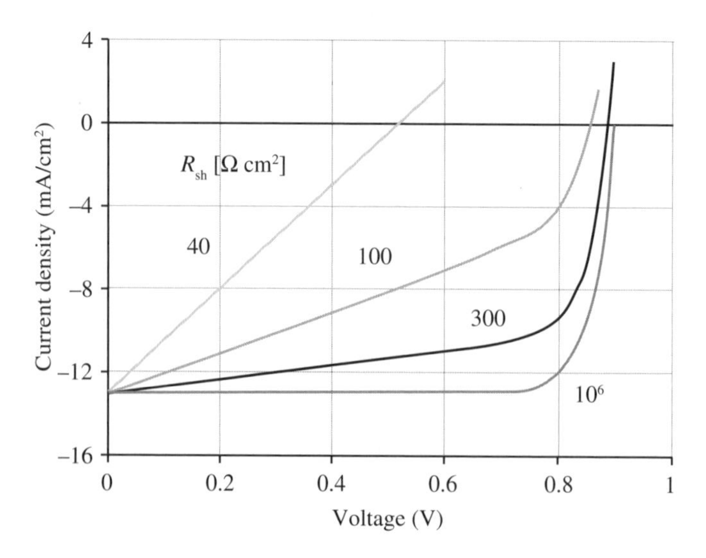


**Fig. 6.53** Calculated $I(V)$ characteristics of *pin* cells shunted by various resistances.

All these local shunts can be made visible with thermographic imaging, as shown for example in Figure 6.54. The bright spots indicate locally high current densities accompanied by higher temperatures.

*Effects of shunting on the fill factor and open-circuit voltage*

The $I(V)$ characteristics shown in Figure 6.52 represent photo-shunts of cells with infinitely high dark shunts, whereas those shown in Figure 6.53 apply to cells with infinitely high photo-shunts (i.e. with $L_c/d_i$ infinitely high) and various dark shunts. In reality, the $I(V)$ characteristics are a result of the combined effect of both types of shunting. Since dark shunts are independent of the illumination intensity, whereas photo-shunts are inversely proportional to it, $I(V)$ characteristics can reveal information particularly on the magnitude of dark shunts. This information is obtained from the fill factor ($FF$) and the open-circuit voltage ($V_{oc}$) as a function of the light intensity. Both quantities significantly decrease if, with decreasing illumination intensity, the intensity-dependent photo-shunt exceeds the intensity-*in*dependent dark shunt.

Based on the model of $I(V)$ characteristics, assumed here for negligible carrier recombination, the fill factor is calculated as a function of a voltage-independent dark shunt resistance for various illumination levels (see Fig. 6.55). Accordingly, to operate solar cells with a satisfactory fill factor ($FF \geq 0.68$), the dark shunt resistance is required to be higher, the lower is the illumination intensity. As a practical consequence it follows that for applications at low light levels, like indoor illumination, single-junction a-Si cells are better suited, because their dark shunt resistances must only stay above the typical photo-shunt value, which is about four times lower than for tandem-junction cells (see above). From Figure 6.55 it can be seen that for satisfactory performance ($FF \geq 0.68$) single-junction *pin* cells will tolerate shunt resistances of 350 $\Omega$cm$^2$ at 120,000 lux, to 5000 $\Omega$cm$^2$ at 10,000 lux. For low-level illumination, shunt resistances must exceed 5000 $\Omega$cm$^2$. For tandem-junction and micromorph cells, these shunt resistance requirements are raised by a factor of about 4 for a-Si/a-Si cells, and about 2 for micromorph a-Si/μc-Si cells, as explained above. Thus, for low-level illumination, single-junction cells may be a better choice for application than tandem cells, if dark shunts tend to be low.

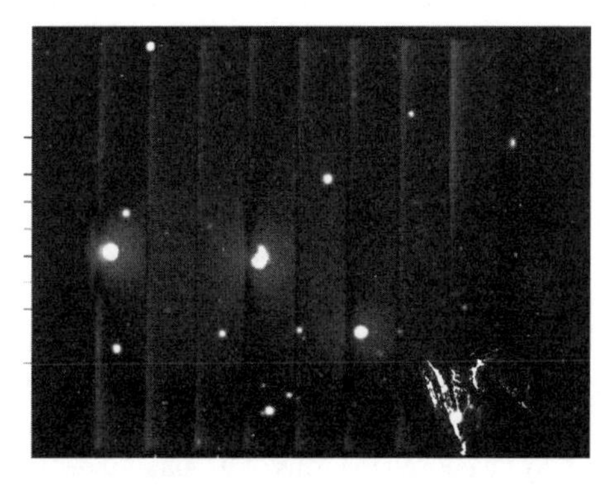

**Fig. 6.54** Thermographic imaging of a portion of an a-Si module being operated under low-level illumination (courtesy of [Rech-2005]).

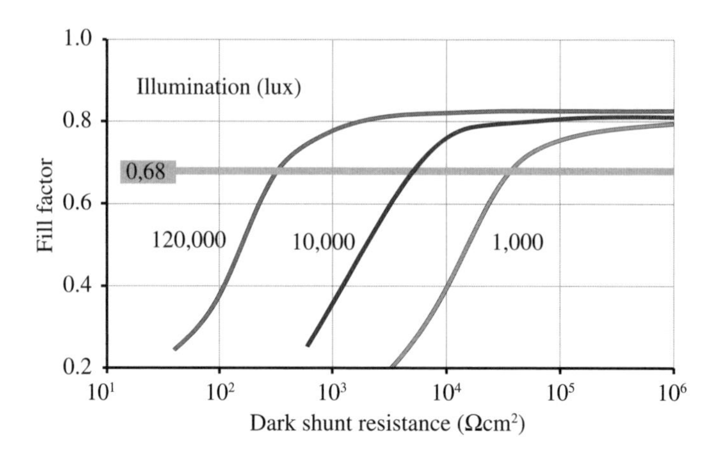


**Fig. 6.55** Fill factor as a function of the dark shunt resistance of single-junction *pin* cells for various illumination intensities (120,000 lux ≈ 1 sun).

As already apparent from Figure 6.53, distinct decreases of the open-circuit voltage occur, which are considerably beyond the inherent decrease as a function of the logarithm of the short-circuit current (and hence illumination intensity, see Sect. 4.5). These decreases are calculated, based again on the diode model used above.

Assuming a voltage-independent photocurrent given by $I_{ph}(L_c /d_i) = I_{sc}$, the $I(V)$ characteristic given in Equation 6.29 is evaluated for $I = 0$, i.e. at $V = V_{oc}$, leading to the relationship

$$I_{sc} = I_0 (\exp(eV_{oc} /nkT) - 1) + V_{oc} /R_{sh}. \tag{6.30}$$

Figure 6.56 shows this relationship for various illumination intensities. For sufficiently large values of $R_{sh}$, $V_{oc}$ linearly decreases with the logarithm of $I_{sc}$, as shown by the limiting straight line. However, for lower dark shunt resistances, $V_{oc}$ deviates from this straight line towards smaller values, starting at the higher values of $V_{oc}$ and $I_{sc}$ (i.e. of the illumination intensity) the lower the shunt resistances are. Without relying on thermographic imaging, this behavior is utilized in practice as a rapid quality control, in which the $V_{oc}$ of solar modules is measured under low-level illumination. $V_{oc}$-values smaller than expected from the logarithmic intensity dependence indicate excessive shunting.

***Removal of shunts***

Without knowing the nature and location of shunts, in practice, it is common to attempt to render shunts inactive by the application of a reverse bias. Thereby high current densities and thus thermal energy are concentrated at the shunts, which may result in oxidation or evaporation of the shunt material. In both cases, the shunt path is eliminated. The effectiveness of this electrical method of shunt removal, also called shunt-busting, is strongly dependent on both the electrical details and the materials being involved. Specifically, the following details must be considered:

- The applied bias may be chosen in terms of its magnitude (typically –4 to –8 V per cell), and also of its time dependence (shape and duration of pulses). Furthermore, shunt-busting may be applied cell-by-cell, or for portions of a module.

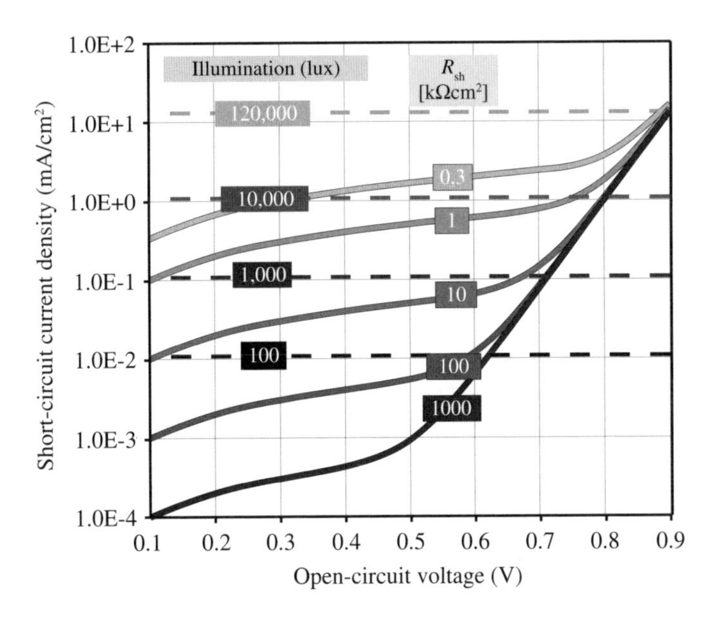


**Fig. 6.56** Open-circuit voltage as a function of the short-circuit current (proportional to the illumination intensity) for *pin* cells with various dark shunt resistances.

- Distribution of shunts: For many shunts of a cell or module area, the extra shunt current density is subdivided, and the individual shunt current densities may not be sufficient to be effective at the maximum tolerable bias. This reverse bias must not exceed a critical value, above which the *pin* junction of the actual cell structure would irreversibly break down.
- The material combination, particularly involving materials of the front and back contact layers, react quite differently in terms of the resulting local resistance.

The interplay of these influences sketched above is usually found only empirically, and may not necessarily lend itself to shunt-busting. Nevertheless, in practice shunt-busting methods are often applied (but rarely publicized) to further improve the yield of a production line.

## 6.7 MODULE FINISHING

So far, all the elements that essentially determine the function of thin-film silicon modules have been described. In order to put modules into practical use, additional elements are required, as follows:

- The electrical connection to the external load, i.e. the attachment of connection cables. Without going into specific details, this cable connection is established typically with lead-free solder to the contact ribbons on the backside of the module, and it is, in general, tightly sealed within a junction box that is attached to the module.

- The overall protection of the module from exposure to environmental conditions, which mainly involves an encapsulation.
- A formal qualification that certifies for any given module type not only the electrical output, but also that certain environmental and mechanical tests have been successfully passed, according to exact standardized procedures.

The latter two points will be discussed below in more detail.

### 6.7.1 Encapsulation

Besides the legally required guarantees on workmanship, which typically last for 2 years from the date of purchase, module manufacturers generally offer long-term power output guarantees. These guarantees slightly vary, depending on the manufacturer; typically they state that the module will furnish at least 90 % of the specified power after 10 years, and at least 80 % after 20 or 25 years. Since the specified power may imply an allowable range of ± 5 %, the long-term power guarantee may be further relaxed accordingly.

Without protection, the functional layer sequence of the module would be adversely affected, predominantly by interaction with environmental humidity, possibly additionally aggravated by elevated operational temperatures, and by the inherent presence of operational voltages that may give rise to electro-corrosive effects. All such influences must be impeded by an encapsulation that basically reduces, if not prevents, moisture from penetrating to the functional layers of the module.

There are generally two types of encapsulation, namely glass/glass and glass/foil lamination. Glass/glass lamination typically applies to superstrate-based modules, whereas glass/foil lamination (or simply foil lamination) is used for both superstrate- and substrate-based modules. In the latter case the front-side foil must additionally fulfil the condition of having high and stable optical transmission.

Regarding protection against moisture, for glass/glass lamination the moisture ingress must be impeded along the edge regions of the laminate ("horizontal" ingress), while for glass/foil laminates moisture ingress must be impeded across the entire area of the laminating foil ("vertical" ingress). These two basic encapsulation arrangements are illustrated in Figure 6.57.

An additional possibility, which can be applied particularly for building-integrated PV (BIPV), is derived from the construction of insulating glazing. Here, the front glass carrying the PV module is combined with the rear-side glass panel by an edge seal that consists of a thermoplastic butyl spacer tape.

In all cases, the laminating material must be characterized by a low water vapor transmission rate (*WVTR*). This quantity, $WVTR = D \times C / \ell$, is related to the diffusivity $D$, to the saturation concentration $C$ of water vapor, and to the diffusion distance $\ell$, which is determined by the thickness of the laminating material measured in the direction of moisture penetration, or by the distance of the functional layers from the module edges (see Fig. 6.57).

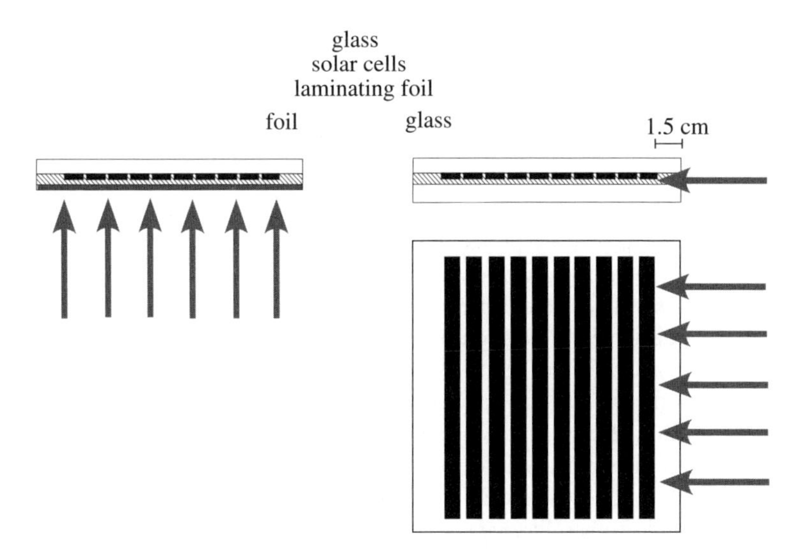


**Fig. 6.57** Encapsulation by glass/foil and glass/glass lamination for superstrate modules.

Besides providing adequate protection against moisture, additional considerations determine the choice between these two types of lamination, which are further described here:

### *Glass/glass laminates*

These are mostly used for superstrate-based modules, where the front glass carrying the functional layers is laminated to a rear-side pane of glass. The lamination is accomplished by a laminating foil that acts as an embedding and adhesive medium. The foil materials, partly derived from the safety glass industry, are ethylene vinyl acetate (EVA), polyvinyl butyral (PVB), or thermoplastic polyurethane (TPU). Their choice is governed by the specific features of these materials and by cost considerations. Particularly for building-integrated PV (see Chap.7), legal requirements, set by various building regulations, must be fulfilled; for laminates used for overhead glazing, for example, normally only PVB is approved as the laminating foil. Except for typical water vapor transmission rates listed below, further details on laminating foils and their processing go beyond the scope of this book.

However, some other aspects of glass/glass lamination should be pointed out here, namely the weight and the mechanical stability of the glass/glass laminate. Adding the rear-side glass to the "raw" module results in approximately doubling the weight; this obviously complicates handling, transportation, and mounting of modules. On the other hand, the mechanical load resistance is increased, and the requirements for wind and snow loads, as well as for impact resistance against projectiles like hail, may be satisfied by choosing an adequate thickness of the rear-side glass. If the rear-side glass thickness is sufficient, the need for using heat-strengthened or tempered front glass may be avoided.

***Foil lamination***

This method typically involves a combination of different foils starting with an adhesive foil material. While sufficient adherence to the functional module layer avoids the ingress of moisture along the interface between the module and the foil ("horizontal" ingress), the penetration of moisture through the foil ("vertical" ingress) across the entire module area is determined by the *WVTR*-value of the foil material (see Table 6.7). Besides providing protection from moisture, the foils must also exhibit sufficiently high electrical insulation to avoid any shunting of the module. Common foil materials satisfying the above properties are combinations of polyethylene terephthalate (PET, trade name Mylar, about 75 $\mu$m) and polyvinyl fluoride (PVF, tradename Tedlar, typically about 30 $\mu$m thick). More recently, the *WVTR* values of these foils or combinations thereof have been further reduced by the incorporation of inorganic films of Al or $SiO_X$. Representative values are compiled in Table 6.7. For a proper comparison, the values must be referred to the same materials thickness and diffusion temperature. Values for EVA and PVB are also included; although these appear to be orders of magnitude higher than for the foils, their effect in terms of the actual amount of moisture reaching the cells close to the module edge is lower. This is determined by the "horizontal" penetration, as opposed to the "vertical" foil penetration, for which the length and cross section of penetration appreciably differ: the penetration path length is about 1.5 cm "horizontally", as shown in Figure 6.57, compared to about 30 to 80 $\mu$m "vertically" through the foil thickness. The relevant cross sections are given by the product of the length of the module circumference and the laminating foil thickness (order of $4\ m \times 10^{-4}\ m$) vs. the module area (order of $1\ m^2$) for horizontal and vertical moisture ingress, respectively.

For superstrate-based modules, the foils are "behind" the module and, thus, their optical properties are not relevant to the proper functioning of the module. On the other hand, one generally uses the substrate-based version with front-side encapsulation for flexible modules (see Sect. 6.1.9). Here, high optical transmission across the visible and near-infrared spectral range is an additional requirement for the encapsulating foil material. For this application, particularly fluoropolymer films, such as ethylene tetrafluoroethylene (ETFE, also known under the trade name Tefcel), are frequently applied.

In a similar manner as for glass/glass laminates, the consequences of foil lamination have to be judged with respect to the resulting module weight, and the resulting mechanical properties. The advantages of a lower weight translate into lower handling, transportation and mounting costs. Furthermore, the process of foil encapsulation may typically be faster than the process of glass/glass lamination. However, these advantages must be balanced against the need to use heat-strengthened or tempered glass in order to satisfy the mechanical load requirements.

**Table 6.7** *WVTR* for various encapsulants, related to the thickness of 140 $\mu$m, at 25 °C.

| Encapsulant | EVA | PVB | PET (Mylar) | PVF (Tedlar) | PVF/PET/PVF | PVF/$SiO_X$/PVF |
|---|---|---|---|---|---|---|
| WVTR [$g/m^2\ d$] | 43 | 70 | 1.7 | 2.7 | 1.5 | 0.3 |

### 6.7.2 Module certification

Solar modules on the market are required to be certified for a number of performance criteria that include the power output as well as construction features, both designed for long-term stability during environmental exposure. These criteria serve as basic inputs to the designer of the solar power installation, and also to the investing consumer. The criteria are formulated on the basis of various standards defined by organizations, such as IEC (International Electrotechnical Commission), IEEE (Institute of Electrical and Electronics Engineers), ASTM International (originally American Society for Testing and Materials). Modules are certified by third-party, independent institutions, such as UL (Underwriters Laboratories), NREL (National Renewable Energy Laboratory), TÜV Rheinland, AIST (Japanese Institute of Advanced Science and Technology).

To illustrate the scope of the tests, that are required to be passed for certification, the main items prescribed by the standard IEC 61646, "Thin-film terrestrial photovoltaic (PV) modules - Design qualification and type approval", are summarized below [TÜV-2009]. This standard is in many aspects identical to standard IEC 61215, which is applied for crystalline-silicon modules. The main difference between these two standards lies in additional test procedures required in the case of thin-film modules, due to the degradation behavior of a-Si under light exposure. The qualification tests according to IEC 61215/IEC 61646 are performed with a total of eight modules of a given type. Not all modules are subjected to the same tests, but they are allocated to essentially four different test sequences that partly consist of different tests. The resulting certification will then refer to that particular module type. Without describing the specific test sequences and procedures, the most important tests comprise:

- power output at standard test conditions (STC, i.e. irradiance 1000 W/m$^2$, spectral distribution AM 1.5, cell temperature 25 °C);
- insulation test: 1000 $V_{dc}$ plus twice the open-circuit voltage of the module at STC for 1 min, isolation resistance x module area > 40 MΩ m$^2$ at 500 V$_{dc}$;
- temperature coefficients of short-circuit current, $(dI_{sc}/I_{sc})/dT$, open-circuit voltage, $(dv_{oc}/v_{oc})/dT$, and maximum power $(dP_{mp}/P_{mp})/dT$, within a temperature interval of 30 °C;
- determination of the normal operating cell temperature, NOCT, with irradiance of 800 W/m$^2$ (AM 1.5), and wind speed of 1 m/s;
- power output at NOCT, with irradiance of 800 W/m$^2$ (AM 1.5);
- power output at low irradiance of 200 W/m$^2$ (AM 1.5), and at 25 °C;
- thermal cycling, i.e. 200 or 50 cycles between –40 °C and 85 °C, depending on the specific test sequence;
- humidity/freeze, i.e. 10 cycles between –40 °C and 85 °C at 85 % relative humidity;
- damp heat, i.e. 85 °C at 85 % relative humidity for 1000 hours;
- mechanical load, i.e. 3 cycles under a uniform load of 2400 Pa, applied for 1 hour to the front and rear side in turn;
- impact resistance (hail test), i.e. ice balls of 25 mm diameter to impact the front surface at 11 locations with a speed of 23 m/s;

- light-soaking, i.e. light exposure cycles with at least 43 kWh/m$^2$ (i.e. 43 - 54 h at 1000 - 800 W/m$^2$) at 50 °C ± 10 °C, until $P_{mp}$ is stable to within 2 %. This test concludes each of the different test sequences.

*Pass criteria* [TÜV-2009]: A module design shall be judged to have passed the qualification tests, and therefore to be IEC-type approved, if each sample meets the following criteria:

- the degradation of the maximum power output at STC does not exceed 5 % after each test, nor 8 % after each test sequence;
- the requirements for the insulation tests are met;
- no major damage (breakage or cracks, delamination of any kind) is visible;
- no sample has exhibited any open circuit or ground fault during the tests;
- the maximum power output after final light-soaking shall not be less than 90 % of the minimum value specified by the manufacturer.

It is important to note that a test certificate is issued for a certain module construction type that has passed the described tests. The construction type is considered to be the same, and hence the certificate valid, as long as all materials and module components remain the same, the same number or less cells are bridged by a by-pass diode, and the module dimensions and rated power output do not deviate by more than + 10 % from the originally certified module type.

### 6.7.3 Long-term stability

The properties of finished modules, as described in the previous sections, are designed to fulfill the certification criteria, and ultimately to satisfy the long-term performance guarantees. The latter are fundamentally determined by the environmental exposure of the modules. Although the qualification procedures address this aspect by a sort of accelerated testing (thermal cycling, humidity/freeze cycles), the corresponding certification cannot be viewed as an assurance for satisfactory long-term performance within the granted guarantees. Furthermore, the IEC test procedures do not specifically include climatic tests in combination with electrical stress – this is a combination, that may give rise to electro-corrosive phenomena.

So far, the situation exists that thin-film module manufacturers inherently cannot, or at most can only partially, grant long-term guarantees based on true experience, since thin-film module installations were widely put into operation only less than 20 years ago, i.e. less than the usually stated duration of the guarantee. By now many examples, particularly large-area installations in BIPV, have been installed worldwide [Maurus-2004] (see Sect. 7.1). However, published, well-documented performance data monitored over many years of operation are relatively scarce. Figure 6.58 shows as an example the maximum power output of representative a-Si/a-Si tandem modules, measured under STC conditions after extended periods of outdoor exposure during a time span of 11 years.

There does not exist any documented experimental proof of the guaranteed power of thin-film silicion modules after 25 years of operation. However, the following estimates do lend some support to the long-term performance guarantees which

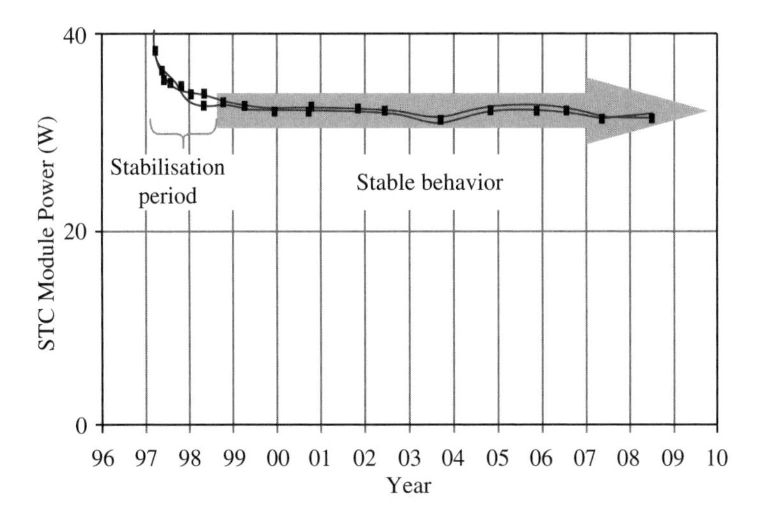


**Fig. 6.58** Time dependence of the maximum power output of a-Si/a-Si tandem modules, measured at STC after extended periods of outdoor exposure over a total time span of 11 years (Copyright © [2008] IEEE, with permission[Lechner-2008]).

**Table 6.8** Projected relative power decreases due to the Staebler-Wronski effect.

| cell type | $\Delta P/P$ (1000 h) (Light-soak) [%] | $FF$ (1 h) | $FF$ (1000 h)[1] | $\Delta FF/$ 1000 h | $\Delta FF/$ decade [h] | $FF$ (25 years)[1] | $\Delta P/P$ (25 years)[2] | $\Delta P/P$ per year (%/year) |
|---|---|---|---|---|---|---|---|---|
| a-Si | −23 | 0.75 | 0.58 | −0.17 | −0.058 | 0.49 | −16 | −0.64 |
| a-Si/a-si | −18 | 0.74 | 0.61 | −0.13 | −0.044 | 0.54 | −12 | −0.47 |
| a-Si/ c-Si | −11 | 0.73 | 0.65 | −0.08 | −0.027 | 0.61 | −7 | −0.26 |

[1] Assumption $\Delta P/P = \Delta FF/FF$

[2] Assumption:1000 h STC irradiance = 0.7 years outdoor exposure, i.e. 25 years = 35700 h, 1000 h to 35700 h corresponds to 1.6 decades [h].

most thin-film module manufacturers give. These estimates are based on the following two influences that affect module performance.

### *Staebler-Wronski effect*

This effect (see Sect. 4.5.16) is predominantly determined by a decrease of the fill factor ($FF$) and by its time dependence, which has been shown to be logarithmic [Smith-1985]. It is, when plotted on a logarithmic time scale, characterized by a constant *absolute* $FF$-decrease, $\Delta FF$, per time-decade (measured in hours, i.e. the first decade is from 1 hour to 10 hours, the second decade from 10 hours to 100 hours, etc.). For the following estimate (see Table 6.8), it is assumed that the *relative* $FF$-decrease, $\Delta FF/FF$, is equal to the relative decrease of the maximum power. Depending on the cell type, this decrease is derived from the *relative* decrease during the photostabilization period of 1000 hours (i.e. 3 time decades, when measured in hours [h]). From typical initial fill factors of different cell types listed in Table 6.8, designated as *FF*(1 h) in conjunction with a logarithmic time scale, the stabilized fill factors, *FF*(1000 h), are determined. The latter are representative of data specified in product data sheets, and thus must satisfy the long-term guarantees. In order to project the data to be reached after 25 years, the constant *absolute* *FF*-decrease per time decade is obtained from *FF*(1 h) and *FF*(1000 h), i.e.

$$\Delta FF/\text{time-decade[h]} = (FF(1\text{ h}) - FF(1000\text{ h}))/3 \qquad (6.31)$$

From this *FF*-decrease per time decade one may deduce the *FF* after 25 years [*FF*(25 years), as shown in Table 6.8].

A correlation between 1000-h irradiance at STC and outdoor illumination follows from global insolation data, as obtainable, for example, from [Šúri-2007]. Accordingly, the average annual insolation for central Europe is in the range of 1200 to 1600 kWh/$m^2$ year. This corresponds to 1200 to 1600 h/year of the irradiance of 1 kW/$m^2$, i.e. the 1000-h irradiance at STC is about equivalent to an average of 0.7 years outdoor illumination. The outdoor exposure of 25 years then corresponds to approximately 36,000 h of STC irradiance, i.e. to an additional 1.6 time-decades [h], counted from the specified "stabilized" value *FF*(1000 h). The obtained relative decreases are assumed to represent the power decreases to be expected, over the 25 year guarantee period, due to the Staebler-Wronski effect.

Besides these estimated decreases based on the Staebler-Wronski effect, additional effects possibly caused by the encapsulation may adversely affect the module power output during long-term operation. This aspect is briefly discussed below.

### *Encapsulation*

While the Staebler-Wronski effect is an inherent phenomenon of the actual cell structure, additional effects that lead to a decrease in the power output may be caused by certain failures of the encapsulating materials. For example, the ingress of moisture may corrode the back metal reflector and, thus, reduce the generated current; or corrosive effects may lead to higher series resistances at contact connections. Since many of the encapsulating materials are used for both crystalline and thin-film technologies, one may refer to the long-term performance also of crystalline-silicon modules, which is much more widely documented than that of thin-film silicon modules; this will provide an approximate guide. In a review on field-aged modules or module arrays, monitored at several locations for periods of typically 10 years [Dunlop-2005], power degradation rates between 0.2 and 0.7 % per year were obtained. Obviously, these rates depend on the generation (year of fabrication) of the module technology, on the encapsulating materials used, and on the climatic conditions of the outdoor exposure. For the purpose of obtaining a lifetime estimate that can be used to judge the guaranteed output power of modules after long time spans, one may, nevertheless, assume an average annual power decrease of 0.4 % per year, as attributed predominantly to the influence of the encapsulants.

Based on the two contributions from the Staebler-Wronski effect (see Table 6.8) and encapsulation (0.4 %/year), a total annual power decrease may be projected,

- for single-junction a-Si modules: –1 % per year;

- a-Si/a-Si tandem modules: –0.9 % per year; and
- for a-Si/μc-Si micromorph modules –0.7 % per year.

Thus, the usual power guarantee of 80 % after 25 years (i.e. 76 % of the specified power, considering the specified lower tolerance of – 5 %) is likely to be well satisfied, or even surpassed, depending on the cell type.

## 6.8 CONCLUSIONS

The practical use of solar cells generally implies the fabrication of entire solar modules, which represent the basic units for most applications. For thin-film technologies the fabrication of modules relies on two main features:

- uniform deposition of the cell layers onto large-area substrates (order of magnitude: 1 $m^2$); in combination with
- areal definition of cells and their integral series connection.

For silicon-based thin-film technologies, the semiconductor layers mainly comprise amorphous and microcrystalline silicon (a-Si:H and μc-Si:H). Their method of deposition is central to the technology and, to date, has been up-scaled to large-area deposition only on the basis of plasma-enhanced chemical vapor deposition (PECVD). This up-scaling does not solely result from geometric and volumetric up-scaling of laboratory-size conditions, but involves additional fundamental implications, such as, for example, the deposition uniformity as a function of the plasma excitation frequency and the corresponding wavelength.

The transition from small-area laboratory cells to large-area modules leads to inherent losses in efficiency, which are predominantly determined by photo-active area losses, as well as by series resistances of the contact layers as a consequence of the integral series connection of cells.

In particular, the series resistance of the TCO front contact affects the module efficiency. The corresponding losses are proportional to the current density of the cells and are, therefore, less pronouced in modules with tandem or triple-junction cells than in modules with single-junction cells.

Besides the series resistance, the optical properties of the transparent conductive oxide (TCO) play an essential role in the light management for most effective absorption of the insolation within the *i*-layer(s) of the cell structure. The optical properties refer mainly to the absorbance, which is inversely proportional to the sheet resistance (and hence to the TCO series resistance). Moreover, a suitable surface texture of the TCO is required to induce light scattering and light trapping. Therefore, the material properties of the TCOs constitute a key factor for the fabrication of modules with reasonable efficiencies.

Modules are evaluated not only in terms of their efficiencies under standard test conditions (STC), but also in terms of their energy yield in actual field deployment. The energy yield is characterized, for a given deployment site, by the annual energy generation per module power (kWh/k$W_p$year). Due to their favorable spectral response, and due also to the low values of their temperature coefficients, silicon-based thin-film modules have been shown to exceed typical energy yields of conventional crystalline silicon modules.

In order to offer economically viable products to the market, the long-term performance (typically 20 years) must be assured. Light-induced degradation (Staebler-Wronski effect) takes place mainly during the first 1000 hours of operation. It is also less pronounced if the operating temperature of the modules is high. Apart from that, *all* PV modules generally exhibit an efficiency degradation of approximately 1 % (relative) per year. Generally, the long-term performance of *all* PV modules is largely dependent on the encapsulation, and it may be projected by following generally accepted certification procedures that are defined, for example, in IEC standards.

## 6.9 REFERENCES


[Abdel-Fatah-2003] Abdel-Fatah, E., Sugai, H., "Electron heating mode transition observed in very high frequency capacitive discharge", (2003), *Applied Physis Letters*, Vol. **83**, pp. 1533-1535

[Alpuim-2008] Alpuim, P., Junior, G. M., Filonovich, S. A., Roca i Cabarrocas, P., Bourée, J. E., Johnson, E. V., Soro, Y. M., "Polymorphous and nanocrystalline silicon thin-film solar cells deposited at 150 °C on plastic substrates", (2008), *Proc. 23rd European Photovoltaic Solar Energy Conference*, pp. 2455-2458

[Alpuim-2009] Alpuim, P., Gonçalves, L. M., Marins, E.S, Viseu, T. M. R., Ferdov, S., Bourée, J. E., "Deposition of silicon nitride thin films by hot-wire CVD at 100 °C and 250 °C", (2009), *Thin Solid Films*, Vol. **517**, pp. 3503-3506

[Amanatides-2005] Amanatides, E., Mataras, D., Rapakoulias, D., van den Donker, M. N., Rech, B., "Plasma Emission Diagnostics for the transition from microcrystalline to amorphous silicon solar cells", (2005), *Solar Energy Materials & Solar Cells,* Vol. **87,** pp 795–805

[Asari-2008] Asari, S., Fujinaga, T., Takagi, M., Hashimoto, M., Saito, K., Harada, M., Ishikawa, M., "ULVAC research and development of Cat-CVD applications", (2008), *Thin Solid Films*, Vol. **516**, pp. 541-544

[Ballif-2006] Ballif, C., Bailat, J., Dominé, D., Steinhauser, J., Fay, S., Python, M., Feitknecht, L., "Fabrication of high efficiency microcrystalline and micromorph thin film solar cells on LPCVD ZnO coated glass substrates", (2006), *21st European Photovoltaic Solar Energy conference, Dresden*, pp. 1552-1555

[Ballif-2007] Ballif, C., Terrazzoni-, V., Haug, F.-J., Fischer, D., Soppe, W., Löffler, J., Andreu, J., Fahland, M., Schlemm, H., Topic, M., Wurz, M., "Flexcellence: Towards roll-to-roll mass-production of low cost silicon solar cells" (2007), *22nd European Photovoltaic Solar Energy Conference, Milan,* pp. 1835-1838

[Bartl-2006] Bartl, R., Schlatmann, R., Stannowski, B., Gordijn, A., van den Donker, M. N., Finger, F., Rech, B., "TCO development for thin film silicon solar cells", (2006), *21st European Photovoltaic Solar Energy Conference, Dresden*, pp. 1666-1668

[Beneking-1990] Beneking, C., "Power dissipation in capacitively coupled rf discharges", (1990), *J. Appl. Phys.*, Vol. **68**, pp. 4461-4473

[Beneking-1999] Beneking, C., Rech, B., Wieder, S., Kluth, O., Wagner, H., Frammelsberger, W., Geyer, R., Lechner, P., Rübel, H., Schade, H., "Recent developments of silicon thin film solar cellson glass substrates", (1999), *Thin Solid Films,* Vol. **351**, pp. 241-246

[Bennet-1989] Bennet, J. M., Mattson, L., "Introduction to Surface Roughness and Scattering", (1989), *Optical Society of America, Washington, DC,* p. 51

[Bourée-2003] Bourée, J.E., Guillet, J., Grattepain, C., Chaumont, J., "Quantitative analysis of tungsten, oxygen and carbon concentrations in the microcrystalline silicon films deposited by hot-wire CVD", (2003), *Thin Solid Films*, Vol. **430**, pp. 110-115

[Branz-2008] Branz, H.M., Teplin, C.W., Young, D.L., Page, M.R., Iwaniczko, E., Roybal, L., Bauer, R., Mahan, A.H., Xu, Y., Stradins, P., Wang, T., Wang, Q., "Recent advances in hot-wire CVD R and D at NREL: From 18 % silicon heterojunction cells to silicon epitaxy at glass-compatible temperatures", (2008), *Thin Solid Films*, Vol. **516**, pp. 743-746

[Bubenzer-1990] Bubenzer, A., Schmitt, J. P. M., "Plasma processes under vacuum conditions" (1990), *Vacuum,* Vol. **41,** pp. 1957-1961

[Chittick-1969] Chittick, R. C., Alexander, J. H., Sterling, H. F., (1969), *J. Electrochem. Soc.,* Vol. **116,** p.77

[Cody-1981] Cody, G. D., Tiedje, T., Abeles, B., Moustakas, T. D., Brooks, B., Goldstein, Y., "Disorder and the optical absorption edge of hydrogenated amorphous silicon", (1981), *Journal Physique, Colloque C 4 supplément au no. 10*, Vol. **42**, pp. C4-301 - C4-304

[Collins-2002] Collins, R. W., Ferlauto, A. S.,"Advances in plasma-enhanced chemical vapor deposition of silicon films at low temperatures", (2002), *Current Opinion in Solid State and Materials Science*, Vol. **6**, pp. 425-437

[Comerford-2009] Comerford, D. W., Smith, J. A., Ashfold, M. N. R., Mankelevich, Y. A., "On the mechanism of H atom production in hot filament activated $H_2$ and $CH_4/H_2$ gas mixtures", (2009), *The Journal of Chemical Physics*, Vol. **131**, pp.044326 044326-12

[Crandall-1982] Crandall, R., "Transport in hydrogenated amorphous silicon p-i-n solar cells", (1982), *J. Appl. Phys.,* Vol. **53**, pp. 3350-3352

[Curtins-1987a] Curtins, H., Favre, M., Wyrsch, N., Brechet, M., Prasad, K., Shah, A. V., "High-rate deposition of hydrogenated amorphous silicon by the VHF-GD method", (1987), *Proc. 19th IEEE Photovoltaic Specialists Conference, New Orleans, LA*, pp. 695-698

[Curtins-1987b] Curtins, H., Wyrsch, N., Favre, M., Shah, A.V.,"Influence of plasma excitation frequency for a-Si:H thin film deposition", (1987), *Plasma Chem. Plasma Process*. Vol. **7**, pp. 267-273

[Daudrix-2000] Daudrix, V., Droz, C., Wyrsch, N., Ziegler, Y., Niquille, X., Shah, A., "Development of more stable amorphous silicon thin film solar cells deposited at „moderately high" temperature", (2000), 16*th European Photovoltaic Solar Energy Conference, Glasgow,* pp. 385-388

[Dekker-2005] Dekker,T., Metselaar, J. W., Schlatmann, R., Stannowski, B., van Swaaij, R. A. C. M. M., Zeman, M., "Correlation between surface-textured tin- and zinc oxide substrates and curren enhancement in amorphous silicon solar cells", (2005), 20*th European Photovoltaic Solar Energy Conference, Barcelona,* pp. 1517-1520

[Dorier-1992] Dorier, J.-L., Hollenstein, Ch., Howling, A. A., Kroll, U., "Powder dynamics in very high frequency silane plasmas", (1992), *Journal of Vacuum Science and Technology*, Vol. **A 10,** pp. 1048-1052

[Doyle-1988] Doyle, J., Robertson, R. Lin, G.H., He, M.Z., Gallagher, A., "Production of high-quality amorphous silicon films by evaporative silane surface decomposition", (1988), *Journal of Applied Physics*, Vol. **64**, pp. 3215-3223

[Droz-2003] Droz, C., "Thin Film Microcrystalline Silicon Layers and Solar Cells: Microstructure and Electrical Performances", (2003), *Thesis, University of Neuchâtel*

[Droz-2004] Droz, C., Vallat-Sauvain, E., Bailat, J., Feitknecht, L., Meier, J., Shah, A., "Relationship between Raman crystallinity and open-circuit voltage in microcrystalline silicon solar cells", (2004), *Solar Energy Materials and Solar Cells,* Vol. **81,** pp. 61-71

[Duan-2001] Duan, H.L., Zaharias, G.A., Bent, S.F., "Probing radicals in hot wire decomposition of silane using single photon ionization", (2001), *Applied Physics Letters*, Vol. **78**, pp. 1784-1786

[Dunlop-2005] Dunlop, E. D., Halton, D., Ossenbrink, H. A., "20 years of life and more: where is the end of life of a PV module?", (2005), *Proc. 31st IEEE Photovoltaic Solar Energy Specialists Conference, Lake Buena Vista, Florida,* pp. 1593-1596

[Dutta-1992] Dutta, J., Kroll, U., Chabloz, P., Shah, A., Howling, A.A., Dorier, J.-L., Hollenstein, Ch., "Dependence of Intrinsic Stress in Hydrogenated Amorphous Silicon on Exitation Frequency in Plasma-Enhanced Chemical Vapor Deposition Process", (1992), *J. Appl. Phys.*, Vol. **72**, pp. 3220-3222

[Faughnan-1984] Faughnan, B., Crandall, R., "Determination of Carrier Collection Length and Prediction of Fill Factor in Amorphous Silicon Solar Cells", (1984), *Appl. Phys. Lett.*, Vol. **44**, pp. 537-539

[Finger-2009] Finger, F., Astakhov, O., T., Bronger, Carius, R., Chen, T., Dasgupta, A. Gordijn, A., Houben, L., Huang, Y., Klein, S. Luysberg, M., Wang, H., Xiao, L., "Microcrystalline silicon carbide alloys prepared with HWCVD as highly transparent and conductive window layers for thin film solar cells", (2009), *Thin Solid Films*, Vol. **517**, pp. 3507-3512

[Frammelsberger-2000] Frammelsberger, W., Geyer, R., Lechner, P., Rübel, H., Schade, H., Müller, J., Schöpe, G., Kluth, O., Rech, B., "Effects of TCO surface texture on light absorption in thin film silicon solar cells", (2000), *16th European Photovoltaic Solar Energy Conference, Glasgow,* pp. 389-392

[Frammelsberger-2004] Frammelsberger, W., Geyer, R., Lechner, P., Lundszien, D., Psyk, W., Rübel, H., Schade, H., "Entwicklung und Optimierung von Prozessen zur Herstellung hocheffizienter grossflächiger Si-Dünnschicht-PV-Module", (2004), *Abschlussbericht, Bundesministerium für Umwelt, Naturschutz und Reaktorsicherheit*, Förderkennzeichen 0329810A

[Gerhardinger-1996] Gerhardinger, P. F., McCurdy, R. J., "Float line deposited transparent conductors - implications for the PV industry", (1996), *Mat. Res. Soc. Symp. Proc.*, Vol. **426,** pp. 399-410

[Golay-2000] Golay, S., Meier, J., Dubail, S., Faÿ, S., Kroll, U., Shah, A., "Laser scribing of p-i-n/p-i-n micromorph (a-Si:H/μc-Si:H) tandem cells", (2000), *Proc. 28th IEEE Photovoltaic Sprecialists Conference, Anchorage, Alaska*, pp. 1456-1459

[Hamers-2008] Hamers, E. A. G., Lenssen, J. M. T. Borreman, A., Ammerlaan, J., Broekhof, S., Dubbeldam, G. C., Périn, S., Scherder, W., Sportel, E., Stigter, L. A., Welling, F., Schlatmann, R., Gordijn, A., Jongerden, G. J., "Manufacturing of large area thin film silicon flexible solar cell modules employing a temporary superstrate foil", (2008), *23rd European Photovoltaic Solar Energy Conference, Valencia*, pp. 2065-2068

[Hegedus-1995] Hegedus, S. S., Kampas, F., Xi, J., "Current transport in amorphous silicon n/p junctions and their application as "tunnel" junctions in tandem solar cells", (1995), *Appl. Phys. Lett.,* Vol. **67**, pp. 813-815

[Herrmann-1998] Herrmann, W., Adrian, M., Wiesner, W., Operational behaviour of commercial solar cells under reverse biased conditions, (1998), *2nd World Conference on Photovoltaic Solar Energy Conversion, Vienna,* pp. 2357-2359

[Heya-2008] Heya, A., Minamikawa, T., Niki, T., Minami, S., Masuda, A., Umemoto, H., Matsuo, N., Matsumura, H., "Cat-CVD SiN passivation films for OLEDs and packaging", (2008), *Thin Solid Films*, Vol. **516**, pp. 553-557

[Holt-2002] Holt, J.K., Swiatek, M., Goodwin, D.G., Atwater, H.A., "The aging of tungsten filaments and its effect on wire surface kinetics in hot-wire chemical vapor deposition", (2002), *Journal of Applied Physics*, Vol. **92**, pp. 4803-4808

[Honda-2008] Honda K., Ohdaira K., Matsumura H., "Study of silicidation process of tungsten catalyzer during silicon film deposition in catalytic chemical vapour deposition", (2008), *Japanese Journal of Applied Physics*, Vol. **47**, pp. 3692-3698

[Howling-1992] Howling, A. A., Dorier, J.-L., Hollenstein, Ch., Kroll, U.,Finger, F., "Frequency effects in silane plasmas for plasma enhanced chemical vapor deposition", (1992), *J. Vac. Sci.Technol.* Vol. **A 10**, 1080-1085

[Howling-2000] Howling, A. A., Sansonnens, L., Ballutaud, J., Grangeon, F., Delachaux, T., Hollenstein, Ch.,, V., Kroll, U., "The influence of plasma chemistry on the deposition of microcrystalline silicon for large area photovoltaic solare cells"(2000), *16th European Photovoltaic Solar Energy Conference, Glasgow, UK*, pp. 518-521

[Howling-2005] Howling, A. A., Derendinger, L., Sansonnens, L. Schmidt, H., Hollenstein, C., Sakanaka, E., Schmitt, J. P. M., "Probe measurements of plasma potential nonuniformity due to edge asymmetry in large-area radio-frequency reactors: The telegraph effect", (2005), *Journal of Applied Physics*, Vol. **97**, pp. 123308-123308-13

[Huang-2007] Huang, Y., Dasgupta, A., Gordijn, A., Finger, F., Carius, R., "Highly transparent microcrystalline silicon carbide grown with hot wire chemical vapor deposition as window layers in *n-i-p* microcrystalline silicon solar cells", (2007), *Applied Physics Letters*, Vol. **90**, pages 203502

[Ishibashi-2003] Ishibashi, K., Karasawa, M., Xu, G., Yokokawa, N., Ikemoto, M., Masuda, A., Matsumura, H., "Development of Cat-CVD apparatus for 1-m-size large area deposition", (2003), *Thin Solid Films*, Vol. **430**, pp. 58-62

[Izu-2003] Izu M.; Ellison T., "Roll-to-roll manufacturing of amorphous silicon alloy solar cells with in situ cell performance diagnostics", (2003), *Solar Energy Materials and Solar Cells*, Vol. **78**, pp. 613-626

[Jansen-2006] Jansen, K. W., Kadam, S. B., Groelinger, J. F.,"The high energy yield of amorphous silicon modules in a hot coastal climate", (2006), *21st European Photovoltaic Solar Energy conference, Dresden*, pp. 2535-2538

[Jardine-2002] Jardine, C., Lane, K., "PV COMPARE: Relative performance of PV technologies in northern and southern Europe", (2002), *Proc. PV in Europe Conference, Rome,* pp. 1057-1060

[Kasper-1992] Kasper, W., Böhm, H., Hirschauer,B., "The influence of electrode areas on radio frequency glow discharge", (1992), *J. Appl. Phys.,* Vol. **71,** pp. 4168-4172

[Katsuma-2007] Katsuma, K., Hayakawa, S., Masuda, A., Matsui, T., Kondo, M., "Fabrication of superstrate-type thin-flm silicon solar cells on textured plastic substrates", (2007), *22nd European Photovoltaic Solar Energy Conference, Milan,* pp. 1831-1834

[Kawamura-2006] Kawamura, K., Mashima, H., Takeuchi, Y., Takano, A., Noda, M., Yonekura, Y., Takatsuka, H., "Development of large-area a-Si:H films deposition using controlled VHF plasma", (2006), *Thin Solid Films*, Vol. **506-507**, pp. 22-26

[Klein-2003] Klein, S., Finger, F., Carius, R., Dylla, T., Rech, B., Grimm, M., Houben, L., Stutzmann, M., "Intrinsic microcrystalline silicon prepared by hot-wire chemical vapour deposition for thin film solar cells", (2003), *Thin Solid Films*, Vol. **430**, pp. 202-207

[Klein-2007] Klein, S., Repmann, T., Wieder, S., Müller, J., Buschbaum, S., Rohde, M., "a-Si:H/μc-Si:H tandem cell development on 1.4 $m^2$ substrate size in a vertical in-line reactor", (2007), *22nd European Photovoltaic Solar Energy Conference, Milan*, pp. 1791-1794

[Kluth-2003] Kluth, O. Schöpe, G., Hüpkes, J., Agashe, C., Müller, J., Rech, B., "Modified Thornton model for magnetron sputtered zinc oxide: film structure and etching behaviour", (2003), *Thin Solid Films,* Vol. **442**, pp. 80-85

[Kluth-2004] Kluth, O., Zahren, C., Stiebig, H., Rech, B., Schade, H., "Surface morphologies of rough transparent conductive oxide films applied in silicon thin-film solar cells", (2004), *19th European Photovoltaic Solar Energy Conference, Paris,* pp. 1587-1590

[Kondo-2006] Kondo, M., Matsui, T., Nasuno, Y., Sonobe, H., Shimizu, S., "Key issues for fabrication of high quality amorphous and microcrystalline silicon solar cells", (2006), *Thin Solid Films,* Vol. **501**, pp. 243.246

[Kroll-1995] Kroll, U.,. "VHF-Plasmaabscheidung von amorphem Silizium: Einfluss der Anregungsfrequenz, der Reaktorgestaltung sowie Schichteigenschaften", (1995), *Ph.D. Thesis, University of Neuchâtel,* Hartung-Gorre Verlag Konstanz, ISBN 389191-905-0

[Kroll-1999] Kroll, U., Fischer, D., Meier, J., Sansonnens, L., Howling, A., Shah, A., (1999), *Proc. Mat. Res. Soc. Symp.,* Vol. **557**, pp. 121-126

[Lau-2008a] Lau, K.K.S., Gleason, K. K., "Applying HWCVD to particle coating and modelling the deposition mechanism", (2008), *Thin Solid Films*, Vol. **516**, pp. 674-677

[Lau-2008b] Lau, K.K.S., Gleason, K.K., "Initiated chemical vapor deposition (iCVD) of copolymer thin films", (2008), *Thin Solid Films*, Vol. **516**, pp. 678-680

[Lechner-2000] Lechner, P., Geyer, R., Schade, H., Rech, B., Müller, J., "Detailed accounting for quantum efficiency and optical losses in a-Si:H based solar cells", (2000), *Proc. 28th IEEE Photovoltaic Specialists Conference, Anchorage, Alaska,* pp. 861-864

[Lechner-2004] Lechner, P., Geyer, R., Schade, H., Rech, B., Kluth, O., Stiebig, H., Optical TCO properties and quantum efficiencies in thin-film silicon solar cells", (2004), *19th European Photovoltaic Solar Energy Conference, Paris,* pp. 1591-1594

[Lechner-2008] Lechner, P., Frammelsberger, W., Psyk, W., Geyer, R., Maurus, H., Lundszien, D., Wagner, H., Eichhorn, B., "Status of performance of silicon thin film solar cells and modules", (2008), *23rd European Photovoltaic Solar Energy Conference, Valencia,* pp. 2023-2026

[Lehmann-2006] Lehmann, J.-C., "Glass and glass products", (2006), *Europhysics News*, Vol. **37**, pp. 23-27

[Mahan-2000] Mahan, A.H., Mason, A., Nelson, B.P., Gallagher, A.C., "The influence of W filament alloying on the electronic propertiesof HWCVD deposited a-Si:H films", (2000), *Materials Research Society Symposium Proceedings*, Vol. **609**, paper A6.6.1

[Mahan-2002] Mahan, A.H., Xu, Y., Iwaniczko, E., Williamson, D.L., Nelson, B.P., Wang, Q., "Amorphous silicon films and solar cells deposited by HWCVD at ultra high deposition rates", (2002), *Journal of Non-Crystalline Solids*, Vol. **299-302**, pp. 2-8

[Mahan-2003] Mahan, A.H., "Hot wire chemical vapour deposition of Si containing materials for solar cells", (2003), *Solar Energy Materials and Solar Cells*, Vol. **78**, pp. 299-327

[Masuda-2006] Masuda, A., Umemoto, H., Matsumura, H., "Various applications of silicon nitride by catalytic chemical vapour deposition for coating, passivation and insulating films", (2006), *Thin Solid Films*, Vol. **501**, pp. 149-153

[Mataras-1997] Mataras, D., Rapakoulias, D., (1997), "Improvements in control and understanding of radio frequency silane discharges", *High Temp. Material Processes,* Vol. **1,** pp. 383-391

[Matsuda-2004] Matsuda, A., "Microcrystalline silicon, growth and device application", (2004), *Journal of Non-Crystalline Solids*, Vol. **338-340**, pp. 1-12

[Matsumura-1986] Matsumura, H., "Catalytic Chemical Vapor Deposition (Cat-CVD) Method Producing High Quality Hydrogenated Amorphous silicon", (1986), *Japanese Journal of Applied Physics*, Vol. **25**, pp. L949-L951

[Matsumura-2008] Matsumura, H., Ohdaira, K., "Recent situation of industrial implementation of Cat-CVD technology in Japan", (2008), *Thin Solid Films*, Vol. **516**, pp. 537-540

[Maurus-2004] Maurus, H., Schmid, M., Blersch, B., Lechner, P., Schade, H., "PV for buildings", (2004), *Refocus, Nov./Dec.*, pp. 22-27

[Meier-1994] Meier, J., Dubail, S., Flückiger, R., Fischer, D., Keppner,H., Shah, A., "Intrinsic microcrystalline silicon (μc-Si:H) - a promising new thin film solar cell material", (1994), *1st World Conference on Photovoltaic Energy Conversion, Hawai,* pp. 409 - 412

[Meier-2004] Meier, J., Kroll, U., Vallat-Sauvin, E., Spitznagel, J., Graf, U., Shah, A., "Amorphous solar cells, the micromorph concept and the role of the VHF-GD deposition technique", (2004), *Solar Energy*, Vol. **77**, pp. 983-993

[Meier-2005] Meier, J., Kroll, U., Spitznagel, J., Benagli, S., Roschek, T., Pfanner, G., Ellert, C., Androutsopoulos, G., Hügli, A., Nagel, M., Bucher, C., Feitknecht, L., Büchel, G., Büchel, A., "Progress in up-scaling of thin film silicon solar cells by large-area PECVD KAI systems", (2005), *Proc. 31st IEEE Photovoltaic Specialists Conference, Buenavista, Florida,* pp. 1464-1467

[Meillaud-2005] Meillaud, F., Vallat-Sauvain, E., Niquille,X., Dubey, M., Bailat, J., Shah, A., Ballif, C., "Light-induced degradation of thin film amorphous and microcrystalline silicon solar cells", (2005), *Proc. 31st IEEE Photovoltaic Specialists Conference, Lake Buena Vista, Florida,* pp. 1412-1415

[Meillaud-2006] Meillaud, F., Shah, A., Bailat, J., Vallat-Sauvain, E., Roschek, T., Rech, B., Didier, D., Söderström, T., Python, M., Ballif, C., "Microcrystalline silicon solar cells: Theory and diagnostic tools", (2006), *4th World Conference on Photovoltaic Solar Energy Conversion, Kona Island, Hawaii*, pp. 1572 -1575

[Meillaud-2009] Meillaud, F., Feltrin, A., Dominé, D., Buehlmann, P., Python, M., Bugnon, G., Billet, A., Parascandola; G., Bailat, J., Fay, S., Wyrsch, N., Ballif, C., Shah, A., "Limiting factors in the fabrication of microcrystalline silicon solar cells and microcrystalline/amorphous ('micromorph') tandems", (2009),*Philosophical Magazine*, Vol. **89**, pp. 2599-2621

[Middelman-1998] Middelman, E., van Andel, E., Schropp, R. E. I.., de Jonge-Meschaninova, L. V., Peters, P. M. G. M., Severens, R. J., Meiling, H., Zeman, M., van den Sanden, M. C. M., Kuijpers, A., Spee, C. I. M. A., Jongerden, G. J.,"New superstrate process for roll-to-roll production of thin film solar cells", (1998), *2nd World Conference on Photovoltaic Solar Energy Conversion, Vienna,* pp. 816-819

[Molenbroek-1991] Molenbroek, E., Waddington, D. W., Emery, K. A., "Hot spot susceptibility and testing of PV modules", (1991), *Proc. 22nd IEEE Photovoltaic Specialists Conference, Las Vegas, Nevada,* pp. 547-552

[Molenbroek-1996] Molenbroek, E.C., Mahan, A.H., Johnson, E.J., Gallagher, A.C., "Film quality in relation to deposition conditions of a-Si:H films deposited by the "hot wire" method using highly diluted silane", (1996), *Journal of Applied Physics*, Vol. **79**, pp. 7278-7292

[Müller-2001] Müller, J., Schöpe, G., Kluth, O., Rech, B., Szyszka, B., Höing, T., Sittinger, V., Jiang, X., Bräuer, G., Geyer, R., Lechner, P., Schade, H., Ruske, M., "Large area mid-frequency sputtered ZnO films as substrates for siliconthin-film solar cells", (2001), *17th European Photovoltaic Solar Energy Conference, Munich,* pp. 2876-2879

[Nebel-2000] Nebel, C. E., (2000), *Walter Schottky Institute, Munich,* private communication

[Ogawa-2008] Ogawa, Y., Ohdaira, K., Oyaidu, T., Matsumura, H., "Protection of organic light-emitting diodes over 50000 hours by Cat-CVD $SiN_x/SiO_xN_y$ stacked thin films", (2008), *Thin Solid Films*, Vol. **516**, pp. 611-614

[Oku-2008] Oku, T., Kamo, Y., Totsuka, M., "AlGaN/GaN HEMTs passivated by Cat-VCVD SiN Film", (2008), *Thin Solid Films*, Vol. **516**, pp. 545-547

[Oshinowo-2009] Oshinowo, J., Riva, M., Pittroff, M., Schwarze, T., Wieland, R., "Etch performance of Ar2/N2/F2 for CVD/ALD chamber clean", (2009), *Solid State Technology*, Vol. **52**, pp 22-24

[Ossadnik-1995] Ossadnik, C., Frammelsberger, W., Psyk, W., Lechner, P., Rübel, H., Schade, H., "Plasma impedances and voltage distributions in silane rf glow discharges - correlations with amorphous silicon film properties", (1995), *13th European Photovoltaic Solar Energy Conference, Nice*, pp. 210 - 213

[Parascandolo-2010] Parascandolo, G., Bugnon, G., Feltrin, A., Ballif, C., "High-rate deposition of microcrystalline silicon in a large-area pecvd reactor and integration in tandem solar cells", (2010), *Prog. Photovolt: Res. Appl.*, Vol. **18**, pp. 1-8

[Pellaton Vaucher-1997] Pellaton Vaucher, N., Rech, B., Fischer, D., Dubail, S., Götz, M., Keppner, H., Wyrsch, N., Beneking, C., Hadjad, O., Shklover, V., Shah, A., "Controlled nucleation of thin microcrystalline layers of the recombination junction in a-Si stacked cells", (1997), *Solar Energy Materials and Solar Cells*, Vol. **49**, pp. 27-33

[Perrin-1988] Perrin, J., Roca i Cabarrocas, P., Allain, B., Friedt, J.-M., "a-Si:H deposition from $SiH_4$ and $Si_2H_6$ rf-discharges: pressure and temperature dependence of film growth in relation to the $\alpha$-$\gamma$ discharge transition", (1988), *Jap. J. Appl. Phys.,* Vol. **27**, pp. 2041-2052

[Perrin-1991] Perrin, J., "Plasma and surface reactions during a-Si:H film growth", (1991),*Journal of Non-Crystalline Solids*, Vol. **137&138**, pp. 639-644

[Platz-1998] Platz, R., Hof, C., Wieder, S., Rech, B., Fischer, D., Shah, A., Payne, A., Wagner, S., "Comparison of VHF, RF and DC plasma excitation for a-Si: H deposition with hydrogen dilution", (1998), *Mater. Res. Soc. Symp. Proc.* Vol. **507**, pp. 565-570

[Rech-2005] Rech, B. (2005), *Research Centre Julich, Institute of Photovoltaics,* private communication

[Roschek-2002] Roschek, T., Repmann, T., Müller, J., Rech, B., Wagner, H., "Comprehensive study of microcrystalline silicon solar cells deposited at high rate using 13.56 MHz plasma-enhanced chemical vapor deposition", (2002), *J. Vac. Sci. Technol.* Vol. **A 20(2),** pp. 492-498

[Sansonnens-1997] Sansonnens, L., Pletzer, A., Magni, D., Howling, A. A., Hollenstein, Ch., Schmitt, J. P. M., "A voltage uniformity study in large-area reactors for RF plasma deposition", (1997), *Plasma Sources Science & Technology*, Vol. **6**, pp. 170-178

[Sansonnens-1998] Sansonnens, L., Howling, A. A., Hollenstein, Ch., "Degree of dissociation measured by FTIR absorption spectroscopy applied to VHF silane plasmas", (1998), *Plasma Sources Sci. Technol.*, Vol. **7**, pp. 114-118

[Sansonnens-2005] Sansonnens, L., Hollenstein, Ch., "Ladder electrode principle for VHF uniform plasma generation and comparison with a solid electrode", (2005), *Presentation Mannheim Experts Meeting*

[Sansonnens-2006] Sansonnens, L., Schmidt, H., Howling,A. A., Hollenstein, Ch., Ellert,Ch., Buechel, A., "Application of the shaped electrode technique to a large area rectangular capacitively coupled plasma reactor to suppress standing wave nonuniformity", (2006), *J. Vac. Sci. Technol. A,* Vol. **24**, pp. 1425-1430

[Sato-1990] Sato, K., Gotoh, Y., Hayashi, Y., Adachi, K., Nishimura, H., "Improvement of textured $SnO_2$:F TCO films for a-Si solar cells", (1990), *Reports Res. Lab. Asahi Glass Co, Ltd.*, Vol. **40**, pp. 233-241

[Schade-1984] Schade, H., Smith, Z E., Thomas III, J. H., Catalano, A., "Hydrogen plasma interactions with tin oxide surfaces", (1984), *Thin Solid Films,* Vol. **117,** pp. 149-155

[Schade-1985a] Schade, H., Smith, Z E., "Optical properties and quantum efficiency of a-$Si_{1-x}C_x$:H/a-Si:H solar cells", (1985), *J. Appl. Phys.,* Vol. **57**, pp. 568-574

[Schade-1985b] Schade, H., Smith, Z E., "Mie scattering and rough surfaces", (1985), *Applied Optics*, Vol. **24**, pp. 3221-3226

[Schade-1989] Schade, H., Chao, H., "Microcrystalline silicon-carbon p-layers prepared by photo-CVD and glow discharge", (1989), *Proc. Amorphous Silicon Subcontactors' Review Meeting, Solar Energy Research Institute, Golden, CO*, p.199

[Schade-1998] Schade, H., Lechner, P., Geyer, R., Frammelsberger, W., Rübel, H., Schmid, M., Maurus, H., Hoffmann, W., "Application-related features of a-Si based PV technology", (1998), *2nd World Conference on Photovoltaic Solar Energy Conversion, Vienna*, pp. 2054-2057

[Schade-2001] Schade, H., (2001), *RWE Solar, Phototronics*

[Schade-2005] Schade, H., Lechner, P., Geyer, R., Stiebig, H., Rech, B., Kluth,O., "Texture properties of TCO uniquely determining light trapping in thin-film solar cells", (2005), *Proc. 31st IEEE Photovoltaic Specialists Conference, Lake Buena Vista, Florida*, pp. 1436-1439

[Schottler-2008] Schottler, M., de Wild-Scholten, M., "The carbon footprint of PECVD chamber cleaning using fluorinated gases", (2008), *23th European Photovoltaic Solar Energy Conference, Valencia*, pp. 2505-2509

[Schroeder-2003] Schroeder B., "Status report: solar cell related research and development using amorphous and microcrystalline silicon deposited by HW(Cat)CVD", (2003), *Thin Solid Films*, Vol. **430**, pp. 1-6

[Schroeder-2008] Schroeder B., Kupich, M., Kumar, P., Grunsky, D., "Recent contributions of the Kaiserslautern research group to thin silicon solar cell R and D applying the HW(Cat)CVD", (2008), *Thin Solid Films*, Vol. **516**, pp. 722-727

[Shah-2004] Shah, A. V., Schade, H., Vanecek, M., Meier, J., Vallat-Sauvin, E., Wyrsch, N., Kroll, U., Droz, C., Bailat, J., "Thin-film silicon solar cell technology", (2004), *Progr. Photovolt.: Res. Appl.*, Vol. **12**, pp. 113-142

[Shah-2006] Shah, A., Meier, J., Buechel, A., Kroll, U., Steinhauser, J., Meillaud, F., Schade, H., Dominé, D., "Towards very low-cost mass production of thin-film silicon photovoltaic (PV) solar modules on glass", (2006), *Thin Solid Films*, Vol. **502**, pp. 292-299

[Smith-1985] Smith, Z E., Wagner, S., "A carrier lifetime model for the optical degradationof amorphous silicon solar cells", (1985), *Mat. Res. Soc. Symp. Proc.*, Vol. **49**, pp. 331-338

[Spear-1975] Spear, W. E., LeComber, P. G., (1975), *Sol. State Comm.,* Vol.**17,** p.1193

[Stiebig-2006] Stiebig, H.; Schulte, M.; Zahren, C.; Haase, C.; Rech, B.; Lechner, P., "Light trapping in thin-film silicon solar cells by nano-textured interfaces", (2006), *Photonics Europe: Optoelectronics and Photonic Materials - Proc. of the SPIE,* Vol. **6197**, pp. 1-9

[Stolk-2008] Stolk, R. L., Li, H., Franken, R. H., Schüttauf, J. W. A., van der Werf, C. H. M., Rath, J. K., Schropp, R. E. I., "Improvement of the efficiency of triple junction n-i-p solar cells with hot-wire CVD proto- and microcrystalline silicon absorber layers", (2008), *Thin Solid Films*, Vol. **516**, pp. 736-739

[Strahm-2007a] Strahm, B., "Investigations of radio-frequency capacitively coupled large-area industrial reactor: cost-effective production of thin-film microcrystalline silicon for solar cells", (2007), *Thesis Nr. 3895, École Polytechnique Fédérale de Lausanne,* pp. 18-20, 52-59

[Strahm-2007b] Strahm, B., Howling, A. A., Sansonnens, L., Hollenstein, Ch., Kroll, U., Meier, J., Ellert, Ch., Feitknecht, L., Ballif, C., "Microrystalline silicon deposited at high rates on large areas from pure silane with efficient gas utilization", (2007), *Solar Energy Materials and Solar Cells,* Vol. **91**, pp. 495-502

[Strobel-2008] Strobel, C., Zimmermann, T., Albert, M., Bartha, J. W., Beyer, W., Kuske, "Dynamic high-rate-deposition of silicon thin film layers for photovoltaic devices", (2008), *23th European Photovoltaic Solar Energy Conference, Valencia*, pp. 2497-2504

[Šúri-2007] Šúri, M., Huld, T.A., Dunlop, E.D., Ossenbrink, H.A., "Potential of solar electricity generation in the European Union member states and candidate countries", (2007), *Solar Energy*, Vol. **81**, pp. 1295–1305, http://re.jrc.ec.europa.eu/pvgis/

[Takagi-2008] Takagi T., Yamamoto N., Murayama A., Miyata, N., Kawasaki, Y., "Microcrystalline deposition by array antenna VHF-PECVD", (2008), *23rd European Photovoltaic Solar Energy Conference, Valencia*, pp. 2125-2128

[Takatsuka-2004] Takatsuka, H., Noda, M., Yonekura, Y., Takeuchi, Y., Yamauchi, Y., "Development of high efficiency large area silicon thin film modules using VHF-PECVD", (2004), *Solar Energy,* Vol. **77**, pp. 951-960

[Takatsuka-2006] Takatsuka, H., Yamauchi, Y., Fukagawa, M., Mashima, H., Kawamura, K., Yamaguchi, K., Nishimiya, T., Takeuchi, Y., "High efficiency thin film solar cells, (2006), *21st European Photovoltaic Solar Energy Conference, Dresden,* pp. 1531-1534

[Taneda-2008] Taneda, N., Masumo, K., Kambe, M., Oyama, T., Sato, K., "Highly textured $SnO_2$ films for a-Si/μc-Si tandem solar cells", (2008), *23rd European Photovoltaic Solar Energy Conference, Valencia,* pp. 2084-2087

[Tauc-1966] Tauc, J., Grigorivici, A, Vancu, R., (1966), *Phys. Stat. Sol.,* Vol. **15**, p. 627

[Terrazzoni-2006] Terrazzoni-, V., Haug, F.-J., Ballif, C.,Fischer, D., Soppe, W., Andreu, J., Fahland, M., Roth, K., Topic, M., Willford, T., "The European project Flexcellence roll to roll technology for the production of high efficiency low cost thin film solar cells", (2006), *21st European Photovoltaic Solar Energy Conference, Dresden,* pp. 1669-1672

[TÜV-2009] TÜV Rheinland, "Design qualification and type approval of PV modules acc. to IEC 61215:2005 / 61646:2008", (2009), *TÜV Rheinland Immissionsschutz und Energiesysteme GmbH, Renewable Energies,* pp. 1-5

[Umemoto-2002] Umemoto, H., Ohara, K., Morita, D., Nozaki, Y., Masuda, A., Matsumura, H., "Direct detection of H atoms in the catalytic chemical vapor deposition of the $SiH_4/H_2$ system", (2002), *Journal of Applied Physics*, Vol. **91**, pp. 1650-1656

[Vallat-Sauvin-2001] Vallat-Sauvin, E., Fay, S., Dubail, S., Meier, J., Bailat, J., Kroll, U., Shah, A., "Improved interface between front TCO and microcrystalline silicon p-i-n solar cells, (2001), *Proceedings of the MRS Symp., Spring Meeting, San Francisco,* Vol. **664**, pp. A15.3.1-A15.3.5

[van der Werf-2009] van der Werf, C.H.M., Li, H., Verlaan, V., Oliphant, C.J., Bakker, R., Houweling, Z.S., Schropp, R.E.I., "Reversibility of silicidation of Ta filaments in HWCVD of thin film silicon", (2009), *Thin Solid Films*, Vol. **517**, pp. 3431-3434

[Verlaan-2008] Verlaan, V., Houweling, Z.S., van der Werf, C.H.M., Romijn, I.G., Weeber, A.W., Goldbach, H.D., Schropp, R.E.I., "deposition of device quality silicon nitride with ultra high deposition rate (>7 nm/s) using hot-wire CVD", (2008), *Thin Solid Films*, Vol. **516**, pp. 533-536

[Villar-2009] Villar, F., Antony, A., Escarré, J., Ibarz, D., Roldan, R., Stella, M. Munoz, D., Asensi, J.M., Bertomeu, J., "Amorphous silicon thin film solar cells deposited entirely by hot-wire chemical vapour deposition at low temperature (<150 °C)", (2009), *Thin Solid Films*, Vol. **517**, pp. 3575-3577

[Wang-2004] Wang, Q., Ward, S., Gedvilas, L., Keyes, B., Sanchez, E., Wang, S., "Conformal thin-film silicon nitride deposited by hot-wire chemical vapor deposition", (2004), *Applied Physics Letters*, Vol. **84**, pp. 338-340

[Yamauchi-2005] Yamauchi, Y., Takatsuka, H., Kawamura, K., Yamashita, N., Fukagawa, M., Takeuchi, Y., "Development of a-Si/microcrystalline-Si tandem-type photovoltaic solar cell", (2005), *Mitsubishi Heavy Industries Technical Review,* Vol. **42**, No. 3, pp. 1-5

[Zedlitz-1992] Zedlitz, R., Heintze, M., Bauer, G. H., "Analysis of VHF Glow Discharge of a-Si:H over a wide frequency range", (1992), *Proc. MRS Symp.* Vol. **258**, pp. 147.

[Zheng-2006] Zheng, W., Gallagher,A., "Hydrogen dissociation on high-temperature tungsten", (2006), *Surface Science*, Vol. **600**, pp. 2207-2213

CHAPTER 7

# EXAMPLES OF SOLAR MODULE APPLICATIONS

*Horst Schade*

## 7.1 BUILDING-INTEGRATED PHOTOVOLTAICS (BIPV): ASPECTS AND EXAMPLES

Thin-film solar cell technologies based on large areas are particularly well-suited to applications in the building industry. These applications of BIPV consist mainly of photovoltaic roofs and facades, and generally are grid-connected.

Compared to other thin-film modules (such as those based on CdTe and CIGS), silicon-based thin-film modules presently are most widely spread in BIPV installations. This is the case in spite of generally lower stabilized efficiencies compared to the other technologies. But this drawback is outweighed by a number of unique features which Si-based thin-film technology has to offer.

- The use of silicon as base material implies a non-toxic and abundant materials supply, free from heavy metals like cadmium, or from relatively rare materials like indium and tellurium.
- Low process temperatures (around 200 °C) are employed. This facilitates the use of low-cost substrates like float-glass. It also leads to moderate energy consumption, as well as to shorter energy payback times.
- Upscaling of the deposition processes to very large areas has been realized, leading to modules in sizes up to 5.7 $m^2$ [Tanner-2008].
- Low temperature coefficients of the power output (see Sect. 4.4.3), i.e. –0.1 % per K to –0.2 %/K for a-Si, and –0.25 %/K for a-Si/μc-Si, compared to around –0.4 %/K for c-Si.
- High energy yields (see Sect. 6.6.2) due to the low temperature coefficients, and the spectral response, which is more closely matched to outdoor illumination than that of c-Si.

BIPV installations combine several functions, namely

- electricity generation;
- thermal insulation;
- shading and glare protection;
- architectural design.

Depending on the particular application, these functions may be weighted differently. Unlike applications that solely provide electricity generation, and usually are valued in terms of price/Wp, BIPV products are often more suitably valued in terms of price/m$^2$. Here the following aspects are taken into account:

- energy yield; this determines the amount of energy generated by the installation, and hence the monetary return, that can be obtained from feeding electricity into the grid, as well as the savings that arise from correspondingly lower consumption of conventional electricity;
- cost comparison with conventional roof or façade building elements;
- possibilities for architectural design, and the perceived image and reputation derived from the chosen installation.

As a particularly attractive design option, thin-film superstrate technology offers the unique possibility of providing semitransparent modules. By laser scribing, the opaque cell layers are removed along narrow lines perpendicular to the scribe lines of the integral series connection of the cells. Thereby a grid-like rectangular pattern with a color-neutral transmission of typically 10 % is created (see Fig. 7.1 [Phototronics-2002]), resulting in a "see-through" effect, from which the trade name ASI THRU® for semitransparent modules was derived.

Semitransparent modules were created as early as 1991 [Ricaud-1991]. They are now being used with grid connection, mainly as double-glazing PV windows. Obviously,

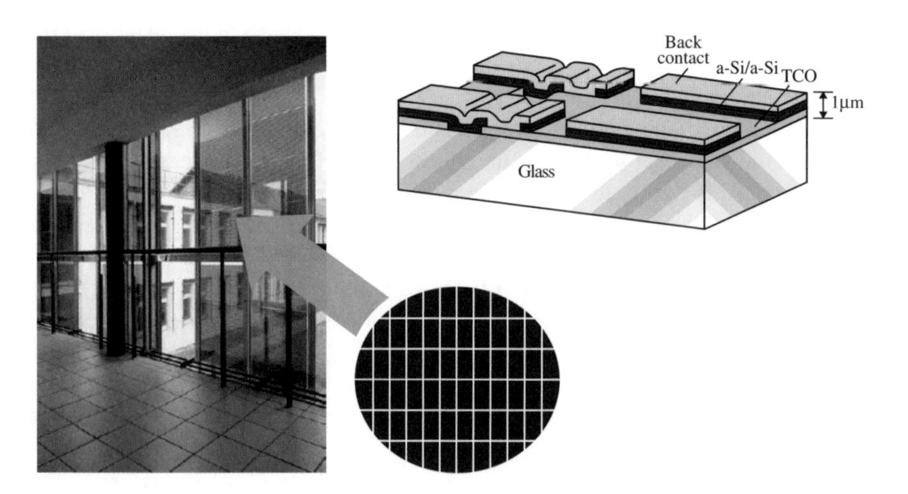


**Fig. 7.1** Principle and appearance of semitransparent modules. (adapted from [Shah-2004], and credit RWE SCHOTT Solar)

the efficiency of semitransparent modules is reduced by the percentage of their optical transmission. However, besides electricity generation, they serve additional functions in terms of light and temperature management in buildings, i.e. by providing shading and thermal insulation. These aspects are generally characterized by the total energy transmittance ($g$-value [%]), and the heat transfer coefficient ($U$-value [$W/m^2K$]).

The use of a-Si modules in BIPV is manifold and extends to all elements of a building shell, namely roofs, façades, windows, doors and awnings. Although the substrate size for the semiconductor deposition is fixed by the particular deposition equipment, the final module sizes can be adapted to the specific size of the desired building element by cutting and formation of laminates. Building elements exceeding the individual module size may also be formed by laminating a number of modules side-by-side onto a panel of the required size of the building element. A high degree of flexibility in design is thus obtained. The installation of BIPV elements further benefits from established methods and procedures that are routinely used in the conventional building industry. But by the same token, the various building codes and regulations, concerning e.g. static requirements and mechanical load tolerances, must also be satisfied.

The applicability of BIPV is inseparably linked to the long-term stability of the installation, as discussed in Section 6.7.3. Although there are many examples of thin-film Si-based BIPV worldwide [Maurus-2003], it is only in relatively few cases that evaluation of their long-term performance has been published or is otherwise available.

### 7.1.1 PV Façade in Munich (Germany)

The PV façade of the Bavarian State Ministry of the Environment and Public Health in Munich, Germany, is one of the oldest PV façades (see Fig. 7.2) based on a-Si modules, and was completed in August 1993. In architectural terms, it is a curtain-wall

(a)

(b)

**Fig. 7.2** PV facade of the Bavarian State Ministry of the Environment and Public Health, in Munich: (a) as originally installed in 1993 (© H. Schade [1993], with permission [Shah-2004]); (b) after building renovation in 2003 (credit RWE SCHOTT Solar).

facing with airspace, also described as a rear-ventilated cold façade, suspended in front of a flagstone façade. In 2003 the hull of the entire building was renovated. At that time, the PV façade was preserved; however, for architectural reasons the lowest row of panels, i.e. $6 \times 4$ modules, was omitted. The main features of the façade as originally installed are summarized in Table 7.1.

The rated power given in the table below refers to the entire PV façade system, comprising 240 modules, wiring, and power conditioning and monitoring equipment. The performance of the façade, oriented towards the south, was monitored [FfE-2002] from its start of operation, over a period of almost 9 years (see Fig. 7.3).

After an initial phase of light-induced degradation, the power output remained stable for the rest of this period until the end of 2002. From the continued retention of this PV façade after the renovation of the building, it must be concluded that the preceding history of stable performance was confidently projected to continue into

**Table 7.1** Main features of the PV façade of the Bavarian State Ministry of the Environment and Public Health, in Munich, as originally installed.

| | |
|---|---|
| Modules | a-Si/a-Si tandem-junction cells on 0.6 $m^2$ modules, fabricated by Phototronics, RWE SCHOTT Solar, 60 laminates, arranged in 10 rows and 6 columns, $2 \times 2$ modules/laminate, module size $0.6 \times 1.0$ $m^2$, nominal power 32 W |
| Mounting system | aluminum profile system, clamping system Schüco FW 50 |
| Rated power | 6.5 kW |
| Inverter | single-phase, efficiency 84-89 % |
| Year of installation | 1993 |

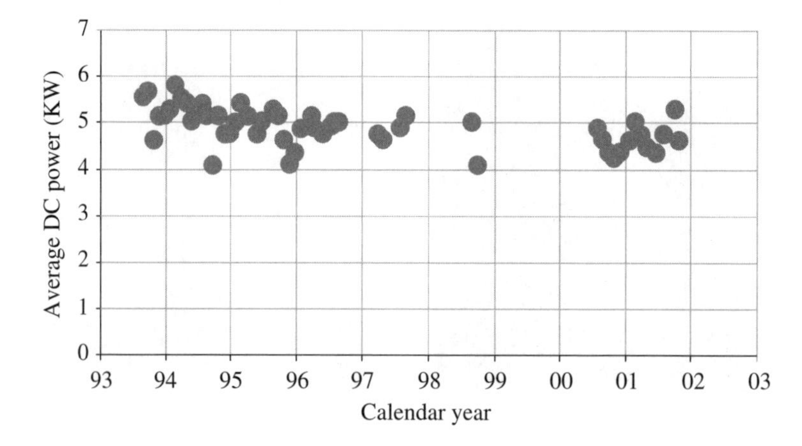


**Fig. 7.3** Long-term performance of the original PV façade at the Bavarian State Ministry of the Environment and Public Health, Munich, shown left in Figure 7.2. (with permission [Shah-2004])

the future. Reportedly, the façade is still in operation and renewed data collection is being planned.

### 7.1.2 Alpine roof integrated PV (by Luc Feitknecht)

The grid-connected 1.5 kW PV test plant "Mungg" in Bellwald, Switzerland, is an example of a BIPV system recently implemented in an alpine environment. Modules of the type Nova®-T were taken from a pilot production of ersol Thin Film GmbH (now BOSCH Solar). The underlying module technology features a-Si single-junction cells deposited on commercial TCO, and uses a dielectric white-paint back-reflector. Further details are summarized in Table 7.2.

The main reason to build this PV test plant (see Fig. 7.4) was to replace the existing aging roof tiles. In the planning phase, the aim was to replace as many roof tiles as possible with PV modules. This roof area allowed for 18 modules of 1.4 $m^2$ area (1.1 m width and 1.3 m height), leaving enough space for servicing. To minimize the effects of partial shading that might occur at the lower portions of the roof, the modules were mounted in a "portrait" orientation, i.e. with their widths horizontally.

**Table 7.2** Main features of the roof-integrated test plant Mungg in Bellwald, Switzerland.

| | |
|---|---|
| Modules | a-Si single-junction cells on 1.1 × 1.3 $m^2$ modules, fabricated by ersol Thin Film GmbH, framed glass/glass laminates, 6 strings with 3 modules each |
| Mounting system | Solrif D from Ernst Schweizer Metallbau AG replaces the roof tiles |
| Power | |
| initially | 1994 W |
| nominally | 1535 W |
| array yield | 2008: 1060 Wh/kWp, 2009: 1095 Wh/kWp |
| Inverter | Fronius, IG 15, Datalogger |
| Sensors: | 2 × temperature, radiation |
| Year of installation | 2007 |

**Fig. 7.4** PV roof integration Mungg in Bellwald (1600 m), Switzerland.

Only portions of cell stripes or cell segments running parallel to the module length can thereby be shaded.

Module mounting was accomplished using a well-proven solar roof integration frame system [SOLRIF, Haller-2006]. A key advantage of the SOLRIF system over conventional PV framing is the free-standing lower glass edge, which allows for improved self-cleaning, since no dirt or dust accumulates, and snow can slide off markedly sooner. In that context, it must be noted that the Mungg PV system is in a geographical location where snow can reach a height of 1 to 2 m in winter. The potential for high snow loads was taken into account by providing additional supports for the module mounting structure.

Snow coverage obviously substantially reduces the energy yield of this particular installation. After the first 18 months of operation, the energy yield was considerably lower than that to be expected from computer simulations [Spyce-2006]. According to these, the annual amount of energy calculated to be generated is about 1920 kWh/1.53 kW, whereas the actual energy produced by the array was only 1675 kWh/1.53 kW, as shown in Table 7.2. The main reason for this shortcoming was snow coverage. Figure 7.5 shows that during winter months (October to March) the measured power output is lower than the simulation prediction. Whereas without snow coverage simulations predict 39 % of the calculated annual energy to fall into the winter period, only 20 % of the measured annual energy could actually be produced during the snow-rich period (see Fig. 7.5, the simulated and measured array outputs, respectively).

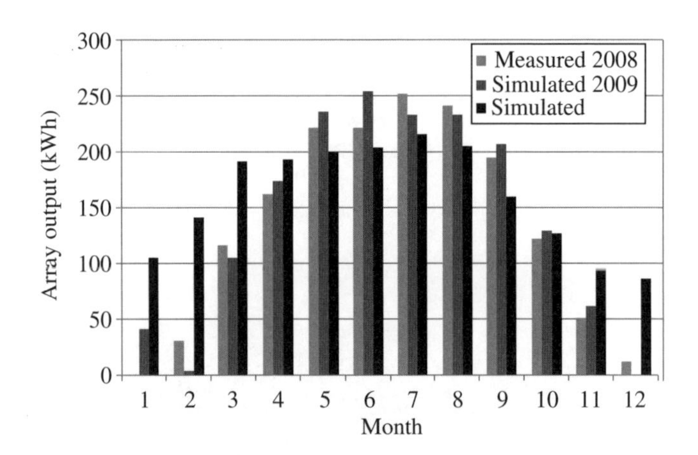


**Fig. 7.5** Comparison of simulated and measured energy output data in alpine climate. A considerable number of days without array output is observed in winter due to snow covering the modules.

### 7.1.3 PV Roof at Auvernier, Switzerland (by Reto Tscharner)

Unlike the BIPV examples which are based on superstrate technology, the application described here employs substrate-based modules on stainless steel. The modules are installed shingle-like on a roof facing south. As shown in Figure 7.6, this resulted in a very esthetical roof integration, replacing the former fibro-cement corrugated roof elements.

There is no ventilation at the rear side of the modules, thus they operate at high temperatures during the summer. This does not adversely affect the performance of the plant, because the low temperature coefficient of the modules (see Sect. 5.3.3) is compensated by partly annealing the Staebler-Wronski effect. The main details of the installation are summarized in Table 7.3.

The plant was commissioned in July 1997. It showed excellent reliability and good performance over 12 years of operation. No breakdown or system failure was observed. The performance of the installation is shown in Figure 7.7.

The annual system performance ratio has decreased from initially 78 % to 70 % at present, due to the following reasons:

- initial light-induced degradation during the first months of operation;
- dirt accumulation; the building is located near a highway, whereby a considerable amount of carbon particles are deposited on the modules. The modules were

**Fig. 7.6** Front view of PV roof "La Galère", Auvernier, Switzerland (with permission [Shah-2004]).

**Table 7.3** Main features of the PV installation in Auvernier, Switzerland.

| | |
|---|---|
| Location | Port of Auvernier, Canton of Neuchâtel, Switzerland |
| Modules | a-Si/a-SiGe double-junction cells *n-i-p/n-i-p*, on modules of 18.5 Wp fabricated by United Solar (formerly USSC), USA, 148 modules in 21 strings of series-connected modules, total array area 60 $m^2$ |
| Mounting system | PV shingles directly mounted on the roofing structure |
| Rated power (DC) | 2.74 kWp, system voltage 100 V DC |
| Inverter | Top Class Grid III TCG 2500/6, rated power 2500 W(AC), Manufacturer: ASP-Aton Sunpower AG |
| Year of installation | 1997 |

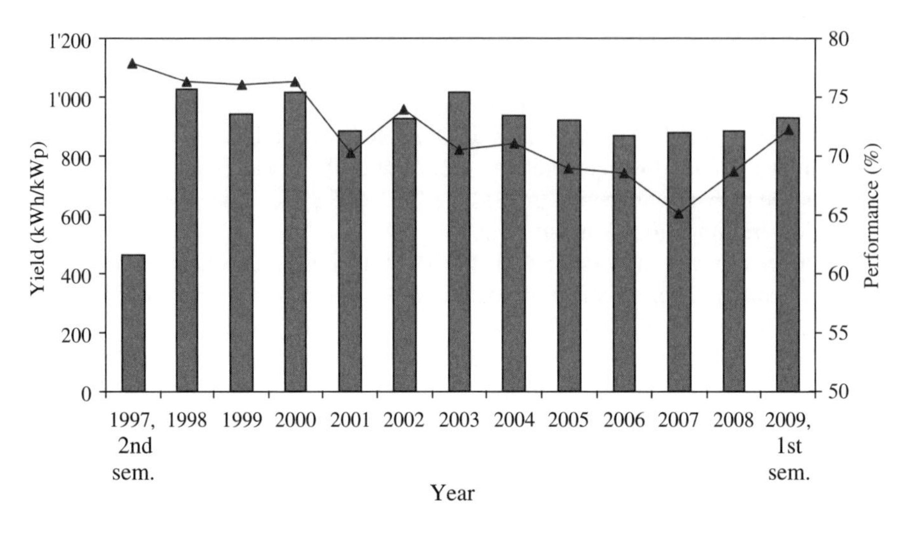


**Fig. 7.7** Performance 1997 to 2009 of PV roof integration "La Galère", Auvernier, Switzerland. (Annual yield shown as histogram, performance ratio as curve)

cleaned by a brush and detergent in 2002 and 2008, after which the performance recovered for some time;

- partial shading of the modules by vegetation growing next to the building.

The most recent annual relative performance is 880 kWh/kW$_p$, which is above the average of around 800 kWh/kW$_p$ for PV installations in Switzerland.

### 7.1.4 PV installation in Brazil (by Ricardo Rüther)

***Description***

The first grid-connected building-integrated thin-film PV system to operate in Brazil was installed in 1997 [Rüther-1998]. The choice of using a-Si was made because of the low temperature coefficient of power for this PV technology which, combined with the temperature-activated annealing of the light-induced degradation (Staebler-Wronski Effect [Staebler-1977], see Sect. 4.5.16), renders a-Si a good performer in warm climates. The system was installed in the facilities of the Solar Energy Research Laboratory (LABSOLAR) at Universidade Federal de Santa Catarina (UFSC), in Florianópolis, Brazil. The PV installation consists of a 2 kWp a-Si array, plus DC/AC inverter, irradiance and temperature measurement instrumentation, and data logging system. The array comprises 54 opaque and 14 semitransparent modules; the semitransparency is about 13 %. The decision to use both opaque and semitransparent a-Si modules was made to draw attention to the architectural and esthetic features of both module types. The unframed modules were installed onto a simple steel structure retrofitted as an overhang to the existing building. Further details of the installation are summarized in Table 7.4. Figures 7.8 and 7.9 show front and lateral views of the PV installation.

**Table 7.4** Main features of the PV installation in Florianópolis, Brazil.

| | |
|---|---|
| Location | Florianópolis, 27 °S, 48 °W, Brazil (sea level)<br>Climate: moist maritime, with warm winters and hot summers. |
| Modules | a-Si/a-Si tandem-junction cells *p-i-n/p-i-n*, same band gap, on 0.6 $m^2$ modules, ASE-30-DG-UT, fabricated by Phototronics Solartechnik GmbH (RWE SCHOTT Solar), 54 opaque and 14 semitransparent glass/glass laminates, module size $0.6 \times 1.0$ $m^2$, total array surface area ≈ 40 $m^2$. |
| Mounting system | Galvanized steel profiles, overhang bolted to existing concrete façade |
| Rated power | 2078 Wp |
| Inverter and data acquisition | single-phase inverter, 2500 W, SMA model Sunny Boy SB2500, ambient and back-of-module temperature, horizontal and tilted (27 °) solar radiation sensors, DC and AC electrical parameters measured at 4-minute intervals |
| Year of installation | 1997 |

**Fig. 7.8** Front view of the 2-kWp a-Si thin-film PV installation in Florianópolis, Brazil. (with permission [Rüther-1998])

### *Operation experience and performance*

Over nearly 12 years of continuous operation, the PV installation has performed consistently well, with minimum downtime and virtually no maintenance. In year 10 (2007), the four original low-power and low-input voltage inverters were replaced by a state-of-the-art inverter. It has been demonstrated that a-Si is a good performer in warm climates [Rüther-1999], [Rüther-2004],[Rüther-2008a], and a most appropriate PV technology for operation in a tropical country like Brazil. Figures 7.10 and 7.11

**Fig. 7.9** Lateral view of the 2 kWp a-Si thin-film PV installation in Florianópolis, Brazil (with permission [Rüther-1998]).

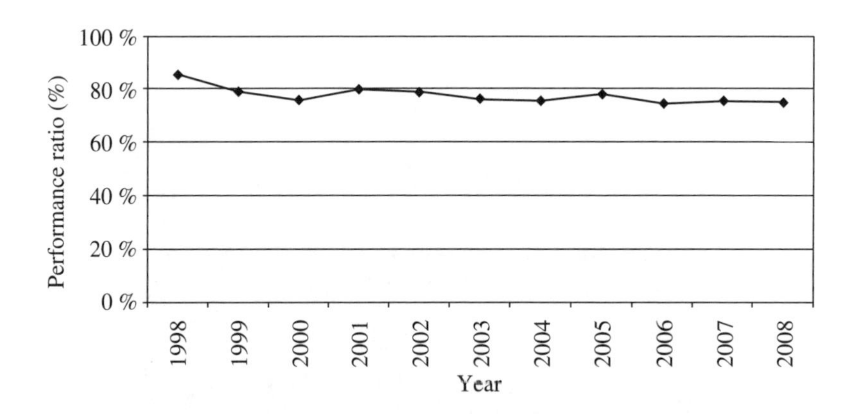


**Fig. 7.10** Measured annual performance ratio of the 2 kWp a-Si installation in Florianópolis, Brazil, over more than 10 years of continuous operation (adapted from [Rüther-2008b]).

show the performance measurements over more than 10 years of continuous operation of the 2 kWp a-Si PV system. The performance ratio (in %) is defined as the ratio of the actually generated annual energy to the energy that would be generated under STC conditions from the annual irradiation received at that particular location. This ratio is equal to the ratio of the effective and STC efficencies defined in Section 6.6.2.

***Temperature coefficients, and the advantages of a-Si PV in warm climates***

It has been argued that a-Si modules represent an appropriate technology for operation in warm climates [Rüther-1999]. This is due to two temperature effects that affect

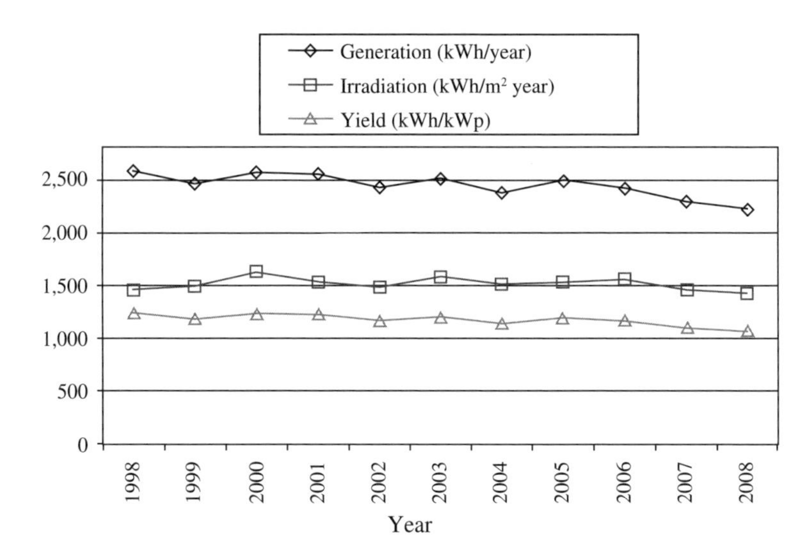


**Fig. 7.11** Measured annual electrical generation, plane-of-array irradiation, and energy yield of the 2 kWp a-Si installation in Florianópolis, Brazil, over more than 10 years of continuous operation (for a comparison of these different data with those of other installations see Table 6.4) (adapted from [Rüther-2008b]).

this technology in a peculiar manner in comparison with the more traditional crystalline Si PV technology.

- The temperature coefficient of the power output from a-Si cells is typically only about one half of that for c-Si cells (see Sect. 4.4.3), leading to smaller power losses at high operating temperatures in warm climates.
- The higher operating temperatures in warm climates offset the Staebler-Wronski effect, stabilizing the a-Si module output at a higher performance level, in comparison with the same a-Si module deployed in colder climates [Rüther-2003].

In this context it is interesting to further discuss the effect of climate on the performance of a-Si modules deployed in different climates. A round-robin outdoor exposure experiment with identical kits of a-Si modules from different manufacturers was conducted over a period of 4 years. The modules were exposed in different locations, namely in Arizona and Colorado (USA) and in Florianópolis (Brazil), but the deployment sites were annually permutated [Rüther-2008a]. In this experiment, higher output performance parameters were always measured when modules were moved from a colder to a warmer site and, conversely, modules that were stabilized at a higher temperature site always presented some further degradation in output when moved to a colder site. After the fourth year (the final round of outdoor deployment), a selected sample of the modules was submitted to simulator measurements at different temperatures, in order to determine the temperature coefficients of the stabilized efficienies. Compared to the values typically specified in data sheets for a-Si modules (about 0.2 %/K), the temperature coefficients of the efficiency derived from this round-robin experiment turned out to be substantially lower, as shown in Table 7.5. It might be concluded that temperature coefficients of stabilized modules are generally lower than those of modules in the initial state, resulting in better performance in warm climates.

**Table 7.5** Temperature coefficients of the efficiency of a-Si modules after four years of outdoor deployment. [Rüther-2008a]

| Manufacturer | Efficiency (%/K) |
|---|---|
| A | – 0.105 |
| B | – 0.087 |
| C | – 0.079 |
| D1 | – 0.101 |
| D2 | – 0.053 |

### 7.1.5 Stillwell Avenue Station, New York City

Coney Island's Stillwell Avenue Terminal is the largest above-ground station of the New York City subway system. The 90-year old station was completely rebuilt in 2003-2005 according to the design by Kiss & Cathcart (Architects G. Kiss and T. Daniels). The glass and steel structure of the roof incorporates one of the largest thin-film BIPV installations in the United States (see Fig. 7.12).

The roof, covering four platforms and eight tracks, consists of three arched sections with a total area of around 7000 m$^2$; 30 % (2060 m$^2$) of this area accounts for the

**Fig. 7.12** Stillwell Avenue Train Station, Coney Island, New York City (credit SCHOTT Solar AG).

steel construction of the roof, and about 55 % (3850 $m^2$) for the PV-active area. The remaining 15 % (1100 $m^2$) is incorporated into the laminates as clear glass. Design criteria had called for sufficient daylight transmission through the roof in order to avoid electrical lighting of the platforms during daytime. The main details of the installation are summarized in Table 7.6 (see also Fig. 7.13).

The performance of the entire installation has been designed to be monitored by a central Supervisory Control and Data Acquisition (SCADA) system [Bing-2006]. Although no detailed data are available at this time, the overall performance of the system has been communicated quite recently [Abdallah-2009]. Accordingly, the annual electricity production is given as 167, 000 to 199, 000 kWh/year. Combined with the rated power of 200 $kW_p$, the relative annual performance amounts to 835 to 995 kWh/$kW_p$. The effective efficiency (see Sect. 6.6) is derived from the annual insolation of 1380 kWh/$m^2$ year at New York City [Marion-1994] and results in 3.2 to 3.8 %. It must be noted that these values of effective efficiency include the

**Table 7.6** Main features of the PV roof of the Stillwell Avenue Station, New York City.

| | |
|---|---|
| Modules | a-Si/a-Si tandem-junction cells on custom-sized modules, fabricated by Phototronics, RWE SCHOTT Solar, and combined into laminates of 1.8 $m^2$ (1.4 $m^2$ PV-active), based on the PVB process, and fabricated by Glaswerke Arnold, Germany; total number of laminates 2730, total PV area 3843 $m^2$ |
| Mounting system | 5 laminates mounted into steel frames; 2 × 210 plus 126 frames for the two outer and the middle roof sections, respectively |
| Rated power | 200 kWp; total-area efficiency 5.2 % (STC) |
| Inverter | 2 Xantrex 150 kW inverters |
| Year of installation | 2003-2005 |

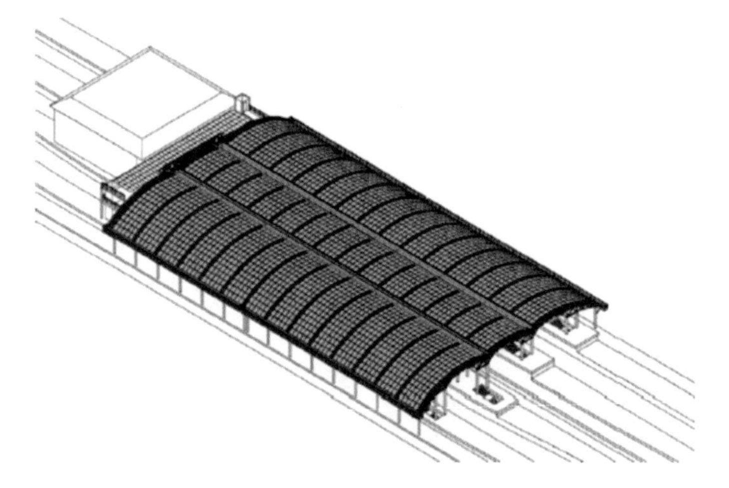

**Fig. 7.13** Schematic view of the Stillwell Avenue Train Station; the individual small rectangles represent the 1.8 $m^2$ laminates, see text (with permission [Miras-2003]).

effects of temperature, various tilts of mounting, inverter efficiency and wiring, and thus are not truly comparable to the nominal STC efficiency of 5.2 %. Even including these effects which must be individually taken into account, the performance ratio (effective/STC efficiency), based on the present information, is estimated to be 62 to 73 %.

## 7.2 STAND-ALONE AND PORTABLE APPLICATIONS

PV solar cells have been used for many years to provide electricity to isolated consumers and remote electrical installations that are not connected to a grid.

For example, stand-alone installations are provided for:

- irrigation pumps in tropical countries;
- lighting and television in remote villages, e.g., in Africa, Indonesia or India;
- holiday homes in remote locations;
- telecommunication repeater stations in inaccessible areas;
- lighthouses on small islands or rocks.

Portable applications are manifold; they may be found, e.g., on:

- sailing boats, to power refrigerators and lighting;
- temporary traffic lights on roads under repair;
- devices used for camping and other outdoor activities.

Thus far, PV systems for such applications have been mostly equipped with crystalline silicon cells. However, since several years, the use of amorphous silicon cells has become increasingly popular for these applications.

With the possible exception of irrigation pumps, the usefulness of these PV installations is tied to electricity storage, in order to extend their use beyond the periods of illumination of the solar cells. Irrigation is generally needed whenever the sun is shining. If some amount of irrigation water is required on cloudy but rainless days, it is easily possible to have this water stored in an elevated location. Nevertheless, sometimes it is found useful to have a small amount of electricity stored in a battery in order to start the pumping system early in the morning, before the PV system can deliver sufficient electricity to directly power the irrigation pump.

For all the other applications, electricity storage in batteries is essential. So far, almost all batteries used for PV systems have been lead-acid batteries, similar to the batteries used in automobiles. "Solar batteries", as they are called, have a slightly different design; they do not have thick electrodes (as car batteries do), and they have better sealing so that they can remain for several years in an outdoor location at elevated temperatures, without the need for adding water. But lead-acid batteries suffer from a number of disadvantages, such as limited lifetime, damage from a complete discharge, and special disposal requirements. Recently, a number of other battery types have come on the market, although they are still considerably more expensive than lead-acid batteries, especially for large storage capacity.

There are many different products on the market with a-Si-based panels in the range of 10 to 200 $W_p$. A few selected examples of stand-alone PV systems with a-Si modules are described below.

Figure 7.14 illustrates an irrigation pumping system, powered by framed a-Si modules on stainless steel (by UNI-SOLAR). The application of PV modules for water pumping systems, particularly for irrigation purposes, is very important in many "economically emergent" countries, such as China, India, Thailand, Indonesia, Brazil, etc. Although at present most water pumping systems are implemented with wafer-based crystalline silicon modules, one may predict that in view of expected trends in future module manufacturing costs a large part of such water pumping systems will be based, from the next decade onwards, on thin-film silicon modules. Note that in India, e.g., more than 25 % of all electricity is presently used for irrigation purposes. Obviously if a large part of these irrigation pumps could become solar-driven during the next few decades, this would provide a major development opportunity, opening up a huge market for photovoltaics, and liberating the Indian electricity grid from a substantial part of its present load, i.e. from many hundreds of GWh per year, which would then be available for industrial applications.

The second example (see Fig. 7.15) is a portable solar lamp, which is basically sufficient to illuminate a room during a period of 7 hours once the battery is fully charged. For this purpose, the solar panel must be placed outdoors in the sun. Solar lamps are an important application in remote areas of the "Third World". Often they are the only alternative to kerosene lamps for providing light to villagers. Solar lamps based on crystalline silicon panels are, for example, already extensively used in the Indian Himalayas. These panels are relatively heavy and cumbersome. It is to be expected that solar lamps based on flexible light-weight thin-film silicon panels, in conjunction with LED lamps, will become popular in the near future.

The third example features a modern innovative application: the design and implementation of an autonomous portable PV system (28 $W_p$). Presently this system

**Fig. 7.14** "Historical" water pumping system in Thailand with ground-mounted framed a-Si modules. (courtesy of United Solar Ovonic)

**Fig. 7.15** Portable solar lamp (7 W compact fluorescent tube) with Solarex 12 V/5 Wp module and lead acid battery (12V; 3.4 Ah) (Credit Muntwyler Energietechnik AG).

is commercialized under the brand name "iland®" [Flexcell-2009]. It contains flexible modules and a storage battery, and is described below.

The initial motivation for the development of the iland® came from the "Decentralized Rural Electrification Program" of Morocco. The following three criteria were to be considered.

- provide 200 Wh/day of electricity for lighting and for charging electronic appliances (at a later stage this could be extended to a higher storage capacity, so that the user could include a portable computer, a solar pump and a refrigerator);
- for storage, use batteries that are light-weight, have a long life and allow for many thousands of cycles;
- as PV panel, use a light-weight panel which can be folded and stowed away within a "container" of minimal volume. A 28 $W_p$ Flexcell® panel, produced by VHF Technologies S.A., Yverdon, Switzerland, was chosen.

The daily electricity consumption of a rural household in Morocco was estimated as follows:

- four 10 W lamps (low-consumption fluorescent lamps) during 3 hours a day: 120 Wh;
- TV with parabolic antenna, 15 W during 3 hours per day: 45 Wh;
- radio, 10 W during 4 hours per day: 40 Wh;
- mobile telephone: 5 Wh;
- total electricity consumption per day: 210 Wh.

In the future, LED lamps may replace the fluorescent lamps, so that a power supply for 200 Wh per day will be sufficient.

For the "iland®" system, consisting of the above-mentioned solar panel and a battery with a storage capacity of 200 Wh, the required electricity can be provided by charging the battery during approximately seven hours of full sunlight.

For the battery, a lithium-ion battery with a lithium iron phosphate ($LiFePO_4$) cathode was chosen. This type of a battery offers a rapid charging possibility and allows for a full battery discharge without any reduction of battery lifetime. The lifetime of such a battery is given as six to seven years (3000 charge-discharge cycles, on the basis of one cycle per day); however, this needs to be further qualified, since these batteries may lose about 20 to 30 % per year of their capacity, especially at high ambient temperatures. But these batteries are considered to be ecologically more acceptable than conventional types of such batteries as lead-acid or nickel-cadmium batteries.

The specifications of the Flexcell panel are:

- power (STC) 28 W;
- open-circuit voltage 23 V;
- short-circuit current 2400 mA;
- dimensions 642 mm × 1310 mm;
- active area 572 mm × 1220 mm;
- weight 1500 g.

Apart from the technical criteria, the iland® system must provide the following features:

- easy to assemble;
- easy to transport, either by vehicle or by a person on foot;
- sealing of the whole system, to prevent ingress of humidity, water and sand;
- mechanical stability;
- easy to dismantle into individual elements for individual replacements, containing only non-polluting materials that can easily be recycled.

The whole iland® system, shown in Figure 7.16, has a tubular form about 1 m long, with a diameter of 20 cm.

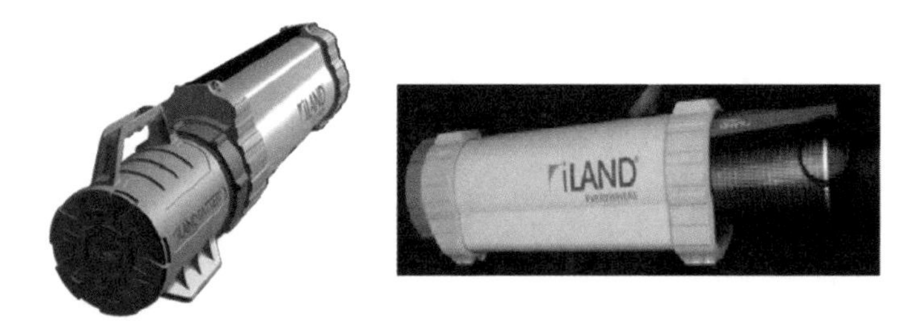


**Fig. 7.16** iland® power pack by Flexcell (Courtesy Daniel Oppizzi).

## 7.3 INDOOR APPLICATIONS OF AMORPHOUS SILICON SOLAR CELLS (by Julien Randall)

Solar electronic calculators are ubiquitous in most offices and homes. Many of these devices rely on series-connected amorphous silicon solar cells ("mini-modules").

Using a-Si cells for indoor applications is supported by some basic technical reasons explained in this Section. Furthermore, recommendations for electronic applications and a simple categorization of indoor solar cell products will be presented.

### 7.3.1 Why is amorphous silicon well suited for indoor applications?

The selection of a-Si rather than other photovoltaic materials can, in the first instance, be associated with issues related to specifications of the application. The low indoor levels of radiant energy (1-100 W/m$^2$) require series-connected solar cells in order to deliver sufficient voltage for electronic devices. Achieving series connection with crystalline silicon, whilst not impossible, is not realistic for mass production. Thin-film solar cell technologies have the advantage of facilitating modules with monolithically series-connected cells (see Sect. 6.5).

Furthermore, most thin-film solar cells exhibit a significantly higher sensitivity to low light levels, compared to wafer-based crystalline silicon solar cells. This is clearly illustrated in Figure 7.17, showing stronger decreases of the efficiency towards low light intensities for the c-Si samples.

The advantageous performance of thin-film cells, particularly a-Si cells, is a consequence of several factors, namely:

- a-Si cells have, in general, a higher parallel resistance (recombination shunt resistance) than conventional wafer-based c-Si cells.
- The energy gap is higher, and thus better matched to the spectrum of the indoor radiant energy sources as, shown in Figures 7.18 and 7.19.
- Due to the much lower indoor irradiance, light-induced degradation (Staebler-Wronski effect, see Sect. 4.5.16) is a lesser issue, and thus does not significantly affect indoor operation.

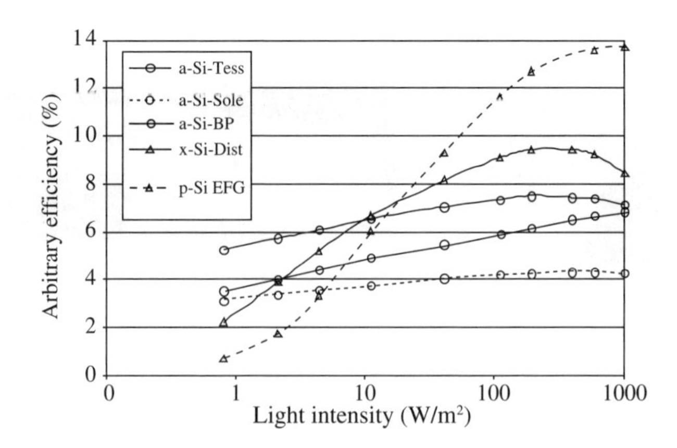


**Fig. 7.17** Typical light intensity dependences for different types of silicon solar cells: amorphous silicon (a-Si), polycrystalline silicon (p-Si, from edge-defined film-fed growth, EFG) and monocrystalline silicon (xSi), from 1/1000 sun- to 1 sun-equivalent illumination levels (with permission [Randall-2005a]).

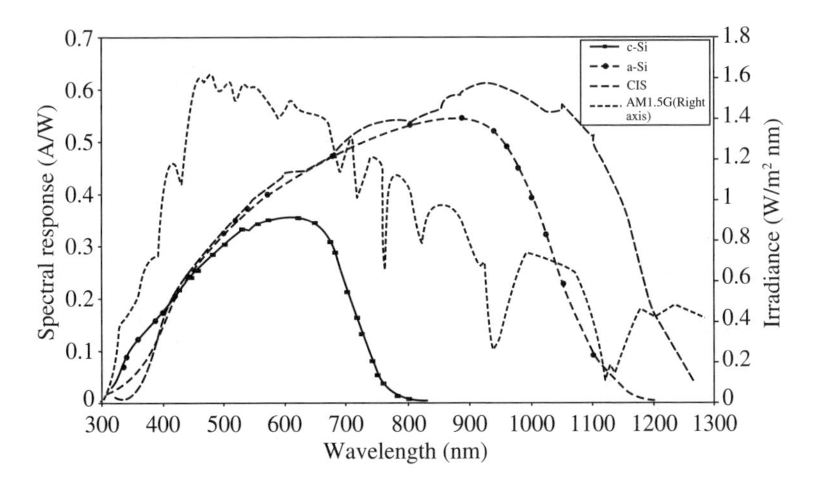


**Fig. 7.18** Solar irradiance with AM 1.5 g spectrum, and spectral response of solar cells of different PV technologies: c-Si, a-Si and CIS. (with permission [Kenny-2006])

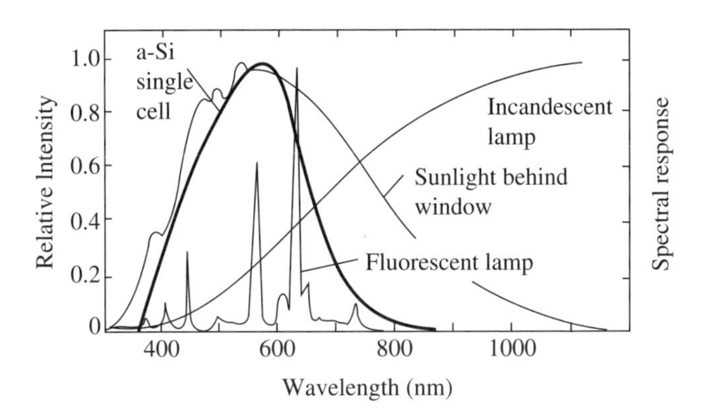


**Fig. 7.19** Qualitative comparison of light source spectra, and the spectral response of an a-Si cell (with permission [Roth-1990]).

## 7.3.2 Design guidelines for solar powering of indoor applications

The surface areas of solar cells, as determined by the size of most electronic products, generally cannot deliver sufficient energy to replace their main "competitor", the battery. Nevertheless, significant numbers of satisfactory indoor solar products have been produced that have freed users from the inconvenience of either changing or (re-) charging batteries. The product designer can recognize products that are well suited to indoor solar power by considering whether a target product specification lies within one of the following two categories of successful solar products:

- *Low or no storage*: calculators, e.g., where the energy is used immediately and storage is limited. In the case of calculators, a small capacitance maintains the

**Fig. 7.20** Various hand-held photovoltaic applications (with permission [Randall-2005b]).

display for some seconds. For such products, the surface of the solar cell(s) must be correlated with the product power requirements.

- *Higher storage capacity*: torch/flashlight, e.g., where the short cycles of use are assured by collecting energy over relatively longer time periods.

Figure 7.20 illustrates both categories of indoor solar products. Examples of the low or no storage products include calculators, weighing scales, and wall-mounted thermometers. The higher storage capacity products include the alarm clock (top left), stapler (top right), whisk, and various lights.

Assuming that the target product falls within one of these two categories, further factors will then require consideration. For example, an assessment of the available radiant energy should include all light sources, including natural light (daylight) and man-made (artificial) light. This assessment should also consider the orientation of the radiant energy vectors, in order to optimally align the solar cells where possible. An example of such alignment is the wall-mounted sensor shown in Figure 7.21, where the solar cells are tilted in order to face upward, i.e. toward the expected light source.

## 7.4 SPACE APPLICATIONS (by Nicolas Wyrsch)

### 7.4.1 Introduction

Exploration of space has been very much enabled by the development of photovoltaics. The first high-efficiency silicon cells developed by Bell Labs in 1954 were rapidly followed by the launch of the first satellite, powered by PV modules (Vanguard I) in 1958. The development of PV technologies for space applications has been so successful that today more than 99 % of the missions are powered by

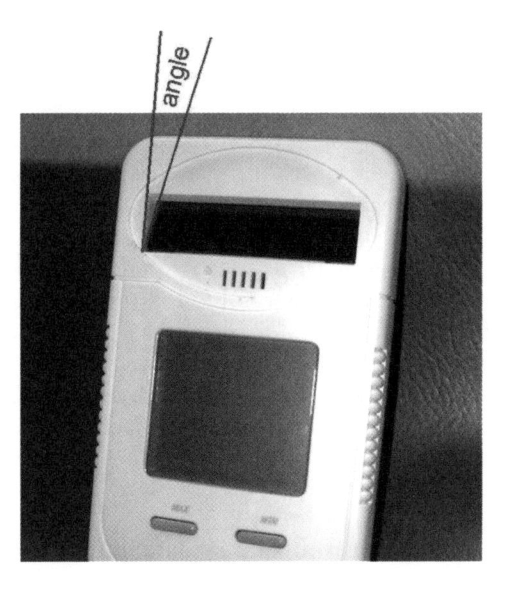


**Fig. 7.21** Example of a wall-mounted sensor mounted so as to face the direction of the incoming light. (with permission [Randall-2005b])

PV modules. In contrast to terrestrial applications, where cost reduction is one of the main objectives, development of space cells is mainly driven by the following requirements:

- high resistance to the "harsh" space environment (particle radiation, vacuum, thermal cycling, meteorite bombardment, space plasma, atomic hydrogen, etc);
- high end-of-life (EOL) efficiency;
- high specific power density.

These requirements, together with the high reliability necessary for space missions, have generally led to very high prices for the corresponding PV cells and modules: the prices are in the order of 1000 $/W. Up to now the development of PV for space has been concentrated on three bulk crystalline technologies,

- c-Si: derived from the standard and broadly used terrestrial PV technology; this option is widely used, in space applications, for commercial missions, or for very large solar generators, where cost is an issue;
- InP: interesting, because of its high radiation hardness; nevertheless, this technology is slowly being abandoned, due to the low industrial base and its high cost;
- GaAs-based solar cells: mainly in the form of multi-junction cells on Ge substrates; such a cell design is used to increase the cell efficiency.

Recent developments of the satellite market have led to a demand for increasing PV generator power; such a demand is beginning to be difficult to satisfy with the current double-(DJ) or triple-junction (TJ) high efficiency GaAs-based solar cells. Limitations are here mainly given by the launch weight and stowing size, as well as by the deployment system. In this context, one of the important evaluation parameters is the ***specific power***

***density***. In order to increase the specific power density, one may work on improving the efficiency (for example by replacing DJ by TJ), or on reducing the weight by using thinner cells and/or thinner substrates. Nevertheless, this strategy has only a reduced effect on the total specific power density of the power generator, because a large fraction of the weight is due to the protective glass sheets of the cells (needed for the mitigation of radiation effects), and to the mechanical support of the cells (e.g. the honeycomb core).

A radical approach to considerably improve the specific power density is to switch to thin-film technologies. Thereby, the specific power density is basically given by the substrate thickness. Using roll-to-roll processing technologies and/or temporary carriers for the processing of the cells, the total thickness can be considerably reduced and the specific power density enhanced. Furthermore, most thin-film technologies, such as CIGS (copper indium gallium diselenide), CdTe and a-Si:H, are significantly more radiation-resistant than c-Si or GaAs based cells. These issues will be discussed in Section 7.4.3.

### 7.4.2 Satellite power generators and specific power density

A typical satellite solar generator (as seen in Fig. 7.22) consists of wings (or arrays), attached to the main body of the spacecraft. The wings include a rigid honeycomb core or a flex blanket, on which the individual solar cell as well as the electrical cell interconnect are attached, and the deployment system.

As seen in the cross-section view of an array (Fig. 7.23), additional layers are used to isolate the cells and interconnects, dissipate the excess heat, and attach all components together. Finally, a protective cover glass is put in front of the cell in

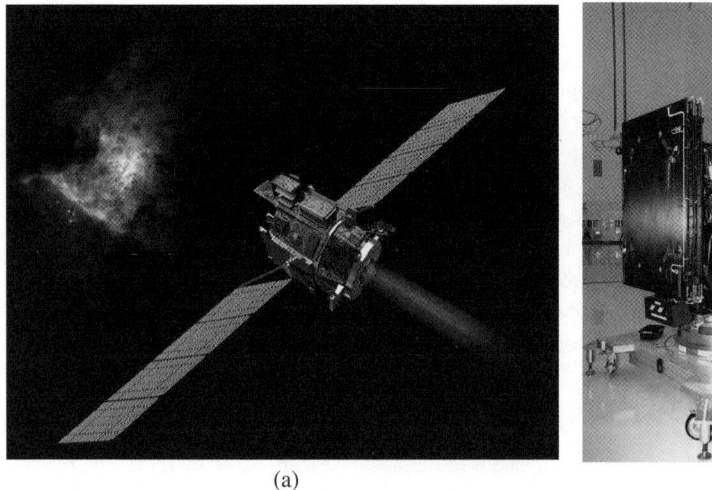

(a)

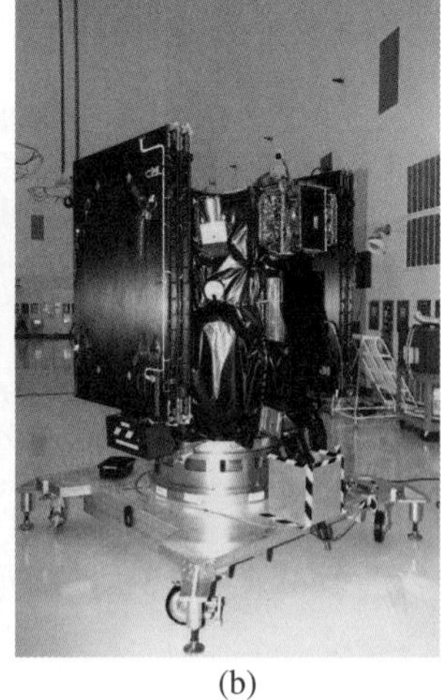

(b)

**Fig. 7.22** (a) Artist's picture of Deep Space 1 spacecraft; (b) the two wings are the two parts of the solar generator powering the spacecraft and are folded for launch, as seen in the photograph (credit NASA-JPL).

order to mitigate degradation due to proton and electron irradiation. The role of this cover glass is to stop damaging particles in the glass rather than in the cell. Therefore, specific power density values at the array level are much smaller than those obtained at the cell level. In this context, the use of thin-film PV technologies, and especially of a-Si:H, can considerably simplify the structure of a PV array (see Fig. 7.24).

- The cell substrate (when using an isolated substrate) can be used as the array blanket, thereby eliminating the need for any additional layers or for a honeycomb core.
- The high radiation hardness eliminates the need for a relatively heavy cover glass.
- The monolithic serial interconnect of thin-film PV modules simplifies the array interconnect.

Table 7.7 summarizes the specific power densities achieved by today's best commercially available cells, compared to values obtained by a-Si:H cells, and compares them to the specific power densities of the entire generator. It can be seen that despite a rather low efficiency, a-Si:H can achieve lower specific power densities, already at the cell level. Compared to standard crystalline technologies, and given the much simpler array design, it is expected that a-Si:H-based power generators will have specific power densities which are at least one order of magnitude lower than those of state-of-art solutions based on c-Si or TJ-GaAs cells [Beernink-2002, Wyrsch-2007].

As a-Si:H cells or modules for space applications should preferably be fabricated on thin flexible polyimide (PI) foils, the entire array can then be rolled or folded into a much smaller volume than a conventional PV array for space applications, which uses standard cell technology. This smaller stowing volume will again positively affect the weight budget of the spacecraft.

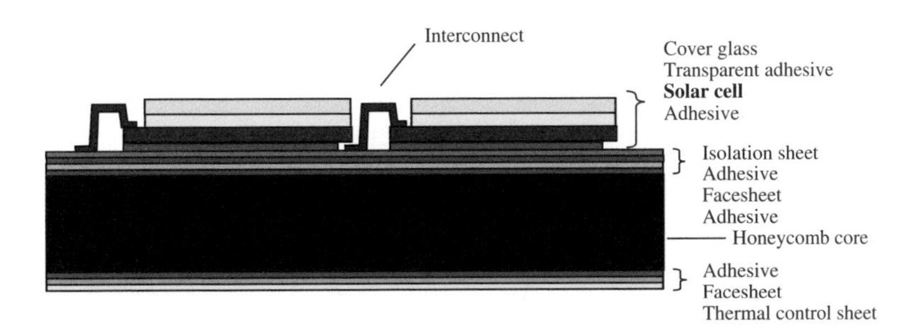


**Fig. 7.23** Cross-section of a conventional space PV array for crystalline cells.

**Fig. 7.24** Cross-section of a thin-film a-Si:H PV array. Light enters the arrays on the protective coating (e.g. $SiO_x$ layer), but the array could be designed to enter the cell through the polyimide substrate when using ultra-thin and transparent types of polyimide foils.

**Table 7.7** Specific power density at the cell level of commercially available space cells and for a-Si:H on polyimide (PI) as a function of the substrate thickness, as well as the achieved (or expected) specific power at the solar array level.

| Technology | Efficiency | Substrate/wafer thickness | Cell-specific power density [W/kg] | Array-specific power density [W/kg] |
|---|---|---|---|---|
| c-Si | 16 % | c-Si, 130 $\mu$m | 676 | <100 |
| DJ ($GaInP_2$/GaAs) | 22 % | Ge, 140 $\mu$m | 354 | <100 |
| TJ ($GaInP_2$/GaAs/Ge) | 30 % | Ge, 140 $\mu$m | 483 | <100 |
| a-Si:H | 6 % | PI, 50 $\mu$m | 940 | >500 |
| a-Si:H | 6 % | PI, 5 $\mu$m | 3680 | >1000 |

### 7.4.3 Radiation resistance of a-Si:H and other PV technologies

As mentioned above, any PV technology used for space applications must sustain the very harsh space environment and has, in particular, to be very "radiation-hard". In crystalline materials, interaction of energetic particles with the material tends to create crystallographic defects by displacing atoms; at high radiation levels it furthermore leads to an amorphization of the structure. In Figure 7.25 the effect of irradiation on the cell efficiency is plotted for various PV technologies as function of the displacement damage dose. The formalism of the displacement damage dose was developed by Summers et al. to predict irradiation damage for a given cell technology in any radiation environment, and to compare it to other technologies [Messenger-2002]. As one can observe, thin-film compound semiconductors (CdTe, CIGS,...), as well as a-Si:H, have a much higher "radiation hardness". Two sets of data are presented here for a-Si:H; the discrepancies originate in the values of the non-ionizing energy loss, as used by authors to determine the amount of displacement in a-Si:H and to calculate the displacement damage dose (see Refs. in the caption of Fig. 7.25). A high hydrogen concentration in the material also benefits the radiation hardness, as hydrogen helps to mitigate defect creation (i.e. dangling bond creation).

Defects created in a-Si:H by energetic particles are similar to those created by light soaking. In both cases, thermal annealing allows full recovery of the initial undegraded state. Therefore, the possible thermal annealing of the created defects in solar cells should be considered. As seen in Figure 7.26, a-Si:H degrades significantly less than μc-Si:H (upon proton irradiation with a broad energy spectrum of 0 - 5 MeV), and a full recovery is obtained after thermal annealing of the cell at 100 °C. Micromorph (a-Si:H/μc-Si:H tandem) cells exhibit a behavior similar to that of μc-Si:H cells.

As a solar cell mounted on a satellite experiences large thermal cycling, and thereby attains operating temperatures in the order of 100 °C, almost no degradation is expected from proton irradiation. Similar results have been obtained with a-Si:H based triple-junction devices, showing that defects created by electron and proton irradiation can be fully annealed at temperatures as low as 70 °C [Grigorieva-2003]. Note here that the low thermal coefficient of a-Si:H cells (see Sect. 4.4.3) offers reduced efficiency losses for a-Si:H compared to other PV technologies.

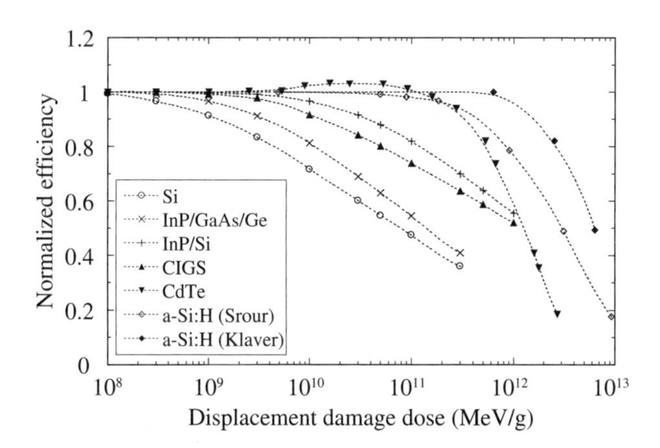


**Fig. 7.25** Effect of irradiation, expressed as displacement damage dose, on the normalized solar cell efficiency of several PV technologies (from [Bätzner-2004], [Srour-1998], [Klaver-2007]).

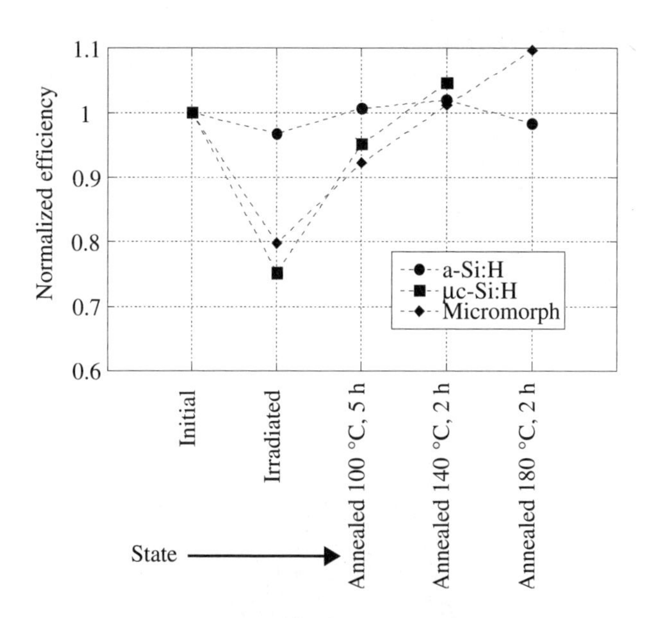


**Fig. 7.26** Normalized efficiency of a-Si:H and μc-Si:H single-junction cells, as well as of a micromorph tandem device, in the initial state, in the degraded state after irradiation by $10^{13}$ protons/cm$^2$ with a broad energy spectrum, and after three different steps of thermal annealing (from [Kuendig-2003]).

### 7.4.4 a-Si:H based cells for space

Two different paths are followed to fabricate a-Si:H solar cells on very-thin substrates. The first uses conventional roll-to-roll processes developed for terrestrial solar cells on stainless steel (SS) or polyimide (PI) foils. An example of such a development is shown in Figure 7.27. The other one is based on the use of a temporary glass (or metallic) carrier with a thin polyimide coating; the former is released from the carrier at the end of the cell processing, as shown in Figure 7.28 [Wyrsch-2007].

In order to achieve very high specific power densities, the use of PI foils as substrate is preferred to SS. The use of isolating substrates allows the implementation of a monolithic serial interconnection, and avoids the need for an additional blanket to attach the individual cells (see Fig. 7.24). Depending on the type of substrate and on its thickness, a-Si:H solar cells can be deposited in the *pin* (for thin transparent substrate) or *nip* configuration. Due to the high radiation hardness, no cover glass is required. However, additional thin isolating and protecting layers (such as $SiO_2$ layers) can be used to encapsulate the cell.

**Fig. 7.27** Roll of a-Si:H triple-junction cells developed for space applications (courtesy of S. Guha, United Solar Ovonic).

**Fig. 7.28** Photograph showing the release of a 6 $\mu$m PI film, carrying an experimental a-Si:H cell, from the underlying glass carrier.

### 7.4.5 Space applications of a-Si:H modules

For the time being, only prototypes of a-Si:H modules have flown, both on the MIR orbital station [Kagan-2000], and on the international space station, as part of comparison tests for various PV technologies (forward technology solar cell experiment – FTSCE) [Walters-2007]. a-Si:H based triple-junction devices deposited on thin SS foils have already successfully undergone space qualification tests [Kagan-2000]. With their high specific power density and small stowing volume, a-Si:H based PV cells on thin substrates are very attractive candidates for all those space applications which require large total PV array power, such as:

- space habitats (including both orbital as well as moon-or Mars-based);
- spacecrafts powered by ion thrusters;
- space power satellites;
- large telecommunication satellites.

The emerging space applications mentioned above cannot be powered simply by an extension of existing standard space PV technologies because of weight and size constraints. Cost is at present an increasingly important aspect; thus, a-Si:H solar cell technology (using production solutions close to those developed for terrestrial applications) is at an advantage.

The deployment of large a-Si:H PV blankets will require new systems, which are quite different from the ones used so far in spacecrafts. Nevertheless, the deployment systems used for "solar sails" (i.e. for spacecrafts employing solar winds for their propulsion) appears to be well suited for the deployment of thin a-Si:H PV cells, as the material utilized in both cases is rather similar (thin coated polyimide films) (see Fig. 7.29).

**Fig. 7.29** 400 m$^2$ solar sail prototype deployment test. The sail consists of an approx. 5 $\mu$m thick aluminized PI foil. (credit NASA-JPL)

As a more convenient and cheaper replacement for monitoring or telecommunication satellites, several companies are developing high-altitude aircrafts (HAA). These aircrafts look like large dirigibles or blimps, which should cruise or be kept in stationary positions at very high altitudes, well above 20,000 m, in order to avoid atmospheric perturbations (see Fig. 7.30). For such platforms, very light-weight PV modules with specific power densities above 1000 W/kg are also required; here thin-film PV technologies, including a-Si:H, are the best candidates.

**Fig. 7.30** Artist's picture of a high altitude airship. (credit Lockeed-Martin)

## 7.5 CONCLUSIONS

Applications of silicon-based thin-film modules range at present from replacing or charging batteries in portable consumer products such as calculators or lamps, to providing off-grid electricity in remote locations. Increasingly, silicon-based thin-film modules are used for replacing and/or supplementing grid-connected electricity, such as in solar farms and in BIPV (Building-Integrated Photovoltaics). In addition, due to the radiation hardness of amorphous silicon, applications of silicon-based thin-film modules even extend to generators used in satellites.

Among these areas of applications, BIPV already is, and in future will be, of ever-increasing importance if measured in terms of the produced power. The availability of large-area thin-film modules, tailored to the various requirements of the building industry, provides a wide range of design possibilities. Many building-integrated installations worldwide, both roof- and façade-mounted, have already demonstrated long-term stable performance. This track record supports further confidence and warrants increasing investments.

Furthermore, and above all, silicon-based thin-film modules have potentially no problems with respect to the

- abundance and availability of the raw materials involved;
- ecological acceptance of materials and fabrication processes to be used in future manufacturing plants;
- production energy invested for their fabrication (i.e. relatively short energy recovery times).

For all these reasons, silicon-based thin-film modules constitute one of the main technological priorities for the future application of photovoltaics.

## 7.6 REFERENCES


[Abdallah-2009] Abdallah, T., (August 2009), *New York City Transit Authority,* private communication.

[Bätzner-2004] Bätzner, D.L., Romeo, A., Terheggen, M., Döbeli, M., Zogg, H., Tiwari, A.N., "Stability aspects in CdTe/CdS solar cells", (2004), *Thin Solid Films*, Vol. **451–452**, pp. 536–543.

[Beernink-2002] Beernink, K.J., Pietka, G., Nochl, J., Wolf, D., Banerjee, A., Yang, J., Guha, S., Jones, S.J., "High specific power amorphous silicon alloy photovoltaic modules", (2002), *Proc. of the 29th IEEE Photovoltaic Specialist Conference, New Orleans, Louisiana*, pp. 998-1001.

[Bing-2006] Bing, J., Kern, E., Hassan, F., "Redundant industrial grade DC switching in a 200 kW BIPV system", (2006), *SOLAR 2006 Conference American Solar Energy Society, Denver, Colorado*, pp. 1-5.

[FfE-2002] FfE, private communication, (2002), *Research Institute for Energy Technology (Forschungsstelle für Energiewirtschaft), Munich.*

[Flexcell-2009] Press release, (2009), *VHF-Technologies, Yverdon, Switzerland.*

[Grigorieva-2003] Grigorieva, G., Kagan, M., Nadorov, V., Zvyagina, K., "Perspectives for application of amorphous silicon thin-film solar cells in conditions of higher-level space radiation", (2003), *3rd World Conference on Photovoltaic Energy Conversion, Osaka*, pp. 734-736.

[Haller-2006] Haller, A., Hartwig,H., "Long term experiences with a versatile PV in roof system", (2006), *21st European Photovoltaic Solar Energy Conference, Dresden*, pp. 2073-2077].

[Kagan-2000] Kagan, M., Nadorov, V., Guha, S., Yang, J., Banerjee, A., "Space qualification of amorphous silicon alloy lightweight modules", (2000), *Proc. of the 28th IEEE Photovoltaic Specialist Conference, Anchorage, Alaska*, pp. 1261-1264.

[Kenny-2006] Kenny, R. P., Ioannides, A., Müllejans, H., Zaaiman, W., Dunlop, E. D., "Performance of thin film PV modules", (2006), *Thin Solid Films,* Vol. **511-512**, pp. 663-672.

[Klaver-2007] Klaver, A., "Irradiation-induced degradation of amorphous silicon solar cells", (2007), Ph.D. Thesis, TU Delft.

[Kuendig-2003] Kuendig, J., Goetz, M., Shah, A., Gerlach, L., Fernandez, E., "Thin film silicon solar cells for space applications: Study of proton irradiation and thermal annealing effects on the characteristics of solar cells and individual layers", (2003), *Solar Energy Materials & Solar Cells*, Vol. **79**, pp. 425–438.

[Marion-1994] Marion, W., "Solar radiation data manual for flat-plate and concentrating collectors", (1994), *National Renewable Energy Laboratory, Renewable Resource Data Center*, WBAN No. 94728 (New York City).

[Maurus-2003] Maurus, H., Schmid, M., Blersch, B., Schade, H., "BIPV installations worlwide in ASI® technology", (2003), *3rd World Conference on Photovoltaic Energy Conversion, Osaka,* pp. 2375-2378.

[Messenger-2002] Messenger, S.R., Summers, G.P., Burke, E.A., Walters, R.J., Xapsos, M.A., "Modeling Solar Cell Degradation in Space: A Comparison of the NRL Displacement Damage Dose and the JPL Equivalent Fluence Approaches", (2001), *Prog. in Photovoltaics*, Vol. **9**, pp. 103–121.

[Miras-2003] Miras R., Hassan, F., Daniels, T., "Incorporating Building Integrated Photovoltaic Technology into New York City Transit's BMT Stillwell Avenue Terminal Train Shed", (2003), *Proc. of the 2003 Annual Conference of the American Railway Engineering and Maintenance of Way Association, Chicago, Illinois,* pp. 1-32.

[Phototronics-2002] Phototronics, "Building integration with ASI® thin-film solar modules", (2002), *RWE SCHOTT Solar.*

[Randall-2005a] Randall, J.F., "Designing Indoor Solar Products – Photovoltaic Technologies for AES", (2005), *John Wiley and Sons, Chichester, U.K.*, p. 82, Figure 5.2

[Randall-2005b] Randall, J.F., "Designing Indoor Solar Products – Photovoltaic Technologies for AES", (2005), *John Wiley and Sons, Chichester, U.K.*, p. 58, Figure 4.6

[Ricaud-1991] Ricaud, A. M., Schmitt, J. P. M., Siefert, J. M., Méot, J., Roelen, E., Bubenzer, A., Kümmerle, W., Häussler, W., Böttger, M., "Progress in manufacturing at the megawatt level of a-Si based PV technology", (1991), *10th European Solar Energy Conference, Freiburg,* pp. 1184-1187.

[Roth-1990] Roth W., Steinhüser A., "Photovoltaische Energieversorgung von Geräten im kleinen und mittleren Leistungsbereich", (1990), *Photovoltaik: Strom aus der Sonne; Technologie, Wirtschaftlichkeit und Marktentwicklung, 2. Auflage,* pp. 92-109.

[Rüther-1998] Rüther, R., "Experiences and operational results of the first grid-connected, building-integrated, thin-film photovoltaic installation in Brazil", (1998), $2^{nd}$ *World Conference on Photovoltaic Solar Energy Conversion, Vienna, Austria,* pp. 2655-2659.

[Rüther-1999] Rüther, R., "Demonstrating the superior performance of thin-film, amorphous silicon for building-integrated PV systems in warm climates", *(1999), International Solar Energy Society's 1999 Solar World Congress, ISES, Jerusalem*, pp. 221-226.

[Rüther-2000] Rüther, R., Dacoregio, M., "Performance assessment of a 2 kWp grid-connected, building-integrated, amorphous silicon solar photovoltaic installation in Brazil", (2000), *Progress in Photovoltaics: Research and Applications*, Vol. **8**, pp. 257-266.

[Rüther-2003] Rüther, R., Mani, G., del Cueto, J., Adelstein, J., Montenegro, A, von Roedern, B., "Performance test of amorphous silicon modules in different climates: higher minimum operating temperatures lead to higher performance levels", (2003), $3^{rd}$ *World Conference on Photovoltaic Energy Conversion, Osaka*, pp. 501-504.

[Rüther-2004] Rüther, R., Beyer, H.G., Montenegro, A.A.,.Dacoregio, M.M., Salamoni, I.T., Knob, P., " "erformance Results of the First Grid-Connected, Thin-Film PV Installation in Brazil: Temperature Behaviour and Performance Ratios over Six Years of Continuous Operation", (2004), *19th European Photovoltaic Solar Energy Conference, Paris,* pp. 1487-1490.

[Rüther-2008a] Rüther, R., del Cueto, J., Mani, G., Montenegro, A., Rummel, S., Anderberg, A., von Roedern, B., "Performance Test of Amorphous Silicon Modules in Different Climates – Year Four: Progress in Understanding Exposure History

Stabilization Effects", (2008), *Proc. 33rd IEEE Photovoltaic Specialists Conference, San Diego, California,* pp. 423-427.

[Rüther-2008b] Rüther, R., Viana, T. S., Salamoni, I. T., "Reliability and long term performance of the first grid-connected, building-integrated amorphous silicon PV installation in Brazil", (2008), *Proc. 33rd IEEE Photovoltaic Specialists Conference, San Diego, California.*

[Shah-2004] Shah, A. V., Schade, H., Vanecek, M., Meier, J., Vallat-Sauvin, E., Wyrsch, N., Kroll, U., Droz, C., Bailat, J., "Thin-film silicon solar cell technology", (2004), *Progr. Photovolt.: Res. Appl.*, Vol. **12**, pp. 113-142.

[Spyce-2006] Spyce, Satellite photovoltaic yield control and evaluation, (2006), *Meteotest, Berne, Switzerland*, www.spyce.ch.

[Srour-1998] Srour, J.R., Vendura, G.J., Lo, D.H., Toporow, C.M.C., Dooley, M., Nakano, R.P., King, E.E., "Damage Mechanisms in Radiation-Tolerant Amorphous Silicon Solar Cells", (1998), *IEEE Transactions on Nuclear Science*, Vol. **45**, pp. 2624-2631.

[Staebler-1977] Staebler,D.L., Wronski, C.R., "Reversible conductivity changes in discharge-produced amorphous silicon", (1977), *Appl. Phys. Lett.*, Vol. **31,** pp. 292-294.

[Tanner-2008] Tanner, D., Mei, F., Le, M., Su, J., Luu, C., Lu, W., Frei, M., Prabhu, G., Chae, Y. K., Eberspacher, C., "Fabrication and performance of large area thin film solar modules", (2008), *23rd European Photovoltaic Solar Energy Conference, Valencia*, pp. 2489-2491.

[Walters-2007] Walters, R.J., Garner, J.C., Lam, S.N., Vasquez, J.A., Braun, W.R., Ruth, R.E., Warner, J.H., Lorentzen, J.R., Messenger, S.R., Bruninga, R., Jenkins, P.P., Flatico, J.M., Wilt, D.M., Piszczor, M.F., Greer, L.C., Krasowski, M.J., "Forward Technology Solar Cell Experiment First On-Orbit Data", (2007), *Proc. of the 19th Space Photovoltaic Research and Technology Conference*, pp. 79-94.

[Wyrsch-2007] Wyrsch, N., Dominé, D., Freitas, F., Feitknecht, L., Bailat, J., Ballif, C., Poe, G., Bates, K., Reed, K., "Ultra-Light Amorphous Silicon Cell for Space Applications", (2007), *Proc. of the 4th World Conference on Photovoltaic Energy Conversion, Kona Island, Hawaii, 2006*, pp. 1785-1788.

CHAPTER 8

# THIN-FILM ELECTRONICS

*Nicolas Wyrsch*

## 8.1 THIN-FILM TRANSISTORS AND DISPLAY TECHNOLOGY

### 8.1.1 Introduction

Thin-film transistors (TFT) are field-effect transistors fabricated from thin-film semiconductors deposited directly on a substrate. The TFT was the first application of a-Si:H and it is still the most important one in the context of flat panel displays.

Historically, the first functional TFT was reported by P.K. Weimer in 1962 [Weimer-1962] using microcrystalline cadmium sulfide (CdS) for the semiconductor layer. TFTs rapidly attracted a lot of attention and in the early 60's were in competition with single-crystal silicon MOSFETs for display and electronic applications. The rapid progress in the development of the MOSFET limited the field of application of TFTs to display. The first active-matrix liquid crystal display (AMLCD) was demonstrated in 1973 [Brody-1973]. It was composed of CdS TFTs and a nematic liquid crystal cell.

Until the early 80's, several examples of TFTs were developed from various other materials, e.g CdSe, Te, InSb, Ge, whereas CdS remained the candidate of choice, but without much industrial success. The situation changed quite drastically with the fabrication of the first a-Si:H TFT by Spear and LeComber in 1979 [LeComber-1979]. Tremendous improvements in the deposition process and large-area manufacturing of a-Si:H TFTs led in the late 80's to the first mass production of a-Si:H-based AMLCDs, definitely removing the chance of success for CdS.

Today, AMLCD is a market worth tens of billions of euros per year, and a-Si:H is still the dominating technology for flat panel displays. Alternative materials such as μc-Si:H, "conventional" poly-silicon (obtained by thermal annealing), low-temperature poly-silicon (LTPS obtained by laser crystallization of a-Si:H), metal-oxide semiconductors and organic semiconductors are all being investigated with the goal of replacing a-Si:H for TFTs in the active matrix (AM). Several issues concerning the use of other materials, and the motivation for looking at them, will also be briefly presented in this section. At the same time, new display technologies are being introduced which also make use of a-Si:H-based AM. In this chapter, we will focus on thin-film silicon-based TFTs.

### 8.1.2 TFTs and flat panel displays

A TFT is a particular kind of field-effect transistor. It consists of a thin semiconductor layer in which a conducting channel is formed by applying a potential on a gate electrode; the gate is separated from the channel by a dielectric layer. Current is drawn through the channel from the source to the drain. A typical TFT structure is given in Figure 8.1. In display applications, TFTs are used to control the voltage (as in LCD display technology) or the current (in the case of OLED displays) applied on each pixel. By using thin films, these field-effect transistors can be deposited on the display plate next to each pixel.

In AMLCD, each pixel of the display is composed of a liquid crystal (LC) cell which is controlled by applying an electric field between a common electrode and the pixel electrode. The voltage on the latter is usually controlled by a TFT acting as a switch. Depending on the field applied on the LC cell, the polarization is changed and the transmission of the light blocked by the front or rear polarizer. The schematic layout of an AMLCD using TFTs is illustrated in Figure 8.2, while the complete display configuration is given in Figure 8.3.

The main advantage of AMLCDs is the simplicity of the circuit with only one active device (TFT) per pixel (Note that to achieve redundancy or improve performance, multiple TFTs or double-gate TFTs are nevertheless sometimes implemented.) A further advantage is that the TFT is used "only" as a switch. The characteristics of the TFT have therefore just a marginal effect on the performance of the display. The main disadvantage of AMLCDs is the need for a light source to achieve reasonable brightness and contrast. For this purpose, several types of light source (e.g. fluorescent lamp, LED, electroluminescent panel) can be implemented at the back of the display.

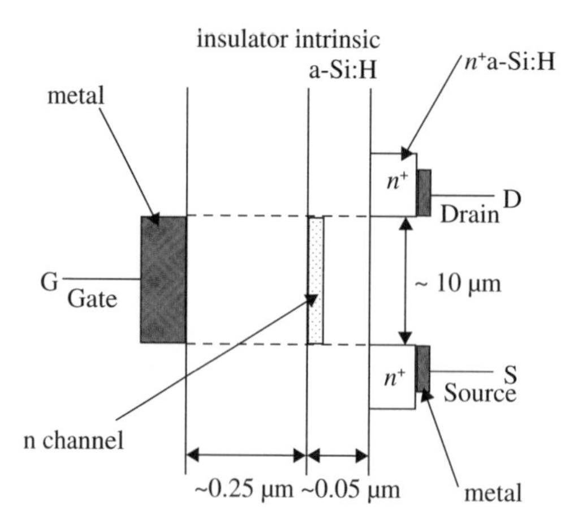


**Fig. 8.1** Schematic structure of a typical a-Si:H TFT.

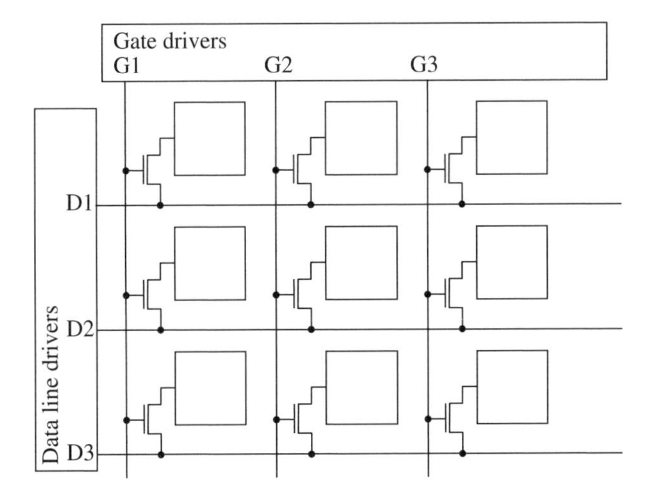


**Fig. 8.2** Schematic layout of an AMLCD.

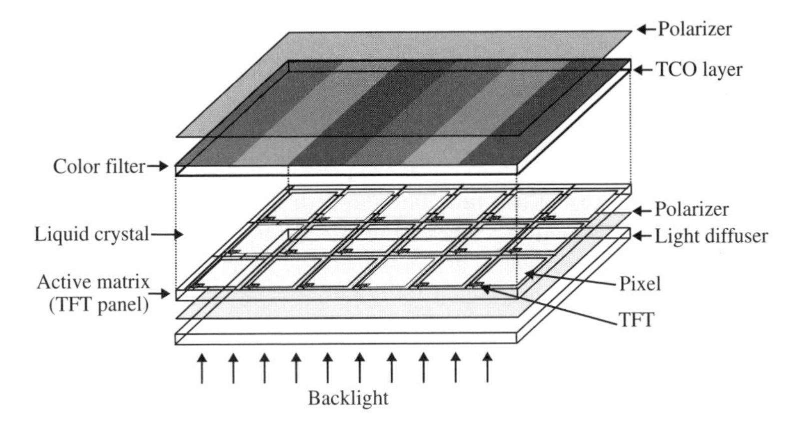


**Fig. 8.3** Cross-section of an AMLCD.

AMLCDs are available in a very broad range of display and pixel sizes, e.g. from mobile phones up to very large TV sets.

For monochrome and slow applications, ***electrophoretic displays*** are emerging as an alternative to AMLCDs. The applications of such display technology comprise price tags, control screens and, increasingly, also so-called "electronic paper", as electrophoretic display technology is also relatively well suited for the fabrication of flexible displays. Here, each pixel is composed of a cell filled with black-colored beads embedded in a white viscous liquid. By applying a voltage on the cell, the beads can either rise to the surface of the liquid, resulting in a black or colored pixel (ON state), or be immersed in the liquid, resulting in a white pixel (OFF state). A more recent and sophisticated implementation of the technology (used now in several digital book readers) involves two types of colored beads immersed in a transparent liquid (see Figure 8.4). This type of display also requires an active matrix (AM) which can be identical to that in an AMLCD.

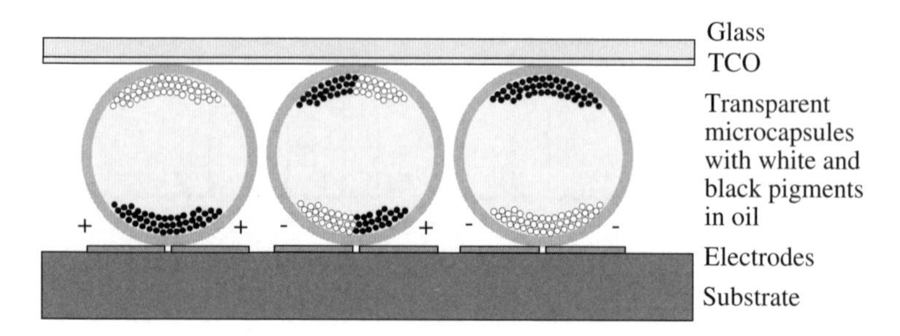


**Fig. 8.4** Schematic cross-section of an electrophoretic display cell in the ON, "grey" and OFF states. The positively charged black pigments and negatively charged white pigments are attracted according to the voltage present on the electrodes.

In order to enhance the brightness of display screens, research in display technologies using intrinsic light emission schemes is being actively pursued. Here organic light emitting diodes (OLED) are one of the main candidates. In OLED displays, each pixel consists of an OLED whose current is controlled by its own circuit. The latter necessitates several TFTs (a minimum of 3 and up to 7) and capacitances to program and control the current flowing through the OLED. The design and performance of the display are very critically related to the performance of the TFTs. The main issues here are:

- Mobility: High current can flow through both the diode and the driver TFT. A high mobility is therefore required in order to limit the size of this transistor and leave the maximum pixel area to the OLED, especially in the case of small pixels (high pixel density).
- Uniformity of the TFT performance: A high uniformity of the TFT performance allows for a simpler design of the circuit controlling the pixel and reduces the area needed for this circuit.
- Stability: TFTs with stable performance, especially in terms of the threshold voltage, are required for simple circuit design and optimal display performance.

For current display technologies such as AMLCDs or OLEDs, a-Si:H is used when a relatively low mobility is acceptable for the display and when there is a low pixel density. For displays with high pixel density, e.g. for mobile phones as well as for OLED displays, LTPS (low temperature poly-Si) is, in general, the material of choice because of its higher mobility. In order to reduce the number of connections to the display, line-driver and shift-register chips are bonded directly to the display substrate. Research also aims at directly integrating these functionalities with LTPS TFTs deposited on the glass substrate of the display. In the long term, graphic interfaces and even processors and memories could be integrated on to the same substrate.

Several other display technologies (e.g. field-emission) are at present receiving attention. In order to obtain high-performance displays, all these technologies require the implementation of an active matrix (AM). The requirements here are, in most cases, similar either to those for AMLCD or to those for OLED displays.

### 8.1.3 TFT configurations and basic characteristics

TFTs are found in four basic configurations, as presented in Figure 8.5. Depending on the material used as a semiconductor, one of the configurations is preferred (see Table 8.1). Inverted configurations are also described as "bottom-gate" transistors while the non-inverted ones are described as "top-gate".

In the case of a-Si:H TFTs, the inverted staggered (bottom gate) configuration is usually preferred because of the sequence of deposition temperatures. In thin-film technologies, one should avoid depositing a layer at a temperature higher

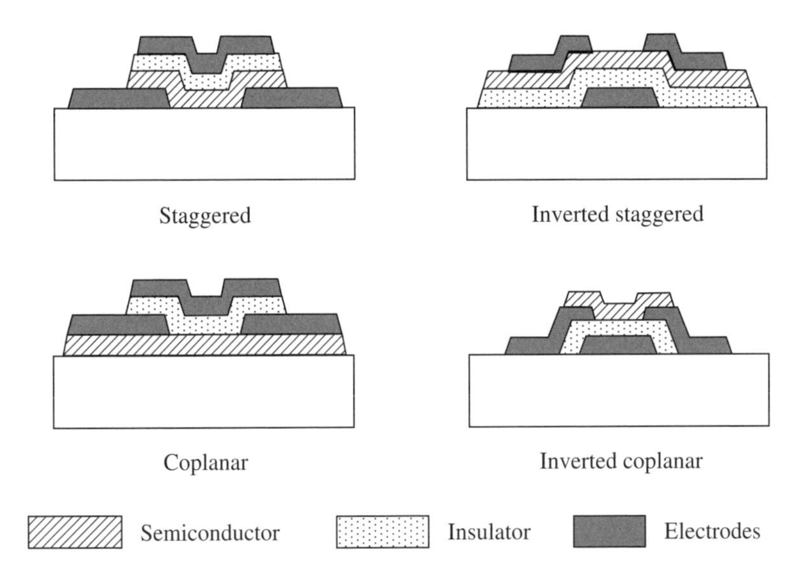


**Fig. 8.5** Main possible TFT configurations. Staggered configurations (inverted for a-Si:H TFT and non-inverted for poly-Si) are the most common configurations.

**Table 8.1** Preferred TFTs configuration and dielectric material for several kinds of thin-film silicon semiconductors, as well as basic device performance and deposition technology.

| | a-Si:H | μc-Si:H | poly-Si |
|---|---|---|---|
| Configuration | Bottom-gate (inverted staggered) | Top-gate (staggered) | Top-gate (staggered) |
| Dielectric layer /insulator | a-$SiN_x$ | $SiO_x$ | $SiO_x$ |
| Channel type | *n*-type | *n*- and *p*-type | *n*- and *p*-type |
| μ [cm2/Vs] | 0.1 - 2 | 1-5 (–200) | 200 - 1000 |
| $I_{on}$ [A] | $\leq 10^{-5}$ | $\leq 10^{-5}$ | $\leq 10^{-4}$ |
| $I_{off}$ [A] | $\leq 10^{-11}$ | $\leq 10^{-10}$ | $\leq 10^{-10}$ |
| Deposition technology | PE-CVD, HW | PE-CVD, HW | PE-CVD or sputtering + thermal annealing or laser crystallization |

than the deposition temperature of an underlying layer. For a-Si:H TFTs, the best semiconductor/insulator combination (best interface properties for optimal performance) is obtained with silicon nitride ($Si_3N_4$) deposited at temperatures between 300-350 °C. As the best a-Si:H is obtained at temperatures around 200 °C, it must be deposited after the $Si_3N_4$ layer. These inverted staggered a-Si:H TFTs are usually found in two versions: in the back-channel-etched (BCE) structure or in the channel-passivated (CHP) structure. The process flow of a BCE TFT (for an AMLCD) as well as its structure is illustrated in Figure 8.6. In CHP TFT, the channel-passivation layer is deposited and patterned between the deposition of the intrinsic and the doped a-Si:H layers. Typical thicknesses for the different layers of a TFT structure are indicated in Table 8.2.

**Table 8.2** Typical thicknesses for the different layers of a TFT strutcture, from [Nathan-2004].

| Layer | Thickness [nm] |
|---|---|
| Gate metal | 130 |
| a-$SiN_x$ gate dielectric | 250 |
| a-Si:H | 50 |
| a-$SiN_x$ passivation layer | 250 |
| $n^+$ μc-Si:H contact layer | 30 |
| Metal contact layer | 500 |

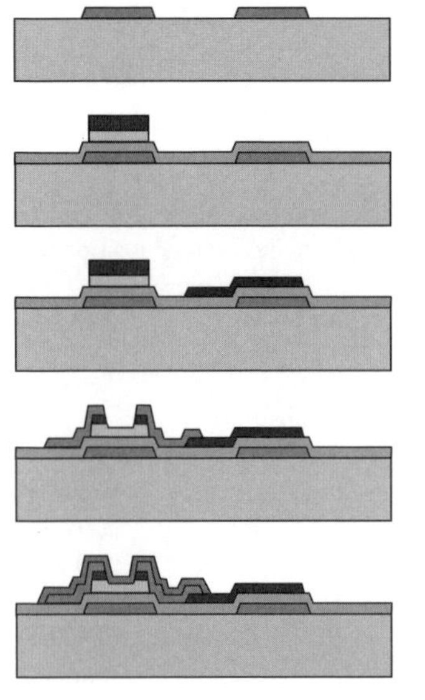

1. Gate electrode deposition and patterning
2. Gate dielectric deposition ($SiN_x$) deposition, a-Si:H stack (intrinsic and n-doped) deposition and patterning
3. ITO (pixel contact) deposition and patterning
4. Source/drain metal deposition and patterning
5. Passivation layer deposition and patterning

**Fig. 8.6** Process flow chart of a back-channel-etched (inverted staggered) a-Si:H TFT for AMLCD. The ITO layer corresponds here to the pixel electrode of the liquid crystal cell.

### 8.1.4 a-Si:H TFT operation

An a-Si:H TFT is a metal oxide semiconductor field-effect transistor (MOSFET) working in electron accumulation mode. As the voltage on the gate increases, electrons accumulate in the a-Si:H semiconductor close to the interface between a-Si:H and gate insulator (see Figure 8.7) in the dangling bond states, in the bandtail states and in the conduction band states, Electrons can also be captured by interface states, and the density of the latter must therefore be kept as low as possible by careful process optimization. When a sufficiently high voltage $V_G$ is applied to the gate electrode, i.e. when the voltage exceeds a threshold voltage $V_{th}$, the charge accumulated in the conduction band will form a ***conduction channel***. For the semiconductor material, the formation of this channel corresponds to a change from an intrinsic to an *n*-type character. Note that in the c-Si field-effect transistor, we rely on the formation of an ***inversion channel*** by changing the semiconductor character from *p*- to *n*-type, or from *n*- to *p*-type.

The total charge present in the channel is given by the capacitor defined by the gate electrode and the conduction channel (see also Figure 8.1) and is equal to $Q_{tot} = C \times V$. The total charge $Q_{tot}$ is the sum of three charge contributions: $Q_f$ from free electrons in the conduction band, $Q_{droit}$ from electrons in the conduction bandtail and $Q_{db}$ from electrons in dangling bonds: $Q_{tot} = Q_f + Q_{bt} + Q_{db}$. As only electrons in the extended states of the conduction band (i.e. the free electrons) are mobile, the conductivity of the channel is directly related to $Q_f$ (or to the ratio $Q_f/Q_{tot}$). In a-Si:H, free electrons possess a finite, but low mobility $\mu_0^n$, while electrons populating the localized states (dangling bond states and tail states) have zero mobility. The mobility that can be measured in a typical device is an average of the mobility of all electrons (free electrons, electrons in tail states and electrons in deep defects or dangling bonds). In

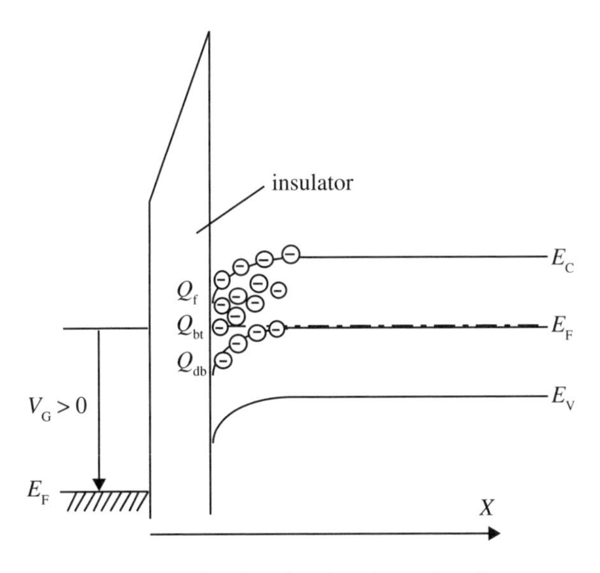


**Fig. 8.7** Band diagram of the metal/insulator/semiconductor interfaces.

a-Si:H TFTs, the average or effective mobility $\mu^{n}_{eff}$ of the electron population in the channel (as seen in Figure 8.7) can be defined as:

$$\mu^{n}_{eff} = \frac{Q_f}{Q_f + Q_{bt} + Q_{db}} \mu^{n}_{0} \tag{8.1}$$

In a-Si:H TFTs, most of the excess charge induced by the positive voltage on the gate is located in the conduction band tail. The effective mobility is $\mu^{n}_{eff} \approx 0.1\mu^{n}_{0}$, with values between 0.5 and 1 $cm^2V^{-1}s^{-1}$. Due to the much gentler slope of the valence bandtail in a-Si:H, the ratio $Q_f/Q_{tot}$ for holes, and therefore the effective hole mobility $\mu^{p}_{eff}$, will remain very small for all practical values of negative gate voltages. *p*-type a-Si:H TFTs therefore cannot achieve any useful performance, and practically only *n*-type TFTs are used.

Due to the voltage difference between the drain and the source (as seen in Figure 8.8), the voltage (and the corresponding charge) will vary as a function of the position in the channel. The relationship between voltage and charge is given here by the capacitance of the gate dielectric. Due to the variation of the total charge along the channel, the free charge, as well as the effective mobility (average mobility of all carriers), is also a function of the position. To simplify the calculation, it is, however, customary to assume that the mobility is constant along the channel and that it corresponds to an average mobility value known as the ***field-effect mobility*** $\mu^{FET}_{n}$. The operation of a TFT can be theoretically analyzed in a similar way as the operation of CMOS FETs. The interested reader will find several textbooks on the topic [Kanicki-2003, Nathan-2004].

Figure 8.9 shows typical output and transfer characteristics of a TFT, i.e. (a) the drain-source current $I_{DS}$ as a function of drain-source voltage $V_{DS}$, and (b) the

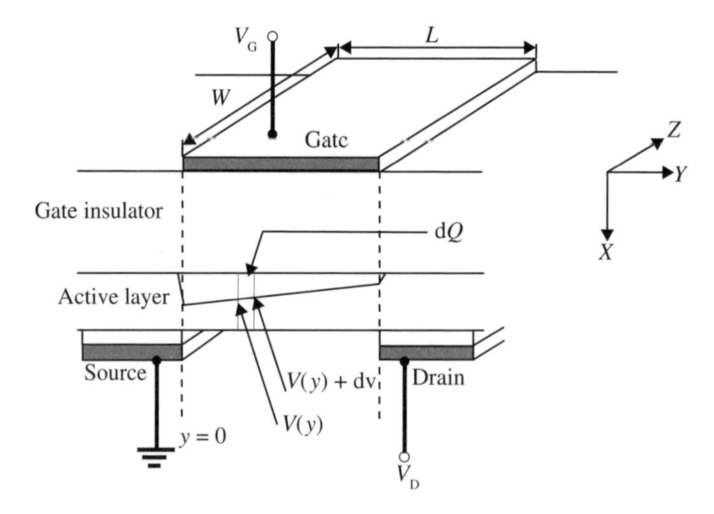


**Fig. 8.8** Schematic view of a thin-film transistor (TFT). *W* and *L* are, respectively, the width and length of the channel. The voltage profile and the corresponding charge in the channel are also indicated.

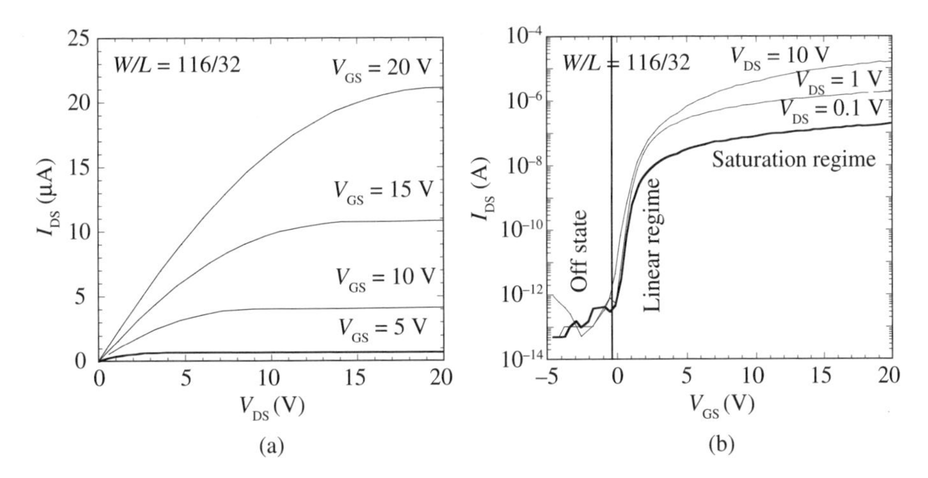


**Fig. 8.9** Example of (a) the TFT output characteristics (drain source current $I_{DS}$ as a function of drain source voltage $V_{DS}$) and (b) TFT transfer characteristics (drain source current $I_{DS}$ as a function of gate source voltage $V_{GS}$) of an a-Si:H TFT. The regions corresponding to the various TFT regimes are also indicated on the transfer characteristic plot (from [Chen-1996]).

drain-source current $I_{DS}$ as a function of gate-source voltage $V_{GS}$. On this latter graph one can observe three different regimes which will be briefly discussed below:

- off state regime;
- linear regime;
- saturation regime.

***Off-state regime***

The gate voltage $V_G$ is negative and the channel is fully depleted. $I_{DS}$ is very low and controlled mainly by the conductivity of the back-channel interface region, i.e. by the region of the semiconductor far away from the dielectric/semiconductor interface. The electronic states present at this back interface will therefore significantly affect the OFF $I_{DS}$ current.

***Linear regime***

The application of a small positive polarization on the gate creates a conduction layer in a channel of length $L$ and width $W$ (see Figure 8.8). When the drain voltage is much smaller than the gate voltage, i.e. when $V_D \ll V_G$, the charge per surface unit at each point of the conduction channel can be assumed to be constant. Using the global average field effect mobility $\mu_n^{FET}$ introduced above, one can show [Kanicki-2003] that the source-drain current $I_{DS}$ is then given by:

$$I_{DS} \approx \mu_n^{FET} \frac{W}{L} C'_{nitride} (V_G - V_{th}^{mean}) V_{DS} \tag{8.2}$$

where $V_{th}^{mean}$ is the mean value of the threshold voltage along the $y$-axis. Equation 8.2 gives $I_{SD}$ as a function of $V_G$ with the drain-source voltage difference $V_{DS}$ as a parameter.

From Equation 8.2, we can extract the expression for the ***transconductance***:

$$\frac{\partial I_{DS}}{\partial V_G} \approx \mu_n^{FET} \frac{W}{L} C'_{nitride} V_{DS} \tag{8.3}$$

This expression allows the determination of the mobility value $\mu_n^{FET}$ from the transfer characteristics of an a-Si:H TFT. Equations 8.2 and 8.3 are only valid for small values of polarization ($V_{DS} \ll V_G$).

### *Saturation regime*

When $V_{DS} = V_G - V_{th}$ the electron channel becomes completely pinched off (see Figure 8.8) and the source-drain current saturates. In this saturation regime ($V_{DS} > V_G - V_{th}$) we must take into account the variation of the channel (potential) along the *y*-axis. One can show [Kanicki-2003] that:

$$I_{DS} \approx \mu_n^{FET} \frac{W}{2L} C'_{nitride} (V_G - V_{th}^{mean})^2 \tag{8.4}$$

Equation. 8.4 can be used for the determination of the field-effect mobility in the saturation regime.

### *TFT performance*

Besides the field-effect mobility, other important transistor parameters are the ON and OFF currents, as well as the gate voltage swing *S*. The latter is defined as the voltage required to increase the source-drain current by a factor of 10. *S* is given by:

$$S = \frac{dV_G}{d(\log I_{DS})} \tag{8.5}$$

The dominant factors influencing a-Si:H TFT performance (such as ON and OFF currents, field-effect mobility $\mu_n^{FET}$ and voltage swing) are summarized in Table 8.3.

### *a-Si:H TFT instabilities*

a-Si:H TFTs usually exhibit instabilities related to shifts in the threshold voltage after the extended application of a gate voltage (bias stress). This instability may be due to:

- charging of the defects in the insulator (usually a-$SiN_x$:H) (this charging affects the polarization of the channel);
- creation of defects in the a-Si:H layer (in the channel); this effect is similar to the Staebler-Wronski effect in solar cells.

Examples of the effect of such instabilities on the transfer characteristics, as well as on the threshold voltage, are shown in Figures 8.10 and 8.11. Instabilities related to the silicon nitride can be reduced by improving the a-$SiN_X$:H quality, as well as by a reduction of the defects at the Si:H/a $SiN_X$:H interface (e.g. by plasma treatments). Note that these instabilities are not too inconvenient for LCD

**Table 8.3** Factors influencing a-Si:H TFT performance [Kanicki-2003].

| TFT performance | Dominant factor |
|---|---|
| ON-current | *W/L* |
| | Drift mobility |
| | Interface states (a-Si:H/SiNx) |
| | Ohmic contact |
| | Gap state density |
| | Back interface states |
| OFF-current | W/L |
| | Fermi level (a-Si:H) |
| | Interface states (a-Si:H/SiNx) |
| | Back surface charge |
| | $n^+$ contact ($n^+$ a-Si:H) |
| | Band gap |
| Field-effect mobility $\mu_n^{FET}$ | Width of band tails (disorder) |
| | Interface states (a-Si:H/SiNx) |
| Gate voltage swing *S* | Gap states (defect states) |
| | Interface states (a-Si:H/SiNx) |

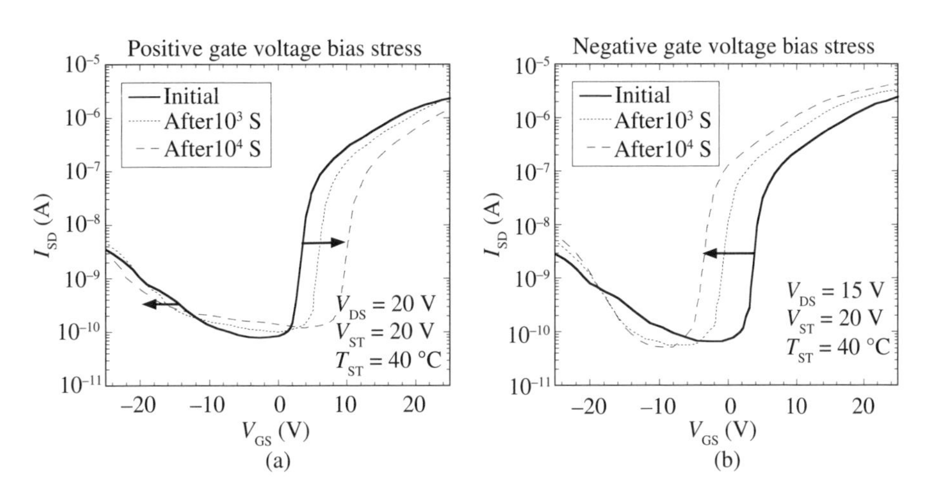


**Fig. 8.10** Examples of instabilities in a-Si:H TFTs: The extended application of a negative (left) or positive (right) gate bias voltage usually results in a shift of the transfer characteristics and threshold voltage; the magnitude of this shift is a function of the gate bias voltage (bias stress) application time (from [Chiang-1998]).

display applications, where TFTs are used as simple switching devices. However, these instabilities are much more problematic for OLED display applications, where TFTs are used to control the voltage flowing through the LEDs. Optimization of the device materials and of the interfaces is necessary to limit the drift of $V_{th}$ and to obtain satisfactory long-lifetime operation of OLED displays based on a-Si:H TFTs.

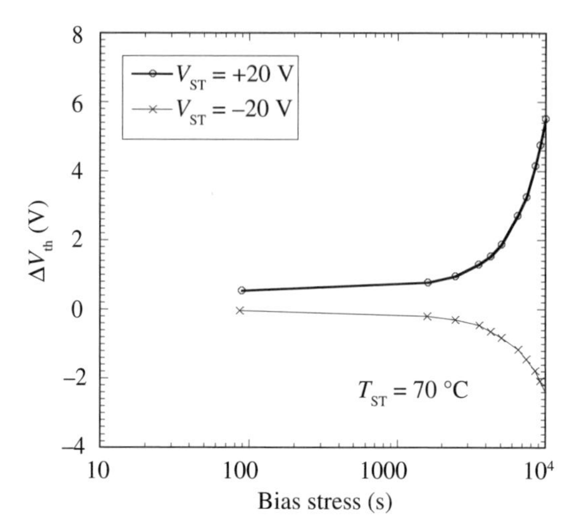


**Fig. 8.11** Threshold voltage shift $\Delta V_{th}$ of an a-Si:H TFT as a function of bias stress time (application time of the bias voltage applied on the gate), for a positive gate stress voltage $V_{ST}$ = 20 V and a negative gate stress voltage $V_{ST}$ = −20 V (from [Chiang-1998]).

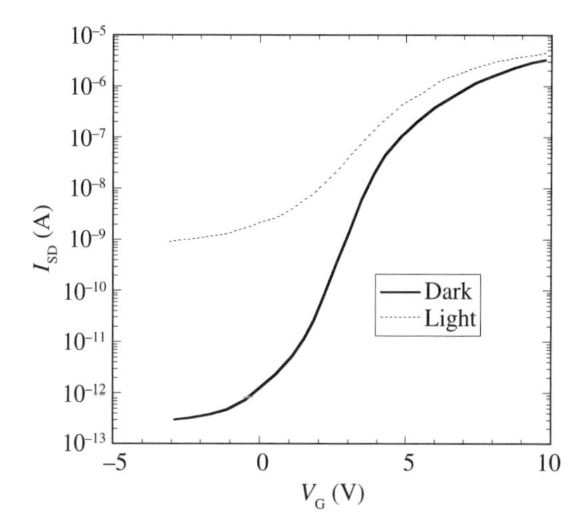


**Fig. 8.12** Transfer characteristics of an a-Si:H in the dark as well as under illumination (at $2 \times 10^4$ Lux) for a drain voltage $V_D$ = 10 V (from [Powell-1989]).

### *Photosensitivity of a-Si:H TFT*

As a-Si:H is a strongly photosensitive material, the transfer characteristics of an a-Si:H TFT also vary with light illumination, as illustrated in Figure 8.12. In order to avoid the disturbing effect of the light (e.g. from the back illumination in AMLCDs), light-shields have to be implemented within the active matrix (AM). If light is only coming from one side, the metallic gate of the TFT is usually sufficient for light shielding of the channel.

### 8.1.5 μc-Si:H and poly-Si TFT performance and other issues

The main drawback of a-Si:H TFT technology is the absence of *p*-type TFTs. No complementary MOS circuit can therefore be built. This considerably limits the possible circuits that can be designed. For display applications it would, for example, be desirable to integrate the line drivers and multiplexers on the same substrate, in order to simplify the connections between the display and the control electronics. However, these rather simple circuits require the availability of both *p*- and *n*-type TFTs. For this purpose, poly-Si TFTs are being developed; they offer the possibility to implement also *p*-type, and generally exhibit higher mobility values than a-Si:H TFTs (see also Table 8.1). Further development of poly-Si technology as well as the achievement of higher mobility values would allow the integration of other electronic functions directly on the display panel. In the long term, systems incorporating even processors and memories are envisaged.

Poly-Si TFTs are fabricated by crystallization of an a-Si layer deposited by sputtering or PE-CVD. The crystallization is obtained by thermal annealing at temperatures above 600 °C (requiring the use of special glass substrates) or by laser annealing at lower temperatures. The main issue for active matrixes (AMs) with poly-Si TFTs is to obtain uniform TFT performances. As the field-effect mobility is mainly limited by semiconductor defects in the channel, control of these defects in terms of density and localization (with regards to TFT geometry) is of paramount importance. For example, sophisticated procedures and complex TFT designs have been developed to control the crystallization process and ensure that the TFT channel is located within a single-grain [Voutsas-2003].

Fabrication of active matrixes (AMs) using poly-Si requires dedicated and rather costly equipment (especially in the case of laser crystallization). One of the problems is thereby the uniformity of performance, and this is one the drivers for the present very active development of μc-Si:H TFTs. μc-Si:H permits the fabrication of complementary electronics whilst using the same lower-cost equipment as used for a-Si:H. However, μc-Si:H TFTs exhibit performances significantly inferior to those of poly-Si TFTs: performances, which are, in general, comparable to those of (n-type) a-Si:H TFTs. Nevertheless, careful control of the growth of μc-Si:H could overcome these limitations and enable the fabrication of μc-Si:H TFTs with performances similar to those of poly-Si [Lee-2006].

## 8.2 LARGE-AREA IMAGERS

### 8.2.1 Introduction and device configuration

The development of flat panel displays has paved the way for the emergence of new applications, such as large-area imagers. By combining an active matrix (as used in a display) with an a-Si:H photodiode array, one obtains a large-area imager, i.e. an active-matrix flat panel imager (AMFPI). Full integration is thereby obtained by depositing the photodiode array directly on top of the active matrix. A schematic view of such a large-area imager is shown in Figure 8.13. A schematic cross-section of one pixel with the TFT switch and the photodiode is presented in Figure 8.14.

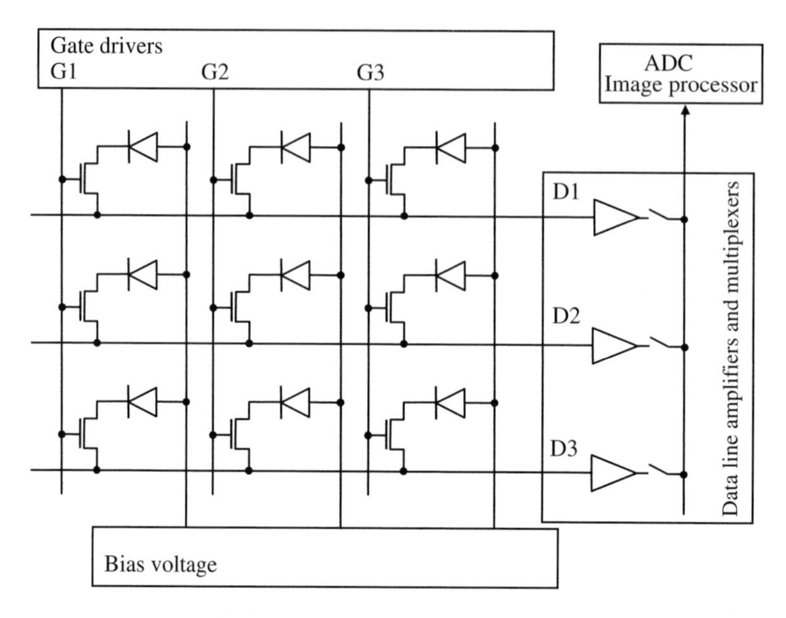


**Fig. 8.13** Schematic layout of a large-area imager. Each pixel consists of an individual photodiode with a top common electrode.

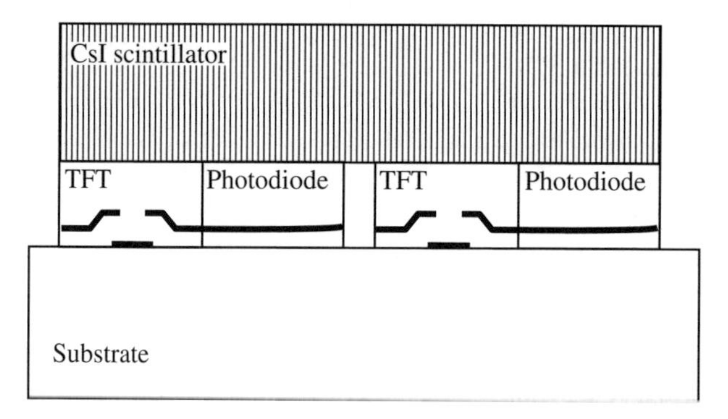


**Fig. 8.14** Schematic cross-section of a single pixel of a large-area X-ray imager (from [Kim-2008]).

Even though such imagers could be used to replace visible light scanners (in FAX machines or photocopiers), their main interest lies in X-ray imaging, mainly for medical applications. By covering the photodiode array with a scintillating layer (usually thallium-doped cesium iodide, CsI:Tl) one obtains a large-area X-ray imager. Nowadays several products are available with various sizes and resolutions, for different applications of X-ray imaging, ranging from dental radiography to full-chest radiography. The versatility of such imagers enables one to perform not only digital radiography but also fluoroscopy (live moving radiography), with reduced radiation exposure of the patient. Besides medical imaging, other applications such as non-destructive testing or security are also targeted. The largest X-ray imager so far developed (prototype) measures $45 \times 46\ \text{cm}^2$ with a resolution of $3072 \times 3072$ pixels [Kim-2008]. Commercial products with dimensions in excess of 40 cm are available, and the development of even larger imagers on Gen 4 glass substrates ($73 \times 92$ cm$^2$) is already in progress.

### 8.2.2 Performance and limitations

A measure of the performance of an imaging system is usually given by the ***detective quantum efficiency DQE*** [Antonuk-2004] defined as:

$$DQE = \frac{(SNR_{out})^2}{(SNR_{in})^2} \quad (0 \leq DQE \leq 1) \tag{8.6}$$

$SNR_{in}$ is the signal-to-noise ratio of the incoming beam (the beam to be detected) and $SNR_{out}$ is the signal-to-noise ratio at the output of the system. $DQE$ is a measure of the capacity of the system to transfer the information present in the incoming beam or incoming illumination to the output; it should be as close to unity as possible. $DQE$ values depend on the conversion layer (e.g. scintillator layer), on the detector at the pixel level, on the TFT's performances, and finally on amplifier and array design.

An important characteristic of any imager is the electronic noise, which defines the minimum charge that still can be detected. In large-area imagers, the electronic noise is mainly given by the minimum noise arising when collecting the charge located on a pixel photodiode. This noise is thereby primarily governed by the data line capacitance, which increases with imager size. This effect represents one of the main limiting factors for the performance of flat panel imagers [Antonuk-2004, Weisfield-1998].

In order to improve the overall device performance and reduce the complexity of the connections, integration of more driving circuits (such as line drivers and amplifiers) using LTPS (low temperature poly-Si) is also being pursued. Implementation of amplification circuit at the pixel level using a-Si:H-based circuits has already been studied, but no noticeable gains in the signal-to-noise ratios have been obtained.

As an alternative to flat panel X-ray imagers using a-Si:H photodiodes and a scintillating layer, X-ray imagers using a-Se photoconductors are also being developed. Thereby, a direct conversion of the X-ray radiation takes place in the a-Se layer, and one can thus avoid the use of a scintillator coupled with an a-Si:H photodiode. Direct detection enables a slightly better spatial resolution and could possibly offer better performance for applications in mammography [Kasap-2002]. This technology also offers a cost advantage (compared to the indirect detection scheme using a-Si:H diodes and a scintillator), due to the fact that a-Se can be deposited directly on top of a commercially available active matrix (AM) as used for flat panel displays. Detailed review of AMFPI fabrication, operation and performances can be found in [Antonuk-2004].

## 8.3 THIN-FILM SENSORS ON CMOS CHIPS

### 8.3.1 Introduction

Active pixel sensors (APS) in CMOS technology are widely used for many imaging applications. In these sensors, each pixel of the array comprises a photodiode and an electronic circuit for local signal processing (at the pixel level). Thus, there is a trade-off between the sensitivity of the sensor (given by the area of the photo diode) and the complexity of the electronics that can be implemented on the pixel area. In order to avoid this trade-off, the concept of "vertical integration" has been introduced: Thereby, direct deposition of an a-Si:H photodiode on top of the readout electronics is carried out as shown in Figure 8.15 [Schneider-1999,Wyrsch-2005]. This technology is known as ***TFA technology*** ("thin-film on application-specific integrated circuits"); it is also called "thin-film on CMOS (TFC)" technology or "diode on top" ("DOT") technology. The TFA concept offers several advantages, such as:

- high sensitivity;
- high dynamic imaging;
- high geometrical fill-factor (ratio between the active area and the total sensor area) close to 100 %;
- no dead area between the pixels;
- high radiation hardness (when using a-Si:H diodes);
- easier coupling of the incoming light with the photodiode (in contrast to CMOS APS, where the light has to go through the many metal layers of a CMOS chip);
- clear separation between the fabrication of the electronic circuitry and the sensor design and optimization.

TFA technology was pioneered in the late 1990s by the University of Siegen (Germany, [Schneider-1999]) and has since been successfully introduced for applications such as vision sensors with high sensitivity [Benthien-2000, Sterzel-2002] and/or high dynamic range [Lulé-1999]. This concept has attracted world wide a large interest for imaging [Theil-2003] and color detection [Neidlinger-1999], but also for other applications such as infrared light vision [Syllaios-2001], particle detection [Wyrsch-2004a, Wyrsch-2004b, Despeisse-2008] and MEMS (micro electro-mechanical systems) [Adrega-2006]. Bio-chips (such as chips for DNA strand detection or for studying biological activity) [Fixe-2004, Moridi-2009] and "lab-on-chip" [Schäfer-2008] are also possible target applications for TFA technology. Although vertical

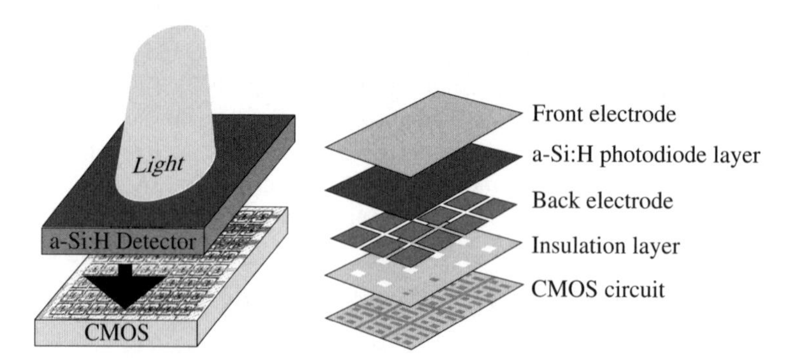


**Fig. 8.15** Schematic view of an array of sensors in TFA technology. In most cases, the CMOS circuit passivation layer is used as the insulation layer. Either the top metal layer of the CMOS chip is used as the back electrode of the a-Si:H diode layer, or an additional metal layer is evaporated on top of the chip. The array is defined by the patterning of the back electrode.

integration may basically involve various types of materials and circuits, the following sections will be dedicated to the discussion of a-Si:H based devices on CMOS circuits.

### 8.3.2 a-Si:H sensor integration

For TFA technology, a-Si:H offers two key advantages:

- It offers a low deposition temperature (around 200 °C) which is fully compatible with the CMOS chip processing and allows direct deposition of the a-Si:H photodiode on top of the chip.
- It also offers a relatively large bandgap of a-Si:H (around 1.75 eV, much higher than the bandgap of c-Si) leading to a very low value of dark conductivity. This very low value of dark conductivity, in turn, results in a low leakage current for the photodiodes (allowing a high sensitivity of the TFA sensors) as well as in a reduced current flow to adjacent pixels ( thus reducing cross-talk).

As mentioned in Section 2.7, a-Si:H can be easily alloyed with Ge in order to obtain smaller bandgap values, or with C or O in order to obtain larger bandgap values. The same alloying possibilities can be used in TFA technology for tuning performances and wavelength sensitivity. For the same purpose of modifying the bandgap, a-Si:H can be replaced by μc-Si:H (which has a significantly lower bandgap, 1.1 eV).

Photodiode arrays usually consist of a stack of *n*-doped, intrinsic and *p*-doped a-Si:H layers covered with a transparent conductive oxide (TCO) layer, which is used as the common electrode for all pixels [see Figure 8.16.a]. In order to limit cross-talk, the *n*-layer should be optimized for low conductivity, or patterned as shown in Figure 8.16(b). Alternative solutions involve e.g. the implementation of ditches in the CMOS chip between the pixels [Schneider-1999] or the removal of the *n*-layer [Miazza-2006]. The active area of a given pixel is defined by the extension of the *n*-layer (when patterned) or by the metal contact of the CMOS chip. In standard CMOS technology, a passivation layer is present at the surface of the chip with openings through this passivation layer (insulation layer seen in Figure 8.15) to allow access to the bonding pads. The *nip* layers of the photodiode array are then deposited directly on this isolation layer. The walls of these openings and any other asperities or steps influence the deposition of the a-Si:H diode array. They usually lead to additional defect formation in the material and, as a consequence, to higher leakage currents. For

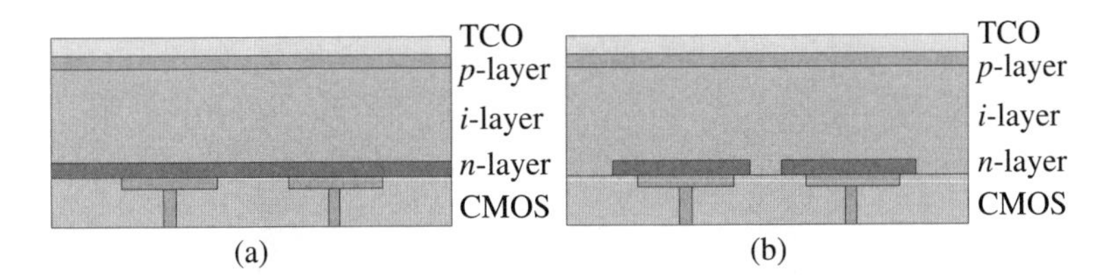


**Fig. 8.16** Schematic view of the cross-section of two pixels of a TFA sensor. Compared to the standard design (a), the *n*-layer can be patterned (b) in order to reduce cross-talk.

this reason, planar surface morphologies are required in order to obtain lowest leakage currents [Miazza-2006, Despeisse-2008].

For particle detection, two different designs can be used. As for X-ray imagers (see Sect. 8.2), the sensor can be coupled to a scintillator to convert the particle into UV or visible light. As an alternative, thick a-Si:H diodes can be used to directly detect the electron-hole pairs generated by the particle traversing the diode. A thick diode (with thickness in the range of 10-30 $\mu$m) is required, as the weak interaction between a high-energy particle and an a-Si:H layer generates only in the order of 100 electron-hole pairs per micrometer of material. An example of such a particle detector is shown in Figure 8.17. Deposition of such diodes with an *i*-layer thickness in the range of 10-30 $\mu$m requires specially adapted deposition conditions in order to reduce the internal mechanical stress and thus avoid peeling of the thick diode.

### 8.3.3 Performance and limitations

One of the critical device characteristics is (as already seen above) the dark leakage current, i.e. the reverse saturation current under dark conditions. It primarily depends on the semiconductor material used and on the interface properties of the a-Si:H diode: on glass substrates the "best" *nip* a-Si:H diodes (best in the sense of having the lowest leakage current) can attain leakage currents that are as low as 1 pA/cm$^2$ [Wyrsch-2005]. However, the surface morphology and the contact geometry of the underlying CMOS chip generally lead to additional leakage currents [Miazza-2006, Despeisse-2008]. For the "best" diodes, the leakage current is primarily governed by thermal activation of carriers from the defect states in the bulk of the intrinsic layer. With light-soaking or degradation of the TFA sensor, the density of defects increases and consequently the leakage current also increases.

**Fig. 8.17** Photograph of a particle sensor in TFA technology with 64 × 64 pixels (picture supplied by EPFL-IMT).

### *Light imagers*

The continuous distribution of states in the bandgap of a-Si:H has another detrimental effect on the performance of a-Si:H devices. Any change in the illumination level and or in the photodiode disturbs the prevailing equilibrium in the occupation of gap states and induces current transients. This results in an "image lag" (i.e. the persistence in time of an image after switching the light. This is due to the charging of the deep defects, which require a certain time to return to equilibrium). The "image lag" is an effect which tends to become even more pronounced with the light-induced degradation of the a-Si:H diode. Therefore TFA sensors are not very well suited for applications requiring high speeds combined with a high dynamic range (i.e. large difference in illumination levels between pixels) [Wyrsch-2008]. Detailed review of TFA sensor performance can be found in [Schneider-1999, Wyrsch-2008].

#### *Particle detectors*

For optimum detection, full depletion of the a-Si:H diode should be achieved. This means that the electric field applied should be high enough to avoid a full screening of this field in any part of the device by the charge carriers. Any part of the device which is not fully depleted will, in fact, reduce the collection of the charge carriers (generated by the incoming particles). Full depletion is therefore particularly important in the case of particle detection, in order to maximize the number of collected electron-hole pairs. However, the voltage required for full depletion $V_D$ increases with the square of the thickness $L$, as seen in the following formula [Wyrsch-2004b]:

$$V_D = \frac{qN_{DB}^* L^2}{2\varepsilon_0 \varepsilon_{Si}} \tag{8.7}$$

where $q$ is the elementary charge, $N_{DB}^*$ is the density of ionized dangling bonds and $\varepsilon_0\varepsilon_{Si}$ the dielectric constant of Si. For a 20 $\mu$m thick diode, assuming a defect density of $5 \times 10^{15}$ cm$^{-3}$ and an ionization fraction of 30 %, a reverse bias voltage of $\approx$ 200 V is thus needed. In order to improve the sensitivity of the sensor, one should increase the thickness (to create more electron-hole pairs). However, thicker devices require larger bias voltages, which in turn lead to much higher leakage current. This ultimately limits the maximum useful thickness to values between 10 and 30 $\mu$m [Wyrsch-2004b], [Despeisse-2008]. Compared with standard 300 $\mu$m thick c-Si sensors, this is a relatively small thickness. Therefore a very sensitive and very low-noise readout chip is required, in the case of a-Si:H sensors, so that one is able to detect the very few electron-hole pairs which are generated here. On the other hand, such a small sensor thickness allows for fast detection as well as for very high spatial resolution.

One of the very attractive features of TFA sensors based on a-Si:H is the very high radiation hardness of this material [Wyrsch-2006]. As also mentioned in the context of its application for space solar cells (see Sect. 7.4.3), a-Si:H is among those semiconductor materials that have the highest radiation resistance. This type of device is therefore of interest for applications as particle trackers in the core of detectors of next-generation particle colliders.

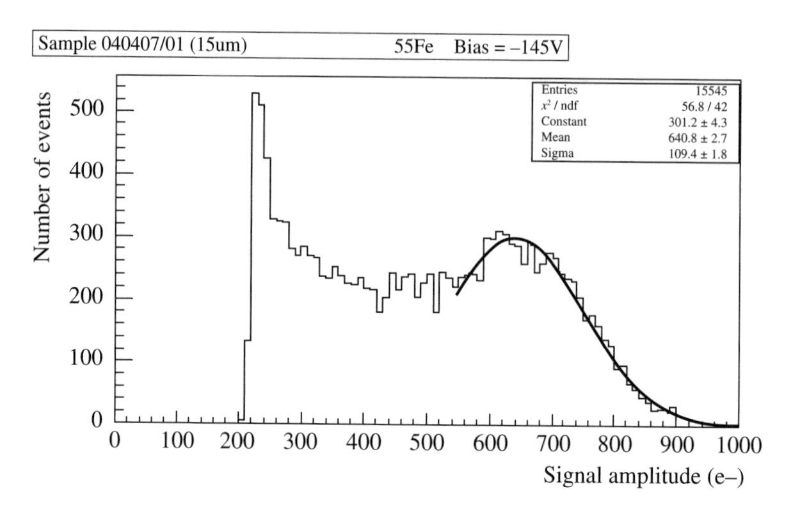


**Fig. 8.18** Histogram of X-ray particle (energy of 5.9 keV) detection from a $^{55}$Fe source by one of the pixels of a TFA sensor comprising a 15 μm thick *nip* a-Si:H diode (source EPFL-IMT/CERN).

TFA sensors have already been successfully tested for detection of beta particles, protons, muons and X-ray particles of various energies. An example for high-energy X-ray detection is given in Figure 8.18. The development is, at present, focused on the detection of particles at the minimum ionizing energy (MIP) (i.e. very energetic particles, which have the lowest interaction with the material.). Even though such particles have already been detected, the signal-to-noise ratio is still not high enough to insure the detection of single MIPs [Wyrsch-2004b, Despeisse-2008].

## 8.4 CONCLUSIONS

Active Matrix *flat panel displays* are today the main application of amorphous silicon (a-Si:H). The total surface area of substrates coated every year for the fabrication of a-Si:H thin-film transistors exceeds by far the area coated for thin-film silicon solar modules. With the introduction of increasingly larger displays, and despite the rapid growth of thin-film silicon PV production, the preponderance of the display market for a-Si:H technology will remain for years to come.

Nevertheless, the emergence of new display technologies such as OLED displays requires enhanced TFT performance; this factor may allow new materials to gradually replace a-Si:H. Low temperature poly-silicon, but also metal-oxide based semiconductors [Jackson-2009] are potential candidates for large-scale manufacturing of OLED displays. Organic semiconductors could also play an important role for specific applications, such as flexible displays. On the other hand, due to the large industrial base and thanks to the technical maturity of a-Si:H technology, its long-term future remains bright.

The situation is rather similar in the case of *large-area imagers*. A total replacement of a-Si:H for both the active matrix and for the sensor array is unlikely in the short- and medium-term. For X-ray detectors, the combination of an a-Si:H photodiode array and a scintillating layer remains by far the leading solution, despite the technical advantages of direct X-ray detection using amorphous selenium (a-Se).

The introduction of *vertically integrated sensors and imagers* has been forecast for many years. The increasing number of metal layers and the reduction in the feature size in modern CMOS imager technology render the coupling of incoming light with photodiodes more and more difficult. It would thus be an obvious evolution to implement such a photodiode on top of the readout chip, and on top of all metal layers, using the TFA (thin-film on ASIC) concept. However TFA technology has not yet been successfully introduced commercially. One of the clear obstacles to this application of a-Si:H is the metastability of the material, which can lead to fluctuations in device performance which are judged to be inacceptable by device manufacturers. Nevertheless, with the increasing integration of manifold devices into microsystems using hybrid technology, the direct deposition of a-Si:H sensors on various substrates is becoming more and more interesting with respect to integration and performance.

## 8.5 REFERENCES


[Adrega-2006] Adrega, T., Prazeres, D.M.F., Chu, V., Conde, J.P., "Thin-film silicon MEMS DNA sensors", (2006), *Journal of Non-Crystalline Solids*, Vol. **352**, pp. 1999-2003.

[Antonuk-2004] Antonuk, L.E., "a-Si:H TFT-Based Active Matrix Flat-Panel Imagers For Medical X-Ray Applications", in "Thin film transistors, Materials and Processes", Vol. 1, edited by Kuo, Y., Kluwer Academic Publishers, 2004, pp. 395-481.

[Benthien-2000] Benthien, S., Lulé, T., Schneider, B., Wagner, M., Verhoeven, M., Böhm, M., "Vertically Integrated Sensors for Advanced Imaging Applications", (2000), *IEEE Journal of Solid-State Circuits*, Vol. **35**, pp. 939-945.

[Brody-1973] Brody, T.P., Asars, J.A., Dixon, G.D., "A 6 × 6 Inch 20 Lines-per-Inch Liquid-Crystal Display Panel", (1973), *IEEE Transactions on Electron Devices*, Vol. **ED-20**, pp. 995-1001.

[Chen-1996] Chen, C.-Y., Kanicki, J., "High Field-Effect-Mobility a-Si:H TFT Based on High Deposition-Rate PECVD Materials", (1996), *IEEE Electron Device Letters*, Vol. **17**, pp. 437-439.

[Chiang-1998] Chiang, C.-S., Kanicki, J., Takechi, K., "Elecrtical Instability of Hydrogenated Amorphous Silicon Thin-Film Transistors of Active-Matrix Liquid-Crystal Displays", (1998), *Japanese Journal of Applied Physics*, Vol. **37**, pp. 4704-4710.

[Despeisse-2008] Despeisse, M., Anelli, G., Jarron, P., Kaplon, J., Moraes, D., Nardulli, A., Powolny, F., Wyrsch, N., "Hydrogenated Amorphous Silicon Sensor Deposited on Integrated Circuit for Radiation Detection", (2008), IEEE Transactions on Nuclear Science, Vol. **55,** pp. 802-811.

[Fixe-2004] Fixe, F., Chu, V., Prazeres, D.M.F., Conde, J.P., "An on-chip thin film photodetector for the quantification of DNA probes and targets in microarrays", (2004), *Nucleic Acids Research*, Vol. **32**, pp. e70.

[Jackson-2009] Jackson, W.B., "Flexible Transition Metal Oxide Electronics and Imprint Lithography", in "Flexible Electronics: Materials and Applications", edited by Wong, W.S., Salleo, A., Springer, 2009, pp.107-142.

[Kanicki-2003] Kanicki, J., Martin, S., "Hydrogenated Amorphous Silicon Thin-Film Transistors", in "Thin-Film Transistors", edited by Kagan, C.R., and Andry, P., Marcel Dekker Inc, 2003, pp. 71-137.

[Kasap-2002] Kasap, S.O., Rowlands, J.A., "Direct-Conversion Flat-Panel X-Ray Image Sensors for Digital Radiography", (2002), *Proceedings of the IEEE*, Vol. **90**. pp. 591-604.

[Kim-2008] Kim, H. K., Cunningham, I.A., Yin, Z., Cho, G., "On the Development of Digital Radiography Detectors : A Review", (2008), *International Journal of Precision Engineering and Manufacturing*, Vol. **9**, pp. 86-100.

[LeComber-1979] LeComber, P.G., Spear, W.E., Ghaith, A., "AMORPHOUS-SILICON FIELD-EFFECT DEVICE AND POSSIBLE APPLICATION", (1979), *Electronics Letters*, Vol. **15**, pp. 179-1981.

[Lee-2006] Lee, M C.-H., Sazonov, A., Nathan, A., Robertson, J., "Directly deposited nanocrystalline silicon thin-film transistors with ultra high mobilities", (2006), *Applied Physics Letters*, Vol. 89, pp. 252101.

[Lulé-1999] Lulé, T., Schneider, B., Böhm, M., "Design and Fabrication of a High-Dynamic-Range Image Sensor in TFA Technology", (1999), *IEEE Journal of Solid-State Circuits*, Vol. **34**, pp. 704-711.

[Miazza-2006] Miazza, C., Wyrsch, N., Choong, G., Dunand, S., Ballif, C., Shah, A., Blanc, N., Kaufmann, R., Lustenberger, F., Moraes, D., Despeisse, M., Jarron, P., "Image Sensors Based on Thin-film on CMOS Technology: Additional Leakage Currents due to Vertical Integration of the a-Si:H Diodes", (2006), *Material Research Society Proceedings*, Vol. 910, pp. A17-03.

[Moridi-2009] Moridi, M., Tanner, S., Wyrsch, N., Farine, P.-A., Rohr, S., "A Highly Sensitive a-Si Photodetector Array with Integrated Filter for Optical Detection in MEMS", (2009) *Procedia Chemistry*, Vol. **1**, 1367–1370.

[Nathan-2004] Nathan, A., Servati, P., Karim, K.S., Striakhilev, D., Sazonov, A. , "Device Physics, Compact Modeling, and Circuit Applications of a-Si:H TFTs", in "Thin film transistors, Materials and Processes", Vol. 1, edited by Kuo, Y., Kluwer Academic Publishers, 2004, pp. 79-175.

[Neidlinger-1999] Neidlinger, T., Harendt, C., Glockner, J., Schubert, M.B., "Novel Device Concept for Voltage-Bias Controlled Color Detection in Amorphous Silicon Sensitized CMOS Cameras", (1999), *Material Research Society Proceedings*, Vol. **558**, pp. 285-290.

[Powell-1989] Powell, M.J., "The Physics of Amorphous-Silicon Thin-Film Transistors", (1989), IEEE Transactions on Electron Devices, Vol. **36**, pp. 2753-2763.

[Schäfer-2008] Schäfer, H., Lin, H., Schmittel, M., Böhm, M., "A labchip for highly selective and sensitive electrochemiluminescence detection of $Hg^{2+}$ ions in aqueous solution employing integrated amorphous thin film diodes", (2008), *Microsystem Technologies*, Vol. **14**, pp. 589–599.

[Schneider-1999] Schneider, B., Rieve, P., Böhm, M., Jähne, B., Haußecker, H., Geißler, P., "Image Sensors in TFA (Thin Film on ASIC) Technology", in "Handbook on Computer Vision and Applications", Vol. 1, edited by , Jähne, B., Haussecker, H., Geissler, P., Academic Press, Boston, 1999, pp. 237-270.

[Sterzel-2002] Sterzel, J., Blecher, F., Hillebrand, M., Schneider, Böhm, B.M., "TFA Image Sensors For Low Light Level Detection", (2002), *Material Research Society Proceedings*, Vol. **715**, pp. A.7.1.1.

[Syllaios-2001] Syllaios, A.J., Schimert, T.R., Gooch, R.W., McCardel, W.L., Ritchey, B.A., Tregilgas, J.H., "Amorphous Silicon Microbolometer Technology", (2000), *Material Research Society Proceedings*, Vol. **609**, pp. A14.4.1.

[Theil-2003] Theil, J., "Advances in elevated diode technologies for integrated circuits: progress towards monolithic instruments". (2003), *IEEE Proceedings Circuits, Devices & Systems*, Vol. **150**, pp. 235-249.

[Voutsas-2003] Voutsas, A.T., Hatalis, M.K., "Technology of Polysilicon Thin-Film Transistors", in "Thin-Film Transistors", edited by Kagan, C.R., and Andry, P., Marcel Dekker Inc, 2003, pp. 139-207.

[Weimer-1962] Weimer, P.K., "The TFT-a new thin-film transistor",(1962), *Proceedings of the Institute of Radio Engineers*, Vol. **50**, pp. 1462-1469.

[Weisfield-1998] Weisfield, R.L., "AMORPHOUS SILICON TFT X-RAY IMAGE SENSORS", (1998), *Electron Devices Meeting, IEDM '98 Technical Digest*, pp. 21-24.

[Wyrsch-2004a] Wyrsch, N., Miazza, C., Dunand, S., Shah, A., Moraes, D., Anelli, G., Despeisse, M., Jarron P., Dissertori, G., Viertel, G., "Vertically Integrated Amorphous Silicon Particle Sensors", (2004), *Material Research Society Proceedings*, Vol. **808**, pp. 441-446.

[Wyrsch-2004b] Wyrsch, N., Dunand, S., Miazza, C., Shah, A., Anelli, G., Despeisse, M. Garrigos, A. Jarron, P., Kaplon, J., Moraes, D., Commichau, S.C., Dissertori, G., Viertel, G.M., "Thin-film silicon detectors for particle detection", *Physica Status Solidi (c)*, Vol. **1**, pp. 1284-1291.

[Wyrsch-2005] Wyrsch, N., Miazza, C., Ballif, C., Shah, A., Blanc, N., Kanfmann, R., Lustenberger, F., Jarron, P., "Vertical integration of hydrogenated amorphous silicon devices on CMOS circuits," *Material Research Society Proceedings*, Vol. **869,** pp. 3-14.

[Wyrsch-2006] Wyrsch, N., Miazza, C., Dunand, S., Ballif, C., Shah, A., Despeisse, M., Moraes, D., Powolny, F., Jarron, P., "Radiation hardness of amorphous silicon particle sensors",(2006), *Journal of Non-Crystalline Solids*, Vol. **352**, pp. 1797-1800.

[Wyrsch-2008] Wyrsch, N., Choong, G., Miazza, C., Ballif, C., "Performance and Transient Behavior of Vertically Integrated Thin-film Silicon Sensors", (2008), *Sensors*, Vol. **8**, pp. 4656-4668.

# INDEX